NONLINEAR OPTICS

NONLINEAR OPTICS

PHENOMENA, MATERIALS, AND DEVICES

George I. Stegeman
Robert A. Stegeman

WILEY

A JOHN WILEY & SONS, INC., PUBLICATION

For general information on our other products and services or for technical support, please contact our Customer Care Department within the United States at (800) 762-2974, outside the United States at (317) 572-3993 or fax (317) 572-4002.

Wiley also publishes its books in a variety of electronic formats. Some content that appears in print may not be available in electronic formats. For more information about Wiley products, visit our web site at www.wiley.com.

Library of Congress Cataloging-in-Publication Data:

ISBN: 978-1-118-07272-1

Printed in the United States of America

10 9 8 7 6 5 4 3 2 1

■ CONTENTS

The field of nonlinear optics came into being in the 1960s, stimulated essentially by the invention of the laser. Its impact has been acknowledged by the award of the Nobel Prize in physics to one of its pioneers Nicholas Bloembergen in 1981 (1) and by other Nobel prizes for work enabled by nonlinear optics in chemistry and physics and multiple awards by the American Institute of Physics, the Optical Society of America, and other scientific organizations.

The fundamental principles of nonlinear optics are now well known and have been elucidated in excellent nonlinear optics textbooks by Ron Shen, Doug Mills, Robert Boyd, Govind Agrawal, and others over the last 20 years. These books served two purposes: to discuss basic principles and to give an overview of interesting applications and experiments. Of the two branches of nonlinear optics—nonlinear phenomena and nonlinear materials—that have evolved over the years, the latter has accounted for most of the progress in the field over the last few decades whereas the former has been the subject of most textbooks. In fact, the nonlinear materials evolution has been spectacular. Since the early days of nonlinear optics, the requirements for some nonlinear phenomena have been reduced from kilowatt lasers to milliwatt lasers. Our particular goal for this textbook, in addition to elucidating the fundamentals of nonlinear optics from our own perspective, was to discuss nonlinear materials, new nonlinear phenomena developed in the last few decades such as solitons, and the practical aspects of the most common nonlinear devices.

This textbook is based on a nonlinear optics course initially developed at CREOL (Center for Research in Electro-Optics and Lasers, now a part of the College of Optics and Photonics), University of Central Florida, in the 1980s by Eric Van Stryland and David Hagan. After George I. Stegeman joined CREOL in 1990, he took over this course, expanded and continuously revised it, put it into the PowerPoint format, and added new problems every year. Robert A. Stegeman, the coauthor, has used this course in his professional career, which involves current applications of nonlinear optics, primarily in the mid-infrared region of the spectrum, and he has contributed most of the application discussions to this text.

This course deals with the physics and applications of optical phenomena that occur at intensity levels at which optical processes become dependent on optical intensity or integrated flux. It is designed for graduate students, postdoctoral fellows, and others with prior knowledge of electromagnetic wave propagation in materials as well as for professionals in the field who are looking for newly developed fields and concepts. Although a rudimentary knowledge of quantum mechanics would be helpful, it is not a requirement. When quantum mechanics is used, it is reviewed at the level needed for the course.

Nonlinear optics is *not* just a simple extension of linear optics. A keystone concept in linear optics is that electromagnetic waves do not interact with one another. Solutions to Maxwell's equations lead to "orthogonal" eigenmodes, i.e., summing the fields due to two or more overlapping field solutions and calculating the intensity leads to a net intensity that is the sum of the intensities of the individual waves. Nonlinear optics is all about interactions that occur between light and matter at high intensities. Hence, the solutions to the nonlinear wave equations do not lead to eigenmodes, just nonlinear modes. As will be discussed in the later part of the textbook, the modes of nonlinear optics are solitons, spatial solitons for continuous waves that do not diffract in space, temporal solitons for noncontinuous waves that do not disperse (spread) in time and are confined in some kind of a waveguide that inhibits spatial diffraction, and spatiotemporal solitons that spread neither in time nor in space. Such modes, in general, exchange energy when they interact so that the reader should be prepared to give up notions such as the superposition principle, which may be satisfied only approximately.

It proves useful to explain the philosophy adopted here. There are two approaches to discussing nonlinear optics. One is to introduce macroscopic nonlinear susceptibilities at a phenomenological level. These susceptibilities are measured in the laboratory and used to quantify phenomena, devices, and so forth. The second approach starts at a more physical level with the electric dipole interaction of isolated atoms and molecules with radiation fields to find the response at the atomic level and the molecular level. Although it is satisfying from a physical insight perspective, difficulties occur in going to the condensed matter limit where most experiments are done. Because there are dipolar fields induced in neighboring molecules, the "local" fields experienced by an atom or a molecule are not just the fields given by the macroscopic Maxwell's equations, called the "Maxwell" fields. The "local" field is the sum of the Maxwell fields and all the induced dipole fields at the site of an atom or a molecule. A rigorous and satisfactory approach to accurately estimating the local fields has been a subject of continuing discussion for many years. Hence this approach does not necessarily yield reliable numbers for measured nonlinear susceptibilities, but does give fairly accurately many key features, including the frequency spectrum of the nonlinear response functions, i.e., the nonlinear susceptibilities.

Here we use a combination of these approaches. Whenever possible, the nonlinear optics response at the molecular level is treated first, approximate susceptibilities are derived, and then measured susceptibilities are used in discussing applications. To facilitate the separation of microscopic and macroscopic parameters, the isolated molecule parameters are identified by a "bar," e.g., the transition dipole moment between energy levels i and j ($\bar{E}_i$ and $\bar{E}_j$) is $\bar{\mu}_{ij}$, the mass of an electron is $\bar{m}_e$, the reduced mass for the βth vibration in a molecule is $\bar{m}_\beta$, and so on.

This book also includes appendices in which the fundamentals of a number of concepts such as Raman scattering and two- and three-level models are presented, as well as some tables of the relation between nonlinear susceptibilities, their conversion between different systems of units, and crystal symmetry.

G.S. thanks his colleagues at CREOL for their helpful discussions over the years, especially Demetrius Christodoulides, Eric Van Stryland, and David Hagan, as well as the many graduate students who diligently corrected lecture notes and asked probing questions.

REFERENCE

1. N. Bloembergen, Encounters in Nonlinear Optics: Selected Papers of Nicolaas Bloembergen with Commentary (World Scientific Press, Singapore,1996).

CHAPTER 1

Introduction

1.1 WHAT IS NONLINEAR OPTICS AND WHAT IS IT GOOD FOR?

In general, nonlinear optics takes place when optical phenomena occur in materials that *change optical properties with input power or energy* and/or *generate new beams or frequencies*. Examples are power-, intensity-, or flux-dependent changes in the frequency spectrum of light, the transmission coefficient, the polarization, and/or the phase. New beams can also be generated either by a shift in frequency from the original frequency or by travelling in different directions relative to the incident beam. Although one frequently refers to the intensity or power dependence of phenomena as being signatures of nonlinear optics, there are many cases characterized by a flux dependence, i.e., changes in beam properties that are cumulative in the illumination time, usually accompanied by absorption.

A frequently asked question is: "How do I really know when nonlinear optics is occurring in my experiment?" Some examples of commonly observed phenomena are shown in Figs 1.1 and 1.2. Figure 1.1a shows harmonic generation, a second- or third-order nonlinear effect. Figure 1.1b shows nonlinear transmission, essentially a third-order nonlinear effect. In an interference experiment an increase in the input intensity can lead to a shift in fringes due to second- or third-order nonlinear optics (see Fig. 1.1c). A very common effect—self-focusing of light—is illustrated in Fig. 1.2, in which a beam narrows with an increase in the input intensity due to propagation through a sample, forming a soliton at high intensities that propagates without change in size or shape and then breaks up into "noise" filaments, i.e., multiple nondiffracting beams, at very high intensities.

The second most frequently asked question is: "What is nonlinear optics good for?" A collage of applications is shown schematically in Fig. 1.3. Probably the most frequently used nonlinear optics device is the second harmonic generator, which doubles the frequency of light, as shown in Fig. 1.1a. Along the same lines are optical parametric devices, also based on second-order nonlinearities, which include amplifiers (optical parametric amplifiers) and frequency-tunable generators (optical parametric generators and optical parametric oscillators), and the last two are commonly used as sources of tunable radiation (see Fig. 1.3a). Nonlinear absorption

Nonlinear Optics: Phenomena, Materials, and Devices, George I. Stegeman and Robert A. Stegeman.
© 2012 John Wiley & Sons, Inc. Published 2012 by John Wiley & Sons, Inc.

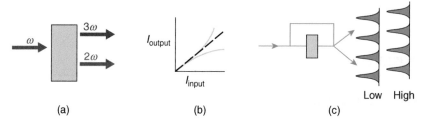

FIGURE 1.1 (a) Second and third harmonic generation. (b) Nonlinear transmission. (c) Nonlinear fringe shift between low and high intensity inputs.

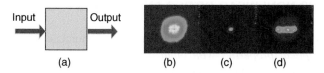

FIGURE 1.2 (a) Beam input and output geometry. (b) Diffracted output beam at low intensity input. (c) Self-focused output beam corresponding to a spatial soliton. (d) Multiple filaments in the output beam at very high intensity.

that depends on intensity is used for the localized activation of drugs or imaging inside media (Fig. 1.3b). A third example is an all-optical control of optical signals, e.g., for communications (Fig. 1.3c).

1.2 NOTATION

The diversity of notations used for optical fields, nonlinear susceptibilities, and so on, is a frequently confusing aspect of this field. A perusal of the nonlinear optics literature shows that there is little consistency, especially when dealing with third-order nonlinear coefficients. Here a concentrated effort has been made to be consistent and to introduce more descriptive notations. The assumptions and notation used here are as follows:

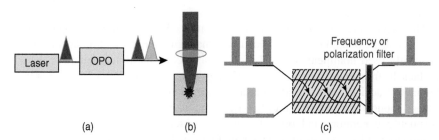

FIGURE 1.3 (a) An optical parametric oscillator (OPO) for producing two tunable frequencies. (b) Two-photon absorption activation of chemistry. (c) All-optical control of routing in a nonlinear coupler. A control beam (lower input arm) is used to isolate (switch out) a single pulse from the input pulse train.

1. *Plane waves* will be explicitly assumed to elucidate nonlinear phenomena in the simplest fashion. Whenever finite beams are considered, which is normally necessary to discuss devices and applications, this will be clearly stated.

2. *Continuous-wave fields* are explicitly assumed unless otherwise stated. The electromagnetic fields are written as

$$\vec{E}(\vec{r}, t) = \frac{1}{2}\vec{E}(\omega)\, e^{-i\omega t} + \text{c.c.} = \frac{1}{2}\vec{\mathcal{E}}(\omega)\, e^{i[kz - \omega t]} + \text{c.c.} \tag{1.1}$$

3. The unit vector is written as $\hat{e}$ and has components $\hat{e}_i$, where $i = x, y, z$.

4. The "Einstein" notation is used for summations over repeated indices; e.g., $a_i b_i c_i = a_x b_x c_x + a_y b_y c_y + a_z b_z c_z$.

5. Quantities with a "bar" above, e.g., $\vec{\bar{\mu}}$, refer to individual molecular properties in the absence of interaction with other molecules as well as parameters in a single molecule's frame of reference.

6. Quantities with a "tilde" above, e.g., $\tilde{\chi}_{ijk}^{(2)}$, identify parameters and coefficients in the "zero (nonresonant) frequency" limit (Kleinman limit) $\omega \ll \bar{\omega}_r$, i.e., at frequencies much smaller than any resonant frequency $\bar{\omega}_r$ of the material.

7. SI units are used throughout. Here intensity is used to mean power per unit area, usually in units of watts per square centimeter. It is equivalent to irradiance. In the cases of pulses of light, flux per unit area is defined as the integrated intensity of a pulse over time, typically over the duration of the pulse. Flux is defined as the energy of a pulse integrated over both time and cross section.

It is important to realize that in this textbook $\vec{E}(\omega)$ is *not* the Fourier transform of $\vec{E}(t)$ and its use is restricted to Eq. 1.1. For a monochromatic wave of frequency ω_a, $E_i(\omega_a)$ is the notation used for the Fourier transform of the field and $E_i(\omega_a) \neq \mathbf{E}_i(\omega_a)$. The relations between the two can be derived easily from the unitary Fourier transform equations:

$$E(t) = \sqrt{\frac{1}{2\pi}} \int_{-\infty}^{\infty} E(\omega)\, e^{-i\omega t}\, d\omega; \qquad E(\omega) = \sqrt{\frac{1}{2\pi}} \int_{-\infty}^{\infty} E(t')\, e^{i\omega t'}\, dt'$$

$$\delta(t - t') = \sqrt{\frac{1}{2\pi}} \int_{-\infty}^{\infty} e^{-i\omega(t - t')}\, d\omega; \qquad \delta(\omega - \omega_a) = \sqrt{\frac{1}{2\pi}} \int_{-\infty}^{\infty} e^{i(\omega - \omega_a)t')}\, dt'. \tag{1.2}$$

Substituting Eq. 1.1 for $E(t')$ into the $E(\omega)$ equation in Eq. 1.2 gives

$$E_i(\omega) = \sqrt{\frac{1}{2\pi}} \frac{1}{2} \left\{ \int_{-\infty}^{\infty} \mathbf{E}_i(\omega_a)\, e^{i(\omega - \omega_a)t}\, dt + \int_{-\infty}^{\infty} \mathbf{E}_i^*(\omega_a)\, e^{i(\omega + \omega_a)t}\, dt \right\}$$

$$= \frac{1}{2} \left\{ \mathbf{E}_i(\omega_a) \sqrt{\frac{1}{2\pi}} \int_{-\infty}^{\infty} e^{i(\omega - \omega_a)t}\, dt + \mathbf{E}_i^*(\omega_a) \sqrt{\frac{1}{2\pi}} \int_{-\infty}^{\infty} e^{i(\omega + \omega_a)t}\, dt \right\}$$

$$\to E_i(\omega) = \frac{1}{2} [\mathbf{E}_i(\omega_a)\delta(\omega - \omega_a) + \mathbf{E}_i(-\omega_a)\delta(\omega + \omega_a)]. \tag{1.3}$$

If fields have a distribution of frequencies, then the δ functions are replaced by $g(\omega - \omega_a)$ and $g(\omega + \omega_a)$, normalized so that their integrals over frequency are unity. Additional notation will be introduced as needed in succeeding chapters.

1.3 CLASSICAL NONLINEAR OPTICS EXPANSION

The simplest and most general expansion of the nonlinear polarization induced by the mixing of optical fields is

$$
P_i(\vec{r}, t) = \varepsilon_0 \left[\int_{-\infty}^{\infty} \int_{-\infty}^{t} \hat{\chi}_{ij}^{(1)}(\vec{r}-\vec{r}'; t-t') E_j(\vec{r}', t') \, d\vec{r}' \, dt' \right.
$$

$$
+ \int_{-\infty}^{\infty} \int_{-\infty}^{\infty} \int_{-\infty}^{t} \int_{-\infty}^{t} \hat{\chi}_{ijk}^{(2)}(\vec{r}-\vec{r}', \vec{r}-\vec{r}''; t-t', t-t'') E_j(\vec{r}', t') E_k(\vec{r}'', t'') \, d\vec{r}' \, d\vec{r}'' \, dt' \, dt''
$$

$$
+ \int_{-\infty}^{\infty} \int_{-\infty}^{\infty} \int_{-\infty}^{\infty} \int_{-\infty}^{t} \int_{-\infty}^{t} \int_{-\infty}^{t} \hat{\chi}_{ijk\ell}^{(3)}(\vec{r}-\vec{r}', \vec{r}-\vec{r}'', \vec{r}-\vec{r}'''; t-t', t-t'', t-t''')
$$

$$
\left. \times E_j(\vec{r}', t') E_k(\vec{r}'', t'') E_\ell(\vec{r}''', t''') \, d\vec{r}' \, d\vec{r}'' \, d\vec{r}''' \, dt' \, dt'' \, dt''' + \cdots \right] \quad \text{with}
$$

$$
\int_{-\infty}^{\infty} d\vec{r}' \equiv \int_{-\infty}^{\infty} \int_{-\infty}^{\infty} \int_{-\infty}^{\infty} dx' \, dy' \, dz' \tag{1.4}
$$

To understand the physical implications of this formula, consider the first nonlinear term due to the second-order susceptibility, i.e., $\hat{\chi}_{ijk}^{(2)}(\vec{r}-\vec{r}', \vec{r}-\vec{r}''; t-t', t-t'')$. The polarization $P_i(\vec{r}, t)$ is created at time t and position $\vec{r}$ by two separate interactions of the total electromagnetic field at time t' and position $\vec{r}'$ and at time t'' and position $\vec{r}''$ in a material in which $\chi^{(2)} \neq 0$. This form also includes nonlocal-in-space effects, such as thermal nonlinearities, in which the refractive index changes due to absorption, e.g., diffuses. In most cases encountered on optics, the response is local in space and so

$$
\hat{\chi}_{ijk}^{(2)}(\vec{r}-\vec{r}', \vec{r}-\vec{r}''; t-t', t-t'') \quad \rightarrow \quad \hat{\chi}_{ijk}^{(2)}(t-t'; t-t'') \delta(\vec{r}-\vec{r}') \delta(\vec{r}-\vec{r}'') \tag{1.5}
$$

and

$$
P_i^{(2)}(\vec{r}, t) = \varepsilon_0 \int_{-\infty}^{\infty} \int_{-\infty}^{\infty} \hat{\chi}_{ijk}^{(2)}(t-t', t-t'') E_j(\vec{r}, t') E_k(\vec{r}, t'') \, dt' \, dt''. \tag{1.6}
$$

Only near a "resonance" does a noninstantaneous response typically occur for Kerr-type nonlinearities. A noninstantaneous time response translates into a frequency dependence for all the susceptibilities. An example of how a noninstantaneous response occurs is shown in Fig. 1.4 for a simple two-level model.

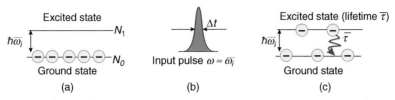

FIGURE 1.4 (a) Two-level model with all electrons initially in the ground state N_0. (b) Incidence of a short pulse ($\Delta t \ll \bar{\tau}$) causes many electron transitions to the excited state. (c) Excited-state population after the pulse passes.

As the excited-state electrons relax back to the ground state, the induced polarization relaxes back to the ground-state polarization, leading to time evolution in both the refractive index and the absorption coefficient, as illustrated in Fig. 1.5. The Fourier transform of this time evolution gives the frequency response.

Equation 1.4 is not the one normally used because of its complexity. Assuming plane waves of the form

$$P_i(\vec{r}, t) = \frac{1}{2} P_i(\omega, z)\, e^{-i\omega t} + \text{c.c.}, \qquad E_i(\vec{r}, t) = \frac{1}{2} \sum_m E_i^m(\omega_m, z)\, e^{-i\omega_m t} + \text{c.c.},$$

$$E_j^m(-\omega_m) = E_j^{*m}(\omega_m) \tag{1.7}$$

and expanding again in terms of the total field gives

$$
\begin{aligned}
P_i(\omega, z)\, e^{-i\omega t} = \varepsilon_0 \Bigg[&\hat{\chi}_{ij}^{(1)}(-\omega;\, \omega_m) \sum_m E_j^m(\omega_m, z)\, e^{-i\omega_m t} \\
&+ \frac{1}{2} \sum_m \sum_p \hat{\chi}_{ijk}^{(2)}(-\omega;\, \pm\omega_m,\, \pm\omega_p) E_j^m(\pm\omega_m, z) E_k^p(\pm\omega_p, z)\, e^{-i(\pm\omega_m \pm\omega_p)t} \\
&+ \frac{1}{4} \sum_m \sum_p \sum_q \hat{\chi}_{ijk\ell}^{(3)}(-\omega;\, \pm\omega_m, \pm\omega_p, \pm\omega_q) E_j^m(\pm\omega_m, z) E_k^p(\pm\omega_p, z) \\
&\times E_\ell^q(\pm\omega_q, z)\, e^{-i(\pm\omega_m \pm\omega_p \pm\omega_q)t} + \cdots \Bigg]
\end{aligned}
\tag{1.8}
$$

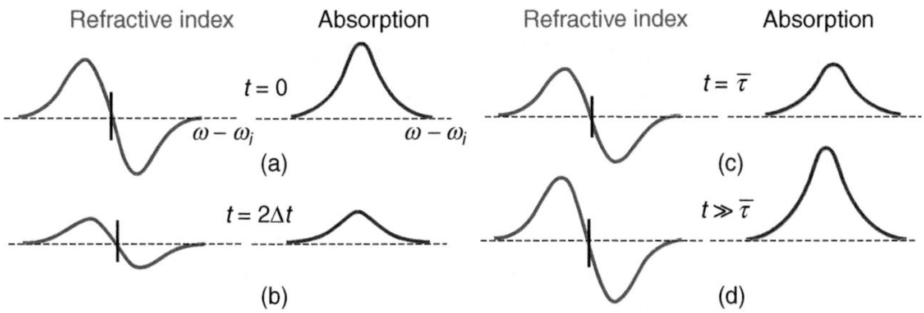

FIGURE 1.5 (a) Spectral distribution of refractive index and absorption before the incidence of the pulse ($t = 0$). (b)–(d) Time evolution of refractive index and absorption: $t = 2\Delta t$ (b); $t = \bar{\tau}$ (c), and $\Delta t \gg \bar{\tau}$ (d).

In each case, $\omega = \pm\omega_m \pm \omega_p \pm \omega_q$ is the output frequency generated by the interaction. The hat (roof) superscript is meant to emphasize that the quantity underneath is a complex number.

A key question is the order of magnitude of the nonlinear susceptibilities. The simplest atom is hydrogen. Its structure and spectrum of excited states is well known and is simple to calculate since it has only one electron, the minimum needed for the interaction of electromagnetic radiation with matter. The atomic Coulomb field binding the electron to the proton in its orbit of Bohr radius r_B is given by

$$E_{\text{atomic}} = \frac{\bar{e}}{4\pi\varepsilon_0 r_B^2}, \qquad r_B = \frac{4\pi\varepsilon_0 \hbar^2}{\bar{m}_e \bar{e}^2}, \qquad (1.9)$$

in which $\bar{e}$ is the charge on the electron (-1.6×10^{-19} C), ε_0 is the permittivity of free space (8.85×10^{-12} F/m), $\bar{m}_e$ ($=9.11 \times 10^{-31}$ kg) is the electron mass, and $h = 2\pi\hbar = 6.63 \times 10^{-34}$ J s is Planck's constant. Equation 1.9 gives the order of magnitude of $E_{\text{atomic}} = 10^{12}$ V/m. It is reasonable to adopt this field as an approximate field at which nonlinear optics becomes important. Since $\chi^{(1)} = n^2 - 1$ (where n is the refractive index of the order of unity) for a perturbation expansion in terms of products of electric fields to be valid, $P^{(1)} \geq 10 \times P^{(2)}$:

$$\frac{P^{(1)}}{P^{(2)}} = \frac{\chi^{(1)}}{\chi^{(2)}E} \approx \frac{1}{\chi^{(2)}E_{\text{atomic}}} \approx 10 \quad \rightarrow \quad \chi^{(2)} \approx 10^{-13} \text{ m/V}. \qquad (1.10)$$

This is a reasonable estimate for the lower limit value of the second-order susceptibility, especially since the field was based on hydrogen, which has only a single electron and proton. Following the same approximations but now assuming that

$$\frac{P^{(1)}}{P^{(3)}} = \frac{\chi^{(1)}}{\chi^{(3)}E^2} \approx \frac{\chi^{(1)}}{\chi^{(3)}E_{\text{atomic}}^2} \approx 10 \quad \rightarrow \quad \chi^{(3)} \approx 10^{-25} \text{ m}^2/\text{V}^2. \qquad (1.11)$$

As will become clear later, these approximate values are close to the minimum values found for these susceptibilities.

1.4 SIMPLE MODEL: ELECTRON ON A SPRING AND ITS APPLICATION TO LINEAR OPTICS

There are many physical mechanisms that lead to nonlinear optical phenomena. Initially, the focus here is on transitions between the electronic states associated with atoms and molecules in matter. Although the appropriate treatment (Chapter 8) for completely describing the interaction of radiation with atoms and molecules involves quantum mechanics, initially a simpler classical approach that provides a useful description of the *linear* (and as it turns out exact) susceptibility is adopted.

As an example of this approach, consider the molecule O_2 and its electron cloud, as illustrated in Fig. 1.6. This molecule has inversion symmetry (i.e., a center of

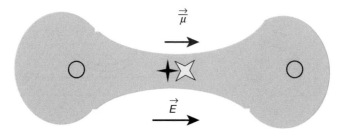

FIGURE 1.6 The oxygen molecule O_2, its center of mass ✕, and positive and negative charges in the absence of the field. After the field is applied, **+** is the center of the negative charge and the electron cloud is shifted (and distorted) from its original position.

symmetry halfway between the oxygen atoms) and hence has no permanent dipole moment since the centers of positive (nuclei) and negative charges are coincident. When an electric field

$$\vec{E}(\vec{r}, t) = \frac{1}{2}\mathbf{E}_x(\omega_a)\, e^{i(kz - \omega_a t)} + \text{c.c.} \tag{1.12}$$

is applied along the molecular axis ($+x$-axis), the negative and positive charges and their centers of charge are displaced in opposite directions by the Coulomb forces, giving rise to the forces $\bar{m}_{e,n}\ddot{x}_{e,n}$, where $\bar{m}_e$ and x_e are the electron mass and its displacement and $\bar{m}_n$ and x_n are the nuclear mass and its displacement, respectively. Since $\bar{m}_n \gg \bar{m}_e$, only the displacements of the electrons are important for inducing dipoles.

The electrons are bound to the nucleus (atoms) or nuclei (molecules) by Coulomb forces and, for isolated atoms or molecules, exist in discrete states m with an energy $\hbar(\bar{\omega}_m - \bar{\omega}_g) = \hbar\bar{\omega}_{mg}$ above the ground state and an excited-state lifetime $\bar{\tau}_{mg}$. They move in "orbits" around nuclei described by probability density functions $\bar{\psi}_m(\vec{r}, t)$, with $|\bar{\psi}_m(\vec{r}, t)|^2\, dx\, dy\, dz\, dt$ giving the probability that the electron "exists" at time t in the volume element $dx\, dy\, dz$ at position $\vec{r}$. Since "optics" usually deals with the spectral region longer in wavelength (smaller in frequency) than the low frequency absorption edge of the material determined by the transitions between electronic states, the electron in the lowest lying energy level is normally the prime participant when radiation interacts with matter; i.e., it is the electron with the largest displacement. With these approximations, the dipole moment induced by an electromagnetic field is $\bar{\mu}_x = \bar{e}_e x_e$, as shown in Fig. 1.6. For the most general case, $\bar{\mu}_i = \bar{\alpha}_{ij}E_j$, where $\bar{\alpha}_{ij}$ is the polarizability tensor, and the induced dipole and the electric field are not necessarily collinear.

In linear optics, it is possible to diagonalize the polarizability tensor. The deflections $\vec{q} = \hat{e}_x\bar{q}_x + \hat{e}_y\bar{q}_y + \hat{e}_z\bar{q}_z$ of this representative electron are defined in terms of these axes and so $\vec{\bar{\mu}} = -\bar{e}\vec{q}$. (Note, however, that for very anisotropic crystal classes the coordinate system may be nonorthogonal and/or frequency dependent.)

The Coulomb interaction between the net positive and negative charges provides a restoring force that oscillates at the frequency of the applied field, and so the motion of

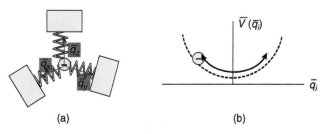

(a)　　　　　　　　　　　　(b)

FIGURE 1.7　(a) Electron connected via three springs oriented along the axes that diagonalize the polarizability. (b) One-dimensional cut of the three-dimensional parabolic potential well inside which the electron oscillates at the frequency of an applied field.

the electron can be described as a simple harmonic oscillator. In three dimensions, this can be visualized as the electron attached to three orthogonal springs, as illustrated in Fig. 1.7a, and the electron motion can be described as oscillation in a harmonic potential well.

The equation of motion of an electron is described by a simple harmonic oscillator with the potential

$$\bar{V}^{(m)}(\bar{q}) = \frac{1}{2}\bar{k}_{ii}^{(m)}\bar{q}_i^{(m)}\bar{q}_i^{(m)}. \tag{1.13}$$

From classical mechanics, the restoring force is given by

$$\bar{F}_i^{(m)} = -\frac{\partial \bar{V}^{(m)}}{\partial \bar{q}_i^{(m)}} = -\frac{1}{2}\bar{k}_{ii}^{(m)}\left[\frac{\partial \bar{q}_i^{(m)}}{\partial \bar{q}_i^{(m)}}\bar{q}_i^{(m)} + \bar{q}_i^{(m)}\frac{\partial \bar{q}_i^{(m)}}{\partial \bar{q}_i^{(m)}}\right] = -\bar{k}_{ii}^{(m)}\bar{q}_i^{(m)}, \tag{1.14}$$

in which the spring constant $\bar{k}_{ii}^{(m)}$ is defined in terms of the excited state's energy by $\bar{\omega}_{mg} = \sqrt{\bar{k}_{ii}^{(m)}/\bar{m}_e}$ and the restoring force is given by $\bar{F}_i^{(m)} = \bar{m}_e\bar{\omega}_{mg}^2\bar{q}_i^{(m)}$. The inertial force is $\bar{m}_e\ddot{\bar{q}}_i^{(m)}$. Therefore, the force balance equation describing the electron motion in a simple harmonic oscillator model is

$$\bar{m}_e\left[\ddot{\bar{q}}_i^{(m)} + 2\bar{\tau}_{mg}^{-1}\dot{\bar{q}}_i^{(m)} + \bar{\omega}_{mg}^2\bar{q}_i^{(m)}\right] = -\bar{e}E_i$$

$$\rightarrow \left[\bar{\omega}_{mg}^2 - \omega_a^2 - 2i\omega_a\bar{\tau}_{mg}^{-1}\right]\bar{q}_i^{(m)}(t) = -\frac{\bar{e}}{\bar{m}_e}E_i(t). \tag{1.15}$$

Assuming

$$\bar{q}_i^{(m)} = \frac{1}{2}\bar{Q}_i^{(m)}e^{-i\omega_a t} + \text{c.c.} \quad \rightarrow \quad \bar{Q}_i^{(m)} = -\frac{\bar{e}E_i(\omega_a)}{\bar{m}_e\bar{D}_i^{(m)}(\omega_a)}, \tag{1.16}$$

where $D_i^{(m)}(\omega_a) = -\omega_a^2 - 2i\omega_a \bar{\tau}_{mg}^{-1} + \bar{\omega}_{mg}^2$ is the resonance denominator. When $\omega_a \approx \bar{\omega}_{mg}$, the amplitude of the displacement is enhanced. Note that $\bar{e}_e = -\bar{e}$ and that in the zero-frequency limit, $D_i^{(m)}(\omega_a) = \bar{\omega}_{mg}^2$ and $\bar{Q}_i^{(m)} = -\bar{e}E_i(\omega_a)/\bar{m}_e\bar{\omega}_{mg}^2$ is just a net steady-state displacement of the electron.

For a dilute medium with N noninteracting atoms (molecules) per unit volume, the induced linear polarization and the first-order susceptibility $\bar{\chi}_{ii}^{(1)}(-\omega_a; \omega_a)$ are given as follows:

$$P_i(t) = -N\bar{e} \sum_m \left\{ \frac{1}{2}\bar{Q}_i^{(m)} e^{-i\omega_a t} + \text{c.c.} \right\} = \frac{1}{2}P_i(\omega_a) e^{-i\omega_a t} + \text{c.c.}$$

$$P_i(\omega_a) = \frac{N\bar{e}^2}{\bar{m}_e} E_i(\omega_a) \sum_m \frac{1}{D_i^{(m)}(\omega_a)} = \varepsilon_0 \hat{\chi}_{ii}^{(1)}(-\omega_a; \omega_a) E_i(\omega_a) \tag{1.17}$$

$$\rightarrow \quad \hat{\chi}_{ii}^{(1)}(-\omega_a; \omega_a) = \frac{N\bar{e}^2}{\varepsilon_0 \bar{m}_e(\omega_a)} \sum_m \frac{1}{D_i^{(m)}(\omega_a)}.$$

The fact that $\hat{\chi}_{ij}^{(1)}(-\omega_a; \omega_a)$ is a diagonal tensor is a direct consequence of choosing a coordinate system in which the polarizability tensor is diagonal.

The first-order susceptibility $\hat{\chi}_{ii}^{(1)}(-\omega_a; \omega_a)$ can easily be defined in terms of $\hat{\chi}_{ii}^{(1)}(\vec{r} - \vec{r}'; t - t')$. From Eq. 1.4,

$$P_i^{(1)}(t) = \varepsilon_0 \left[\int \hat{\chi}_{ij}^{(1)}(t-t')E_j(t')\,d(t-t') \right] = \frac{1}{2}P_i^{(1)}(\omega_a) e^{-i\omega_a t} + \text{c.c.} \tag{1.18}$$

Substituting for the field $E_j(t')$ gives

$$P_i^{(1)}(t) = \frac{\varepsilon_0}{2} \left[\int \hat{\chi}_{ij}^{(1)}(t-t')[E_j(\omega_a) e^{-i\omega_a t'} + \text{c.c.}]\,d(t-t') \right]$$

$$= \frac{1}{2}\varepsilon_0 E_j(\omega_a) e^{-i\omega_a t} \int_{-\infty}^{\infty} \hat{\chi}_{ij}^{(1)}(t-t') e^{i\omega_a(t-t')}\,d(t-t') + \text{c.c.} \tag{1.19}$$

$$\rightarrow \quad \hat{\chi}_{ij}^{(1)}(-\omega_a; \omega_a) = \int_{-\infty}^{\infty} \hat{\chi}_{ij}^{(1)}(t-t') e^{i\omega_a(t-t')}\,d(t-t'),$$

i.e., $\hat{\chi}_{ij}^{(1)}(-\omega_a; \omega_a)$ is the Fourier transform of $\hat{\chi}_{ij}^{(1)}(t-t')$.

Decomposing $\hat{\chi}_{ii}^{(1)}(-\omega_a; \omega_a)$ into its real and imaginary components yields

$$\hat{\chi}_{ii}^{(1)}(-\omega_a; \omega_a) = \frac{N\bar{e}^2}{\bar{m}_e\varepsilon_0} \sum_m \frac{(\bar{\omega}_{mg}^2 - \omega_a^2) + 2i\omega_a\bar{\tau}_{mg}^{-1}}{(\bar{\omega}_{mg}^2 - \omega_a^2)^2 + 4\omega_a^2\bar{\tau}_{mg}^{-2}}. \tag{1.20a}$$

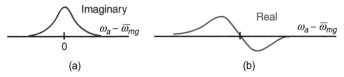

FIGURE 1.8   Spectral dispersion of the (a) imaginary and (b) real parts of $\hat{\chi}_{ii}^{(1)}(-\omega; \omega)$.

This equation is always valid. It can be simplified near and on resonance $(\omega_a \approx \bar{\omega}_{mg})$ to give

$$\hat{\chi}_{ii}^{(1)}(-\omega_a; \omega_a) = \frac{N\bar{e}^2}{2\bar{\omega}_{mg}\bar{m}_e\varepsilon_0} \sum_m \frac{\bar{\omega}_{mg} - \omega_a + i\bar{\tau}_{mg}^{-1}}{(\bar{\omega}_{mg} - \omega_a)^2 + \bar{\tau}_{mg}^{-2}} \tag{1.20b}$$

and off resonance $(|\omega_{mg} - \omega_p|\bar{\tau}_{mg} \gg 1)$ to give

$$\hat{\chi}_{ii}^{(1)}(-\omega_a; \omega_a) = \frac{N\bar{e}^2}{2\bar{\omega}_{mg}\bar{m}_e\varepsilon_0} \sum_m \frac{\bar{\omega}_{mg} - \omega_a + i\bar{\tau}_{mg}^{-1}}{(\bar{\omega}_{mg} - \omega_a)^2}. \tag{1.20c}$$

Figure 1.8 shows the frequency dispersion in the imaginary and real parts of $\hat{\chi}_{ii}^{(1)}(-\omega_a; \omega_a)$ for a single excited state.

The refractive index and the absorption coefficient (for the field) are defined in the usual way by $n^2(\omega) = 1 + \mathfrak{Real}\{\hat{\chi}_{ii}^{(1)}(-\omega; \omega)\}$ and $\alpha(\omega) = k_{vac}\mathfrak{Imag}\{\hat{\chi}_{ii}^{(1)}(-\omega; \omega)\}/2n(\omega)$, respectively. Note that the absorption spectrum, i.e., $\alpha(\omega)$, has contributions only from transitions from the ground state that are electric dipole allowed. For symmetric molecules in which the states are described by wave functions that are either symmetric or antisymmetric in space, the linear absorption spectrum does not contain contributions from the even-symmetry excited states because electric dipole transitions from the even-symmetry ground state are not dipole allowed.

As stated previously, optics normally refers to electromagnetic waves in the spectral region defined by frequencies below the lowest lying electronic resonance due to electric dipole transitions. Assuming that $|\bar{\omega}_{mg} - \omega_a|\bar{\tau}_{mg}^{-1} \gg 1$, $\mathfrak{Imag}\{\hat{\chi}_{ii}^{(1)}(-\omega_a; \omega_a)\}$ decreases faster with increasing frequency difference from the resonance $|\bar{\omega}_{mg} - \omega_a|$ than does $\mathfrak{Real}\{\hat{\chi}_{ii}^{(1)}(-\omega_a; \omega_a)\}$. This will also be the case for the real and imaginary parts of the nonlinear susceptibilities.

1.5 LOCAL FIELD CORRECTION

Although local field correction is discussed in most introductory textbooks on optics, it will prove useful to repeat it here since the transition to nonlinear optics is not straightforward. The preceding analysis for the linear susceptibility was for a single isolated atom or molecule and, to a good approximation, for a dilute gas. The situation is more complex in dense gases or condensed matter (liquids and solids) where the atoms and molecules interact with one another via the dipole fields induced by an applied optical field.

Experiments are usually performed with an optical field incident onto a nonlinear material from another medium, typically air. Maxwell's equations in the material and the usual boundary conditions at the interface are valid for spatial averages of the fields over volume elements small on the scale of a wavelength, but large on the scale of a molecule. The "averaged" quantities also include the refractive index, the Poynting vector, and the so-called Maxwell field, which has been written here as $\vec{E}(\vec{r}, t)$. It is the Maxwell field that satisfies the wave equation for a material with the averaged refractive index n.

However, at the site of a molecule the situation can be quite complex since the dipoles induced by the Maxwell electric fields on all the molecules create their own electric fields, which must be added to the "averaged" field to obtain the total ("local") field $\vec{E}_{\text{loc}}(\vec{r}, t)$ acting on a molecule, as shown in Fig. 1.9. In the low density limit, the dipolar fields decay essentially to zero with distance from their source dipole and so $\vec{E}_{\text{loc}}(\vec{r}, t) \cong \vec{E}(\vec{r}, t)$ and the single molecule result is converted to a macroscopic polarization by multiplying the molecular result by N, the number of molecules per unit volume.

The situation is more complex in condensed matter. It is very difficult to calculate the "local" field accurately because it depends on crystal symmetry, intermolecular interactions, and so on. Standard treatments such as Lorenz–Lorenz are only approximately valid even for isotropic and cubic crystal media. Nevertheless, they are universally used. Here the usual formulation found in standard electromagnetic textbooks will be followed. The dipole moments of the molecules induced by the Maxwell field $\vec{E}(\vec{r}, t)$ produce a Maxwell polarization $\vec{P}(\vec{r}, t) = \varepsilon_0 \vec{\chi}^{(1)} : \vec{E}(\vec{r}, t)$ in the material. Consider a spherical cavity around the molecule of interest to find the local field acting on the molecule (see Fig. 1.9c). Assuming that the effects of the induced dipoles inside the cavity average to zero, the polarization field outside the cavity induces charges on the walls of the cavity, which produce an additional electric field on the molecule in the cavity (see standard texts on electrostatics):

$$\left\langle \sum \vec{E}_{\text{dipoles}}(\vec{r}, t) \right\rangle = \frac{1}{3\varepsilon_0} \vec{P}(\vec{r}, t) \quad \rightarrow \quad \vec{E}_{\text{loc}}(\vec{r}, t) = \vec{E}(\vec{r}, t) + \frac{1}{3\varepsilon_0} \vec{P}^{(1)}(\vec{r}, t).$$

$$(1.21)$$

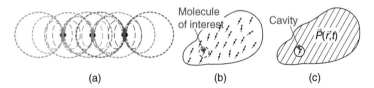

(a)   (b)   (c)

FIGURE 1.9 (a) The local fields created by the induced dipoles in a medium. (b) Dipoles induced everywhere in the material. The average gives the Maxwell polarization $\vec{P}(\vec{r}, t)$. (c) Artificial spherical cavity assumed around the molecule of interest, embedded in a uniform medium with polarization $\vec{P}(\vec{r}, t)$.

The induced dipole on a molecule at the center of the cavity is now given by

$$\vec{p}(1)(\vec{r},t) = \stackrel{\leftrightarrow}{\alpha} \cdot \left[\vec{E}(\vec{r},t) + \frac{1}{3\varepsilon_0} \vec{P}^{(1)}(\vec{r},t) \right]$$

$$\rightarrow \vec{P}^{(1)}(\vec{r},t) = N\stackrel{\leftrightarrow}{\alpha} \cdot \left[\vec{E}(\vec{r},t) + \frac{1}{3\varepsilon_0} \vec{P}^{(1)}(\vec{r},t) \right]$$

$$\rightarrow \left[1 - \frac{1}{3}N\stackrel{\leftrightarrow}{\alpha} \right] \cdot \vec{P}^{(1)}(\vec{r},t) = N\stackrel{\leftrightarrow}{\alpha} \cdot \vec{E}(\vec{r},t). \tag{1.22}$$

From the Clausius–Mossotti relation that connects the macroscopic relative dielectric constant to the molecular polarizability,

$$\frac{\varepsilon_r - 1}{\varepsilon_r + 2} = \frac{1}{3}\langle N\stackrel{\leftrightarrow}{\alpha} \rangle \quad \rightarrow \quad \vec{P}^{(1)}(\vec{r},t) = \frac{\varepsilon_r + 2}{3}N\stackrel{\leftrightarrow}{\alpha} \cdot \vec{E}(\vec{r},t), \tag{1.23}$$

and so the local field and the local field correction $f^{(1)}$ is defined as

$$\vec{E}_{\text{loc}}(\vec{r},t) = \frac{\varepsilon_r(\omega_a) + 2}{3}\vec{E}(\vec{r},t), \qquad f^{(1)} = \frac{\varepsilon_r + 2}{3}, \tag{1.24}$$

respectively, where $\varepsilon_r(\omega_a) = \varepsilon(\omega_a)/\varepsilon_0$. Since the field driven displacement is now

$$\bar{Q}_i^{(m)} = -f^{(1)} \frac{\bar{e}E_i(\omega_a)}{\bar{m}_e D_i^{(m)}(\omega_a)}, \tag{1.25}$$

the linear susceptibility from Eq. 1.20a, including the local field correction, becomes

$$\chi_{ii}^{(1)}(-\omega_a; \omega_a) = \frac{N\bar{e}^2 f^{(1)}(\omega_a)}{\bar{m}_e \varepsilon_0} \sum_m \frac{(\bar{\omega}_{mg}^2 - \omega_a^2) + 2i\omega_a \bar{\tau}_{mg}^{-1}}{(\bar{\omega}_{mg}^2 - \omega_a^2)^2 + 4\omega_a^2 \bar{\tau}_{mg}^{-2}}. \tag{1.26}$$

PROBLEMS

1. The purpose of this problem is to show that absorption decreases much faster than refractive index with frequency difference from a resonance. Consider an isolated molecule with a single excited state with a transition frequency $\bar{\omega}_i$ and a phenomenological decay constant $\bar{\tau}_i^{-1}$.

   (a) Assuming that $\bar{\tau}_i^{-1}\bar{\omega}_i \gg 1$ and $\omega \approx \bar{\omega}_i$, show that $\chi^{(1)}$ can be written as

   $$\hat{\chi}_{ii}^{(1)}(-\omega; \omega) = \Re\text{eal}\{\hat{\chi}_{ii}^{(1)}\} + i\,\Im\text{mag}\{\hat{\chi}_{ii}^{(1)}\}$$

   $$= \frac{Ne^2(\bar{\omega}_i - \omega)}{2\bar{\omega}_i\bar{m}\varepsilon_0[(\bar{\omega}_i - \omega)^2 + \bar{\tau}_i^{-2}]}$$

   $$+ i\frac{Ne^2\omega\bar{\tau}_i - 1}{2\bar{\omega}_i^2\bar{m}\varepsilon_0[(\bar{\omega}_i - \omega)^2 + \bar{\tau}_i^{-2}]}.$$

(b) Find the maximum change in the real part of the susceptibility and show that the ratio R of change at frequency ω to maximum change occurs at a frequency shift given by $\bar{\omega}_i - \omega \cong 2\bar{\tau}_i^{-1}/R$ for $|\bar{\omega}_i - \omega| \gg 2\bar{\tau}_i^{-1}$.

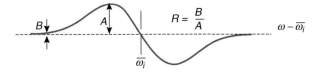

(c) Find the maximum change in the imaginary part of the susceptibility and show that the ratio P of the change at frequency ω to maximum change is given by

$$P = \frac{\bar{\tau}_i^{-2}}{\left[(\bar{\omega}_i - \omega)^2 + (\bar{\tau}_i^{-2})\right]}.$$

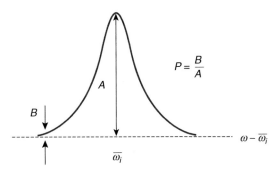

(d) How small is P for values of R equal to 10%: 1%? Although you have calculated this difference for the linear susceptibility, the results are also typical of what is obtained for higher order susceptibilities. A 10% or less "remnant" in the susceptibility is the upper limit for calling the value "nonresonant."

SUGGESTED FURTHER READING

B. I. Bleaney and B. Bleany, Electricity and Magnetism, 2nd Edition (Oxford University Press, London, 1968).

M. Born and E. Wolf, Principles of Optics, 7th Edition (Cambridge University Press, Cambridge, UK, 1999).

R. W. Boyd, Nonlinear Optics, 3rd Edition (Academic Press, Burlington, MA, 2008).

F. A. Hopf and G. I. Stegeman, Applied Classical Electrodynamics, Volume 1: Linear Optics (John Wiley & Sons, New York, 1985).

D. L. Mills, Nonlinear Optics: Basic Concepts, 2nd Edition (Springer, New York, 1998).

Y. Ron Shen, Principles of Nonlinear Optics (John Wiley & Sons, New York, 1984).

SECOND-ORDER PHENOMENA

Second-Order Susceptibility and Nonlinear Coupled Wave Equations

It will be shown in Chapter 8 that the electron on a spring model does not accurately give the frequency dispersion of nonlinear susceptibilities. Nevertheless, for nonlinear optics in second-order materials, it does provide an *approximate* spectral distribution of susceptibilities but *not* their actual magnitudes. This is adequate at this stage since the magnitudes of the coefficients are normally obtained experimentally and they are used near the nonresonant limit, i.e., suitably far from the resonances associated with the electronic molecular states.

The simple harmonic oscillator model for the linear susceptibility obtained in Chapter 1 is now extended by adding an additional (cubic) term to the potential well. An expression for the second-order susceptibility is then derived by using an anharmonic oscillator model. The exact quantum mechanical derivation of second-order susceptibilities in terms of measurable molecular parameters will be given in Chapter 8. The anharmonic oscillator model is sufficiently accurate in the spectral regions far away from the resonances associated with transitions between energy levels. For applications, the measured susceptibilities are used since there is no reliable way to calculate the nonlinear force constant required in the anharmonic oscillator model.

The additional nonlinear term is obtained by taking the product of a nonlinear force constant with the product of two linear displacements obtained from the linear oscillator model. Assuming that the contribution of the nonlinearity to the total electron displacement is small, the nonlinear polarizations are derived, leading to the second-order susceptibilities $\chi^{(2)}$. The slowly varying phase and amplitude equation is formulated for cases involving second harmonic generation (SHG) and sum and difference frequency generation. The slowly varying phase and amplitude equation allows the generated harmonic fields to be calculated from the nonlinear polarizations. These equations in the Kleinman (zero-frequency) limit are then used to demonstrate the Manley–Rowe relations, which verify energy conservation and the photon flux relations.

Nonlinear Optics: Phenomena, Materials, and Devices, George I. Stegeman and Robert A. Stegeman.
© 2012 John Wiley & Sons, Inc. Published 2012 by John Wiley & Sons, Inc.

2.1 ANHARMONIC OSCILLATOR DERIVATION OF SECOND-ORDER SUSCEPTIBILITIES

The electron potential well in which the electron is bound, as given in Eq. 1.3, is now extended to include a term cubic in the displacement; i.e.,

$$\bar{V}^{(m)}\left(\bar{q}_i^{(m)}\right) = \frac{1}{2}\bar{k}_{ii}^{(m)}\bar{q}_i^{(m)}\bar{q}_i^{(m)} + \frac{1}{3}\bar{k}_{ijk}^{(m)}\bar{q}_i^{(m)}\bar{q}_j^{(m)}\bar{q}_k^{(m)}. \tag{2.1}$$

The shape of the potential well versus displacement needs to be asymmetric about the origin for nonzero second-order susceptibilities; i.e., $\bar{V}^{(m)}(\vec{q}^{(m)}) \neq \bar{V}^{(m)}(-\vec{q}^{(m)})$. This is illustrated in Fig. 2.1 in one dimension.

Materials that have this property are called noncentrosymmetric materials. They also exhibit piezoelectricity, and their molecules typically have permanent dipole moments in the ground state.

Note that the anharmonic "force constant" $\bar{k}_{ijk}^{(m)}$ cannot, in general, be diagonalized in the coordinate system established for polarizability. Furthermore, because the terms $\bar{q}_i^{(m)}\bar{q}_j^{(m)}\bar{q}_k^{(m)}$ can be permuted arbitrarily without changing the potential $\bar{V}^{(m)}(\bar{q})$, $\bar{k}_{xxy}^{(m)} = \bar{k}_{xyx}^{(m)} = \bar{k}_{yxx}^{(m)}$, and so on. Again, using the well-known classical mechanics result that the gradient of the potential along a displacement direction gives the force applied to a mass along that direction, i.e., $\bar{F}_i^{(m)} = -\partial\bar{V}^{(m)}/\partial\bar{q}_i^{(m)}$, and the relations $\partial\bar{q}_j^{(m)}/\partial\bar{q}_i^{(m)} = \delta_{ij}$ and $\partial\bar{q}_k^{(m)}/\partial\bar{q}_i^{(m)} = \delta_{ik}$, we obtain

$$\bar{F}_i^{(m)} = -\frac{\partial\bar{V}^{(m)}}{\partial\bar{q}_i^{(m)}} = -\bar{k}_{ii}^{(m)}\bar{q}_i^{(m)} - \bar{k}_{ijk}^{(m)}\bar{q}_j^{(m)}\bar{q}_k^{(m)}. \tag{2.2}$$

Substituting this additional nonlinear force term into Eq. 1.15 gives

$$\left[\bar{\omega}_{mg}^2 - \omega^2 - 2i\omega\bar{\tau}_{mg}^{-1}\right]\bar{q}_i^{(m)}(t) = -\frac{\bar{e}}{\bar{m}_e}E_i - \frac{1}{\bar{m}_e}\bar{k}_{ijk}^{(m)}\bar{q}_j^{(m)}\bar{q}_k^{(m)}, \tag{2.3}$$

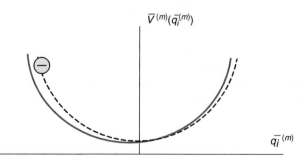

FIGURE 2.1 Comparison between a harmonic potential (dashed line) and an anharmonic potential (solid line) of the form given in Eq. 2.1.

which cannot be solved analytically. The product of the two displacements, both oscillating at frequency ω, leads to nonlinear displacements at the sum frequency 2ω $[\bar{q}_i^{(m)}(2\omega)]$, the difference frequency 0 $[\bar{q}_i^{(m)}(0)]$ with $|\bar{q}_i^{(m)}(\omega)| \gg |\bar{q}_i^{(m)}(2\omega)|, |\bar{q}_i^{(m)}(0)|$; i.e., the nonlinear contributions are small. The total displacement is

$$\bar{q}_i^{(m)} = \bar{q}_i^{(m)}(\omega) + \bar{q}_i^{(m)}(2\omega) + \bar{q}_i^{(m)}(0) + \text{potentially additonal higher order generated fields.} \tag{2.4}$$

Perturbative solutions to Eq. 2.3 involve first setting the nonlinear term to zero, then solving for the linear displacement $\bar{q}_i^{(m)}(\omega)$, inserting this solution into the nonlinear term, and finally solving Eq. 2.3 for the nonlinear displacement. The nonlinear displacement product is

$$\bar{q}_j^{(m)}(\omega)\bar{q}_k^{(m)}(\omega) = \frac{1}{4}\left[\bar{Q}_j^{(m)}(\omega)\,e^{-i\omega t} + \bar{Q}_j^{*(m)}(\omega)\,e^{+i\omega t}\right]$$

$$\times\left[\bar{Q}_k^{(m)}(\omega)\,e^{-i\omega t} + \bar{Q}_k^{*(m)}(\omega)\,e^{+i\omega t}\right]$$

$$= \frac{1}{4}\bar{Q}_j^{(m)}(\omega)\bar{Q}_k^{(m)}(\omega)\,e^{-2i\omega t} + \text{c.c.} + \frac{1}{4}\left[\bar{Q}_j^{(m)}(\omega)\bar{Q}_k^{*(m)}(\omega)\right.$$

$$\left. + \bar{Q}_j^{*(m)}(\omega)\bar{Q}_k^{(m)}(\omega)\right] + \text{c.c.} \tag{2.5}$$

The first term leads to SHG and the second to a zero frequency (DC)-induced electric field. Solving Eq. 2.3 for $\bar{q}_i^{(m)}(2\omega)$ when only a *single-eigenmode* (i.e., field polarized along one of the optical axes) fundamental is assumed, we obtain

$$\bar{q}_i^{(m)}(2\omega) = \frac{1}{2}\bar{Q}_i^{(m)}(2\omega)\,e^{-2i\omega t} + \text{c.c.} = -\frac{1}{4}\bar{k}_{ijk}^{(m)}\frac{\bar{e}^2}{m_e^3}\frac{E_j^2(\omega)}{D_i^{(m)}(2\omega)D_j^{(m)}(\omega)D_k^{(m)}(\omega)}\,e^{-2i\omega t} + \text{c.c.}, \tag{2.6}$$

with

$$D_i^{(m)}(\omega) = -\omega^2 - 2i\omega\bar{\tau}_{mg}^{-1} + \bar{\omega}_{mg}^2,$$

$$D_i^{(m)}(2\omega) = -4\omega^2 - 4i\omega\bar{\tau}_{mg}^{-1} + \bar{\omega}_{mg}^2. \tag{2.7}$$

The widely used notation for the interaction leading to SHG with a single-eigenmode incident is "Type 1" SHG. Defining the macroscopic nonlinear polarization as $P_i^{(2)}(2\omega) = -N\bar{e}\sum_m \bar{Q}_i^{(m)}(2\omega)$, we obtain

$$P_i^{(2)}(2\omega) = \frac{N\bar{e}^3}{2\bar{m}_e^3} \sum_m \bar{k}_{ijj}^{(m)} \frac{E_j^2(\omega)}{D_i^{(m)}(2\omega)D_j^{(m)2}(\omega)} = \frac{\varepsilon_0}{2} \hat{\chi}_{ijj}^{(2)}(-2\omega;\omega,\omega)E_j^2(\omega)$$

$$\rightarrow \quad \hat{\chi}_{ijj}^{(2)}(-2\omega;\omega,\omega) = \frac{Ne^3}{\varepsilon_0\bar{m}_e^3} \sum_m \frac{\bar{k}_{ijj}^{(m)}}{D_i^{(m)}(2\omega)D_j^{(m)2}(\omega)}, \tag{2.8}$$

in which $\hat{\chi}_{ijj}^{(2)}(-2\omega;\omega,\omega)$ is the Fourier transform of the appropriate $\hat{\chi}_{ijk}^{(2)}(t-t',t-t'')$. (This can be easily proven along the lines of the linear susceptibility case in Chapter 1.) Similarly, for the DC term,

$$\bar{q}_j^{(m)}(\omega)\bar{q}_j^{(m)}(\omega) = \frac{1}{4}\left[\bar{Q}_j^{(m)}(\omega)\bar{Q}_j^{*(m)}(\omega) + \bar{Q}_j^{*(m)}(\omega)\bar{Q}_j^{(m)}(\omega) + \text{c.c.}\right]$$

$$\rightarrow \quad P_i^{(2)}(0) = \frac{N\bar{e}^3}{2\bar{m}_e^3} \sum_m \bar{k}_{ijj}^{(m)}\left[\frac{E_j(\omega)E_j^*(\omega)}{D_i^{(m)}(0)D_j^{(m)}(\omega)D_j^{(m)*}(\omega)} + \frac{E_j^*(\omega)E_j(\omega)}{D_i^{(m)}(0)D_j^{(m)*}(\omega)D_j^{(m)}(\omega)}\right]$$

$$= \frac{\varepsilon_0}{2}\left[\hat{\chi}_{ijj}^{(2)}(-0;\omega,-\omega) + \hat{\chi}_{ijj}^{(2)}(-0;-\omega,\omega)\right]E_j(\omega)E_j^*(\omega)$$

$$\rightarrow \quad \hat{\chi}_{ijj}^{(2)}(-0;\omega,-\omega) = \hat{\chi}_{ijj}^{(2)}(-0;-\omega,\omega) = \frac{N\bar{e}^3\bar{k}_{ijj}}{\varepsilon_0\bar{m}_e^3\bar{\omega}_i^2}\sum_m \frac{\bar{k}_{ijj}^{(m)}}{D_i^{(m)}(0)D_j^{(m)*}(\omega)D_j^{(m)}(\omega)} \tag{2.9}$$

The notation adopted for $\hat{\chi}_{ijk}^{(2)}(-2\omega;\omega,\omega)$ and $\hat{\chi}_{ijj}^{(2)}(-0;-\omega,\omega)$ is standard in nonlinear optics. -2ω and -0 refer to the output frequency of the nonlinear polarization, and the two separate values of ω in this case refer to the input beams that are mixed—in this case a fundamental beam with itself.

This approach is easily extended to SHG by using two orthogonally polarized eigenmode fields and to sum and difference frequency cases. Note that this case includes a single incident field not polarized along a crystal axis; i.e., the input beam breaks up into two eigenmodes inside the crystal. The input fields may have difference frequencies, e.g., two being

$$\vec{E}_a(\vec{r},t) = \frac{1}{2}\vec{E}(\omega_a)\,e^{-i\omega_a t} + \text{c.c.}, \qquad \vec{E}_b(\vec{r},t) = \frac{1}{2}\vec{E}(\omega_b)\,e^{-i\omega_b t} + \text{c.c.},$$

$$\rightarrow \quad \vec{E}_c(\vec{r},t) = \frac{1}{2}\vec{E}(\omega_c)\,e^{-i\omega_c t} + \text{c.c.}, \tag{2.10}$$

with the third being the generated field. This *two-eigenmode input* case is labeled as a "Type 2" interaction. Solutions for the linear displacements in this case are

$$\bar{Q}_i^{(m)}(\omega_a) = -\frac{\bar{e}E_i(\omega_a)}{\bar{m}_e D_i^{(m)}(\omega_a)}, \qquad \bar{Q}_j^{(m)}(\omega_b) = -\frac{\bar{e}E_j(\omega_b)}{\bar{m}_e D_j^{(m)}(\omega_b)},$$

$$\bar{Q}_k^{(m)}(\omega_c) = -\frac{\bar{e}E_k(\omega_c)}{\bar{m}_e D_k^{(m)}(\omega_c)}. \tag{2.11}$$

Since the total linear displacement is $\vec{q} = \hat{e}_i \bar{q}_i^{(m)}(\omega_a) + \hat{e}_j \bar{q}_j^{(m)}(\omega_b) + \hat{e}_k \bar{q}_k^{(m)}(\omega_c)$, the fields generated by the interactions will contain many different terms. Usually at this stage it is necessary to isolate the interactions of interest, e.g., $\omega_c = \omega_a + \omega_b$ (sum frequency generation) or $\omega_a = \omega_c - \omega_b$ (difference frequency generation). Assuming the interaction of interest between the frequencies to be $\omega_c = \omega_a + \omega_b$, the sum frequency product $\bar{Q}_j^{(m)}(\omega_a)\bar{Q}_k^{(m)}(\omega_b)$ and the difference frequency products $\bar{Q}_j^{(m)}(\omega_c)\bar{Q}_k^{(m)*}(\omega_b)$ and $\bar{Q}_j^{(m)}(\omega_c)\bar{Q}_k^{(m)*}(\omega_a)$ can be generated efficiently. As an example of the procedure used, consider the case $\omega_a = \omega_c - \omega_b$, in which

$$\bar{Q}_i^{(m)}(\omega_a) = -\frac{\bar{e}^2}{2\bar{m}_e^3}\left[\frac{\bar{k}_{ijk}^{(m)}E_j^*(\omega_b)E_k(\omega_c)}{D_i^{(m)}(\omega_a)D_j^{(m)*}(\omega_b)D_k^{(m)}(\omega_c)} + \frac{\bar{k}_{ijk}^{(m)}E_j(\omega_c)E_k^*(\omega_b)}{D_i^{(m)}(\omega_a)D_j^{(m)}(\omega_c)D_k^{(m)*}(\omega_b)}\right]. \tag{2.12}$$

Since $\bar{k}_{ijk}^{(m)} = \bar{k}_{ikj}^{(m)}$,

$$P_i^{(2)}(\omega_a) = -N\bar{e}\sum_m \bar{Q}_i^{(m)}(\omega_a)$$

$$P_i^{(2)}(\omega_a) = \frac{N\bar{e}^3}{2\bar{m}_e^3}\sum_m \bar{k}_{ijk}^{(m)}\left[\frac{E_j^*(\omega_b)\,E_k(\omega_c)}{D_i^{(m)}(\omega_a)D_j^{(m)*}(\omega_b)D_k^{(m)}(\omega_c)}\right.$$

$$\left. + \frac{E_j(\omega_c)E_k^*(\omega_b)}{D_i^{(m)}(\omega_a)D_j^{(m)}(\omega_c)D_k^{(m)*}(\omega_b)}\right]. \tag{2.13}$$

Noting that $i, j,$ and k are all "dummy" indices, each of which takes the values $x, y,$ and z, respectively, we obtain

$$P_i^{(2)}(\omega_a) = \frac{1}{2}\varepsilon_0\left[\hat{\chi}_{ijk}^{(2)}(-\omega_a; -\omega_b, \omega_c) + \hat{\chi}_{ikj}^{(2)}(-\omega_a; \omega_c, -\omega_b)\right]E_j^*(\omega_b)E_k(\omega_c). \tag{2.14}$$

$$\rightarrow \quad \hat{\chi}_{ijk}^{(2)}(-\omega_a; -\omega_b, \omega_c) = \frac{N\bar{e}^3}{\varepsilon_0\bar{m}_e^3}\sum_m \frac{\bar{k}_{ijk}^{(m)}}{D_k(\omega_c)D_j^*(\omega_b)D_i(\omega_a)} = \hat{\chi}_{ikj}^{(2)}(-\omega_a; \omega_c, -\omega_b). \tag{2.15}$$

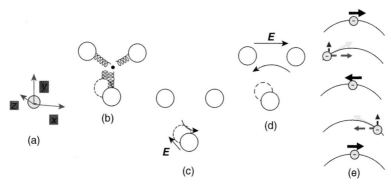

FIGURE 2.2   (a) Axis system used. (b) The tetrahedral arrangement of oxygen atoms and the bonds (springs) connecting to the "active" electron located at the phosphorus site. (c) and (d) The displacement of the electron due to two orthogonally polarized optical fields. (e) The electron displacements over one complete cycle of the applied x-polarized field (black curves). The induced polarization (blue-green arrows) has projections along both the y-axis (blue dashed arrow) and the x-axis (red dashed arrows).

Similarly, for the sum frequency case,

$$P_i^{(2)}(\omega_c) = \frac{1}{2}\varepsilon_0 \left[\hat{\chi}_{ijk}^{(2)}(-\omega_c; \omega_b, \omega_a) + \hat{\chi}_{ikj}^{(2)}(-\omega_c; \omega_a, \omega_b) \right] E_j(\omega_b) E_k(\omega_a) \quad (2.16)$$

$$\rightarrow \hat{\chi}_{ijk}^{(2)}(-\omega_c; \omega_b, \omega_a) = \frac{N\bar{e}^3}{\varepsilon_0 \bar{m}_e^3} \sum_m \frac{\bar{k}_{ijk}^{(m)}}{D_i^{(m)}(\omega_a) D_j^{(m)}(\omega_b) D_k^{(m)}(\omega_c)} = \hat{\chi}_{ikj}^{(2)}(-\omega_c; \omega_a, \omega_b).$$

$$(2.17)$$

It is interesting to investigate how off-diagonal elements in the $\hat{\chi}_{ikj}^{(2)}$ tensor arise physically. KH_2PO_4 is a very common doubling crystal that uses a Type 2 interaction (which is simplified to an effective Type 1 interaction in the following example by discussing the interaction in a coordinate system that does not coincide with the diagonalization of the polarizability tensor for the whole molecule). The origin of the nonlinearity lies in the P–O bond. The electron "responsible for" $\hat{\chi}_{ikj}^{(2)} \neq 0$ can be considered to reside at the center of the PO_4 group, as shown in Fig. 2.2b. Due to the tetrahedral arrangement of oxygen atoms, the electron's displacement path when strong optical fields are applied is shown in Fig. 2.2c and d. These trajectories are plotted as a function of time, as shown in Fig. 2.2e. For an x-polarized optical field, note that the y-polarized component of the induced polarization oscillates at 2ω and that there is a DC polarization offset also along the y-axis (see Figs 2.2e and 2.3). In the axis system assumed, this would give rise to $\hat{\chi}_{yxx}^{(2)}(-2\omega; \omega, \omega)$. In the proper axis

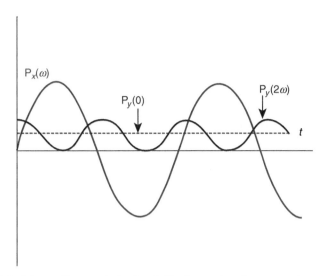

FIGURE 2.3 The nonlinear polarization fields due to the electron motion excited by an oscillating field polarized along the x-axis.

system in which the polarizability is diagonalized, the nonlinear coefficient is actually $\hat{\chi}^{(2)}_{zxy}(-2\omega; \omega, \omega)$.

2.2 INPUT EIGENMODES, PERMUTATION SYMMETRY, AND PROPERTIES OF $\chi^{(2)}$

From Eqs 2.14–2.17, further simplifications are clearly possible. Because all the indices can be permuted in the nonlinear force constant, $\hat{\chi}^{(2)}_{ijk}(-\omega_a; -\omega_b, \omega_c) = \hat{\chi}^{(2)}_{ikj}(-\omega_a; \omega_c, -\omega_b)$. This is called permutation symmetry and is a very powerful tool rooted in the anharmonic oscillator model. Note, however, that both the relevant indices and the frequencies are permuted *together*.

Another important aspect that will surface again later in $\chi^{(3)}$ is the number of input fields that are separate eigenmodes *in the nonlinear material*. To be defined as a separate eigenmode, an input field inside the crystal must be defined as a single polarization along a crystal axis, with a single frequency and a single propagation direction. For example, a single input beam (outside the crystal) polarized at 45° to the optical axes of a crystal excites two eigenmodes, which can give rise to Type 1 and/or Type 2 SHG. Also, $E_i(\omega)$ and $E_i(-\omega) \equiv E_i^*(\omega)$ are considered to be individual eigenmodes *in nonlinear optics* since the products $E_i^2(\omega)$ and $E_i^*(\omega)E_i(\omega)$ give rise to different outputs. Specifically, with permutation symmetry with two input eigenmodes, e.g., for Type 2 SHG, i.e., $j \neq k$,

$$P_i^{(2)}(2\omega) = \varepsilon_0 \hat{\chi}^{(2)}_{ijk}(-2\omega; \omega, \omega)E_j^a(\omega)E_k^b(\omega). \tag{2.18}$$

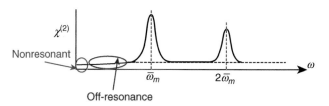

FIGURE 2.4   Definition (location) of the nonresonant and off-resonance regimes for second harmonic generation.

On the other hand, for each of the individual eigenmodes considered as the input separately, $j = k$ and

$$P_i^{(2)}(2\omega) = \frac{1}{2}\varepsilon_0\hat{\chi}_{ijj}^{(2)}(-2\omega;\omega,\omega)E_j^{a2}(\omega), \qquad P_i^{(2)}(2\omega) = \frac{1}{2}\varepsilon_0\hat{\chi}_{ijj}^{(2)}(-2\omega;\omega,\omega)E_j^{b2}(\omega).$$

$$(2.19)$$

Finally, note that DC field generation and, in general, sum and difference frequency generation all have two separate eigenmodes as their input.

To avoid linear losses, efficient parametric frequency conversion usually requires that both the output and the input frequencies be far from the lowest lying absorption peak. This requires operation in the far off-resonance and preferably nonresonant regime (see Fig. 2.4). In the nonresonant regime $\bar{\omega}_m^2 \gg \omega^2$,

$$\tilde{\chi}_{ijk}^{(2)}(-2\omega;\omega,\omega) = \frac{N\bar{e}^3}{\varepsilon_0\bar{m}_e^3\bar{e}^3}\sum_m \frac{\bar{k}_{ijk}^{(m)}}{\bar{\omega}_m^6},$$

which is a real constant, independent of frequency. Here the "tilde" above identifies a parameter in the nonresonant regime. Note that in the nonresonant regime *all* three subscripts and frequency arguments of $\tilde{\chi}_{ijk}^{(2)}(-2\omega;\omega,\omega)$ can be permuted, a property associated with Kleinman symmetry.

As indicated in Eq. 2.9, there is a resonance enhancement in $\hat{\chi}_{ijk}^{(2)}(-2\omega;\omega,\omega)$ near the frequencies $\omega \approx \bar{\omega}_m$ and $\omega \approx \bar{\omega}_m/2$. This occurs because the denominators $D_i(2\omega)D_j(\omega)D_k(\omega)$ are minimized there. Note that just off the resonances, the second-order susceptibility can take on complex values. This is unrelated to loss and just signifies a phase shift (in the generated fields as will be evident later).

In the early days of nonlinear optics when most dielectric second harmonic active crystals contained oxygen bonding to elements such as phosphorous and niobium with comparable permanent dipole moments, it was experimentally found that for a limited number of crystals the second-order susceptibility could be written as

$$\hat{\chi}_{ijk}^{(2)}(-2\omega;\omega,\omega) = \Delta_{ijk}\hat{\chi}_{ii}^{(2)}(2\omega)\hat{\chi}_{jj}^{(2)}(\omega)\chi_{kk}^{(1)}(\omega), \qquad (2.20)$$

with $\Delta_{ijk} = \varepsilon_0^2 \bar{k}_{ijk}/N^2 \bar{e}^3 \approx 1$–$4$ (Miller's coefficient). This coefficient is now known to vary widely in different classes of materials, e.g., taking values up to a thousand in organic crystals.

The local field factor has not been included in the above analysis for practical reasons. It of course should be. However, the discussion of local field effects in nonlinear optics is not a simple extension from linear optics. It will be discussed in Chapter 8. At this point, local field enhancements can be considered to be included in the nonlinear force constant, which will, as a result, vary with the local environment around a molecule. In second-order nonlinear optics, calculations are typically based on the measured values of second-order susceptibilities, which already contain local field effects.

2.3 SLOWLY VARYING ENVELOPE APPROXIMATION

The SVEA (slowly varying envelope approximation, sometimes called the slowly varying phase and amplitude approximation) is a perturbation approach that is widely used for calculating many different wave phenomena in acoustics, plasmas, and so on, as well as in optics. This is a formalism that can be used to calculate from the nonlinear polarization the evolution of an eigenmode field generated at the output frequency of a nonlinear interaction, similar to first-order perturbation theory in quantum mechanics. It is an alternative to finding the generated fields by solving exactly the nonlinear wave equations including the nonlinear polarization terms derived in this chapter. The SVEA becomes exact when the phase velocity of the nonlinear polarization field as determined by the interacting fields matches that of the eigenmode at the generated frequency. This corresponds to the wave-vector-matching condition to be discussed in Chapter 3. The SVEA is derived for plane-wave eigenmodes propagating along the z-axis and leads to an integral that is simple to evaluate in many common geometries or, if necessary, numerically. It works approximately for beams whose width in the transverse beam coordinates x and y is many wavelengths. (For beams confined in waveguides, an equivalent form of the SVEA can be derived.)

The starting point for the derivation of the SVEA is the wave equation to which a perturbation polarization $\vec{P}^p(\vec{r}, t)$ is added. This polarization can be due to a number of different phenomena, such as nonlinear optics, electro-optics, magneto-optics, and acousto-optics.

$$\nabla^2 E_i(\vec{r}, t) - \frac{1}{c^2} \frac{\partial^2 E_i(\vec{r}, t)}{\partial t^2} = \mu_0 \frac{\partial^2 [P_i(\vec{r}, t) + P_i^p(\vec{r}, t)]}{\partial t^2}. \qquad (2.21)$$

Here $P_i(\vec{r}, t)$ is the linear polarization that leads to the refractive index and $P_i^p(\vec{r}, t)$ in nonlinear optics will be called $P_i^{\mathrm{NL}}(\vec{r}, t)$. Assuming that $P_i^p(\vec{r}, t)$ can be written as

$$P_i^p(\vec{r}, t) = \frac{1}{2} \mathbf{P}_i^p(z, t) \, e^{-i\omega_p t} + \text{c.c.} = \frac{1}{2} \boldsymbol{\mathcal{P}}_i^p(z, t) \, e^{i(k_p z - \omega_p t)} + \text{c.c.}, \qquad (2.22)$$

where k_p is the perturbation wave vector, and that the generated field can vary slowly with the propagation distance z, Eq. 2.21 becomes

$$\left\{\frac{\partial^2}{\partial x^2}+\frac{\partial^2}{\partial y^2}+\frac{\partial^2}{\partial z^2}-\frac{n^2}{c^2}\frac{\partial^2}{\partial t^2}\right\}\boldsymbol{\mathcal{E}}_i(z,\omega_p)\,e^{i(kz-\omega_p t)}=-\mu_0\omega_p^2\boldsymbol{\mathcal{P}}_i^P(z,\omega_p)\,e^{i(k_p z-\omega_p t)}+\text{c.c.}$$

$$(2.23)$$

For plane waves or wide beams, the transverse derivatives are zero or negligible, respectively, and k and $\boldsymbol{\mathcal{E}}_i(\omega_p)$ are the eigenmode wave vector and field solution, respectively, for plane waves with frequency ω_p. Furthermore,

$$\left\{\frac{\partial^2}{\partial z^2}-\frac{n^2}{c^2}\frac{\partial^2}{\partial t^2}\right\}\boldsymbol{\mathcal{E}}_i(z,\omega_p)\,e^{i(kz-\omega_p t)}=e^{i(kz-\omega_p t)}\left[\left\{-k^2+\frac{n^2}{c^2}\omega_p^2\right\}+2ik\frac{\partial}{\partial z}+\frac{\partial^2}{\partial z^2}\right]\boldsymbol{\mathcal{E}}_i(z,\omega_p).$$

$$(2.24)$$

The term in the braces on the right-hand side is zero since $\boldsymbol{\mathcal{E}}_i(z,\omega_p)$ is an eigenmode and satisfies the wave equation at ω_p, and assuming slow variation of $\boldsymbol{\mathcal{E}}_i(z,\omega_p)$ with z, the $\partial^2\boldsymbol{\mathcal{E}}_i(z,\omega_p)/\partial z^2$ term can be neglected relative to $2k[\partial\boldsymbol{\mathcal{E}}_i(z,\omega_p)/\partial z]$, finally giving the classical SVEA equation

$$\frac{\partial\boldsymbol{\mathcal{E}}_i(z,\omega_p)}{\partial z}=i\frac{\mu_0\omega_p^2}{2k}\boldsymbol{\mathcal{P}}_i^P(z)\,e^{i(k_p-k)z}.$$

$$(2.25)$$

The wave-vector mismatch between the nonlinear polarization and the freely propagating electric field, both at ω_p, is defined as $\Delta k = k_p - k$.

2.4 COUPLED WAVE EQUATIONS

Since the nonlinear polarization contains the interacting fields, it is used to derive the coupled wave equations for SHG.

2.4.1 Type 1 SHG: Single-Eigenmode Input

The nonlinear polarization term for parametric mixing involving a single-eigenmode input is given by Eq. 2.8. Substituting Eq. 2.8 into Eq. 2.25 gives

$$\frac{d\boldsymbol{\mathcal{E}}_i(z,2\omega)}{dz}=i\frac{k_{\text{vac}}(2\omega)}{4n(2\omega)}\hat{\chi}_{ijj}^{(2)}(-2\omega;\omega,\omega)\boldsymbol{\mathcal{E}}_j^2(\omega)\,e^{i\Delta kz},\quad \Delta k=2k(\omega)-k(2\omega).$$

$$(2.26)$$

To convert this to a scalar equation, write the fields as $\hat{e}_i(2\omega)\boldsymbol{\mathcal{E}}(z,2\omega)$ and $\hat{e}_j(\omega)\boldsymbol{\mathcal{E}}(z,\omega)$, multiply both sides of the equation by $\hat{e}_i^*(2\omega)$, use

$\hat{e}_i^*(2\omega)\hat{e}_i(2\omega) = 1$, and assume that the linear birefringence is small and can be neglected (except in the Δk). Therefore,

$$\frac{d\mathcal{E}(z, 2\omega)}{dz} = i\frac{\omega}{2cn(2\omega)}\hat{\chi}_{\text{eff}}^{(2)}(-2\omega; \omega, \omega)\mathcal{E}^2(z, \omega)\, e^{i\Delta kz}, \qquad (2.27)$$

with

$$\hat{\chi}_{\text{eff}}^{(2)} = \hat{e}_i^*(2\omega)\hat{\chi}_{ijj}^{(2)}(-2\omega; \omega, \omega)\hat{e}_j(\omega)\hat{e}_j(\omega). \qquad (2.28)$$

This represents a projection of the output field onto the nonlinear polarization. Note that this equation is also valid for circularly polarized light. As a matter of notation, the nonlinear coefficient $\hat{d}_{ijk}^{(2)} = \frac{1}{2}\hat{\chi}_{ijk}^{(2)}$ is the one actually tabulated in the literature, which converts Eqs 2.27 and 2.28 into

$$\frac{d\mathcal{E}(z, 2\omega)}{dz} = i\frac{\omega}{cn(2\omega)}\hat{d}_{\text{eff}}^{(2)}(-2\omega; \omega, \omega)\mathcal{E}^2(z, \omega)\, e^{i\Delta kz}, \qquad (2.29)$$

with

$$\hat{d}_{\text{eff}}^{(2)} = \hat{e}_i^*(2\omega)\hat{d}_{ijj}^{(2)}(-2\omega; \omega, \omega)\hat{e}_j(\omega)\hat{e}_j(\omega). \qquad (2.30)$$

If $k(2\omega) \cong 2k(\omega) \to \Delta k \cong 0$, then the harmonic field by the end of a crystal of length L can be large. Since energy must be conserved, the fundamental field must start to deplete; i.e., difference frequency generation via $d_{ijj}^{(2)}(-\omega; 2\omega, -\omega)$ should occur as shown symbolically in Fig. 2.5.

The equation for the down-conversion process is a little different from that for the up-conversion process since there are always two-eigenmode input for down-conversion, namely, the fundamental and the harmonic. Following the previous discussion for Type 2 SHG, the nonlinear polarizations are

$$P_j^{(2)}(z, t; \omega) = \varepsilon_0\hat{\chi}_{ijj}^{(2)}(-\omega; 2\omega, -\omega)\mathcal{E}_i^*(z, \omega)\mathcal{E}_j(z, 2\omega)\, e^{i[k(2\omega)-k(\omega)]z-i\omega t} + \text{c.c.},$$

$$\mathcal{P}_j^{(2)}(z; \omega) = \varepsilon_0\hat{\chi}_{ijj}^{(2)}(-\omega; 2\omega, -\omega)\mathcal{E}_i^*(z, \omega)\mathcal{E}_j(z, 2\omega).$$

$$(2.31)$$

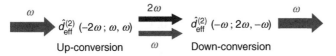

FIGURE 2.5 Schematic of the two processes: up-conversion ($\omega + \omega \to 2\omega$) and down-conversion ($2\omega - \omega \to \omega$), which occur simultaneously during Type 1 second harmonic generation.

Inserting these polarizations into the SVEA and converting to the $\hat{d}_{\text{eff}}^{(2)}$ notation gives

$$\frac{d\mathcal{E}(z,\omega)}{dz} = i\frac{\omega}{n(\omega)c}\hat{d}_{\text{eff}}^{(2)}(-\omega;2\omega,-\omega)\mathcal{E}(z,2\omega)\mathcal{E}^*(z,\omega)\,e^{-i\Delta kz},$$

$$\hat{d}_{\text{eff}}^{(2)}(-\omega;2\omega,-\omega) = \frac{1}{2}\hat{e}_j^*(\omega)\hat{\chi}_{jij}^{(2)}(-\omega;2\omega,-\omega)\hat{e}_i(2\omega)\hat{e}_j^*(\omega).$$
(2.32)

Note that although $k_{ijj} = k_{jij}$,

$$\frac{N\bar{e}^3 k_{ijj}^{(m)}}{\varepsilon_0 \bar{m}_e^3 D_i^{(m)}(2\omega) D_j^{(m)2}(\omega)} \neq \frac{N\bar{e}^3 k_{jij}^{(m)}}{\varepsilon_0 \bar{m}_e^3 D_i^{(m)}(2\omega) D_j^{(m)*}(\omega) D_j^{(m)}(\omega)}$$
(2.33)

$$\rightarrow \quad \hat{\chi}_{jij}^{(2)}(-\omega;2\omega,-\omega) \neq \hat{\chi}_{ijj}^{(2)}(-2\omega;\omega,\omega)$$

because, in general, $D_j^{(m)*}(\omega) \neq D_j^{(m)}(\omega)$. However, far from any of the resonances $(\bar{\omega}_m^2 \gg \omega\bar{\tau}_m^{-1})$, $D_j^*(\omega) \cong D_j(\omega)$ and it is reasonable to set $\tilde{d}_{\text{eff}}^{(2)}(-\omega;2\omega,-\omega) = \tilde{d}_{\text{eff}}^{(2)}(-2\omega;\omega,\omega)$ (Kleinman symmetry); assuming $n(2\omega) \cong n(\omega)$ (a prerequisite for Type 1 efficient conversion, discussed later Section 3.1),

$$\frac{d\mathcal{E}(z,2\omega)}{dz} = i\frac{2\omega}{cn(2\omega)}\tilde{d}_{\text{eff}}^{(2)}(-2\omega;\omega,\omega)\mathcal{E}^2(z,\omega)\,e^{i\Delta kz},$$
(2.34a)

$$\frac{d\mathcal{E}(z,\omega)}{dz} = i\frac{\omega}{cn(\omega)}\tilde{d}_{\text{eff}}^{(2)}(-2\omega;2\omega,\omega)\mathcal{E}(z,2\omega)\mathcal{E}^*(z,\omega)\,e^{-i\Delta kz}.$$
(2.34b)

These two coupled wave equations are universally used to describe the coupling with the distance of the fundamental and the harmonic for a single-eigenmode input.

Although the goal for SHG is to efficiently convert to the harmonic, it is useful to solve for the case of negligible depletion of the fundamental; i.e., $\mathcal{E}(z,\omega) \cong$ constant. Integrating Eq. 2.34a over a distance L gives

$$\mathcal{E}(L,2\omega)-\mathcal{E}(0,2\omega) = i\frac{\omega\tilde{d}_{\text{eff}}^{(2)}L}{n(2\omega)c}\mathcal{E}^2(0,\omega)\exp\left[i\frac{\Delta kL}{2}\right]\text{sinc}\left\{\frac{\Delta kL}{2}\right\}.$$
(2.35)

Under normal circumstances, $\mathcal{E}(0,2\omega) = 0$. Calculating the intensity,

$$I(L,2\omega) = \frac{1}{2}n(2\omega)c\varepsilon_0|\mathcal{E}(L,2\omega)|^2 = \frac{2\omega^2|\tilde{d}_{\text{eff}}^{(2)}|^2}{n^2(\omega)n(2\omega)c^3\varepsilon_0}L^2\,\text{sinc}^2\left(\frac{\Delta kL}{2}\right)I^2(0,\omega).$$
(2.36)

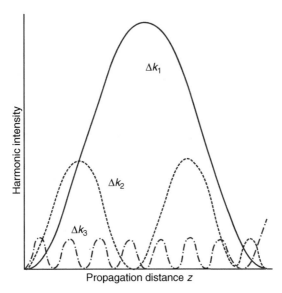

FIGURE 2.6 The evolution with the distance of the second harmonic intensity for $\Delta k_3 > \Delta k_2 > \Delta k_1$.

This formula is useful up to 10% depletion of the fundamental for Type 1. Figure 2.6 quantifies the need to make the phase mismatch ΔkL between the harmonic mode and the nonlinear driving polarization as small as possible. The other parameter that needs to be optimized for efficient SHG is $\tilde{d}_{eff}^{(2)}$. These will be discussed in detail in Chapter 3.

2.4.2   Type 2 SHG: Two-Eigenmode Input

In this case two orthogonally polarized (inside the crystal) fundamental fields are incident along the z-axis. The fields are written as

$$\vec{E}_a(z, t) = \frac{1}{2}\vec{\mathcal{E}}(z, \omega_a)\, e^{i[k_a(\omega)z - \omega t]} + \text{c.c.}, \qquad \vec{E}_b(z, t) = \frac{1}{2}\vec{\mathcal{E}}(z, \omega_b)\, e^{i[k_b(\omega)z - \omega t]} + \text{c.c.}$$

$$(2.37)$$

The fields and interactions involved are shown schematically in Fig. 2.7.

The nonlinear polarizations at 2ω, ω_a (mode a), and ω_b (mode b) with, in reality, $\omega_a = \omega_b = \omega$ are

$$\mathcal{P}_i^{(2)}(z, 2\omega) = \varepsilon_0 \hat{\chi}_{ijk}^{(2)}(-2\omega; \omega_a, \omega)\mathcal{E}_j(z, \omega_a)\mathcal{E}_k(z, \omega_b)\, e^{i[\{k_a(\omega) + k_b(\omega)\}z - 2i\omega t]} + \text{c.c.},$$

$$\mathcal{P}_j^{(2)}(z, \omega_a) = \varepsilon_0 \hat{\chi}_{jik}^{(2)}(-\omega_a; 2\omega, -\omega_b)\mathcal{E}_i(z, 2\omega)\mathcal{E}_k^*(z, \omega_b)\, e^{i[\{k(2\omega) - k_b(\omega)\}z - i\omega t]} + \text{c.c.},$$

$$\mathcal{P}_k^{(2)}(z, \omega_b) = \varepsilon_0 \hat{\chi}_{kij}^{(2)}(-\omega_b; 2\omega, -\omega_a)\mathcal{E}_i(z, 2\omega)\mathcal{E}_j^*(z, \omega_a)\, e^{i[\{k(2\omega) - k_a(\omega)\}z - i\omega t]} + \text{c.c.}$$

$$(2.38)$$

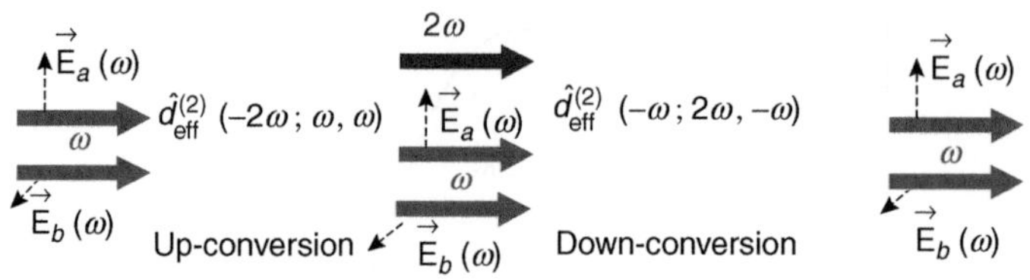

FIGURE 2.7 Schematic of the up-conversion ($\omega + \omega \to 2\omega$) and the two down-conversion ($2\omega - \omega \to \omega$) processes that occur simultaneously during Type 2 second harmonic generation.

Inserting these polarizations into the SVEA and assuming Kleinman symmetry, we obtain

$$\frac{d\mathcal{E}(z, 2\omega)}{dz} = i\frac{2\omega}{n(2\omega)c}\tilde{d}_{\text{eff}}^{(2)}(-2\omega; \omega, \omega)\mathcal{E}(\omega_a)\mathcal{E}(\omega_b)\,e^{i\Delta kz},$$

$$\frac{d\mathcal{E}(z, \omega_a)}{dz} = i\frac{\omega}{n_a(\omega)c}\tilde{d}_{\text{eff}}^{(2)}(-\omega; 2\omega, -\omega)\mathcal{E}(2\omega)\mathcal{E}^*(\omega_b)\,e^{-i\Delta kz}, \qquad (2.39)$$

$$\frac{d\mathcal{E}(z, \omega_b)}{dz} = i\frac{\omega}{n_b(\omega)c}\tilde{d}_{\text{eff}}^{(2)}(-\omega; 2\omega, -\omega)\mathcal{E}(2\omega)\mathcal{E}^*(\omega_a)\,e^{-i\Delta kz},$$

with $\Delta k = k_a + k_b - k(2\omega)$.

It is useful to again consider the small fundamental depletion limit for Type 2:

$$I(L, 2\omega) = \frac{1}{2}n(2\omega)c\varepsilon_0|\mathcal{E}(L, 2\omega)|^2$$

$$= \frac{8\omega^2\left|\tilde{d}_{\text{eff}}^{(2)}\right|^2 L^2}{n_a(\omega)n_b(\omega)n(2\omega)c^3\varepsilon_0}\,\text{sinc}^2\left(\frac{\Delta kL}{2}\right)I_a(0, \omega_a)I_b(0, \omega_b). \qquad (2.40)$$

Comments similar to those of the single-eigenmode input case can be made for optimizing the conversion in this case as well.

Two kinds of mismatch have been introduced here: wave-vector mismatch Δk and phase mismatch ΔkL. The terminology of the field has evolved such that both are frequently used interchangeably when in fact they refer to different physics. The wave-vector mismatch

$$\Delta k = k_{\text{vac}}(\omega)n_a(\omega) + k_{\text{vac}}(\omega)n_b(\omega) - k_{\text{vac}}(2\omega)n(2\omega)$$

$$= 2\omega\left\{\frac{1}{2}\left[\frac{n_a(\omega)}{c} + \frac{n_b(\omega)}{c}\right] - \frac{n(2\omega)}{c}\right\}$$

essentially quantifies the difference between the phase velocities of the nonlinear polarization and harmonic fields, and the phase mismatch ΔkL quantifies the difference in phase accumulated over a distance L due to the wave-vector mismatch, which leads to a reduction in the conversion efficiency.

2.4.3 Sum and Difference Frequency Generation

This case is very similar to the two-eigenmode input SHG case just discussed. Instead of the sum frequency being 2ω as was in the two-eigenmode input SHG case, the sum frequency is generated at $\omega_c = \omega_a + \omega_b$, where all the three frequencies are different. Also the difference frequencies that can be generated are $\omega_b = \omega_c - \omega_a$ and $\omega_a = \omega_c - \omega_b$. For example, for sum frequency generation the processes involved are shown in Fig. 2.8.

The coupled wave equations in the Kleinman limit are given by

$$\frac{d\mathcal{E}(z, \omega_c)}{dz} = i \frac{\omega_c}{n(\omega_c)c} \tilde{d}_{\text{eff}}^{(2)}(-\omega_c; \omega_a, \omega_b)\mathcal{E}(\omega_a)\mathcal{E}(\omega_b)\, e^{i\Delta kz},$$

$$\frac{d\mathcal{E}(z, \omega_a)}{dz} = i \frac{\omega_a}{n(\omega_a)c} \tilde{d}_{\text{eff}}^{(2)}(-\omega_c; \omega_a, \omega_b)\mathcal{E}(\omega_c)\mathcal{E}^*(\omega_b)\, e^{-i\Delta kz}, \qquad (2.41)$$

$$\frac{d\mathcal{E}(z, \omega_b)}{dz} = i \frac{\omega_b}{n(\omega_b)c} \tilde{d}_{\text{eff}}^{(2)}(-\omega_c; \omega_a, \omega_b)\mathcal{E}(\omega_c)\mathcal{E}^*(\omega_a)\, e^{-i\Delta kz},$$

with $\Delta k = k_a + k_b - k_c$.

2.5 MANLEY–ROWE RELATIONS AND ENERGY CONSERVATION

When operating in the Kleinman limit, there are effectively no attenuation mechanisms and power should be conserved between the interacting waves. For the simplest case of Type 1 SHG, the equations for the evolution of the fundamental and second harmonic fields are

$$\frac{d\mathcal{E}(z, 2\omega)}{dz} = i \frac{\omega}{cn(2\omega)} \tilde{d}_{\text{eff}}^{(2)}(-2\omega; \omega, \omega)\mathcal{E}^2(z, \omega)\, e^{i\Delta kz}, \qquad (2.34a)$$

$$\frac{d\mathcal{E}(z, \omega)}{dz} = i \frac{\omega}{cn(\omega)} \tilde{d}_{\text{eff}}^{(2)}(-\omega; 2\omega, -\omega)\mathcal{E}(z, 2\omega)\mathcal{E}^*(z, \omega)\, e^{-i\Delta kz}. \qquad (2.34b)$$

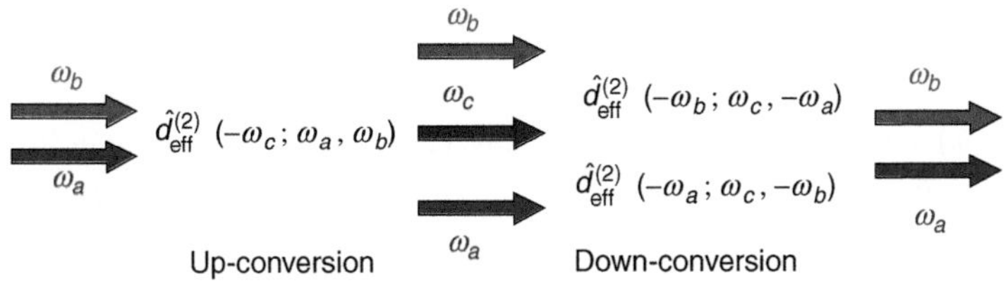

FIGURE 2.8 Schematic of the up-conversion ($\omega_a + \omega_b \rightarrow \omega_c$) and two down-conversion ($\omega_c - \omega_a \rightarrow \omega_b$, $\omega_c - \omega_b \rightarrow \omega_a$) processes that occur simultaneously during sum frequency generation.

Multiplying Eq. 2.34a by $\frac{1}{2}n(2\omega)\varepsilon_0 c\mathcal{E}^*(z, 2\omega)$, taking the complex conjugate of the result, adding it to the original equation, and noting that

$$\frac{dI(z, 2\omega)}{dz} = \frac{1}{2}n(2\omega)\varepsilon_0 c\frac{d[\mathcal{E}^*(z, 2\omega)\mathcal{E}(z, 2\omega)]}{dz} = \frac{1}{2}n(2\omega)\varepsilon_0 c\mathcal{E}^*(z, 2\omega)\frac{d\mathcal{E}(z, 2\omega)}{dz} + \text{c.c.}$$
(2.42)

leads to the following equation for the evolution of the second harmonic intensity:

$$\frac{dI(z, 2\omega)}{dz} = \frac{1}{2}i\omega\varepsilon_0 \tilde{d}_{\text{eff}}^{(2)}\left\{\mathcal{E}^*(z, 2\omega)\mathcal{E}^2(z, \omega)\, e^{i\Delta kz} + \text{c.c.}\right\}.$$
(2.43)

Applying the same procedure to Eqs 2.34a and b gives

$$\frac{dI(z, \omega)}{dz} = -\frac{1}{2}i\omega\varepsilon_0 \tilde{d}_{\text{eff}}^{(2)}\left\{\mathcal{E}^*(z, 2\omega)\mathcal{E}^2(z, \omega)\, e^{i\Delta kz} + \text{c.c.}\right\}.$$
(2.44)

Noting that the right-hand sides of Eqs 2.43 and 2.44 are identical except for sign difference:

$$\frac{d[I(z, 2\omega) + I(z, \omega)]}{dz} = 0 \quad \rightarrow \quad [I(z, 2\omega) + I(z, \omega)] = \text{constant};$$
(2.45)

i.e., the power flow is conserved. Furthermore, defining N_p as the number of photons per square centimeter per second, i.e., $N_p(z, 2\omega) = I(z, 2\omega)/2\hbar\omega$ and $N_p(z, \omega) = I(z, \omega)/\hbar\omega$, Eqs 2.44 and 2.45 give

$$\frac{dN_p(z, 2\omega)}{dz} = -2\frac{dN_p(z, \omega)}{dz}.$$
(2.46)

This simply means that two photons are lost from the fundamental beam for every harmonic photon created.

A similar procedure for the Type 2 case, either for Type 2 SHG or for sum and difference frequency mixing, leads to

$$-\frac{d[I_a(z, \omega_b) + I_b(z, \omega_a)]}{dz} = \frac{dI_c(z, \omega_c)}{dz}, \quad \frac{dN_p(z, \omega_c)}{dz} = -\frac{d[N_p(z, \omega_a) + N_p(z, \omega_b)]}{dz}.$$
(2.47)

Again the interpretation is straightforward; i.e., the annihilation of two photons with energies $\hbar\omega_a$ and $\hbar\omega_b$ is necessary for the creation of a single $\hbar\omega_c$ photon.

PROBLEMS

1. Use the electron on an anharmonic spring model for a noncentrosymmetric medium with additional terms in the potential of the form $\frac{1}{3}\bar{k}_{ijk}\bar{q}_i\bar{q}_j\bar{q}_k + \frac{1}{4}\bar{k}_{ijk\ell}\bar{q}_i\bar{q}_j\bar{q}_k\bar{q}_\ell$ to show that the third-order nonlinear polarizations are given by

(a)

$$P_i(3\omega) = -\frac{e^4 N}{4\bar{m}_e^4}\left\{\bar{k}_{ijk\ell} - \frac{2\bar{k}_{ijm}\bar{k}_{mk\ell}}{\bar{m}_e D_m(2\omega)}\right\}\frac{E_j(\omega)E_k(\omega)E_\ell(\omega)}{D_i(3\omega)D_j(\omega)D_k(\omega)D_\ell(\omega)}.$$

(b)

$$P_i(\omega) = -\frac{e^4 N}{4\bar{m}_e^4}\left\{\left[3\bar{k}_{ijk\ell} - \frac{2\bar{k}_{ijm}\bar{k}_{mk\ell}}{\bar{m}_e D_m(2\omega)}\right]\frac{E_j^*(\omega)E_k(\omega)E_\ell(\omega)}{D_i(\omega)D_j^*(\omega)D_k(\omega)D_\ell(\omega)}\right.$$

$$\left. - \frac{2\bar{k}_{ijm}\bar{k}_{mk\ell}}{\bar{m}_e}\frac{E_j(\omega)E_k^*(\omega)E_\ell(\omega)}{D_i(\omega)D_j(\omega)D_m(0)D_k^*(\omega)D_\ell(\omega)}\right\}.$$

Note that in a noncentrosymmetric medium, two successive $\chi^{(2)}$ responses can also lead to a $\chi^{(3)}$ response. For example, SHG $\propto \bar{q}(\omega)\bar{q}(\omega)$ can lead to a displacement $\bar{q}(2\omega)$ and then sum frequency generation $[\bar{q}(2\omega)\bar{q}(\omega)]$ can lead to $\chi^{(3)}(3\omega)$. Similar considerations apply to (b).

2. Use the electron on an anharmonic spring model for a noncentrosymmetric medium with additional terms in the potential of the form $\frac{1}{3}\bar{k}_{ijk}\bar{q}_i\bar{q}_j\bar{q}_k + \frac{1}{4}\bar{k}_{ijk\ell}\bar{q}_i\bar{q}_j\bar{q}_k\bar{q}$ to show that the third-order nonlinear polarization for DC field-induced second harmonic generation is given by

$$P_i(2\omega) = -\frac{e^4 N}{4\bar{m}_e^4}\left\{\left[3\bar{k}_{ijk\ell} - \frac{2\bar{k}_{ijm}\bar{k}_{mk\ell}}{\bar{m}_e D_m(2\omega)}\right]\frac{E_j(0)E_k(\omega)E_\ell(\omega)}{D_i(2\omega)D_j(0)D_k(\omega)D_\ell(\omega)}\right.$$

$$\left. - \frac{2\bar{k}_{ijm}\bar{k}_{mk\ell}}{\bar{m}_e}\frac{E_j(\omega)E_k(\omega)E_\ell(0)}{D_i(2\omega)D_j(\omega)D_m(\omega)D_k(\omega)D_\ell(0)}\right\}.$$

Note: In a noncentrosymmetric medium, two successive $\chi^{(2)}$ responses can also lead to a $\chi^{(3)}$ response. For example, SHG $\propto \bar{q}(\omega)\bar{q}(\omega)$ can lead to a displacement $\bar{q}(2\omega)$ and then sum frequency generation $[\bar{q}(2\omega)\bar{q}(\omega)]$ can lead to $\chi^{(3)}(3\omega)$. [Hint: Be very careful with the indices.]

3. Given that

$$P_i(3\omega) = -\frac{e^4 N}{4\bar{m}_e^4} \bar{k}_{ijk\ell} \frac{E_j(\omega)E_k(\omega)E_\ell(\omega)}{D_i(3\omega)D_j(\omega)D_k(\omega)D_\ell(\omega)}$$

is the nonlinear-induced polarization in the electron on a spring model for third harmonic generation for the potential term $\frac{1}{4}\bar{k}_{ijk\ell}\bar{q}_i\bar{q}_j\bar{q}_k\bar{q}_\ell$, find the corresponding *third-order* polarizations for $P_i(2\omega)$, $P_i(2\omega_a - \omega_b)$, and $P_i(\omega_a + \omega_b - \omega_c)$.

4. The electro-optic effect is also a second-order nonlinear effect involving a beam at frequency ω and a very low frequency field $\omega \gg \Omega$. Derive the nonlinear polarization $P_i(\omega \pm \Omega)$ and express the electro-optic coefficient r_{ijk} defined by

$$P_i(\omega \pm \Omega) = -\frac{1}{2\varepsilon_0}\varepsilon_{r,ii}\varepsilon_{r,jj}\hat{r}_{ijk}(-[\omega \pm \Omega]; \omega, \pm\Omega)E_j(\omega)E_k(\pm\Omega)$$

in terms of $\hat{\chi}_{ijk}^{(2)}(-\omega; \omega, \Omega \cong 0)$ for materials in which the response is purely electronic. Note that the electro-optics community defines the electro-optic coefficient in the above equation as $\hat{r}_{ijk}$, where the first two terms can be collapsed in the Voigt notation whereas in second harmonic it is the last two terms. As a result, it is, e.g., $\hat{r}_{4,1} \to \hat{\chi}_{1,4}^{(2)}$ that are related.

5. Consider the case of KH$_2$PO$_4$ with incident fields polarized along directions parallel to an axis intersecting two pairs of oxygen atoms, as shown in the figure. This led to a second harmonic polarization orthogonal to the incident field, directed along the axis passing through the center of the molecule as shown in the figure as $\hat{a}$, which corresponds to the x-axis, the direction of the induced second harmonic polarization in the figure. Now consider a circularly polarized field of the form $\vec{E} = (\hat{e}_y + i\hat{e}_z)E_f/\sqrt{2}$ incident along the x-axis so that second harmonic polarizations are induced along the x-axis. In fact, the phosphorus atom sits at the indicated site of the electron in the figure and the nonlinear polarization is induced in the P–O bonds that can be considered to be located at $x - \xi$ and $x + \xi$, where ξ is a molecular distance and x is the center of the PO$_4$ cluster. Hence the induced polarization fields are separated in space along the x-axis by a small submolecular distance 2ξ. For the nonlinear coefficients $\hat{\chi}_{x,yy}^{(2)}(-2\omega; \omega, \omega) = \hat{\chi}_{x,zz}^{(2)}(-2\omega; \omega, \omega) \equiv \hat{\chi}^{(2)}$, show that the induced second harmonic corresponds to $P_x(2\omega) = \varepsilon_0\gamma[\partial E_f^2(\omega, x)/\partial x]$, where $\gamma \equiv \xi\hat{\chi}^{(2)}$ is the nonlinear quadrupole coefficient for SHG. This is just one example of an additional gradient process that leads to a second harmonic polarization. In general,

$$\vec{P}(2\omega) = \varepsilon_0\left\{(\delta - \beta - 2\gamma)[\vec{E}(\omega) \cdot \vec{\nabla}]\vec{E}(\omega) + \beta\vec{E}(\omega)[\vec{\nabla} \cdot \vec{E}(\omega)] + \gamma\vec{\nabla}[\vec{E}(\omega) \cdot \vec{E}(\omega)]\right\},$$

where the coefficients δ, β, and γ can, in principle, be estimated from material parameters or measured experimentally.

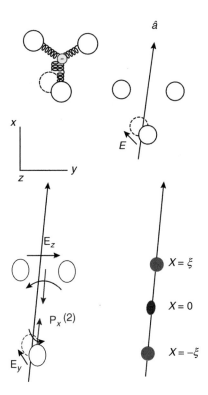

6. Consider a crystal with the following nonzero second-order susceptibilities: $\hat{\chi}^{(2)}_{xxz}$, $\hat{\chi}^{(2)}_{yyz}$, $\hat{\chi}^{(2)}_{zzz}$, $\hat{\chi}^{(2)}_{zyy}$, $\hat{\chi}^{(2)}_{zxx}$ with, of course, $\hat{\chi}^{(2)}_{ijk} = \hat{\chi}^{(2)}_{ikj}$. Light propagates in the x–y plane at $45°$ to the x-axis.

   (a) For an eigenmode of amplitude E_0 polarized in the x–y plane with both $\vec{D}$ and $\vec{E}$ having equal projections along the x- and y-axes, calculate the induced nonlinear polarization vector.

   (b) What is the nonlinear polarization induced for the orthogonally polarized (along the z-axis) eigenmode?

   (c) What is the induced nonlinear polarization when both eigenmodes are present?

7. A continuous-wave, y-polarized, plane-wave beam is incident along the z-axis of a $3m$ noncentrosymmetric crystal which has a "magic" mirror that reflects all wavelengths coated onto its other end, creating oppositely propagating waves in the medium. The input facet of the crystal is also coated with an antireflection layer for all wavelengths. What nonlinear polarizations can be generated within the crystal and what are their appropriate second-order coefficient(s) $\hat{\chi}^{(2)}$?

8. Use the electron on an anharmonic spring model for a noncentrosymmetric medium with nonlinear terms in the potential $\frac{1}{3}\bar{k}_{ijk}\bar{q}_i\bar{q}_j\bar{q}_k + \frac{1}{4}\bar{k}_{ijk\ell}\bar{q}_i\bar{q}_j\bar{q}_k\bar{q}_\ell$ to show that the third-order nonlinear polarization is given by

$$P_i(3\omega) = -\frac{e^4 N}{4\bar{m}_e^4}\left\{\bar{k}_{ijk\ell} - \frac{2\bar{k}_{ijm}\bar{k}_{mk\ell}}{\bar{m}_e D_m(2\omega)}\right\}\frac{E_j(\omega)E_k(\omega)E_\ell(\omega)}{D_i(3\omega)D_j(\omega)D_k(\omega)D_\ell(\omega)}$$

Note: In a noncentrosymmetric medium, two successive $\hat{\chi}^{(2)}$ responses can also lead to a $\hat{\chi}^{(3)}$ response. For example, SHG $\propto \bar{q}(\omega)\bar{q}(\omega)$ can lead to a displacement $\bar{q}(2\omega)$ and then sum frequency generation $[\bar{q}(2\omega)\bar{q}(\omega)]$ can lead to $\hat{\chi}^{(3)}(3\omega)$. [Hint: Be very careful with the indices.]

9. Starting with the following expression for the nonlinear third-order polarization

$$\boldsymbol{\mathcal{P}}_i^{(3)}(\omega) = \frac{3}{4}\varepsilon_0\hat{\chi}_{iiii}^{(3)}(-\omega;\omega,-\omega,\omega)|\boldsymbol{\mathcal{E}}_i(\omega)|^2\boldsymbol{\mathcal{E}}_i(\omega)\exp[ikz],$$

(a) find the slowly varying envelope approximation expression for the evolution of the field $E(\omega,z)$ with the propagation distance where the input field is written as $\frac{1}{2}\boldsymbol{\mathcal{E}}(\omega,z)\exp[i(kz-\omega t)] + $ c.c.

(b) show that in the absence of loss and for real $\chi^{(3)}$, this process leads to a field-dependent contribution to the refractive index.

10. The second harmonic generation process with only a fundamental input is the most common case of second-order nonlinear interaction. This problem examines an example with different input conditions; i.e., *both* the fundamental and the harmonic are incident onto the sample.

(a) For a single-eigenmode fundamental input and Type 1 SHG with small depletion of the fundamental, show that after a small propagation distance L,

$$\mathcal{E}(2\omega, L) = \mathcal{E}(2\omega, 0) + i\kappa L\mathcal{E}^2(\omega, 0)\exp\left[i\frac{\Delta kL}{2}\right]\operatorname{sinc}\left(\frac{\Delta kL}{2}\right),$$

$$\mathcal{E}(\omega, L) = \mathcal{E}(\omega, 0) + i\kappa L\mathcal{E}(2\omega, 0)\mathcal{E}^*(\omega, 0)\exp\left[-i\frac{\Delta kL}{2}\right]\operatorname{sinc}\left(\frac{\Delta kL}{2}\right).$$

(b) In the *phase-matching* limit, show for *fundamental-only input* that the harmonic is $\pi/2$ out of phase with the fundamental. Also show that it in this case the fundamental is depleted by a small amount after ΔL with no change in phase. [Hint: Use the solution of a distance ΔL for the harmonic in the equation for the fundamental.]

(c) Consider again the case where ΔL is an incremental length, i.e., very small; both the fundamental and the harmonic are input; the process is a wave-vector-matched one; and the fundamental and the harmonic input fields are *in phase*. Show that the net result of the interaction between the fields is just to rotate their phase and evaluate the phase rotations $\Delta\phi(\omega)$ and $\Delta\phi(2\omega)$, where $\mathcal{E}(2\omega, \Delta L) = \mathcal{E}(2\omega, 0) \exp[i\Delta\phi(2\omega)]$ and $\mathcal{E}(\omega, \Delta L) = \mathcal{E}(\omega, 0)$ $\exp[i\Delta\phi(\omega)]$.

(d) Use the same procedure to evaluate both fields after propagating another increment ΔL, i.e., $\mathcal{E}(2\omega, 2\Delta L)$ and $\mathcal{E}(\omega, 2\Delta L)$ in terms of $\mathcal{E}(2\omega, 0)$ and $\mathcal{E}(\omega, 0)$. Use this result to show that the interaction continues to produce just phase rotation (i.e., no energy exchange between the fields) only if $2\mathcal{E}^2(2\omega, 0) = \mathcal{E}^2(\omega, 0)$.

Congratulations, you have found the nonlinear modes, i.e., field distributions that do not change on propagation, for SHG. We will discuss this in more detail and in a more elegant way in Chapter 4.

11. The question comes up whether the slowly varying envelope approximation gives the same result as the exact solution to the differential equation for SHG. Exact solutions for plane waves are relatively easy to find for the following case. The differential equation to be solved for the simplest copolarized case with no fundamental depletion is

$$\nabla^2 E(2\omega, z) - \frac{n^2(2\omega)}{c^2} \frac{\partial^2 E(2\omega, z)}{\partial t^2} = \frac{1}{2}\mu_0 \frac{\partial^2 \{\mathcal{P}(2\omega, 0) \, e^{i(2k(\omega)z - 2\omega t)} + \text{c.c.}\}}{\partial t^2}.$$

The complete solution is the sum of a particular field solution $[\mathcal{E}_{\text{inhom}}(2\omega, z)]$ to the polarization-driven wave equation (inhomogeneous wave equation) and a solution $[\mathcal{E}_{\text{hom}}(2\omega, z)]$ to the "homogeneous wave equation," i.e., the equation with $P(2\omega, 0) = 0$. The amplitude $\mathcal{E}_{\text{hom}}(2\omega, z)$ is chosen to satisfy the boundary condition of zero second harmonic field at $z = 0$.

(a) For a wave-vector mismatch $\Delta k = 2k(\omega) - k(2\omega)$, show that the exact result is approximately the same as that obtained from the slowly varying envelope approximation.

(b) Show that in the limit $\Delta k \to 0$, the exact solution field grows linearly with the propagation distance.

12. A frequently encountered case is that a material is effectively lossless at the fundamental frequency, but the second harmonic has absorption with an amplitude absorption coefficient α_1.

(a) Show that in the negligible fundamental depletion limit on phase match, this results in the saturation of the conversion efficiency with the propagation distance, and derive a formula for the maximum harmonic field.

(b) Calculate in the negligible depletion limit the intensity conversion efficiency with the propagation distance in the case of phase mismatch in terms of $\ell_{\text{coh}}\Delta k/\pi = 0, 0.5, 1.0, 2.0$. Find the asymptotic (sample length approaches infinity) conversion efficiency.

13. An occasionally encountered case is that a material is effectively lossless at the harmonic frequency, but the fundamental has absorption with an amplitude

absorption coefficient α_1. Show that in the negligible fundamental conversion to second harmonic limit and when the decay of the fundamental is dominated by the loss term, this results on phase match in the second harmonic field, reaching a maximum value with the distance. Calculate the maximum harmonic conversion efficiency and the propagation distance at which it occurs. Assume Type 1 SHG.

14. In the most general case where a material has an amplitude loss coefficient α_1 at the fundamental frequency and α_2 at the harmonic frequency, calculate a formula for the evolution of the harmonic field with the distance in the limit of negligible depletion of the fundamental to the harmonic. Show that there is a maximum second harmonic field and find the distance at which it occurs. Assume Type 1 SHG. What is distance when $2\alpha_1 = \alpha_2$?

15. In the most general case for difference frequency generation where a material has amplitude loss coefficients α_1 and α_2 at the input frequencies ω_1 and ω_2 ($\omega_1 > \omega_2$) and α_3 at the difference frequency, calculate a formula for the evolution of the difference frequency field with the distance in the limit of negligible depletion of the input fields into radiation at the difference frequency. Show that there is a maximum in the difference frequency field generated and find the distance at which it occurs. What is the distance when $2\alpha_1 = \alpha_2$?

16. In the most general case for sum frequency generation where a material has amplitude loss coefficients α_1 and α_2 at the input frequencies ω_1 and ω_2 and α_3 at the sum frequency, calculate a formula for the evolution of the sum frequency field with the distance in the limit of negligible depletion of the input fields into radiation at the sum frequency. Show that there is a maximum in the sum frequency field generated and find the distance at which it occurs. What is the distance if $\alpha_1 + \alpha_2 = \alpha_3$?

SUGGESTED FURTHER READING

1. R. W. Boyd, Nonlinear Optics, 3rd Edition (Academic Press, Burlington, MA, 2008).

2. F. A. Hopf and G. I. Stegeman, Applied Classical Electrodynamics, Volume 1: Linear Optics (John Wiley & Sons, New York, 1985).

3. F. A. Hopf and G. I. Stegeman, Applied Classical Electrodynamics, Volume 2: Nonlinear Optics (John Wiley & Sons, New York, 1985).

4. Y. R. Shen, Principles of Nonlinear Optics (John Wiley & Sons, New York, 1984).

CHAPTER 3

Optimization and Limitations of Second-Order Parametric Processes

As discussed in Chapter 2, the optimum conversion of energy to different frequencies requires that the wave-vector mismatch Δk be minimized and that the crystal orientation be optimized for maximum $\hat{d}_{\text{eff}}^{(2)}$. Another key factor, since it can limit the effective crystal length, is the walk-off between the fundamental beams and the harmonic beams due to the propagation of these beams in directions away from the crystal axes. Other important practical considerations are the range of input directions for the fundamental in a crystal (angular bandwidth), which maintains reasonable conversion efficiency, and the temperature control needed for efficient and stable second harmonic generation (SHG). These issues are dealt with in this chapter.

3.1 WAVE-VECTOR MATCHING

Wave-vector matching is a linear optics problem, i.e., dispersion in the refractive index. Consider the possibility that i, j, and k in $d_{ijk}^{(2)}$ are all the same polarization. It is clear from Fig. 3.1a, which shows the typical dispersion with frequency in the refractive index, that achieving $\Delta k = 2k_{\text{vac}}(\omega)[n_i(\omega) - n_i(2\omega)] = 0$ is just not possible when i refers to a single polarization. Therefore two refractive indices, each along a different polarization axis, will be necessary for wave-vector matching unless some artificial material modification is done.

One solution is to use uniaxial crystals (two independent refractive indices; see Fig. 3.1b) or biaxial materials (three independent refractive indices). It is also possible to use clever material processing in some ferroelectrics and polymers to periodically reverse the sign of the nonlinearity, which does allow wave-vector matching with a single polarization. Other options are beyond the scope of this textbook and are available in texts discussing waveguide geometries.

Nonlinear Optics: Phenomena, Materials, and Devices, George I. Stegeman and Robert A. Stegeman.
© 2012 John Wiley & Sons, Inc. Published 2012 by John Wiley & Sons, Inc.

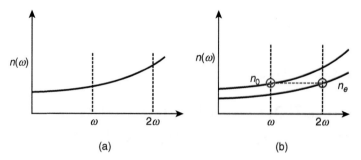

FIGURE 3.1   The dispersion in the refractive index for a (a) single polarization and (b) uniaxial crystal in the visible and near-infrared regions. n_o and n_e are the refractive indices along the ordinary and extraordinary polarizations. The case shown is for noncritical wave-vector matching (defined in text).

3.1.1   Birefringent Wave-Vector Matching: Uniaxial Crystals

The dielectric constant and hence the refractive index of uniaxial crystals is isotropic for propagation along their z-axis. The dielectric constant tensor is of the form shown in Fig. 3.2a, where n_o and n_e are the refractive indices for field polarization in the x–y plane and along the z-axis (optic axis), respectively. Therefore for propagation along the z-axis, wave-vector matching is not possible because there is only one refractive index in the x–y plane and hence both orthogonally polarized waves experience o-polarization, which disperses with frequency. For light propagation in the x–z plane at an angle θ relative to the z-axis (Fig. 3.2b), one eigenmode $\vec{E}_o$ always has its field in the x–y plane and is labeled $\vec{E}_o$. The second eigenmode field $(\vec{E}_e)$ lies in the x–z plane, which is orthogonally polarized to $\vec{E}_o$, with refractive index $n_e(\theta)$ given by

$$n_e(\theta) = \frac{n_o n_e}{\left[n_e^2 \cos^2(\theta) + n_o^2 \sin^2(\theta)\right]^{1/2}}, \tag{3.1}$$

assuming that the birefringence is small so that $\vec{D}$ and $\vec{E}$ are essentially collinear.

Because of the index isotropy in the x–y plane when the incident wave vector $\vec{k}$ is rotated by any arbitrary angle ϕ about the z-axis, $\vec{E}_o$ remains in the x–y plane with $\vec{E}_e$ orthogonal to it. Hence, the refractive indices and $|\Delta\vec{k}|$ are independent of the angle ϕ; i.e., the wave-vector mismatch is a constant along the cone shown in Fig. 3.2c.

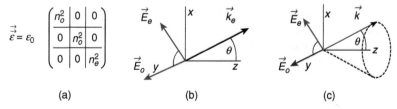

FIGURE 3.2   (a) The dielectric constant tensor in media with uniaxial symmetry. (b) The field geometry for the vector $\vec{k}$ at an angle θ from the z-axis. (c) The cone described by $\vec{k}$ directions for which the linear optical properties and wave vector do not change.

For Type 1 phase matching, $\Delta k = 2k_{\text{vac}}(\omega)[n_j(\omega) - n_i(2\omega)]$, where i and j refer to different polarization axes. Figure 3.1b illustrates an example of how wave-vector matching can be achieved using both polarizations. For $\vec{k}$ at an angle θ from the z-axis, four possibilities for $\Delta k = 0$ exist: two for $n_o > n_e$ (negative uniaxial) and two for $n_e > n_o$ (positive uniaxial).

3.1.1.1 Type 1 Phase Match

3.1.1.1.1 Negative Uniaxial.
For Type 1 negative uniaxial wave-vector matching, $n_o(\omega) = n_e(\theta, 2\omega)$. The appropriate refractive index curves are shown in Fig. 3.3. The refractive index of the e wave is varied between $n_e(2\omega)$ and $n_o(2\omega)$ by changing θ. Thus, the fundamental frequency for doubling can be tuned (see Fig. 3.3a). The lowest possible fundamental frequency for doubling occurs when $n_o(\omega) = n_e(2\omega)$; this is called "noncritical" wave-vector matching for reasons discussed later; see Fig. 3.3b). This occurs for $\theta = \pi/2$; i.e., $\vec{k}$ lies in the x–y plane, $\vec{E}_e$ lies along the z-axis, and $\vec{E}_o$ lies in the x–y plane, orthogonal to $\vec{k}$ and $\vec{E}_e$. Therefore, both fields lie along pure mode axes, and the Poynting vectors are collinear.

The angle θ_{PM} at which wave-vector matching occurs is given by

$$n_o(\omega) = n_e(\theta_{PM}, 2\omega) = \frac{n_o(2\omega)n_e(2\omega)}{\sqrt{n_e^2(2\omega)\cos^2(\theta_{PM}) + n_o^2(2\omega)\sin^2(\theta_{PM})}}$$

$$(3.2)$$

$$\rightarrow \quad \sin(\theta_{PM}) = \frac{n_e(2\omega)}{n_o(\omega)}\sqrt{\frac{n_o^2(2\omega) - n_o^2(\omega)}{n_o^2(2\omega) - n_e^2(2\omega)}}.$$

This case is usually labeled eoo, where the first letter refers to the polarization of the second harmonic and the second and third letters indicate the polarizations of the two fundamental input waves. Note that $e \equiv e(\theta)$ in general, except for the noncritical case. θ_{PM} is called the *phase-match angle*. The SHG efficiency and the angular bandwidth depend on the wave-vector mismatch, as will be discussed in detail in Chapter 4 and section 2.1.4 respectively.

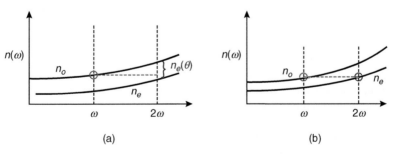

FIGURE 3.3 (a) The dispersion in n_e and n_o for a negative uniaxial crystal. $n_e(\theta, 2\omega)$ can be tuned with angle θ. The dashed red line indicates the wave-vector-matching condition. (b) The noncritical wave-vector-matching condition that can occur at only one frequency.

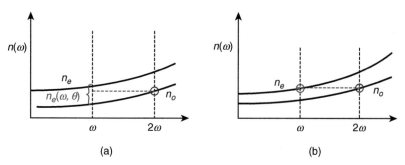

FIGURE 3.4   (a) The dispersion in n_e and n_o for a negative uniaxial crystal. $n_e(\theta, \omega)$ can be tuned with angle θ. The dashed red line indicates the wave-vector-matching condition. (b) The noncritical wave-vector-matching condition that can occur at only one frequency.

3.1.1.1.2 *Positive Uniaxial.* For Type 1 positive uniaxial wave-vector matching, $n_e(\omega) > n_o(\omega)$, and it is the fundamental beam frequency that can be tuned by varying the angle θ and the second harmonic must lie on the n_e curve. Thus, it is possible to obtain tunable SHG as long as $n_o(2\omega) = n_e(\theta, \omega)$ is satisfied. The wave-vector-matching condition is shown in Fig. 3.4a. The lowest fundamental frequency for doubling occurs when $n_o(2\omega) = n_e(\omega)$. The noncritical phase match occurs for $\theta = \pi/2$; i.e., $\vec{k}$ lies in the x–y plane, $\vec{E}_e$ lies along the z-axis, and $\vec{E}_o$ lies in the x–y plane, orthogonal to $\vec{k}$. Therefore, both fields lie along pure mode axes, and the Poynting vectors are collinear.

The analytical solution for the wave-vector-match angle is

$$n_o(2\omega) = n_e(\omega, \theta) = \frac{n_o(\omega)n_e(\omega)}{\left[n_e^2(\omega)\cos^2(\theta_{\mathrm{PM}}) + n_o^2(\omega)\sin^2(\theta_{\mathrm{PM}})\right]^{1/2}}$$

$$\rightarrow \quad \sin(\theta_{\mathrm{PM}}) = \frac{n_e(\omega)}{n_o(2\omega)}\sqrt{\frac{n_o^2(\omega) - n_o^2(2\omega)}{n_o^2(\omega) - n_e^2(\omega)}}.$$

$$(3.3)$$

This case is labeled *oee*, where the first letter refers to the polarization of the second harmonic and the second and third letters indicate the polarization of the single fundamental input.

3.1.1.2 *Type 2 Wave-Vector Match.* The Type 2 case can be considerably more complex than the Type 1 case discussed above, especially the *eoe* geometry just discussed, in which varying θ changes the extraordinary refractive index for wave-vector matching at both ω and 2ω.

3.1.1.2.1 *Negative Uniaxial.* For Type 2 negative uniaxial wave-vector match, labeled *eoe*, the most complex case, the wave-vector mismatch is

$$\Delta k = k_e(\omega) + k_o(\omega) - k_e(2\omega) = k_{\mathrm{vac}}(\omega)\{n_e(\theta, \omega) + n_o(\omega) - 2n_e(\theta, 2\omega)\}.$$

$$(3.4)$$

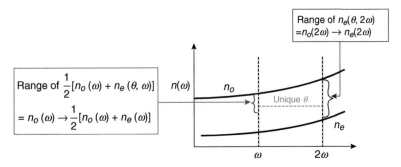

FIGURE 3.5 The dispersion in n_e and n_o for a negative uniaxial crystal. Both $n_e(\theta, \omega)$ and $n_e(\theta, 2\omega)$ are tuned with angle θ. The dashed green line indicates the wave-vector-matching condition. The ranges over which the refractive index for the fundamental and the harmonic can vary are also shown.

The appropriate refractive index curves are shown in Fig. 3.5. The refractive indices of the two e waves change in different ways when θ is varied, and wave-vector matching occurs for a single value of θ. Mathematically,

$$\frac{1}{2}\left[\frac{n_o(\omega)n_e(\omega)}{\sqrt{n_e^2(\omega)\cos^2\theta_{PM}+n_o^2(\omega)\sin^2\theta_{PM}}}+n_o(\omega)\right]=\frac{n_o(2\omega)n_e(2\omega)}{\sqrt{n_e^2(2\omega)\cos^2\theta_{PM}+n_o^2(2\omega)\sin^2\theta_{PM}}}.$$

$$(3.5)$$

3.1.1.2.2 *Positive Uniaxial.* For Type 2 positive uniaxial wave-vector match, labeled *oeo*, the wave-vector mismatch is

$$\Delta k = k_e(\omega) + k_o(\omega) - k_o(2\omega) = k_{vac}(\omega)\{n_e(\theta, \omega) + n_o(\omega) - 2n_o(2\omega)\} = 0.$$

$$(3.6)$$

The dispersion curves are shown in Fig. 3.6, and mathematically, the phase-match angle is calculated from

$$\sin(\theta_{PM}) = \frac{2n_e(\omega)}{2n_o(2\omega) - n_o(\omega)}\sqrt{\frac{n_o(2\omega)\{n_o(\omega) - n_o(2\omega)\}}{n_o^2(\omega) - n_e^2(\omega)}}.$$

$$(3.7)$$

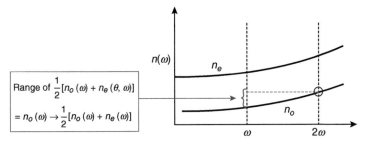

FIGURE 3.6 The dispersion in n_e and n_o for a positive uniaxial crystal. $n_e(\theta, \omega)$ is tuned with angle θ. The dashed green line indicates the wave-vector-matching condition.

3.1.2 Quasi-Phase Matching

Quasi-phase matching (QPM) is a very powerful technique that has made high efficiency, tunable, frequency conversion devices available (2,3). It is commonly used for ferroelectric materials, such as LiNbO$_3$ and LiTaO$_3$, that have relatively large $\hat{d}^{(2)}_{zzz}$ (tens of picometers per volt) coefficients. The principle is relatively straightforward: In linear optics, the refractive index is independent of whether light is polarized along the $+z$-axis or the $-z$-axis. However, in ferroelectric crystals, e.g., LiNbO$_3$, it is possible by selective (usually periodic) optical poling to change the sign of $\hat{d}^{(2)}_{zzz}$ in optical wavelength size crystal domains along the z-axis (see Fig. 3.7). Since the differently oriented domains have identical refractive indices, they cannot be identified with linear optical techniques. Fortunately, there are etching materials that etch different domains with different orientations, allowing the periodic poled regions to be identified.

The physics of this poling process is as follows. The molecular unit cells have a permanent dipole moment that can be oriented either along the $+z$-axis or along the $-z$-axis; i.e., there is a double potential well along the z-axis in the unit cell. During the crystal growth, single domain crystals are formed by controlling the growth conditions; i.e., all dipole moments are aligned. For large electric fields applied to a specific region, the dipole can flip its orientation (see Fig. 3.7) and the sign of $\hat{d}^{(2)}_{zzz}$ reverses. The minimum size of the interface region between neighboring domains limits the smallest periodicity achievable. This periodicity needs to be introduced into the coupled wave equations to understand the effect of the periodic poling. A straightforward Fourier analysis of the structure shown in Fig. 3.7 gives

$$
\begin{aligned}
\hat{d}^{(2)}_{p,zzz} &= (2a - 1)\hat{d}^{(2)}_{zzz}, \qquad p = 0 \\
&= 2\frac{\sin(pa\pi)}{p\pi}d^{(2)}_{zzz}, \quad |p| \geq 1 \\
\rightarrow \quad \hat{d}^{(2)}_{zzz}(x) &= \sum_p d^{(2)}_{p,zzz}\exp\left[i\frac{2p\pi}{\Lambda}x\right].
\end{aligned}
\tag{3.8}
$$

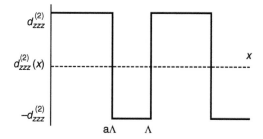

FIGURE 3.7   Variation in the $\hat{d}^{(2)}_{zzz}$ coefficient as a function of the propagation coordinate in poled LiNbO$_3$. The periodicity is Λ, and the mark-to-space ratio a defines the fraction of the period, which is $+\hat{d}^{(2)}_{zzz}$.

Writing $K = 2\pi/\Lambda$, including the periodic $\hat{d}^{(2)}_{zzz}(x)$ in the nonlinear polarization, and inserting it into the slowly varying envelope approximation gives

$$\mathcal{P}^{(2)}_z(2\omega) = \varepsilon_0 \sum_{p=-\infty}^{p=\infty} \hat{d}^{(2)}_{p,zzz} \mathcal{E}^2_z(\omega) \, e^{i[2k(\omega)+pK]x}$$

$$\rightarrow \quad \frac{d\mathcal{E}_z(2\omega, x)}{dx} = i\frac{k_{\text{vac}}(2\omega)}{2n_z(2\omega)} \sum_p \hat{d}^{(2)}_{p,zzz}(-2\omega;\ \omega,\ \omega)\mathcal{E}^2_z(\omega) \, e^{\{i[2k_e(\omega)-k_e(2\omega)+pK]x\}}.$$

$$(3.9)$$

The appropriate wave-vector mismatch is now $\Delta k = 2k_e(\omega) - k_e(2\omega) + pK$, and clearly pK can be chosen to make $\Delta k = 0$ by periodic poling with an appropriate poling period. Note that in the visible and near-infrared regions, $k_e(2\omega) > 2k_e(\omega)$ and p must be positive. Furthermore, using $p > 1$ facilitates wave-vector matching deeper into blue and even the ultraviolet region, provided that the domain reversal can be achieved. However, there is a price that needs to be paid in terms of the magnitude of $\hat{d}^{(2)}_{p,zzz}$ relative to $\hat{d}^{(2)}_{zzz}$ even when the parameter a is varied to optimize the nonlinearity. Clearly, for $p = 1$, the optimum is $\hat{d}^{(2)}_{1,zzz} = (2/\pi)\hat{d}^{(2)}_{zzz}$ for $a = 1/2$. For $p = 2$, the optimum is $\hat{d}^{(2)}_{2,zzz} = (1/\pi)\hat{d}^{(2)}_{zzz}$ for $a = 1/4$ and $3/4$. For $p = 3$ and $a = 1/6$, $1/2$, and $5/6$, the optimum is $\hat{d}^{(2)}_{3,zzz} = (2/3\pi)\hat{d}^{(2)}_{zzz}$.

A detailed evolution of the second harmonic intensity for QPM as well as for the perfect wave-vector matching (which does not exist in this geometry) using $\hat{d}^{(2)}_{zzz}$ is shown in Fig. 3.8, where the coherence length (ℓ_c) is the shortest distance at which the non-wave-vector matching yields a maximum conversion to harmonic, i.e., $\Delta k L = \pi$.

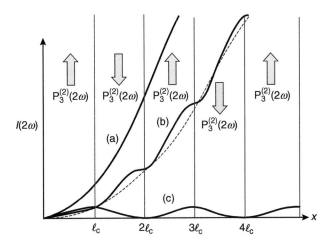

FIGURE 3.8 Second harmonic intensity as a function of the propagation distance for the three cases. (a) Perfect wave-vector matching if there were no dispersion in the refractive index. (b) Quasi-phase matching with $p = 1$. The directions of the nonlinearly induced polarization are indicated by arrows. (c) Phase mismatch $\Delta k L = 2k_{\text{vac}}(\omega)L|n_z(\omega) - n_z(2\omega)|$.

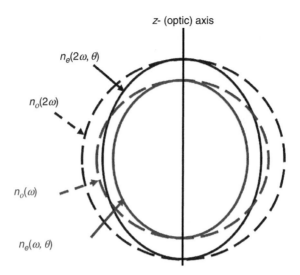

FIGURE 3.9   Cuts of the three-dimensional refractive index surfaces for a negative uniaxial crystal at frequencies ω and 2ω.

3.1.3 Walk-Off

The simplest case is for a negative uniaxial crystal with *eoo* polarizations, and it will be used to illustrate the "critical" and "noncritical" wave-vector-matching cases. For uniaxial crystals, the three-dimensional refractive index surfaces at a given frequency are spheres for n_o and ellipsoids with circular symmetry about the z-axis for n_e. Figure 3.9 shows the cuts of the refractive index surfaces in the z–x plane. The dashed lines correspond to n_o, and the solid line correspond to $n_e(\theta)$. The wave-vector-matching condition $n_o(\omega) = n_e(\theta, 2\omega)$ corresponding to the critical wave-vector-matching case is the intersection of the curves $n_o(\omega)$ and $n_e(\theta, 2\omega)$. Figure 3.10a shows the wave-vector direction for this case. Furthermore, since the direction of the Poynting vectors, i.e., the direction of energy flow, is normal to the index surface, these are not collinear for the fundamental and the harmonic. Thus, for finite width beams with the same phase-velocity direction, i.e., $\vec{k}$, they walk off each other and the conversion efficiency is reduced.

The noncritical wave-vector-matching condition is shown in Fig. 3.10b, which corresponds to $n_o(\omega) = n_e(2\omega)$; i.e., $\theta_{PM} = \pi/2$. The two dispersion curves are tangent to each other and orthogonal to the direction of $\vec{k}$. Hence there is no walk-off between the waves at the two frequencies. This is clearly the optimum case, but it can occur at only one frequency. Limited tuning of the curves can be achieved by changing the temperature since the temperature dependence of the indices n_o and n_e is different.

The walk-off angle δ experienced by the *e* wave from the *o* wave is well known in crystal optics, and at wave-vector match it is given by

$$\tan\delta \cong \delta = \frac{n_e^2(\theta_{PM})}{2}\left(\frac{1}{n_o^2} - \frac{1}{n_e^2}\right)\sin(2\theta), \tag{3.10}$$

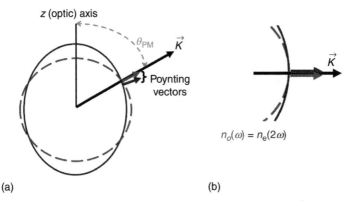

FIGURE 3.10 (a) The phase-match angle θ_{PM} and the direction of the $\vec{k}$ vector for Type 1 critical wave-vector match for a negative uniaxial crystal. The angle between the fundamental and harmonic Poynting vectors is defined as δ. (b) The noncritical wave-vector match condition with collinear Poynting vectors.

which, in the limit of small birefringence, is

$$\delta \cong \frac{n_e - n_o}{n_e} \sin(2\theta). \tag{3.11}$$

For *eoe*, e.g., where there are two extraordinary waves at different frequencies, both waves walk off from the ordinary wave with different walk-off angles.

The QPM cases for z-polarized inputs and outputs, e.g., for LiNbO$_3$ and LiTaO$_3$, can be effectively noncritical wave-vector matched as is clear from Fig. 3.11. For $p = 1$, there is an effective offset $\Delta n_z \cos(\theta)$ of the extraordinary fundamental refractive index introduced, which is maximum for propagation along the x- and y-axes; i.e., $\theta = \pi/2$. For the purpose of wave-vector matching, this offset can be introduced as a distortion of the fundamental index curve so that it can be tangent to the harmonic one in the x–z plane (see Fig. 3.11).

3.1.4 Angular Bandwidth

3.1.4.1 Birefringent Wave-Vector-Matched Uniaxial Crystals. The pertinent question addressed in this section is: How accurately must the phase-match angle be adjusted to obtain efficient SHG? This will of course depend on how "fast" the pertinent index curves, e.g., n_o and $n_e(\theta)$ for Type 1 wave-vector matching, separate with angle θ after their intersection. Clearly, noncritical wave-vector matching will have a larger angular bandwidth than for critical wave-vector matching (see Figs 3.10 and 3.11).

To illustrate this effect, the simplest SHG geometry, namely, Type 1 *eoo*, is examined. The typical $I(2\omega) \propto \mathrm{sinc}^2(\Delta kL/2)$ response falls to ≈ 0.4 when $\Delta kL = \pi$. Therefore a useful definition for the angular bandwidth is the detuning angle $\Delta\theta_{PM}$,

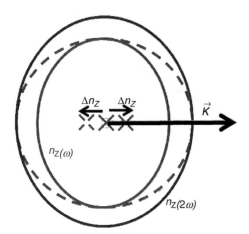

FIGURE 3.11   The outer solid curve is the cut in the x–z plane of the three-dimensional refractive index ellipsoid for the $n_e(2\omega, \theta)$ surface, and the inner solid curve is for $n_e(\omega, \theta)$. The dashed curve is the effective refractive index $n_e(2\omega, \theta)$ augmented by the periodic poling contribution $\Delta n_z = K/2k_{vac}$ for light polarization along the x-axis.

which causes a decrease of 60% in the SHG efficiency from its peak value. This occurs at $\Delta kL = \pi$. For eoo,

$$\Delta k = 2k_{vac}(\omega)[n_o(\omega) - n_e(2\omega, \theta)] = 2\frac{\omega}{c}[n_o(\omega) - n_e(2\omega, \theta)]. \qquad (3.12)$$

The change in the refractive index $n_e(2\omega, \theta)$ from the wave-vector match can be expanded in terms of a power series around the phase-match angle as

$$n_e(2\omega, \theta_{PM} + \Delta\theta_{PM}) = n_e(2\omega, \theta_{PM}) + \frac{\partial n_e(2\omega, \theta)}{\partial\theta}\bigg|_{\theta_{PM}} \Delta\theta_{PM} +$$

$$\frac{1}{2}\frac{\partial n_e^2(2\omega, \theta)}{\partial\theta^2}\bigg|_{\theta_{PM}} \Delta\theta_{PM}^2 + \cdots \qquad (3.13)$$

$$\rightarrow \quad \frac{\Delta k}{2}L = \frac{2\omega[n_o(\omega) - n_e(2\omega, \theta)]}{2c}L = \frac{\omega}{c}L[n_o(\omega) - n_e(2\omega, \theta_{PM}) -$$

$$\frac{\partial n_e(2\omega, \theta)}{\partial\theta}\bigg|_{\theta_{PM}} \Delta\theta_{PM}]. \qquad (3.14)$$

The derivative $\frac{\partial n_e(2\omega, \theta)}{\partial\theta}\big|_{\theta_{PM}}$ is evaluated from Eq. 3.1 and inserted into Eq. 3.14 to give, for $\Delta kL = \pi$,

$$\Delta\theta_{PM} \simeq \frac{\lambda_{vac}(\omega)}{4L}\frac{1}{\{n_e(2\omega) - n_o(2\omega)\}\sin(2\theta_{PM})}. \qquad (3.15)$$

At and near a noncritical wave-vector match, i.e.,$\theta_{PM} = \pi/2$, Eq. 3.15 diverges. It is then necessary to use the next term in the expansion in Eq. 3.13, i.e.,

$$\frac{\Delta k}{2} L = \frac{2\omega[n_o(\omega) - n_e(2\omega, \theta)]}{2c} L = \frac{\omega}{c} L \left[-\frac{1}{2} \frac{\partial^2 n_e(2\omega, \theta)}{\partial \theta^2} \bigg|_{\theta_{PM}} \Delta\theta_{PM}^2 \right], \quad (3.16)$$

which gives

$$\Delta\theta_{PM} \cong \left\{ \frac{\lambda_{vac}(\omega)}{4L|n_o(2\omega) - n_e(2\omega)|} \right\}^{1/2}. \quad (3.17)$$

Note the critical role that birefringence plays in the angular bandwidth, i.e., $\Delta\theta_{PM} \propto [n_o(2\omega) - n_e(2\omega)]^{-1}$. Small birefringence is an advantage. Although many organic crystals have been found with large nonlinearities, invariably they also have large birefringence, which has inhibited their application to date.

The other birefringent wave-vector-matching geometries are also analyzed in the same way. For example, the starting point for Type 2 *eoe* would be

$$\Delta k = k_{vac}(\omega)\{[n_e(\omega, \theta) + n_o(\omega)] - 2n_e(2\omega, \theta)\}$$
$$= k_{vac}(\omega) \left\{ \frac{\partial n_e(\omega, \theta)}{\partial \theta} - 2\frac{\partial n_e(2\omega, \theta)}{\partial \theta} \right\} \bigg|_{\theta_{PM}} \Delta\theta_{PM}. \quad (3.18)$$

3.1.4.2 Quasi-Phase-Matched Crystals.
The final formulas for the birefringent wave-vector match cannot be used here because the poling introduces a contribution to the wave vector and hence the bandwidth. Assuming that the poling direction is along a crystal's z-axis (e.g., $LiNbO_3$), the k-vector contribution at an angle φ from the x- or y-axis can be written as $K(\varphi) = K\cos(\varphi)$. For light polarization and poling along the n_e-axis,

$$\Delta k = 2k_{vac}(\omega) \left[n_e(\omega, \theta) - n_e(2\omega, \theta) + \frac{K(\varphi)}{2k_{vac}(\omega)} \right]$$
$$= 2\frac{\omega}{c} \left[n_e(\omega) - n_e(2\omega, \theta) + \frac{K\cos(\varphi)}{2k_{vac}(\omega)} \right]. \quad (3.19)$$

The refractive indices are again expanded in terms of a power series as

$$n_e(\omega, \theta_{PM} + \Delta\theta) = n_e(\omega, \theta_{PM}) + \frac{\partial n_e(\omega, \theta)}{\partial \theta} \bigg|_{\theta_{PM}} \Delta\theta_{PM} + \frac{1}{2} \frac{\partial n_e^2(\omega, \theta)}{\partial \theta^2} \bigg|_{\theta_{PM}} \Delta\theta_{PM}^2 + \cdots$$

$$(3.20a)$$

and

$$n_e(2\omega, \theta_{PM} + \Delta\theta) = n_e(2\omega, \theta_{PM}) + \frac{\partial n_e(2\omega, \theta)}{\partial\theta}\bigg|_{\theta_{PM}} \Delta\theta_{PM} + \frac{1}{2}\frac{\partial n_e^2(2\omega, \theta)}{\partial\theta^2}\bigg|_{\theta_{PM}} \Delta\theta_{PM}^2 + \cdots,$$

(3.20b)

and $\cos(\varphi)$ is expanded as $1 - \frac{1}{2}\Delta\varphi^2 = 1 - \frac{1}{2}\Delta\theta_{PM}^2$ for the usual QPM case ($\theta_{PM} = \pi/2$) so that

$$\frac{\Delta k}{2}L = \frac{\omega}{c}L\left\{ \left[\frac{\partial n_e(\omega, \theta)}{\partial\theta} - \frac{\partial n_e(2\omega, \theta)}{\partial\theta}\right]\bigg|_{\theta_{PM}} \Delta\theta_{PM} + \right.$$
$$\left. \frac{1}{2}\left[\frac{\partial n_e^2(\omega, \theta)}{\partial\theta^2} - \frac{\partial n_e^2(2\omega, \theta)}{\partial\theta^2} - \frac{K}{2k_{vac}}\right]\Delta\theta_{PM}^2 + \cdots \right\}.$$

(3.21)

Since

$$\frac{\partial n_e(\omega, \theta)}{\partial\theta} - \frac{\partial n_e(2\omega, \theta)}{\partial\theta}\bigg|_{\theta_{PM}} = \frac{1}{2}\sin(2\theta_{PM})\left\{\frac{n_0(\omega)}{n_e^2(\omega)}\left[n_0^2(\omega) - n_e^2(\omega)\right] - \right.$$
$$\left. \frac{n_0(2\omega)}{n_e^2(2\omega)}\left[n_0^2(2\omega) - n_e^2(2\omega)\right]\right\}$$

(3.22)

at $\theta_{PM} = \pi/2 \to \sin(2\theta_{PM}) = 0$, the second-order terms proportional to $\Delta\theta_{PM}^2$ are needed to find the bandwidth. Evaluating this case gives

$$\left[\frac{\partial n_e^2(\omega, \theta)}{\partial\theta^2} - \frac{\partial n_e^2(2\omega, \theta)}{\partial\theta^2}\right] - \frac{K}{2k_{vac}}$$
$$= \cos(2\theta_{PM})\left\{\frac{n_0(\omega)}{n_e^2(\omega)}\left[n_0^2(\omega) - n_e^2(\omega)\right] - \frac{n_0(2\omega)}{n_e^2(2\omega)}\left[n_0^2(2\omega) - n_e^2(2\omega)\right]\right\} - \frac{K}{2k_{vac}}.$$

(3.23)

Therefore,

$$\frac{\Delta k}{2}L = \frac{\omega}{2c}L\left\{\frac{n_0(\omega)}{n_e^2(\omega)}\left[n_0^2(\omega) - n_e^2(\omega)\right] - \frac{n_0(2\omega)}{n_e^2(2\omega)}\left[n_0^2(2\omega) - n_e^2(2\omega)\right] - \frac{K}{2k_{vac}}\right\}\Delta\theta_{PM}^2.$$

(3.24)

For $\dfrac{\Delta kL}{2} = \dfrac{\pi}{2}$ and $\dfrac{K}{2k_{vac}(\omega)} = \dfrac{\lambda_{vac}(\omega)}{2\Lambda}$,

$$\Delta\theta_{PM} = \sqrt{\frac{\lambda_{vac}(\omega)}{2L}\left|\left\{\frac{n_0(\omega)}{n_e^2(\omega)}\left[n_0^2(\omega) - n_e^2(\omega)\right] - \frac{n_0(2\omega)}{n_e^2(2\omega)}\left[n_0^2(2\omega) - n_e^2(2\omega)\right] - \frac{\lambda_{vac}(\omega)}{2\Lambda}\right\}^{-1}\right|},$$

(3.25)

which is a general result for QPM. For small birefringence (the usual case),

$$\Delta\theta_{PM} \cong \sqrt{\frac{\lambda_{vac}(\omega)}{4L} \left| \left\{ [n_o(\omega) - n_o(2\omega)] - [n_e(\omega) - n_e(2\omega)] - \frac{\lambda_{vac}(\omega)}{4\Lambda} \right\}^{-1} \right|}.$$

$$(3.26)$$

3.1.5 Noncollinear Wave-Vector Match

The previous discussion has been limited to the most common second harmonic geometries, namely, collinear wave vectors for the interacting beams. There is an additional method for wave-vector matching that uses copolarized waves. This geometry, called the noncollinear phase matching, is shown in Fig. 3.12. The input $\vec{k}$-vectors are not collinear and arranged so that the output wavevector lies in a preferred direction for phase-matching. Although this clearly expands the wave-vector-matching possibilities, there are some serious drawbacks. The most obvious is that when operating with beams of finite cross section, the interaction volume is limited due to walk-off. Furthermore, the harmonic beam shape no longer has a circular symmetry.

3.1.6 Biaxial Crystals

Although SHG is a more complex problem in biaxial crystals, which in general have lower crystal symmetries than uniaxial crystals, a number of biaxial crystals have been developed for the generation of new frequencies. The dielectric constant tensor is shown in Fig. 3.13a in the coordinate system, which diagonalizes the polarizability tensor as well as the dielectric constant tensor. The typical dispersion in the refractive indices is shown in Fig. 3.13b; the notation used, $n_\gamma > n_\beta > n_\alpha$, is unrelated to the crystal's crystallographic axes (a, b, c). In optics, n_γ is always associated with the largest refractive index, n_α with the smallest, and n_β with the middle one. Because it is possible that two or more axes can cross at some frequency, the definition is normally applied far from resonance. Because of the complications associated with three independent refractive indices that require the full set of Euler angles for defining arbitrary wave vector and field directions, the discussion here is limited to the

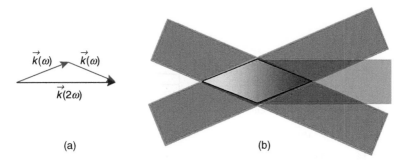

(a)   (b)

FIGURE 3.12   (a) Wave-vector-matching geometry in the noncollinear case. (b) The beam geometry showing the interaction region (diamond shape) and the output harmonic.

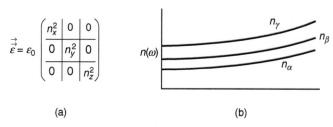

(a)　　　　　　　　　　　　　　　(b)

FIGURE 3.13 (a) The dielectric constant tensor for a biaxial crystal. (b) The dispersion in the three independent refractive indices in a coaxial crystal.

propagation wave vectors $\vec{k}$ lying in one of the principal planes α–β, α–γ, or β–γ. This means that one of the eigenmodes is always along a crystal axis, i.e., orthogonal to one of these planes. For example, for $\vec{k}$ in the α–β plane, one eigenmode lies on the γ-axis and acts like $\vec{E}_o$ in the uniaxial case previously discussed. As a result, the *eoo* and *eoe* as well as Type 1 and Type 2 notations are also used in the same way here; i.e., Type 1 has one eigenmode input and Type 2 has two eigenmode inputs. For simplicity, only the cases associated with $\vec{k}$ lying in the α–β plane are used to illustrate the wave-vector matching. The other cases are straightforward extensions of this case.

The Type 1 *eoo* case is considered first. Bearing in mind that the *o* wave is polarized along the γ-axis, the wave-vector-matching possibilities are explored in Fig. 3.14a. Clearly, wave-vector matching is possible if $n_\beta(2\omega) \geq n_\gamma(\omega) \geq n_\alpha(2\omega)$. Using the analogy with the uniaxial case, with θ measured from the α-axis,

$$\sin(\theta_{PM}) = \frac{n_\beta(2\omega)}{n_\gamma(\omega)} \sqrt{\frac{n_\alpha^2(2\omega) - n_\gamma^2(\omega)}{n_\alpha^2(2\omega) - n_\beta^2(2\omega)}}, \tag{3.27}$$

$$\Delta\theta_{PM} \cong \frac{\lambda(\omega)}{4L} \frac{1}{\{n_\beta(2\omega) - n_\alpha(2\omega)\}\sin(2\theta_{PM})}, \tag{3.28}$$

$$\tan\delta \cong \delta = \frac{n_{\alpha\to\beta}^2(2\omega, \theta_{PM})}{2} \left[\frac{1}{n_\beta^2(2\omega)} - \frac{1}{n_\alpha^2(2\omega)} \right] \sin(2\theta_{PM}) \cong \frac{n_\alpha(2\omega) - n_\beta(2\omega)}{n_\alpha(2\omega)} \sin(2\theta_{PM}). \tag{3.29}$$

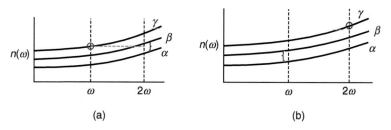

(a)　　　　　　　　　　　　　　　(b)

FIGURE 3.14 (a) The dispersion in the refractive index, showing that Type 1 *eoo* is wave-vector matchable for second harmonic generation with the $\vec{k}$ vector in the α–β plane. (b) The *oee* geometry is not wave-vector matchable.

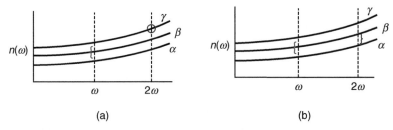

FIGURE 3.15 (a) The dispersion in the refractive index, showing that Type 2 *oeo* is not wave-vector matchable for second harmonic generation. (b) The *eoe* geometry is wave-vector matchable in principle.

For the Type 1 *oee* case, the situation is different, as shown in Fig. 3.14b. This geometry is not wave-vector matchable.

Type 2 *eoe* and *oeo* cases are more complex. The index diagrams are shown in Fig. 3.15. Although the *oeo* can clearly never be wave-vector matched, the *eoe* could be, provided that $n_\gamma(\omega) + n_{\text{eff}}(\theta, \omega) = 2n_{\text{eff}}(\theta, 2\omega)$ with an angle θ measured from the α-axis can be found.

3.2 OPTIMIZING $\hat{d}_{\text{eff}}^{(2)}$

An important problem in optimizing the conversion efficiency is to identify the direction of the incident wave vector inside the crystal that gives the largest possible $\hat{d}_{\text{eff}}^{(2)}$, given an allowed wave-vector-matching condition. The $\hat{d}_{ijk}$ tensor has 27 elements, which are reduced to 18 by permutation symmetry because $\hat{d}_{ikj}^{(2)} = \hat{d}_{ijk}^{(2)}$ (as discussed in Chapter 2). This number is reduced further by the symmetry of the crystal, which also gives relationships between some of the elements. The permutation symmetry allows the information in the 27-element $3 \times 3 \times 3$ tensor to be reduced to a 3×6 tensor constructed using the Voigt notation shown in Fig. 3.16.

jk	J
11	1
22	2
33	3
23, 32	4
13, 31	5
12, 21	6

$\rightarrow \hat{d}_{iJ}^{(2)} =$

$\hat{d}_{11}$	$\hat{d}_{21}$	$\hat{d}_{31}$
$\hat{d}_{12}$	$\hat{d}_{22}$	$\hat{d}_{32}$
$\hat{d}_{13}$	$\hat{d}_{23}$	$\hat{d}_{33}$
$\hat{d}_{14}$	$\hat{d}_{24}$	$\hat{d}_{34}$
$\hat{d}_{15}$	$\hat{d}_{25}$	$\hat{d}_{35}$
$\hat{d}_{16}$	$\hat{d}_{26}$	$\hat{d}_{36}$

FIGURE 3.16 The Voigt notation for transforming the d_{ijk} tensor to the d_{iJ} tensor.

The number of independent elements in the tensors $\hat{\chi}_{ii}^{(1)}$, $\hat{\chi}_{ijk}^{(2)}$, and $\hat{\chi}_{ijk\ell}^{(3)}$ are listed for various crystal classes in Table 3.1. Nonisotropic media are typically characterized by their symmetry rotation axes and mirror planes, which are the basis of the international notation, where it is shown that there are cubic crystal classes such as $\bar{4}3m$ (e.g., GaAs) that are $\hat{\chi}_{ijk}^{(2)}$ active although the crystals are isotropic for linear

TABLE 3.1 Number of Independent Elements in the $\hat{\chi}_{ii}^{(1)}$, $\hat{\chi}_{ijk}^{(2)}$, and $\hat{\chi}_{ijk\ell}^{(3)}$ Tensors for Different Crystal Classes[a]. Note that the Trigonal System Corresponds to Rhombohedral in Crystalographic Notation. Also Different Combinations of Symmetry Operations Are Used to Describe a Given System, e.g. $4/mmm \equiv 4/m\ 2/m\ 2/m$.

| System | Standard Notations | | Number of Independent Elements | | |
	International	Schönflies	$\chi_{ij}^{(1)}$	$\chi_{ijk}^{(2)}$	$\chi_{ijk\ell}^{(3)}$
Triclinic	1	C_1	6*	18	81
	$\bar{1}$	$S_2(c_i)$	6	0	81
Monoclinic	m	C_{1h}	4	10	41
	2	C_2	4*	8	41
	$2/m$	C_{2h}	4	0	41
Orthorhombic	$2mm$	C_{2v}	3	5	21
	222	$D_2(V)$	3*	3	21
	$mmm \equiv 2/mm\ 2/mm\ 2/mm$	$D_{2h}(V_H)$	3	0	21
Tetragonal	4	C_4	2*	4	21
	$\bar{4}$	S_4	2	4	21
	$4/m$	C_{4h}	2	0	21
	$4mm$	C_{4v}	2	3	11
	$\bar{4}2m$	$D_{2h}(V_d)$	2	2	11
	422	D_4	2*	1	11
	$4/mmm \equiv 4/m\ 2/m\ 2/m$	D_{4h}	2	0	11
Trigonal	3	C_3	2*	6	27
	$\bar{3}$	$S_6(C_{3i})$	2	0	27
	$3m$	C_{3v}	2	4	14
	32	D_3	2*	2	14
	$32/m \equiv \bar{3}m$	D_{3d}	2	0	14
Hexagonal	$\bar{6} \equiv 3/m$	C_{3h}	2	2	19
	6	C_6	2*	4	19
	$6/m$	C_{6h}	2	0	19
	$\bar{6}2m \equiv \bar{6}m2$	D_{3h}	2	1	10
	$6mm$	C_{6v}	2	3	10
	622	D_6	2*	1	10
	$6/mm \equiv 6/m\ 2/m\ 2/m$	D_{6h}	2	0	10
Cubic	23	T	1*	1	7
	$2/m\bar{3} \equiv m3$	T_h	1	0	7
	$\bar{4}3m$	T_d	1	1	4
	432	0	1*	0	4
	$m3m \equiv 4/m\bar{3}\ 2/m$	0_h	1	0	4

[a] International and Schönflies are different crystal classifications (5).

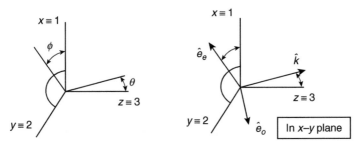

FIGURE 3.17   The axes used, the Euler angles describing the direction of the incident wave vector $\vec{k}$, and the directions of the field unit vectors $\hat{e}_o$ and $\hat{e}_e$.

optics. The important point here is that media with isotropic refractive indices can have nonzero second-order nonlinearities, but isotropic media cannot have second-order nonlinearities.

The problem now is to evaluate $\hat{d}_{\text{eff}}^{(2)}$ as a function of the Euler angles θ and ϕ. For Type 2, $\hat{d}_{\text{eff}}^{(2)} = \hat{e}_i^*(2\omega)\, \hat{d}_{ijk}^{(2)}(-2\omega;\, \omega,\, \omega)\, \hat{e}_j^a(\omega)\, \hat{e}_k^b(\omega)$, where a and b denote the two orthogonally polarized eigenmodes for SHG. The Type 1 case is similar but with $a = b$ and $j = k$. The Euler angles and the definition of $\hat{e}_o$ and $\hat{e}_e$ are shown in Fig. 3.17.

The next step is to evaluate $\hat{d}_{\text{eff}}^{(2)}$. The pertinent equations for the linearly polarized case are as follows:

Type I:

For eoo, $-$ve uniaxial $\rightarrow$ $\hat{e}_{ei}\hat{d}_{ijk}^{(2)}\hat{e}_{oj}\hat{e}_{ok}$

For oee, $+$ve uniaxial $\rightarrow$ $\hat{e}_{oi}\hat{d}_{ijk}^{(2)}\hat{e}_{ej}\hat{e}_{ek}$

$$(3.30)$$

Type II:

For eoe, $-$ve uniaxial $\rightarrow$ $\hat{e}_{ei}\hat{d}_{ijk}^{(2)}\hat{e}_{oj}\hat{e}_{ek}$

For oeo, $+$ve uniaxial $\rightarrow$ $\hat{e}_{oi}\hat{d}_{ijk}^{(2)}\hat{e}_{ej}\hat{e}_{ok}$

Neglecting the difference in the direction between $\vec{D}$ and $\vec{E}$, which is valid for small birefringence, $\hat{e}_o \cdot \hat{e}_e = 0$, $\hat{e}_o \cdot \hat{k} = 0$, and $\hat{e}_e \cdot \hat{k} = 0$. Furthermore,

$$\hat{k} = (\sin\theta\,\cos\phi,\ \sin\theta\,\sin\phi,\ \cos\theta),$$
$$\hat{e}_o = (-\sin\phi,\ \cos\phi,\ 0),$$
$$\hat{e}_e = (\cos\theta\,\cos\phi,\ \cos\theta\,\sin\phi,\ -\sin\theta).$$
$$(3.31)$$

The procedures used to calculate $\hat{d}_{\text{eff}}^{(2)}$ will now be illustrated using the class 32 crystal as the example. The non-zero elements for $\hat{d}_{ijk}^{(2)}$ are listed in Table 3.2 and the appropriate $\hat{d}_{iJ}^{(2)}$ tensor is reproduced in Fig. 3.18.

TABLE 3.2 $\hat{d}_{ij}^{(2)}$ For Different Crystal Classes. The Small Dots Denote Zero Elements. All Other Shapes Are Nonzero. Straight Solid Lines Connect Equal Coefficients. Open Symbols Connected to Solid Symbols Indicate Different Signs. Square Symbols Are Zero Coefficients in the Kleinman Limit. Dashed Lines Connect Elements that Are Equal in the Kleinman Limit (6).

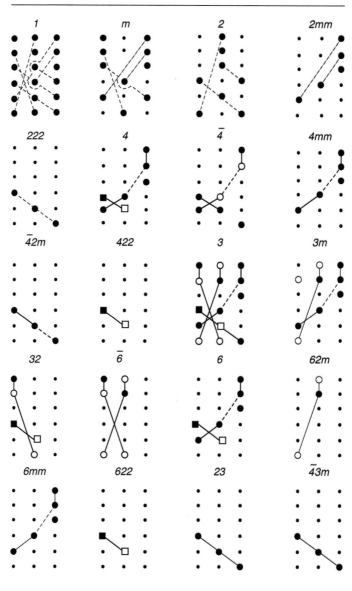

$$\hat{d}_{iJ}^{(2)} = \begin{array}{|c|c|c|}
\hline
\hat{d}_{11}^{(2)} & 0 & 0 \\
\hline
-\hat{d}_{11}^{(2)} & 0 & 0 \\
\hline
0 & 0 & 0 \\
\hline
\hat{d}_{14}^{(2)} & 0 & 0 \\
\hline
0 & -\hat{d}_{14}^{(2)} & 0 \\
\hline
0 & -\hat{d}_{11}^{(2)} & 0 \\
\hline
\end{array}$$

FIGURE 3.18 The d_{iJ} tensor for a 32 crystal.

For Type 1 *eoo*,

$$\hat{d}_{\text{eff}}^{(2)} = \hat{d}_{11}^{(2)}\hat{e}_{e1}\hat{e}_{o1}\hat{e}_{o1} + \hat{d}_{12}^{(2)}\hat{e}_{e1}\hat{e}_{o2}\hat{e}_{o2} + \hat{d}_{14}^{(2)}\hat{e}_{e1}\{\hat{e}_{o2}\hat{e}_{o3} + \hat{e}_{o3}\hat{e}_{o2}\} + \hat{d}_{25}^{(2)}\hat{e}_{e2}\{\hat{e}_{o1}\hat{e}_{o3} + \hat{e}_{o3}\hat{e}_{o1}\}$$

$$+ \hat{d}_{26}^{(2)}\hat{e}_{e2}\{\hat{e}_{o1}\hat{e}_{o2} + \hat{e}_{o2}\hat{e}_{o1}\},$$

$$\hat{e}_3 = 0 \rightarrow \hat{d}_{\text{eff}}^{(2)} = \hat{d}_{11}^{(2)}\{\hat{e}_{e1}\hat{e}_{o1}\hat{e}_{o1} - \hat{e}_{e1}\hat{e}_{o2}\hat{e}_{o2} - 2\hat{e}_{e2}\hat{e}_{o1}\hat{e}_{o2}\},$$

$$\hat{d}_{\text{eff}}^{(2)} = \hat{d}_{11}^{(2)}\{\cos\theta\cos\phi\sin^2\phi - \cos\theta\cos^3\phi + 2\cos\theta\sin^2\phi\cos\phi\},$$

$$\hat{d}_{\text{eff}}^{(2)} = \hat{d}_{11}^{(2)}\cos\theta\{-\cos^3\phi + 3\cos\phi\sin^2\phi\} = -\hat{d}_{11}^{(2)}\cos\theta\cos3\phi\}.$$

$$(3.32)$$

For Type 1 *oee*,

$$\hat{d}_{\text{eff}}^{(2)} = \hat{d}_{11}^{(2)}\hat{e}_{o1}\hat{e}_{e1}\hat{e}_{e1} + \hat{d}_{12}^{(2)}\hat{e}_{o1}\hat{e}_{e2}\hat{e}_{e2} + \hat{d}_{14}^{(2)}\hat{e}_{o1}\{\hat{e}_{e2}\hat{e}_{e3} + \hat{e}_{e3}\hat{e}_{e2}\} + \hat{d}_{25}^{(2)}\hat{e}_{o2}\{\hat{e}_{e1}\hat{e}_{e3} + \hat{e}_{e3}\hat{e}_{e1}\}$$

$$+ \hat{d}_{26}^{(2)}\hat{e}_{o2}\{\hat{e}_{e1}\hat{e}_{e2} + \hat{e}_{e2}\hat{e}_{e1}\},$$

$$\hat{d}_{\text{eff}}^{(2)} = \hat{d}_{11}^{(2)}\{-\sin\phi\cos^2\phi\cos^2\theta + \sin^3\phi\cos^2\theta - 2\cos^2\theta\sin\phi\cos^2\phi\}$$

$$+ \hat{d}_{14}^{(2)}2\{\cos^2\phi + \sin^2\phi\}\sin\theta\cos\theta,$$

$$\hat{d}_{\text{eff}}^{(2)} = -\hat{d}_{11}^{(2)}\sin3\phi\cos^2\theta + \hat{d}_{14}^{(2)}\sin2\theta.$$

$$(3.33)$$

For Type 2 *eoe* and *oeo*,

$$\hat{d}_{\text{eff}}^{(2)} = \hat{d}_{11}^{(2)}\{\hat{e}_{o1}\hat{e}_{e1}\hat{e}_{e1} - \hat{e}_{e1}\hat{e}_{o2}\hat{e}_{e2} - \hat{e}_{e2}[\hat{e}_{e1}\hat{e}_{o2} + \hat{e}_{e2}\hat{e}_{o1}]\}$$

$$+ \hat{d}_{14}^{(2)}\{[\hat{e}_{e1}\hat{e}_{e2}\hat{e}_{o3} + \hat{e}_{e1}\hat{e}_{o2}\hat{e}_{e3}] - [\hat{e}_{e2}\hat{e}_{e3}\hat{e}_{o1} + \hat{e}_{e2}\hat{e}_{o3}\hat{e}_{e1}]\},$$

$$(3.34)$$

$$\hat{e}_3 = 0 \rightarrow \hat{d}_{\text{eff}}^{(2)} = -\hat{d}_{11}^{(2)}\sin3\phi\cos^2\theta + \hat{d}_{14}^{(2)}\sin2\theta$$

and

$$
\begin{aligned}
\hat{d}^{(2)}_{\text{eff}} &= \hat{d}^{(2)}_{11}\hat{e}_{e1}\hat{e}_{o1}\hat{e}_{o1} + \hat{d}^{(2)}_{12}\hat{e}_{o1}\hat{e}_{e2}\hat{e}_{o2} \\
&\quad + \hat{d}^{(2)}_{14}\hat{e}_{o1}\{\hat{e}_{o2}\hat{e}_{e3} + \hat{e}_{o3}\hat{e}_{e2}\} + \hat{d}^{(2)}_{25}\hat{e}_{o2}\{\hat{e}_{o1}\hat{e}_{e3} + \hat{e}_{o3}\hat{e}_{e1}\} \\
&\quad + \hat{d}^{(2)}_{26}\hat{e}_{o2}\{\hat{e}_{o1}\hat{e}_{e2} + \hat{e}_{o2}\hat{e}_{e1}\},
\end{aligned}
$$

$$
\hat{d}^{(2)}_{\text{eff}} = \hat{d}^{(2)}_{11}\cos\theta\{-\cos^3\phi + 3\cos\phi\sin^2\phi\} = -\hat{d}^{(2)}_{11}\cos\theta\cos3\phi,
\tag{3.35}
$$

respectively. Note that the 32 crystal symmetry is reflected in the 3ϕ and 2θ sinusoidal variations in $\hat{d}^{(2)}_{\text{eff}}$. Furthermore, for the Kleinman symmetry, $\tilde{d}^{(2)}_{14} = 0$ and hence the $\hat{d}^{(2)}_{\text{eff}}$'s are identical for *eoo* and *oeo* and also for *oee* and *eoe*. In fact, an approximate Kleinman symmetry is almost always the case in doubling applications; Table 3.3 summarizes all pertinent crystal classes.

Recalling that the wave-vector-matching condition in uniaxials depends only on θ, the optimization is performed on the angle ϕ. For example, for Type 1 *oee* for a 32 crystal,

$$
\hat{d}^{(2)}_{\text{eff}} = -\hat{d}_{11}\sin3\phi\,\cos^2\theta \quad \rightarrow \quad \text{Optima at } \phi = \frac{\pi}{6}, \frac{\pi}{2}, \frac{5\pi}{6}.
\tag{3.36}
$$

Another example is Type 1 *eoo* for $\bar{4}$ symmetry:

$$
\hat{d}^{(2)}_{\text{eff}} = \sin\theta\left[\hat{d}_{15}\cos2\phi - \hat{d}_{14}\sin2\phi\right] \quad \rightarrow \quad \text{Optimum } \phi \text{ depends on the ratio } \hat{d}_{15}/\hat{d}_{14}.
\tag{3.37}
$$

TABLE 3.3 Dependence in the Kleinman Limit of $\tilde{d}^{(2)}_{\text{eff}}$ on the Euler Angles θ and ϕ for All Uniaxial Crystal Classes that Are Noncentrosymmetric[a].

Crystal Class		−ve Uniaxial Type 2 *eoe* +ve Uniaxial Type 1 *oee*	−ve Uniaxial Type 1 *eoo* +ve Uniaxial Type 2 *oeo*
6	4	0	$-d^{(2)}_{15}\sin\theta$
622	422	0	0
6mm	4mm	0	$-d^{(2)}_{31}\sin\theta$
$\bar{6}m2$		$-d^{(2)}_{22}\cos^2\theta\cos3\phi$	$d^{(2)}_{22}\cos\theta\sin3\phi$
3m		$d^{(2)}_{22}\cos^2\theta\cos3\phi$	$d^{(2)}_{31}\sin\theta - d^{(2)}_{22}\cos\theta\sin3\phi$
$\bar{6}$		$\cos^2\theta(-d^{(2)}_{11}\sin3\phi + d^{(2)}_{22}\cos3\phi)$	$\cos\theta(-d^{(2)}_{11}\cos3\phi + d^{(2)}_{22}\sin3\phi)$
3		$\cos^2\theta(-d^{(2)}_{11}\sin3\phi + d^{(2)}_{22}\cos3\phi)$	$-d^{(2)}_{31}\sin\theta + \cos\theta(-d^{(2)}_{11}\cos3\phi + d^{(2)}_{22}\sin3\phi)$
32		$-d^{(2)}_{11}\cos^2\theta\sin3\phi$	$-d^{(2)}_{11}\cos\theta\cos3\phi$
$\bar{4}$		$\sin2\theta(d^{(2)}_{15}\sin2\phi - d^{(2)}_{14}\cos2\phi)$	$\sin\theta(d^{(2)}_{15}\cos2\phi - d^{(2)}_{14}\sin2\phi)$
$\bar{4}2m$		$d^{(2)}_{14}\sin2\theta\cos2\phi$	$d^{(2)}_{14}\sin\theta\sin2\phi$

[a] Ref. 5.

3.3 NUMERICAL EXAMPLES

Consider the case of lithium iodate, which is a uniaxial noncentrosymmetric crystal with the point group symmetry 6. Its refractive indices (with λ in micrometers) are given by

$$n_o^2(\lambda) = 3.4157 + \frac{0.04703}{\lambda^2 - 0.03531} - 0.008801\lambda^2,$$

$$n_e^2(\lambda) = 2.9187 + \frac{0.03515}{\lambda^2 - 0.02822} - 0.003641\lambda^2.$$

(a) Find all possible combinations for Type 1 and Type 2 wave-vector matching. Over what wavelength range is each of these feasible? Can this material be noncritically wave-vector matched? If yes, what are the wavelengths?

(b) What is the wave-vector-matching angle, the acceptance angle, and the walk-off angle for doubling of $1.06\,\mu m$? What is $\tilde{d}_{eff}^{(2)}$? Given $\tilde{d}_{31}^{(2)} = 4.7\,pm/V$ and $\tilde{d}_{33}^{(2)} = 4.8\,pm/V$.

Solutions:

(a) Since lithium iodate is a *negative* uniaxial, the possible wave-vector-matching configurations are *eoo* (Type 1) and *eoe* (Type 2). Let us first consider *eoo* (Type 1):

Type of dispersion in refractive index and wave-vector matching

Field and $\vec{k}$ vector geometry

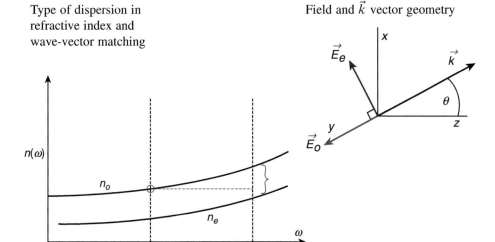

One way to find the range of phase-match angles is to look for the condition $n_e(\lambda/2) - n_o(\lambda) \leq 0$.

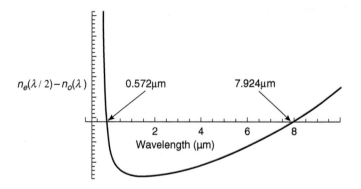

The result shows two noncritical phase-match wavelengths, i.e., the upper and the lower wavelengths for critical phase matching, which occurs for wavelengths between the two noncritical cases.

Now consider *eoe* (Type 2):

Type of dispersion in refractive index and wave-vector matching

Field and $\vec{k}$ vector geometry

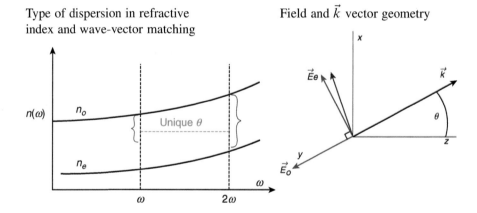

Another way to find the limits of wave-vector matching is to calculate the phase angle θ_{PM} from

$$\frac{1}{2}\left[\frac{n_o(\omega)n_e(\omega)}{\sqrt{n_e^2(\omega)\cos^2(\theta_{\text{PM}})+n_o^2(\omega)\sin^2\theta_{\text{PM}}}}+n_o(\omega)\right]$$

$$=\frac{n_o(2\omega)n_e(2\omega)}{\sqrt{n_e^2(2\omega)\cos^2\theta_{\text{PM}}+n_o^2(2\omega)\sin^2\theta_{\text{PM}}}}.$$

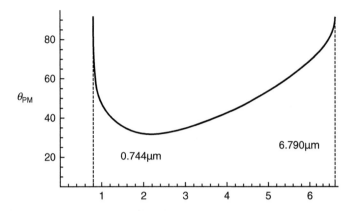

Clearly, the range of wave-vector-matchable wavelengths also has a non-critical wave-vector-matchable wavelength at each end of the range. Although this occurs frequently, it is not always the case. Furthermore, the upper wavelength limit lies in the absorption region for this particular material. The upper wavelength limits are generally not usable because of absorption.

A final word of caution about Sellmeier equations and their coefficients: It may happen that the wavelengths of interest for SHG fall outside the wavelength range of validity of the published coefficients that were obtained for a limited range of wavelengths.

(b) The Kleinman symmetry is valid at 1.06 μm. From Table 3.2, $\tilde{d}_{\text{eff}}^{(2)} \neq 0$ only for Type 1 *eoo*. From wave-vector matching, the angular bandwidth and walk-off equations are

$$\sin(\theta_{\text{PM}}) = \frac{n_e(2\omega)}{n_o(\omega)} \sqrt{\frac{n_o^2(2\omega) - n_o^2(\omega)}{n_o^2(2\omega) - n_e^2(2\omega)}} \quad \Rightarrow \quad \theta_{\text{PM}} = 30.12°,$$

$$\Delta\theta_{\text{PM}}(\text{FWHM}) \cong \frac{\lambda(\omega)}{2L} \frac{1}{\{n_o(2\omega) - n_e(2\omega)\} \sin(2\theta_{\text{PM}})}$$

$$\cong 0.2°/\text{mm (of crystal length)},$$

$$\delta \cong \frac{n_e(2\omega) - n_o(2\omega)}{n_e(2\omega)} \sin(2\theta_{\text{PM}}) \cong 4°.$$

The formula for $d_{\text{eff}}^{(2)}$ is $d_{\text{eff}}^{(2)} = -d_{15}^{(2)} \sin(\theta_{\text{PM}}) = -d_{31}^{(2)} \sin(\theta_{\text{PM}}) = -4.7 \sin(30.1°) = -2.4 \,\text{pm}/\text{V}.$

PROBLEMS

1. For a biaxial crystal, find all possible wave-vector matchable configurations when the "*o* ray" lies along the α-axis where $n_\gamma > n_\beta > n_\alpha$. What are the relations between the indices for each case?

2. **(a)** Find formulas for the acceptance angle for a critical wave-vector match and for a noncritical wave-vector match for birefringent wave-vector matching in uniaxial crystals for Type 1 *oee* and Type 2 *eoe* and *oeo*.

   (b) Find an expression for the acceptance angle for a noncritical wave-vector match in a quasi-phase-matching geometry.

3. One way to reduce the feature size for nanoelectronics photolithography is to use radiation with as short a wavelength as possible. $KBe_2BO_3F_2$ (KBBF) is a negative uniaxial nonlinear crystal with 32 rhombohedral symmetry, which has a "transparency window" from 155 to 3600 nm. Its refractive indices are given by

$$n_o^2(\lambda) = 1 + \frac{1.169725}{\lambda^2 - 0.0062400} - 0.009904\lambda^2,$$

$$n_e^2(\lambda) = 1 + \frac{0.956611}{\lambda^2 - 0.0061926} - 0.027849\lambda^2.$$

   (a) For Type 1 wave-vector match, calculate the variation in the phase-match angle with the second harmonic wavelength.

   (b) A frequency-tripled Nd:YVO4 laser was used to provide 355 nm. This was doubled by a KBBF to 177.3 nm. What was the phase-match angle, the acceptance angle (effective crystal length $= 5.6$ mm), and the walk-off angle for Type 1 second harmonic generation.

4. Find $\hat{d}_{\text{eff}}^{(2)}$ for Type 1 and Type 2 wave-vector match for a uniaxial noncentrosymmetric crystal with the point group symmetry 6 (lithium iodate). Does anything change in the Kleinman limit?

5. Potassium niobate is a negative biaxial crystal with the point group symmetry 2*mm* and nonlinearities $\hat{d}_{31}^{(2)} = 18.3$ pm/V, $\hat{d}_{32}^{(2)} = -15.8$ pm/V, $\hat{d}_{33}^{(2)} = -27.4$ pm/V, $\hat{d}_{15}^{(2)} = -17.1$ pm/V, and $\hat{d}_{24}^{(2)} = -16.5$ pm/V at a fundamental wavelength of 1.064 μm. The refractive indices are given by (with $\gamma \equiv x$, $\beta \equiv y$, $\alpha \equiv z$)

$$n_z^2 = 4.4208 + \frac{0.10044}{\lambda^2 - 0.054084} - 0.019592\lambda^2,$$

$$n_y^2 = 4.8355 + \frac{0.12839}{\lambda^2 - 0.056342} - 0.025379\lambda^2,$$

$$n_x^2 = 4.9873 + \frac{0.15149}{\lambda^2 - 0.064143} - 0.028775\lambda^2.$$

   (a) Comment on how well the Kleinman symmetry is satisfied at 1 μm? Find all possible Type 1 and Type 2 wave-vector-matching geometries and conditions on the refractive indices needed in all three principal planes. Use the frequency dispersion diagrams for the refractive indices.

(b) For Type 1, noncritical wave-vector-matched crystal orientations, find the wavelengths at which these occur for potassium niobate.

(c) Derive formulas for the variation of the acceptance angle and the walk-off angle with the phase-match angle for both Type 1 cases in their respective principal planes. Plot the phase-match, walk-off, and acceptance angles for both cases versus wavelength. Be sure to define the axis from which these angles are being measured.

(d) Derive the variation in $\hat{d}_{eff}^{(2)}$ with angle θ for Type 1 wave-vector match in the two principal planes with large tuning range and plot $\hat{d}_{eff}^{(2)}$ versus wavelength.

(e) Potassium niobate can be Type 1 wave-vector match for doubling of 1.06 μm in a principal plane. What is the wave-vector direction(s)? What are the acceptance and walk-off angles? What are $\hat{d}_{eff}^{(2)}$(s) for those cases?

6. Evaluate $\hat{d}_{eff}^{(2)}$ in terms of θ and ϕ for the two wave-vector-matching geometries of a 3m crystal. Does either change with the Kleinman symmetry?

7. Consider a 3/m uniaxial crystal for propagation along the z-axis whose eigenmodes in the x–y plane can be written as left-circularly polarized and right-circularly polarized light. In this case, the eigenmode fields with $\hat{e}_+ \cdot \hat{e}_+^* = 1$ and $\hat{e}_+ \cdot \hat{e}_+ = 0$ can be written as

Right-circularly polarized : $\vec{E}(\vec{r},t) = \hat{e}_+ E_+(\vec{r},t) = \dfrac{1}{\sqrt{2}}(\hat{e}_x + i\hat{e}_y)E_+(\vec{r},t),$

and

Left-circularly polarized : $\vec{E}(\vec{r},t) = \hat{e}_- E_-(\vec{r},t) = \dfrac{1}{\sqrt{2}}(\hat{e}_x - i\hat{e}_y)E_-(\vec{r},t).$

(a) For an incident field of amplitude $\vec{E}(\vec{r},t) = \frac{1}{\sqrt{2}}\{\hat{e}_x - i\hat{e}_y\}E(\vec{r},t)$, find the nonlinear polarization induced. This corresponds to Type 1.

(b) For Type 2, one fundamental field is right-circularly polarized and the other is left-circularly polarized. Find the nonlinear polarization for this case.

(c) Derive the coupled mode equations for the Type 1 case in (a) in terms of growing circularly polarized second harmonic fields.

(d) Assuming that both harmonic modes can be wave-vector matched, find the intensity of the second harmonic generated.

(e) Can either case be wave-vector matched? Why?

8. Consider a newly invented uniaxial crystal with the following $\hat{d}_{ijk}^{(2)}$ tensor:

0	0	1
0	0	1
0	0	1
0	0	1
0	0	1
0	0	1

(a) Find $\hat{d}_{eff}^{(2)}$
  (i) for propagation along the z-axis,
  (ii) for propagation along the x-axis, and
  (iii) at $45°$ from the z-axis in the x–z plane.
(b) Which cases in (a) can be wave-vector matched? Is the wave-vector matching of Type 1 or Type 2?
(c) Can any of the wave-vector-matching cases be noncritical wave-vector matched? If yes, which one(s)?

9. Consider a 32 uniaxial crystal for propagation along the z-axis whose eigenmodes in the x–y plane can be written as left-circularly polarized and right-circularly polarized light. In this case, the eigenmode fields can be written as

$$\text{Right-circularly polarized}: \vec{E}_+ = \hat{e}_+ E_+ = \frac{1}{\sqrt{2}}(\hat{e}_x + i\hat{e}_y)E_+,$$

and

$$\text{Left-circularly polarized}: \vec{E}_- = \hat{e}_- E_- = \frac{1}{\sqrt{2}}(\hat{e}_x - i\hat{e}_y)E_-.$$

(a) Derive the coupled mode expression for the second harmonic growth in terms of the polarized mode fields. [Hint: Note that $\hat{e}_+\hat{e}_+^* = 1$ and $\hat{e}_+\hat{e}_+ = 0$.]
(b) Find $\hat{d}_{eff}^{(2)}$ for Type 1 eoo for a right-circularly polarized fundamental field and both a right-circularly polarized harmonic and a left-circularly polarized harmonic (two different cases).

10. Consider an *artificial* uniaxial crystal, 1 cm long. Its refractive indices are given by $n_o(\lambda) = 2.0/(1 + 0.05\lambda^2)$ and $n_e = 1.9/(1 + 0.03\lambda^2)$, where λ is in micrometers. Its nonlinear d tensor is given by

0	0	5 pm/V
0	0	5 pm/V
0	0	1 pm/V
2 pm/V	−5 pm/V	0
5 pm/V	2 pm/V	0
0	0	0

(a) Calculate the range of wavelengths over which it can be Type 1 wave-vector matched.
(b) What is $\hat{d}_{eff}^{(2)}$ at the noncritical wave-vector-matched wavelength?

(c) What is the approximate crystal orientation for wave-vector matching for doubling of 1.06 μm. What is $\hat{d}_{\text{eff}}^{(2)}$, the walk-off angle, and the angular bandwidth?

11. For a Type 2 negative uniaxial crystal with the crystal symmetry $\bar{4}2m$, show that
$$\hat{d}_{\text{eff}}^{(2)} = -\hat{d}_{14}^{(2)} \sin(2\theta) \cos(2\phi).$$

12. KH_2PO_4 is a uniaxial noncentrosymmetric crystal with the point group symmetry $\bar{4}2m$, a nonlinearity $\hat{d}_{14}^{(2)} = 0.43$ pm/V, and a transmission range of 0.35–4.5 μm. The refractive indices (with λ in micrometers) are given by

$$n_o^2 = 2.2576 + \frac{0.0101}{\lambda^2 - 0.0142} + 1.7623 \frac{\lambda^2}{\lambda^2 - 57.8984},$$

$$n_e^2 = 2.1295 + \frac{0.0097}{\lambda^2 - 0.0014} + 0.7580 \frac{\lambda^2}{\lambda^2 - 127.0535}.$$

(a) Plot the wavelength, the acceptance angle, and the walk-off angle versus phase-match angle for Type 1 and Type 2 cases. Find the wavelengths at which noncritical wave-vector matching is possible.

(b) Derive the variation in d_{eff} with the wavelength for Type 1 and Type 2 wave-vector match.

(c) KH_2PO_4 can be phase matched for doubling of 1.06 μm. What is the wave-vector direction(s)? What are the acceptance and walk-off angles? What are the $\hat{d}_{\text{eff}}^{(2)}$ for those cases?

13. $AgGaSe_2$ is a tetragonal $\bar{4}2m$ uniaxial crystal, which is transparent from 0.7 to 18 μm. Its relevant nonlinear coefficient is $\hat{d}_{36}^{(2)} = 39$ pm/V for doubling 10.6 μm. The Sellmeier equations (with λ in micrometers) are given by

$$n_o^2 = 4.6453 + \frac{2.2057}{1 - (0.1879/\lambda^2)} + \frac{1.8577}{1 - (1600/\lambda^2)},$$

$$n_e^2 = 5.2912 + \frac{1.3970}{1 - (0.2845/\lambda^2)} + \frac{1.9282}{1 - (1600/\lambda^2)}.$$

(a) Find all possible wave-vector-matching combinations for Type 1 and Type 2 wave-vector matching. Plot the dependence of the phase-match angle on the wavelength. Can this material be noncritical wave-vector matched?

(b) Derive the dependence of $\hat{d}_{\text{eff}}^{(2)}$ on the angles θ and ϕ for both Type 1 and Type 2 wave-vector matching.

(c) What is the phase-match angle, the acceptance angle, and the walk-off angle for Type 1 doubling of 10.6 μm?

14. Calculate $\hat{d}_{eff}^{(2)}$ for all wave-vector-matching geometries for a positive uniaxial crystal with the 4mm symmetry.

15. For a crystal of symmetry class 2mm, find the $\hat{d}_{eff}^{(2)}$ for all possible Type 1 and Type 2 wave-vector-matching conditions for fundamental beam(s) propagating
    (a) along the x-axis of the crystal,
    (b) along the y-axis of the crystal,
    (c) along the z-axis of the crystal,
    (d) at 45° to x- and y-axes in the x–y plane, and
    (e) at 45° to x- and z-axes in the x–z plane.

    Which cases can be noncritical wave-vector matched?

16. **(a)** For a crystal of symmetry class 3, find the $\hat{d}_{eff}^{(2)}$ for all possible Type 1 and Type 2 wave-vector-matching conditions for fundamental beam(s) propagating
    (a) along the x-axis of the crystal,
    (b) along the y-axis of the crystal,
    (c) along the z-axis of the crystal,
    (d) at 45° to x- and y-axes in the x–y plane, and
    (e) at 45° to x- and z-axes in the x–z plane.

    (b) Which of the wave-vector-matching conditions are possible for a positive uniaxial and which for a negative uniaxial?

    (c) Which cases can be noncritical wave-vector matched?

17. **(a)** For a crystal of symmetry class 222, find the $\hat{d}_{eff}^{(2)}$ for all possible Type 1 and Type 2 wave-vector-matching conditions for fundamental beam(s) propagating
    (a) along the x-axis of the crystal,
    (b) along the y-axis of the crystal,
    (c) along the z-axis of the crystal,
    (d) at 45° to x- and y-axes in the x–y plane, and
    (e) at 45° to x- and z-axes in the x–z plane.
    (b) Which cases can be noncritical wave-vector matched?

18. **(a)** For a crystal of symmetry class 4 or $\bar{4}$, find the $\hat{d}_{eff}^{(2)}$ for all possible Type 1 and Type 2 wave-vector-matching conditions for fundamental beam(s) propagating
    (a) along the x-axis of the crystal,
    (b) along the y-axis of the crystal,
    (c) along the z-axis of the crystal,
    (d) at 45° to x- and y-axes in the x–y plane, and
    (e) at 45° to x- and z-axes in the x–z plane.

(b) Which of the wave-vector-matching conditions are possible for a positive uniaxial and which for a negative uniaxial?

(c) Which cases can be noncritical wave-vector matched?

19. Consider a crystal with the following nonzero second-order susceptibilities: $\hat{\chi}_{xxz}^{(2)}$, $\hat{\chi}_{yyz}^{(2)}$, $\hat{\chi}_{zzz}^{(2)}$, $\hat{\chi}_{zyy}^{(2)}$, $\hat{\chi}_{zxx}^{(2)} \hat{\chi}_{zxx}^{(2)}$, with $\hat{\chi}_{ijk}^{(2)} = \hat{\chi}_{ikj}^{(2)}$. Light propagates in the x–y plane at $45°$ to the x-axis.

(a) For an eigenmode of amplitude E_0 polarized in the x–y plane with $\vec{E}$ having unequal projections along the x- and y-axes, what possible crystal symmetries can this correspond to? On the basis of their nonlinear properties, how would you differentiate between them? Is this birefringent crystal uniaxial or biaxial?

(b) Assuming that the birefringence in the x–y plane is small and can be neglected, calculate the induced nonlinear polarization induced for the eigenmode in (a).

(c) What is the induced nonlinear polarization for the orthogonally polarized eigenmode?

(d) What is the induced nonlinear polarization when both eigenmodes are present?

REFERENCES

1. T. Pliska, W.-R. Cho, J. Meier, A.-C. Le Duff, V. Ricci, A. Otomo, M. Canva, G. I. Stegeman, P. Raymond, and F. Kajzar, "Comparative study of nonlinear-optical polymers for guided-wave second-harmonic generation at telecommunication wavelengths," J. Opt. Soc. Am. B, **17**, 1554–1564 (2000).

2. T. Suhara and H. Nishihara, "Theoretical analysis of waveguide second harmonic generation phase-matched with uniform and chirped gratings," IEEE J. Quantum Electron., **26**, 1265–1276 (1990).

3. M. M. Fejer, G. A. Magel, D. H. Jundt, and R. L. Byer, "Quasi-phase-matched second harmonic generation: tuning and tolerances," IEEE J. Quantum Electron., **28**, 2631–2654 (1992).

4. L. E. Myers, G. D. Miller, R. C. Eckardt, M. M. Fejer, R. L. Byer, and W. R. Bosenberg, "Quasi-phase-matched 1.064-mm-pumped optical parametric oscillator in bulk periodically poled LiNbO$_3$," Opt. Lett., **20**, 52–54 (1995).

5. S. Bhagavantam, Crystal Symmetry and Physical Properties (Academic Press, New York, 1966).

6. F. Zernike and J. E. Midwinter, Applied Nonlinear Optics (John Wiley & Sons, New York, 1973).

SUGGESTED FURTHER READING

M. Born and E. Wolf, Principles of Optics, 7th Edition (Cambridge University Press, Cambridge, 1999).

R. W. Boyd, Nonlinear Optics, 3rd Edition (Academic Press, Burlington, MA, 2008).

M. M. Fejer, G. A. Magel, D. H. Jundt, and R. L. Byer, "Quasi-phase-matched second harmonic generation," IEEE J. Quantum Electron., **QE-28**, 2631–2654 (1992).

F. A. Hopf and G. I. Stegeman, Applied Classical Electrodynamics, Volume 1: Linear Optics (John Wiley & Sons, New York, 1985).

F. A. Hopf and G. I. Stegeman, Applied Classical Electrodynamics, Volume 2: Nonlinear Optics (John Wiley & Sons, New York, 1986).

Y. R. Shen, Principles of Nonlinear Optics (John Wiley & Sons, New York, 1984).

F. Zernike and J. E. Midwinter, Applied Nonlinear Optics (John Wiley & Sons, New York, 1973).

CHAPTER 4

Solutions for Plane-Wave Parametric Conversion Processes

In Chapter 2, approximate solutions for second harmonic generation (SHG) valid up to a conversion efficiency of 10% were found by assuming negligible depletion of the fundamental. However, in many cases this is inadequate, and thus solutions for the general case introduced by Armstrong et al. in 1962 are discussed in this chapter (1). Furthermore, the coupled wave equations derived in Chapter 2 are valid for any input, not just the usual zero harmonic input. With the appropriate input of both the fundamental and the harmonic and the relative phase between the two, there exist nonlinear modes, i.e., coupled harmonic and fundamental plane-wave fields, that propagate without change in the amplitude. All these solutions are characterized by additional nonlinear phase shifts, a feature universally present in nonlinear optics.

4.1 SOLUTIONS OF THE TYPE 1 SHG COUPLED WAVE EQUATIONS

In general, the detailed solution of the coupled wave equations depends on the initial conditions as well as the wave-vector mismatch (1). In Chapter 2, the general coupled wave equations were derived for interacting fundamental and harmonic plane waves valid at any frequency and intensity level. However, for fundamental or harmonic frequencies near an electronic resonance, i.e., an electric dipole allowed transition between the ground state and an excited state, linear absorption also exists and hence these regions are avoided if the usual goal of efficient conversion is important. In this case called the Kleinman limit, $\tilde{d}_{\text{eff}}^{(2)}(-\omega; 2\omega, -\omega) = \tilde{d}_{\text{eff}}^{(2)}(-2\omega; \omega, \omega) = \tilde{d}_{\text{eff}}^{(2)}$ and the coupled wave equations for the fundamental $\mathcal{E}_1(z, \omega)$ and the harmonic $\mathcal{E}_3(z, 2\omega)$ simplify to

$$
\begin{aligned}
\frac{d\mathcal{E}_3(z, 2\omega)}{dz} &= i\frac{\omega}{cn(2\omega)}\tilde{d}_{\text{eff}}^{(2)}\mathcal{E}_1^2(z, \omega)\, e^{i\Delta kz} \\
\frac{d\mathcal{E}_1(z, \omega)}{dz} &= i\frac{\omega}{cn(\omega)}\tilde{d}_{\text{eff}}^{(2)}\mathcal{E}_3(z, 2\omega)\mathcal{E}_1^*(z, \omega)\, e^{-i\Delta kz}.
\end{aligned}
\tag{4.1}
$$

Nonlinear Optics: Phenomena, Materials, and Devices, George I. Stegeman and Robert A. Stegeman.
© 2012 John Wiley & Sons, Inc. Published 2012 by John Wiley & Sons, Inc.

Defining for the general case

$$\mathcal{E}_1(z,\omega) = \left[\sqrt{\frac{1}{2}n_1(\omega)c\varepsilon_0}\right]^{-1}\sqrt{I_t(0)}\rho_1(z)\,e^{i\phi_1(z)}$$

$$\mathcal{E}_3(z,2\omega) = \left[\sqrt{\frac{1}{2}n_3(2\omega)c\varepsilon_0}\right]^{-1}\sqrt{I_t(0)}\rho_3(z)\,e^{i\phi_3(z)}$$

(4.2)

$$\rightarrow \quad I_1(z,\omega) + I_3(z,\omega) = [\rho_1^2(z) + \rho_1^2(z)]I_t(0) \tag{4.3}$$

where $\rho_1(z)$ and $\rho_3(z)$ are real quantities, with $\rho_1(0) = 1$ and $\rho_3(0) = 0$, and $I_t(0)$ is the total input intensity, which is conserved. From Eq. 2.45,

$$[\rho_1^2(z) + \rho_3^2(z)] = 1. \tag{4.4}$$

Furthermore, defining

$$\tilde{\kappa} = \frac{\omega}{cn}\tilde{d}_{\text{eff}}^{(2)}(-2\omega;\omega,\omega)|\mathcal{E}(0,\omega)|, \quad \xi = \tilde{\kappa}z, \quad \Delta s = \frac{\Delta k}{\tilde{\kappa}},$$
$$\theta = \Delta s\xi + 2\phi_1(\xi) - \phi_3(\xi), \tag{4.5}$$

where $\tilde{\kappa}$, ξ, Δs, and θ are the coupling coefficient, normalized propagation coordinate, normalized wave-vector mismatch, and global phase, respectively, leads to the equations

$$\frac{d\rho_1(\xi)}{d\xi} + i\rho_1(\xi)\frac{d\phi_1(\xi)}{d\xi} = i\rho_3(\xi)\rho_1(\xi)\,e^{-i\theta} \tag{4.6}$$

and

$$\frac{d\rho_3(\xi)}{d\xi} + i\rho_3(\xi)\frac{d\phi_3(\xi)}{d\xi} = i\rho_1^2(\xi)\,e^{i\theta} \tag{4.7}$$

with all variables real. Separating these equations into real and imaginary parts with $\exp(i\theta) = \cos(\theta) + i\sin(\theta)$ gives

$$\frac{d\rho_1(\xi)}{d\xi} = \rho_1(\xi)\rho_3(\xi)\sin\theta, \tag{4.8}$$

$$\frac{d\phi_1(\xi)}{d\xi} = \rho_3(\xi)\cos\theta, \tag{4.9}$$

$$\frac{d\rho_3(\xi)}{d\xi} = -\rho_1^2(\xi)\sin\theta, \tag{4.10}$$

$$\frac{d\phi_3(\xi)}{d\xi} = \frac{\rho_1^2(\xi)}{\rho_3(\xi)} \cos\theta, \tag{4.11}$$

and

$$\frac{d\theta}{d\xi} = \Delta s + \cos\theta \left[2\frac{d\phi_1(\xi)}{d\xi} - \frac{d\phi_3(\xi)}{d\xi} \right]. \tag{4.12}$$

The last equation can be simplified by inserting Eqs 4.9 and 4.11 for the derivatives of the phases, giving

$$\frac{d\theta}{d\xi} = \Delta s + \cos\theta \left[2\rho_3(\xi) - \frac{\rho_1^2(\xi)}{\rho_3(\xi)} \right]. \tag{4.13}$$

Given that

$$
\begin{aligned}
\frac{d\ln[\rho_1^2\rho_3]}{d\xi} &= \frac{1}{\rho_1^2\rho_3} \left(2\rho_1\rho_3 \frac{d\rho_1}{d\xi} + \rho_1^2 \frac{d\rho_3}{d\xi} \right) \\
&= \frac{\sin\theta}{\rho_1^2\rho_3}(2\rho_1^2\rho_3^2 - \rho_1^4) = \sin\theta \left[2\rho_3 - \frac{\rho_1^2}{\rho_3} \right],
\end{aligned}
\tag{4.14}
$$

Eq. 4.13 becomes

$$\frac{d\theta}{d\xi} = \Delta s + \frac{\cos\theta}{\sin\theta} \frac{d\ln[\rho_1^2\rho_3]}{d\xi}. \tag{4.15}$$

Equations 4.4, 4.8–4.11, and 4.15 along with the boundary conditions at $\xi = 0$ completely describe all Type 1 processes. Simple solutions to these equations can be obtained for $\Delta k = 0 \ (\to \Delta s = 0)$ without resorting to special functions.

4.1.1 SHG with Wave-Vector Match ($\Delta s = 0$)

For normal SHG, only the fundamental wave is input and $\rho_3(0) = 0\,(1)$. Equation 4.15 can then be integrated as shown in Eq. 4.16:

$$\frac{\sin\theta}{\cos\theta} d\theta = d\{\ln[\rho_1^2(\xi)\rho_3(\xi)]\} \quad \to \quad \int \frac{\sin\theta}{\cos\theta} d\theta = \int d\{\ln[\rho_1^2(\xi)\rho_3(\xi)]\},$$

$$\ln\{\cos\theta\rho_1^2(\xi)\rho_3(\xi)\} = \text{constant} \quad \to \quad \cos\theta\rho_1^2(\xi)\rho_3(\xi) = e^{\text{constant}} = C. \tag{4.16}$$

Note that the constant C can be evaluated in terms of the boundary conditions. In this case, $\rho_3(0) = 0$, which gives $C = 0$. Since $\cos\theta\,\rho_1^2(\xi)\rho_3(\xi)$ is a constant for all ξ,

$\cos\theta = 0 \rightarrow \theta = \pm\pi/2$. Because $\dfrac{d\rho_3(\xi)}{d\xi}\big|_{\xi=0}$ must be positive in Eq. 4.10, i.e., the harmonic grows from the input, $\sin\theta$ must be negative, which fixes $\theta = -\pi/2$. Substituting θ into Eq. 4.6 gives $2\phi_1(z) - \phi_3(z) = -\pi/2$. Note that Eqs 4.9 and 4.11 along with the condition $\cos\theta = 0$ show that the phases do not vary on propagation. Since the incident fundamental phase is a free variable, it can be chosen to be zero so that $\phi_3(0) = \pi/2$.

The energy conservation condition $\rho_1^2(\xi) + \rho_1^2(\xi) = 1$ in conjunction with $\theta = -\pi/2$ gives for Eq. 4.10

$$\frac{d\rho_3(\xi)}{d\xi} = \left[1 - \rho_3^2(\xi)\right]. \tag{4.17}$$

Equation 4.17 has a very simple solution, namely,

$$\rho_3(z) = \tanh\left(\frac{z + z_0}{\ell_{pg}}\right), \qquad \ell_{pg} = \frac{cn}{\omega \tilde{d}_{eff}^{(2)} |\mathcal{E}(0, \omega)|} = \tilde{\kappa}^{-1}, \tag{4.18}$$

where ℓ_{pg} is the characteristic parametric gain length and z_0 is a constant of integration, which can be shown to be zero from the boundary condition $\rho_3(0) = 0$. Thus the solutions for the field amplitudes are

$$\rho_3(z) = \tanh\left(\frac{z}{\ell_{pg}}\right) \quad \text{and} \quad \rho_1(z) = \text{sech}\left(\frac{z}{\ell_{pg}}\right). \tag{4.19}$$

The evolution of the fundamental and harmonic intensities is shown in Fig. 4.1. Note that complete conversion is possible for the continuous-wave case on the wave-vector matching condition $\Delta k = 0$. For pulses that have an intensity distribution, the approach to complete conversion requires even higher peak

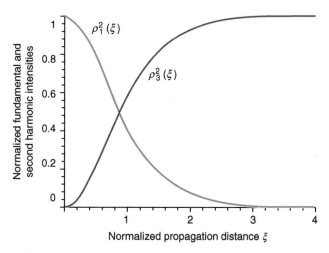

FIGURE 4.1   The evolution of the fundamental and harmonic intensities with the normalized distance.

intensities. In fact, 99% continuous-wave conversion has actually been measured experimentally (2).

The parametric gain length ℓ_{pg} is one of the key parameters in second-order processes, specifically here harmonic generation. When $\ell_{pg} \gg z$, $\tanh(z/\ell_{pg}) \approx z/\ell_{pg}$ and the approximate solutions given by Eqs 2.35 and 2.36 are valid. When $z \geq \ell_{pg}$, the onset of saturation in the conversion efficiency takes place.

4.1.2 SHG with Wave-Vector Mismatch ($\Delta s \neq 0$)

Equation 4.15 can still be integrated for this case by the method of the variation of the parameter in the solution of the homogeneous equation (3), which gives

$$\cos\theta \left[1 - \rho_3^2(\xi)\right]\rho_3(\xi) + \frac{1}{2}\Delta s\rho_3^2(\xi) = C. \tag{4.20}$$

Since $\rho_3(0) = 0$, $C = 0$ and therefore $\cos\theta = -\dfrac{\Delta s\rho_3(\xi)}{2[1 - \rho_3^2(\xi)]}$, which gives

$$\frac{d\rho_3(\xi)}{d\xi} = Sgn\left[1 - \rho_3^2(\xi)\right]\left(1 - \cos^2\theta\right)^{1/2} = Sgn\left[\left\{1 - \rho_3^2(\xi)\right\}^2 - \left(\frac{\Delta s}{2}\right)^2\rho_3^2(\xi)\right]^{1/2}, \tag{4.21}$$

where sgn is the sign of $\sin[2\phi_1(0) - \phi_3(0)]$, i.e., the boundary condition on the phases. The general solution is given in terms of the Jacobi elliptic function $sn(u^{-1}\xi|u^4)$ (4):

$$\rho_1^2(\xi) = 1 - u^2 \, sn^2(u^{-1}\xi|u^4), \qquad \rho_3^2(\xi) = u^2 \, sn^2(u^{-1}\xi|u^4),$$

$$u^{-1} = \Delta s/4 + \sqrt{1 + (\Delta s/4)^2}. \tag{4.22}$$

The response with the propagation distance is now periodic, and the major difference from the negligible fundamental depletion result in Fig. 2.6 is that the harmonic maxima are flattened for $\Delta s = 0.2$ (see Fig. 4.2). There are also differences in the harmonic signal as a function of detuning from the wave-vector match, i.e., harmonic intensity versus phase mismatch (ΔkL) (see Fig. 4.3). Note the following characteristics of these detuning curves with increasing input intensity:

1. The conversion efficiency saturates at unity (as expected).
2. The narrowing of the main ($\Delta k = 0$) peak means that the tuning bandwidth becomes progressively narrower.
3. The side lobes become progressively narrower and their peaks shift to smaller ΔkL.

All these changes make it more difficult to experimentally find the optimum conversion point when tuning the input angle of the incident fundamental.

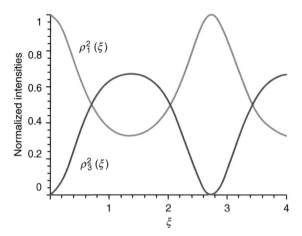

FIGURE 4.2 The evolution of the normalized fundamental and harmonic intensities versus the normalized propagation distance for $\Delta s = 0.2$.

The evolution of the phases with the distance can easily be found from Eqs 4.11, 4.12, and 4.20:

$$\frac{d\phi_1(\xi)}{d\xi} = -\frac{\Delta s \rho_3^2(\xi)}{2[1 - \rho_3^2(\xi)]}, \qquad \frac{d\phi_3(\xi)}{d\xi} = -\frac{\Delta s}{2}. \qquad (4.23)$$

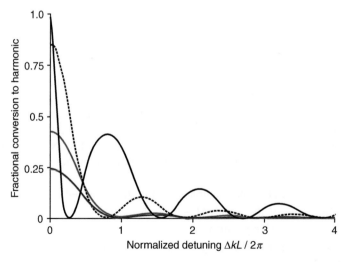

FIGURE 4.3 The fractional conversion to second harmonic generation versus normalized detuning for four different values of the normalized input parameter $\tilde{\kappa}$: $\tilde{\kappa} = 3$ (solid black line); $\tilde{\kappa} = 1.5$ (dashed black line); $\tilde{\kappa} = 0.75$ (red solid line); $\tilde{\kappa} = 0.25$ (solid blue line; curve amplitude multiplied by factor of 4).

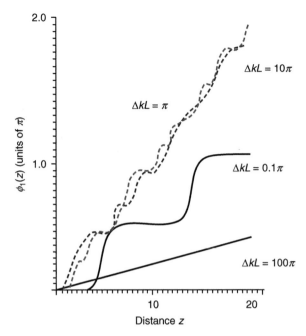

FIGURE 4.4   Calculated variation of the fundamental wave nonlinear phase versus increasing propagation distance for various phase mismatches. Reproduced with permission from the Optical Society of America (5).

The magnitude of both phases increases with ξ, and their sign depends on the sign of Δk. The harmonic phase increases linearly with the distance, whereas the increasing fundamental phase for $\Delta k < 0$ is modulated with the distance since $\rho_3(\xi)$ itself is periodic for the nonlinear contribution to $\phi_1(z)$ (see Fig. 4.4). Note that the biggest contributions in the fundamental phase occur when the fundamental is strongly depleted. A detailed discussion of the physics ("cascading") is given in Chapter 12. All second-order interactions have solutions containing Jacobi elliptic integrals, which are available from most programming software such as MATLAB and MAPLE.

4.1.3   Arbitrary Inputs with $\Delta s = 0$

The preceding case was easily solvable because the constant of integration, C, was zero since $\rho_3(0) = 0$. Discussed now is the case when neither $\rho_1(0)$ nor $\rho_3(0)$ is zero (1). The relative input phase between the fundamental and the harmonic is important since it determines which wave grows initially and which one depletes. Note that by using $\rho_1^2(\xi) = 1 - \rho_3^2(\xi)$, $\sin(\theta) = \pm[1 - \cos^2 \theta]^{1/2}$, and Eq. 4.16, θ and $\rho_1(\xi)$ can still be eliminated from Eq. 4.10 to obtain

$$\frac{d\rho_3(\xi)}{d\xi} = -Sgn\left[1 - \rho_3^2(\xi)\right]\left[1 - \cos^2\theta\right]^{1/2}$$

$$= -Sgn\left[1 - \rho_3^2(\xi)\right]\left[1 - \frac{\cos^2\theta|_{\xi=0}\rho_1^4(0)\rho_3^2(0)}{\left\{1 - \rho_3^2(\xi)\right\}^2\rho_3^2(\xi)}\right]^{1/2}, \tag{4.24}$$

where $\cos^2\theta|_{\xi=0}$ is $\cos^2\theta$ at $\xi = 0$. Multiplying Eq. 4.24 by $\rho_3(\xi)$ and recognizing that $\rho_3(\xi)\dfrac{d\rho_3(\xi)}{d\xi} = \dfrac{1}{2}\dfrac{d\rho_3^2(\xi)}{d\xi}$, Eq. 4.24 now takes the "classic" form, which has solutions in terms of Jacobi elliptic functions:

$$\frac{d\rho_3^2(\xi)}{d\xi} = -Sgn2\left\{[1 - \rho_3^2(\xi)]^2\rho_3^2(\xi) - \cos^2\theta|_{\xi=0}\,\rho_1^4(0)\rho_3^2(0)\right\}^{1/2}. \tag{4.25}$$

The subsequent solution is complicated since there are multiple roots for ρ_3^2 to the cubic equation $\rho_3^2\left[1 - \rho_3^2\right]^2 - C^2$ and they must be found numerically for each individual case. Nevertheless, some insights into the behavior of $\rho_1(\xi)$ and $\rho_3(\xi)$ can be obtained from the general form of the elliptic integral:

$$\xi = \pm\frac{1}{2}\int_{\rho_3^2(0)}^{\rho_3^2(\xi)}\frac{d\rho_3^2}{\sqrt{\rho_3^2\left[1 - \rho_3^2\right]^2 - C^2}}. \tag{4.26}$$

Labeling the two lowest roots as ρ_{3a}^2 and ρ_{3b}^2, with $\rho_{3b}^2 \geq \rho_{3a}^2$, the solutions for $\rho_3^2(\xi)$ are constrained to oscillate periodically between ρ_{3a}^2 and ρ_{3b}^2 with a period given by

$$L_{per} = \tilde{\kappa}\int_{\rho_{3a}^2}^{\rho_{3b}^2}\frac{d\rho_3^2}{\sqrt{\rho_3^2\left[1 - \rho_3^2\right]^2 - C^2}}. \tag{4.27}$$

The fundamental is then given by $\rho_1^2(\xi) = 1 - \rho_3^2(\xi)$. A particular case is shown in Fig. 4.5 from the original paper by Armstrong et al. (1).

The key point is the oscillatory behavior of the solutions, even on the wave-vector match. The individual phases are given by

$$\frac{d\phi_1(\xi)}{d\xi} = \frac{\cos[2\phi_1(0) - \phi_3(0)]\rho_1^2(0)\rho_3(0)}{\rho_1^2(\xi)},$$

$$\frac{d\phi_3(\xi)}{d\xi} = \frac{\cos[2\phi_1(0) - \phi_3(0)]\rho_1^2(0)\rho_3(0)}{\rho_3^2(\xi)}. \tag{4.28}$$

Since all the normalized fields are positive and both $\rho_1^2(\xi)$ and $\rho_3^2(\xi)$ oscillate between two limiting values, both phases grow with the propagation distance. The growth is not linear in ξ, experiencing a periodic modulation with period L_{per}. The sign of the growth is determined by the sign of $\cos[2\phi_1(0) - \phi_3(0)]$.

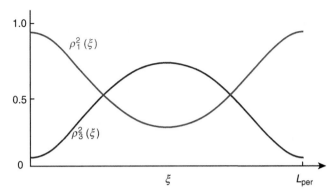

FIGURE 4.5 Variation in the fundamental $[\rho_1^2(\xi)]$ and harmonic $[\rho_3^2(\xi)]$ normalized intensities versus the normalized propagation distance. Reproduced with permission from American Physical Society (1).

4.2 SOLUTIONS OF THE THREE-WAVE COUPLED EQUATIONS

Type 2 SHG is the limiting case of the more general sum frequency generation (1). In the Kleinman limit, there are three independent eigenmode inputs identified by subscripts 1, 2, and 3. (For the Type 2 SHG case, the parameters associated with the harmonic have subscript 3.) The coupled wave equations are (Eq. 2.41)

$$\frac{d\mathcal{E}(z,\omega_3)}{dz} = i\frac{\omega_3}{n(\omega_3)c}\tilde{d}_{\mathrm{eff}}^{(2)}\mathcal{E}(\omega_1)\mathcal{E}(\omega_2)\,e^{i\Delta kz},$$

$$\frac{d\mathcal{E}(z,\omega_1)}{dz} = i\frac{\omega_1}{n(\omega_1)c}\tilde{d}_{\mathrm{eff}}^{(2)}\mathcal{E}(\omega_3)\mathcal{E}^*(\omega_2)\,e^{-i\Delta kz}, \qquad (4.29)$$

$$\frac{d\mathcal{E}(z,\omega_2)}{dz} = i\frac{\omega_2}{n(\omega_2)c}\tilde{d}_{\mathrm{eff}}^{(2)}\mathcal{E}(\omega_3)\mathcal{E}^*(\omega_1)\,e^{-i\Delta kz},$$

with $\Delta k = k_1 + k_2 - k_3$.

Because the frequencies are not necessarily integer multiples of each other, the fields are normalized in a different way from that used for Type 1 (Eq. 4.3); namely,

$$\mathcal{E}_i(z,\omega) = \left[\sqrt{\frac{1}{2\omega_i}n_i(\omega)c\varepsilon_0}\right]^{-1}\sqrt{I_t(0)}\rho_i(z)\,e^{i\phi_i(z)},$$

$$I_t(0) = I_1(z,\omega_1) + I_2(z,\omega_2) + I_3(z,\omega_3). \qquad (4.30)$$

Note that $\rho_i^2(z) = \dfrac{I_i(z)}{\omega_i I_t(0)} \propto N_i(z)$, where $N_i(z)$ is the photon flux at frequency ω_i and $\sum_{i=1}^{3}\rho_i^2(z) \neq 1$.

Defining $\tilde{\kappa} = \tilde{d}_{eff}^{(2)} \sqrt{\dfrac{2\omega_1\omega_2\omega_3}{\varepsilon_0 c^3 n_1(\omega)n_2(\omega)n_3(2\omega)}} \sqrt{I_t}$ and the normalized quantities $\xi = \tilde{\kappa}z$, $\Delta s = \Delta k/\tilde{\kappa}$, and $\theta = \Delta s\xi + \phi_1(\xi) + \phi_2(\xi) - \phi_3(\xi)$, the coupled wave equations can be expressed in terms of six coupled wave equations with only real variables as

$$\frac{d\rho_1(\xi)}{d\xi} = \rho_2(\xi)\rho_3(\xi)\sin\theta, \qquad \frac{d\phi_1(\xi)}{d\xi} = \frac{\rho_2(\xi)\rho_3(\xi)}{\rho_1(\xi)}\cos\theta,$$

$$\frac{d\rho_2(\xi)}{d\xi} = \rho_1(\xi)\rho_3(\xi)\sin\theta, \qquad \frac{d\phi_2(\xi)}{d\xi} = \frac{\rho_1(\xi)\rho_3(\xi)}{\rho_2(\xi)}\cos\theta, \qquad (4.31)$$

$$\frac{d\rho_3(\xi)}{d\xi} = -\rho_1(\xi)\rho_2(\xi)\sin\theta, \qquad \frac{d\phi_3(\xi)}{d\xi} = \frac{\rho_1(\xi)\rho_2(\xi)}{\rho_3(\xi)}\cos\theta.$$

Just as in the preceding Type 1 case, using $d\phi_i(\xi)/d\xi$ from Eqs 4.31, for the global phase

$$\frac{d\theta}{d\xi} = \Delta s + \cos\theta\left[\frac{\rho_3(\xi)\rho_1(\xi)}{\rho_2(\xi)} + \frac{\rho_3(\xi)\rho_2(\xi)}{\rho_1(\xi)} - \frac{\rho_1(\xi)\rho_2(\xi)}{\rho_3(\xi)}\right]$$

$$\rightarrow \quad \frac{d\theta}{d\xi} = \Delta s + \frac{\cos\theta}{\sin\theta}\frac{d\ln[\rho_1\rho_2\rho_3]}{d\xi}. \qquad (4.32)$$

The analysis is now limited to $\Delta s = 0$. (The general case is very complicated.) Equation 4.32 can now be integrated to give

$$\cos\theta\,\rho_1(\xi)\rho_2(\xi)\rho_3(\xi) = C = \cos[\phi_1(0) + \phi_2(0) - \phi_3(0)]\rho_1(0)\rho_2(0)\rho_3(0). \qquad (4.33)$$

Clearly if any one of the inputs is zero, $C = 0$, $\cos\theta = 0$, and $\theta = \pm\pi/2$. Equations 4.31 can now be written as

$$\frac{d\rho_1(\xi)}{d\xi} = \pm\rho_2(\xi)\rho_3(\xi) \;\rightarrow\; \frac{d\rho_1^2(\xi)}{d\xi} = \pm 2\rho_1(\xi)\rho_2(\xi)\rho_3(\xi), \quad \frac{d\phi_1(\xi)}{d\xi} = 0,$$

$$\frac{d\rho_2(\xi)}{d\xi} = \pm\rho_1(\xi)\rho_3(\xi) \;\rightarrow\; \frac{d\rho_2^2(\xi)}{d\xi} = \pm 2\rho_1(\xi)\rho_2(\xi)\rho_3(\xi), \quad \frac{d\phi_2(\xi)}{d\xi} = 0,$$

$$\frac{d\rho_3(\xi)}{d\xi} = \mp\rho_1(\xi)\rho_2(\xi) \;\rightarrow\; \frac{d\rho_3^2(\xi)}{d\xi} = \mp 2\rho_1(\xi)\rho_2(\xi)\rho_3(\xi), \quad \frac{d\phi_3(\xi)}{d\xi} = 0.$$

$$(4.34)$$

The fundamental phases are "locked in" at their input values and $\phi_3(0) = \phi_1(0) + \phi_2(0) \mp \frac{\pi}{2}$.

4.2.1 Sum Frequency Generation

Although the total intensity is still clearly conserved, the photon fluxes $N_i(\xi) = \sqrt{I_t(0)}\rho_i^2(\xi)/\hbar$, even though their total is not a constant, have simple relationships with one another. For sum frequency generation, $\hbar\omega_1 + \hbar\omega_2 \rightarrow \hbar\omega_3$ means that for every ω_3 photon created, one ω_1 and one ω_2 photons are annihilated so that the total number of $\omega_3 + \omega_2$ and $\omega_3 + \omega_1$ photons is a constant; i.e., $\rho_2^2(\xi) + \rho_3^2(\xi) = \rho_2^2(0) + \rho_3^2(0) = m_1$ and $\rho_1^2(\xi) + \rho_3^2(\xi) = \rho_1^2(0) + \rho_3^2(0) = m_2$. Using Eqs 4.34 and imposing the condition that the sum frequency grows from $\xi = 0$, which requires $\theta = -\pi/2$,

$$\frac{d\rho_3^2(\xi)}{d\xi} = 2\big[\{m_1 - \rho_3^2(\xi)\}\{m_2 - \rho_3^2(\xi)\}\rho_3^2(\xi)\big]^{1/2}. \tag{4.35}$$

Assuming that $m_2 = \rho_1^2(0) > m_1 = \rho_2^2(0)$ and defining $\gamma = \rho_2(0)/\rho_1(0)$, the solution for the photon flux with $\sin(\theta) = -1$ is given by

$$N_3(\xi) = N_2(0)\, sn^2(\rho_1(0)\xi|\gamma),$$
$$N_2(\xi) = N_2(0) - N_2(0)\, sn^2(\rho_1(0)\xi|\gamma), \tag{4.36}$$
$$N_1(\xi) = N_1(0) - N_2(0)\, sn^2(\rho_1(0)\xi|\gamma).$$

Defining $\delta = \dfrac{\rho_1^2(0) - \rho_2^2(0)}{\rho_1^2(0) + \rho_2^2(0)}$, a specific example for $\delta = 0.33$ is shown in Fig. 4.6.

Note that the response is periodic, with periodicity L_{per} given by

$$L_{\mathrm{per}} = \frac{2K[(1-\delta)/(1+\delta)]}{\sqrt{1+\delta}}, \tag{4.37}$$

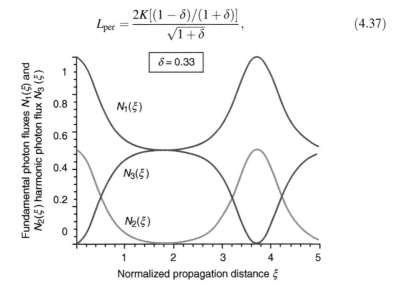

FIGURE 4.6   The evolution of the fundamental and harmonic photon fluxes with the distance.

where K is the elliptic function (also available from some software). There is no asymptotic final state in contrast to the Type 1 behavior, which approaches 100% conversion at large distances. Note that as $\delta \to 0$, $L_{per} \to \infty$ as expected and the results of Section 4.1.1 are obtained. Although the intensity is conserved, the photon flux is not since

$$-\left[\frac{dN_1}{d\xi} + \frac{dN_2}{d\xi}\right] = 2\frac{dN_3}{d\xi}. \tag{4.38}$$

The reason for this is simple: In Fig. 4.6, there are more ω_1 photons than ω_2 photons at the input. Since one photon is required at each frequency to generate a ω_3 photon, the flux N_2 can be completely depleted with both $N_1 \neq 0$ and $N_3 \neq 0$. The relation $d\rho_2(\xi)/d\xi = \rho_1(\xi)\rho_3(\xi)$ in Eqs 4.34 means that N_2 starts to grow again as ω_3 photons are annihilated into ω_1 and ω_2 photons. Similar behavior occurs when N_3 is depleted and ω_3 photons are again generated from ω_1 and ω_2 photons. This cycle then repeats in a periodic fashion.

4.2.2 Difference Frequency Generation

Assume without loss of generality that the inputs are at ω_3 and ω_1, with $\omega_2 = \omega_3 - \omega_1$ being the output. The analysis follows closely that of the sum frequency case but with $m_1 = \rho_3^2(0) > m_2 = \rho_1^2(0)$ and $\rho_2^2(0) = 0$ (6). Furthermore, $\theta = \pi/2$, which gives

$$\frac{d\rho_2^2(\xi)}{d\xi} = -2\left[\{m_1 - \rho_2^2(\xi)\}\{m_2 - \rho_2^2(\xi)\}\rho_2^2(\xi)\right]^{1/2}. \tag{4.39}$$

Solving in terms of the Jacobi elliptic function sn, with $\gamma = \rho_1(0)/\rho_3(0)$, gives

$$N_2(\xi) = -N_1(0)\,sn^2(i\rho_3(0)\xi|\gamma),$$
$$N_3(\xi) = N_3(0) - N_1(0)\,sn^2(i\rho_3(0)\xi|\gamma), \tag{4.40}$$
$$N_1(\xi) = N_1(0) - N_1(0)\,sn^2(i\rho_1(0)\xi|\gamma),$$

and the phases are locked together with $\phi_3(0) - \phi_2(0) = \phi_1(0) + \frac{\pi}{2}$.

4.3 CHARACTERISTIC LENGTHS

The SHG response obtained depends on the values of a number of characteristic lengths. For the plane-wave fields discussed to date and Type 1 SHG, there are three lengths that are important. Of course, the first one is the crystal length L. The second is the parametric gain length ℓ_{pg}, which is the distance required for significant (>40%) conversion to the harmonic for a given input field intensity. The higher the input intensity, the shorter the ℓ_{pg} required for strong conversion on the wave-vector match. There is of course a limit imposed by the material and its quality because material damage can occur at excessively high intensities. Away from $|\Delta k| = 0$, the coherence length ℓ_{coh} is important since it sets the distance over which maximum conversion can be obtained before the harmonic converts back to the fundamental. Furthermore, for a "noncritical" wave-vector match, materials with large birefringence have small angular bandwidths that can make finding and maintaining large ℓ_{coh} difficult unless high quality optical mounts are used. Nevertheless, if $\ell_{coh} > L > \ell_{pg}$, strong conversion is still possible.

There is an additional characteristic distance for Type 2 SHG, namely, the periodic distance over which the evolution repeats itself, even on the wave-vector match when the two orthogonally polarized input beams are not equal in intensity. When dealing with beams that have finite cross sections, there are additional characteristic lengths that must be taken into account. These will be discussed in Chapter 5.

4.4 NONLINEAR MODES

For most of the preceding cases, the boundary condition that the output field was zero at the input was adopted. In this section, it is shown that interesting solutions can be obtained with appropriate boundary conditions on the fundamental and the second harmonic field so that neither field changes magnitude on propagation. These solutions were first found by Baranova (7) and Kaplan (8). The two fields form a nonlinear "mode," defined as a nonlinear field distribution that does not change with the propagation distance. Note that this nomenclature implies neither orthogonality between modes nor the applicability of the superposition principle. For finite cross-section beams that can diffract, similar solutions, i.e., solitons, can be found (9). Solitons are beams that do not spread on propagation.

4.4.1 Type 1 SHG

Steady-state, plane-wave solutions for the field magnitudes require from Eqs 4.8 and 4.10 that

$$\frac{d\rho_1(\xi)}{d\xi} = \rho_1(\xi)\rho_3(\xi)\sin\theta = 0 \quad \text{and} \quad \frac{d\rho_3(\xi)}{d\xi} = -\rho_1^2(\xi)\sin\theta = 0. \quad (4.41)$$

This means that the field amplitudes are independent of ξ, i.e., $\rho_1(\xi) = \rho_1(0)$ and $\rho_3(\xi) = \rho_3(0)$, which require that $\sin(\theta) = 0$ and therefore $\theta = 0$ or $\theta = \pi$. Furthermore, recalling that $sgn = \pm 1$, where $+1$ is identified with $\theta = 0$ and -1 with $\theta = \pi$,

$$\frac{d\phi_1(\xi)}{d\xi} = \rho_3(\xi)\cos\theta = Sgn\,\rho_3(\xi) = \text{constant} \quad \rightarrow \quad \phi_1(\xi) = Sgn\,\rho_3(0)\xi + C_1,$$

$$\frac{d\phi_3(\xi)}{d\xi} = \frac{\rho_1^2(\xi)}{\rho_3(\xi)}\cos\theta = Sgn\frac{\rho_1^2(\xi)}{\rho_3(\xi)} = \text{constant} \quad \rightarrow \quad \phi_3(\xi) = Sgn\frac{\rho_1^2(0)}{\rho_3(0)}\xi + C_3.$$

$$(4.42)$$

The key point is that there is an additional (nonlinear) phase shift that is linear in the propagation distance experienced by both waves. This is called the "cascaded" nonlinear phase shift. Furthermore, since the global phase is also invariant on propagation,

$$\frac{d\theta}{d\xi} = \Delta s + 2\frac{d\phi_1(\xi)}{d\xi} - \frac{d\phi_3(\xi)}{d\xi} = \Delta s + \cos\theta \left[2\rho_3(0) - \frac{\rho_1^2(0)}{\rho_3(0)} \right] = 0$$

$$\rightarrow \quad \Delta s = -Sgn \left[2\rho_3(0) - \frac{\rho_1^2(0)}{\rho_3(0)} \right].$$

(4.43)

4.4.1.1   *On-Wave-Vector Match* ($\Delta s = 0$). In this case, Eq. 4.43 becomes

$$\left[2\rho_3(0) - \frac{\rho_1^2(0)}{\rho_3(0)} \right] = 0 \xrightarrow{\rho_1^2(0)=1-\rho_3^2(0)} \rho_3^2(0) = \rho_3^2(\xi) = \frac{1}{3} \quad \text{and} \quad \rho_1^2(0) = \rho_1^2(\xi) = \frac{2}{3}.$$

(4.44)

Therefore the relative amplitude of the nonlinear mode is fixed. Since $\theta = 2\phi_1(\xi) - \phi_3(\xi) = 0, \pi,$

$$2\phi_1(\xi) - \phi_3(\xi) = (Sgn - 1)\frac{\pi}{2} = \text{constant.}$$

(4.45)

Since the input conditions for one of the fields can still be chosen, the two fields are input either "in phase" or π "out of phase" and remain locked together on propagation.

One might be tempted to assume that there is no energy exchange between the fundamental and the harmonic. However, this is not the case. From Eqs 4.41 and 4.44, it is clear that an equal amount of energy is continuously exchanged between the two waves.

4.4.1.2   *Off-Wave-Vector Match* ($\Delta s \neq 0$). The relative amplitudes in this case are given by

$$-Sgn\,\Delta s = \left[2\rho_3(0) - \frac{\rho_1^2(0)}{\rho_3(0)} \right] = \frac{[3\rho_3^2(0) - 1]}{\rho_3(0)} \quad \rightarrow \quad 3\rho_3^2(0) + Sgn\,\Delta s\rho_3(0) - 1 = 0,$$

(4.46)

which have the following two solutions for $\rho_3(0) > 0$, one associated with $\theta = 0$ and one with $\theta = \pi$:

$$\rho_3(0) = -\frac{\Delta s}{6} + \frac{1}{6}\sqrt{\Delta s^2 + 12},$$

(4.47a)

$$\rho_3(0) = \frac{\Delta s}{6} + \frac{1}{6}\sqrt{\Delta s^2 + 12},$$

(4.47b)

respectively, and $\rho_1(0) = \sqrt{1 - \rho_3^2(0)}$. Since in principle $\infty > \Delta s > -\infty$, the range of Δs is limited for both cases by the condition $1 \geq \rho_3(0)$. These two solutions exist only for $2 \geq \Delta s \geq -2$; otherwise, there is a single solution. The fraction of the harmonic to the fundamental changes with Δs. The discussion on the stability of both cases is beyond the scope of this text.

For $\Delta s \neq 0$, integrating the last term in Eq. 4.43 gives

$$\Delta s \xi = -[2\phi_1(\xi) - \phi_3(\xi)], \tag{4.48}$$

which means that the difference between the nonlinear phase shifts compensates for the phase mismatch.

4.4.2 Type 2 SHG

This case is more complicated than the Type 1 case. The coupled wave equations with real variables were given previously by Eqs 4.31 and 4.33. Again steady-state solutions for the field magnitudes require $\sin(\theta) = 0$ and so $\cos(\theta) = \pm 1$, and from Eqs 4.31,

$$\phi_1(\xi) = Sgn \frac{\rho_2(0)\rho_3(0)}{\rho_1(0)} \xi + \phi_1(0),$$

$$\phi_2(\xi) = Sgn \frac{\rho_1(0)\rho_3(0)}{\rho_2(0)} \xi + \phi_2(0),$$

$$\phi_3(\xi) = Sgn \frac{\rho_1(0)\rho_2(0)}{\rho_3(0)} \xi + \phi_3(0), \tag{4.49}$$

$$Sgn\Delta s + \left\{ \frac{\rho_2(0)\rho_3(0)}{\rho_1(0)} + \frac{\rho_1(0)\rho_3(0)}{\rho_2(0)} - \frac{\rho_1(0)\rho_2(0)}{\rho_3(0)} \right\} = 0.$$

Therefore all nonlinear phases still accumulate linearly with the propagation distance and the field amplitudes adjust, and so the nonlinear phases compensate for the wave-vector mismatch. The steady-state amplitudes depend on Δs via the last equation of Eqs 4.49, in which $d\theta/d\xi = 0$. Again rewriting this last equation after some manipulation, similar to the Type 1 case, leads to

$$\rho_3(0) = \frac{\rho_1(0)\rho_2(0)}{2[\rho_1^2(0) + \rho_2^2(0)]} \left[-Sgn\Delta s \pm \sqrt{\Delta s^2 + 4} \right], \tag{4.50}$$

which can have two physically relevant families of solutions that correspond to $\theta = 0$ and $\theta = \pi$; i.e.,

$$\rho_3(0) = \frac{\rho_1(0)\rho_2(0)}{2[\rho_1^2(0) + \rho_2^2(0)]} \left[-\Delta s + \sqrt{\Delta s^2 + 4} \right] \tag{4.51a}$$

and

$$\rho_3(0) = \frac{\rho_1(0)\rho_2(0)}{2[\rho_1^2(0) + \rho_2^2(0)]} \left[\Delta s + \sqrt{\Delta s^2 + 4}\right], \qquad (4.51b)$$

respectively.

PROBLEMS

1. Consider Type 2 second harmonic generation in the large conversion limit with unequal fundamental inputs; i.e., $\rho_1^2 \neq \rho_2^2$.

   (a) Plot the oscillation period (in units of the normalized distance) as a function of
   $$\delta = \frac{\rho_1^2(0) - \rho_2^2(0)}{\rho_1^2(0) + \rho_2^2(0)}.$$

   (b) For a 1-cm-long crystal with $\hat{d}_{eff}^{(2)} = 5 \, \text{pm/V}$, a fundamental wavelength of $1.06 \, \mu\text{m}$, and a nominal refractive index of 1.5, what input fundamental intensity is required to obtain maximum conversion for $\delta = 0.01$ and $\delta = 0.1$? What conversion efficiencies do these correspond to?

2. The purpose of this problem is to obtain some familiarity with large conversion efficiencies. For Type 1 second harmonic generation, calculate and plot the following quantities:

   (a) The peak second harmonic generation conversion efficiency, i.e., $I(2\omega)/I(\omega)$, as a function of the normalized propagation distance ξ for different normalized wave-vector detuning $\Delta s = 0.05, 0.2, 0.8$, and 3.2.

   (b) The peak second harmonic generation conversion efficiency, i.e., $I(2\omega)/I(\omega)$, as a function of normalized phase detuning $\Delta k L/2\pi$ for a range of $\tilde{\kappa}$ showing the features of the changes.

3. The purpose of this problem is to obtain some familiarity with large conversion efficiencies for Type 2 second harmonic generation on the wave-vector match. Calculate and plot $\rho_1^2(\xi)$ and $\rho_3^2(\xi)$ as a function of the propagation distance ξ for $\rho_1^2(0) = 1$ and $\rho_2^2(0) = 0.1, 0.4, 0.7, 0.9$.

REFERENCES

1. J. A. Armstrong, N. Bloembergen, J. Ducuing, and P. S. Pershan, "Interaction between light waves in a nonlinear dielectric," Phys. Rev., **127**, 1918–1939 (1962).

2. K. R. Parameswaran, J. R. Kurz, R. V. Roussev, and M. M. Fejer, "Observation of 99% pump depletion in single-pass second-harmonic generation in a periodically poled lithium niobate waveguide," Opt. Lett., **27**, 43–45 (2002).

3. W. E. Boyce and R. C. DiPrima, Elementary Differential Equations and Boundary Value Problems, 8th Edition (John Wiley & Sons, Hoboken, NJ, 2005).

4. R. C. Eckardt and J. Reintjes, "Phase matching limitations of high efficiency second harmonic generation," IEEE J. Quantum Electron., **20**, 1178–1187 (1984).

5. G. I. Stegeman, M. Sheik-Bahae, E. W. Van Stryland, and G. Assanto, "Large nonlinear phase shifts in second-order nonlinear-optical processes," Opt. Lett., **18**, 13–15 (1993).

6. Y. R. Shen, Principles of Nonlinear Optics (John Wiley & Sons, New York, 1984).

7. N. B. Baranova, "Adiabatic transition of the pump into second optical harmonic," Pisma Zho Ehsp. Teor. Fiz., **57**, 790–793 (1993).

8. A. E. Kaplan, "Eigenmodes of $\chi^{(2)}$ wave-mixings: cross-induced 2-nd order nonlinear refraction," Opt. Lett., **18**, 1223–1225 (1993).

9. W. Torruellas, Y. Kivshar, and G. I. Stegeman, "Quadratic solitons," in Spatial Solitons, edited by S. Trillo and W. Torruellas (Springer-Verlag, Berlin, 2001), pp. 127–168.

SUGGESTED FURTHER READING

J. A. Armstrong, N. Bloembergen, J. Ducuing, and P. S. Pershan, "Interaction between light waves in a nonlinear dielectric," Phys. Rev., **127**, 1918–1939 (1962).

R. W. Boyd, Nonlinear Optics, 3rd Edition (Academic Press, Burlington, MA, 2008).

Y. R. Shen, Principles of Nonlinear Optics (John Wiley & Sons, New York, 1984).

CHAPTER 5

Second Harmonic Generation with Finite Beams and Applications

The preceding chapters have dealt with the "physics" of second harmonic generation (SHG) with plane-wave and continuous-wave excitation. The reality of experimental nonlinear optics is that finite beams are used and pulsed lasers are often necessary to reach the power levels for interesting nonlinear optics to occur. This chapter deals with some of the practical problems associated with extending the plane-wave solutions to finite beams, e.g., beam walk-off and optimizing conversion by using focusing lenses Here beams wide enough to be considered quasi-plane wave, i.e., beams many wavelengths wide in space, are considered. When pulses in time are considered, they should be many cycles long, and in the first approximation, group velocity delay (which causes pulses to spread in time) can be neglected. Dispersion will be dealt with in detail later in the context of third-order nonlinear effects in fibers.

Material modifications to engineer the wave-vector-matching condition, called quasi-phase matching (QPM), lead to many different application opportunities. For example, using different QPM periods in different segments of a crystal allows the generation of multiple frequencies into spectral regions where lasers do not exist. Doubling bandwidths can be artificially increased, temporally compressed doubled pulses can be generated, and so on. A selection of these applications will be discussed in this chapter.

5.1 SHG WITH GAUSSIAN BEAMS

Real experiments deal with finite beams. Although various beam profiles are used in practice, most theoretical treatments assume Gaussian beams because they have a simple form, are easily integrable in combination with other response functions, and provide a reasonable approximation to laser beams.

Nonlinear Optics: Phenomena, Materials, and Devices, George I. Stegeman and Robert A. Stegeman.
© 2012 John Wiley & Sons, Inc. Published 2012 by John Wiley & Sons, Inc.

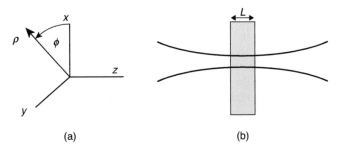

FIGURE 5.1 (a) The parameters used to describe a Gaussian beam's geometry for propagation along the z-axis. (b) Thin sample approximation in which the beam has a planar wave front over the sample and diffraction can be ignored.

The SHG geometry is shown in Fig. 5.1. The equation for a Gaussian field, which includes diffraction, satisfies the wave equation for plane wave-like beams, and conserves power on propagation, is

$$E(\vec{r}, t) = \frac{1}{2}\mathcal{E}_0(\omega)\frac{w_0}{w(z)}\exp\left[i\{kz + \psi(z) - \omega t\} - \frac{\rho^2}{w^2(z)} + i\frac{k\rho^2}{2R(z)}\right] + \text{c.c.}, \quad (5.1)$$

$$\text{Rayleigh range:} \quad z_0(\omega) = \frac{\pi w_0^2(\omega)n(\omega)}{\lambda_{\text{vac}}(\omega)},$$

$$\text{Radius of curvature:} \quad R(z) = z\left[1 + \frac{z_0^2}{z^2}\right], \qquad (5.2a)$$

$$w^2(z) = w_0^2\left[1 + \frac{z^2}{z_0^2}\right], \quad \text{where } 2w(z) = \text{"spot size" (diameter)} \qquad (5.2b)$$

and $2w_0 = $ minimum spot size,

$$\text{Divergence angle (for } z \gg z_0\text{):} \quad \theta_{\text{div}} = \frac{w_0}{z_0}, \qquad \psi(z) = \tan^{-1}\left[\frac{z}{z_0}\right]. \qquad (5.2c)$$

The parameter $z_0(\omega)$ is the distance over which the beam at ω diffracts to $\sqrt{2}w_0$ from its minimum beam radius.

5.1.1 Case I: $L \ll z_0$

For thin enough samples, the radius of curvature of beams is large enough so that the wave front is approximately a plane wave inside the sample and diffraction can be neglected. The input intensity is given by

$$I(\rho, \omega) = I(0, \omega)\exp\left(-\frac{2\rho^2}{w_0^2(\omega)}\right), \quad I(0, \omega) = \frac{1}{2}c\varepsilon_0 n(\omega)|\mathcal{E}_0(\omega)|^2, \qquad (5.3)$$

and the power is given by

$$P(\omega) = I(0,\omega) \int_0^{2\pi} d\phi \int_0^\infty \exp\left[\frac{-2\rho^2}{w_0^2}\right] \rho \, d\rho = \frac{\pi w_0^2(\omega)}{2} I(0,\omega). \tag{5.4}$$

For simplicity, consider the case of Type 1 *eoo*, $\Delta k = 0$, and negligible fundamental depletion. Assuming that the field varies slowly over a wavelength, the second harmonic generated by an incident fundamental beam will have the general form

$$E(\vec{r},t) = \frac{1}{2} \mathcal{E}(z,\rho,2\omega) \, e^{i[k(2\omega)z - 2\omega t]} + \text{c.c.} \tag{5.5}$$

Therefore from the slowly varying phase and amplitude equation, we have

$$\frac{d\mathcal{E}(z,\rho,2\omega)}{dz} = i\frac{\omega \hat{d}_{\text{eff}}^{(2)}}{n(2\omega)c} \mathcal{E}_0^2(\omega) \exp\left[-\frac{2\rho^2}{w_0^2(\omega)}\right]. \tag{5.6}$$

Integrating Eq. 5.6 over the sample length L under the approximations stated gives

$$\mathcal{E}(L,\rho,2\omega) = i\frac{\omega \hat{d}_{\text{eff}}^{(2)}}{n(2\omega)c} \mathcal{E}_0^2(\omega) \exp\left[-\frac{2\rho^2}{w_0^2(\omega)}\right] \int_0^L dz = i\frac{\omega \hat{d}_{\text{eff}}^{(2)}}{n(2\omega)c} \mathcal{E}_0^2(\omega) \exp\left[-\frac{2\rho^2}{w_0^2(\omega)}\right] L. \tag{5.7}$$

Note that $w_0(2\omega) = w_0(\omega)/\sqrt{2}$; i.e., the output harmonic beam is narrowed in space (see Fig. 5.2). This is a *general* feature of nonlinear wave interactions; i.e., the output beam is always narrower than any of the input beams. Furthermore, the "Rayleigh range" of the harmonic is

$$z_0(2\omega) = \frac{\pi w_0^2(2\omega)n(2\omega)}{\lambda_{\text{vac}}(2\omega)} = \frac{\pi w_0^2(\omega)n(2\omega)}{2\lambda_{\text{vac}}(2\omega)} \xrightarrow{n(\omega) \cong n(2\omega)} \frac{\pi w_0^2(\omega)n(\omega)}{\lambda_{\text{vac}}(\omega)} = z_0(\omega). \tag{5.8}$$

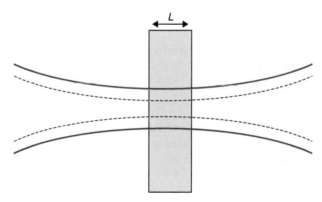

FIGURE 5.2   Gaussian fundamental beam (solid line) and its harmonic (dashed line) in the thin sample approximation.

This means that the Rayleigh range is the same for both beams, although the divergence angle for the harmonic is larger.

Normally, it is the total power in the harmonic that is important. This is obtained by integrating over the beam profile:

$$
P(2\omega) = \frac{1}{2} n(2\omega) c \varepsilon_0 \left| \frac{\omega \hat{d}_{\text{eff}}^{(2)}}{n(2\omega)c} \mathcal{E}_0^2(\omega) \right|^2 L^2 \int_0^{2\pi} d\phi \int_0^\infty \exp\left[-\frac{4\rho^2}{w_0^2(\omega)} \right] \rho \, d\rho
$$

$$
= \frac{1}{2} \varepsilon_0 \frac{\omega^2 \left| \hat{d}_{\text{eff}}^{(2)} \right|^2 L^2}{n(2\omega)c} \frac{\pi w_0^2(\omega)}{4} |\mathcal{E}_0(\omega)|^4 \tag{5.9}
$$

$$
\rightarrow \quad \frac{P(2\omega)}{P(\omega)} = \frac{2\{\omega L\}^2 \left| \hat{d}_{\text{eff}}^{(2)} \right|^2}{\pi w_0^2(\omega) c^3 \varepsilon_0 n(2\omega) n^2(\omega)} P(\omega).
$$

5.1.2 Case II: $L \approx 2z_0(\omega)$ (Optimum Conversion)

Diffraction causes both the fundamental and the harmonic beams to spread on propagation. It is possible to produce higher intensities by strong focusing, but the distance over which this enhancement occurs is reduced to approximately the Rayleigh range z_0 and a priori it is not clear where the optimum will occur. This leads to a trade-off for optimum conversion between minimum beam width and crystal length L (1).

Consider the simplest case of Type 1 eoo, $\Delta k = 0$, and negligibly small birefringence. The total power when diffraction is included is now given by a more complicated integral than in Case I above:

$$
P(2\omega) = \frac{1}{2} n(2\omega) c \varepsilon_0 \left| \frac{\omega \hat{d}_{\text{eff}}^{(2)}}{n(2\omega)c} \mathcal{E}_0^2(\omega) \right|^2 2\pi \int_0^\infty \left\{ \int_{-L/2}^{L/2} \frac{w_0^2(\omega)}{w^2(z,\omega)} \exp\left[-\frac{2\rho^2}{w^2(z,\omega)} \right] dz \right\}^2 \rho \, d\rho.
$$

$$
\tag{5.10}
$$

It proves useful to define a function $h_0(\xi)$ (where $\xi = L/z_0$), so that the power conversion can be written as (1)

$$
\frac{P(2\omega)}{P(\omega)} \cong 4 \frac{\omega^2 \left| \hat{d}_{\text{eff}}^{(2)} \right|^2}{n^2(\omega) n(2\omega) c^3 \varepsilon_0 \lambda_{\text{vac}}(\omega)} h_0(\xi) L P(\omega). \tag{5.11}
$$

The details of the integration are hidden in the function $h_0(\xi)$, which has been evaluated numerically and illustrated graphically in Fig. 5.3 (1).

It is informative to examine this function in different limits. For small ξ, i.e., $\xi \ll 1$, $h_0(\xi) = \xi$. When inserted into Eq. 5.11, the thin sample approximation result is recovered. For large ξ, i.e., $\xi \gg 1$, $h_0(\xi) = \xi^{-1}$, diffraction dominates, and the

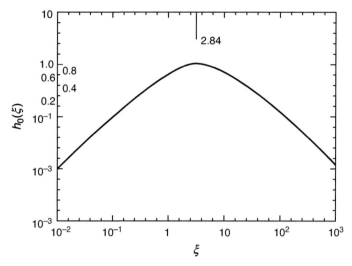

FIGURE 5.3 The function $h_0(\xi)$ as a function of ξ. Reproduced with permission from American Physical Society (1).

second harmonic power reaches a constant value (in a continuously spreading beam):

$$\frac{P(2\omega)}{P(\omega)} \cong 4 \frac{\omega^2 \left| \hat{d}_{\text{eff}}^{(2)} \right|^2}{n^2(\omega)n(2\omega)c^3 \varepsilon_0 \lambda_{\text{vac}}(\omega)} 2z_0 P(\omega). \tag{5.12}$$

Between these two limits there is a broad maximum centered on $L \cong 5.7z_0$, where $h_0(\xi) \cong 1.068$ and

$$\left. \frac{P(2\omega)}{P(\omega)} \right|_{\text{opt}} \cong 4.3 \frac{\omega^2 \left| \hat{d}_{\text{eff}}^{(2)} \right|^2}{n^2(\omega)n(2\omega)c^3 \varepsilon_0 \lambda_{\text{vac}}(\omega)} LP(\omega). \tag{5.13}$$

Note that the power is proportional to the sample length L, not L^2. Since the cost of doubling crystals is approximately linear in their length for reasonable lengths, it is useful to know that by reducing the length by almost a factor of 3 to $L \cong 2z_0, h_0(\xi) \cong 0.8$ and the peak conversion efficiency is reduced only by about $\approx 25\%$. A walk-off between the fundamental and the harmonic beams (which will be discussed in Section 5.1.4) will reduce the conversion efficiency further; however, the shorter the crystal, the smaller the reduction.

5.1.3 Pulsed Fundamental Finite Beam

To achieve high intensities, many nonlinear optics experiments are performed with pulsed lasers and the data are usually taken with some form of an energy meter because the shorter the pulse, the more expensive are the detectors and also because auxiliary apparatus is needed to measure and calibrate the pulse profiles. In this section, a Gaussian pulse in time will be assumed for the input fundamental field and a Gaussian spatial profile in order to calculate the energy in the second harmonic in the absence of a temporal walk-off. The fundamental power is written as

$$P(\omega, t) = P(\omega)\, e^{-2t^2/\tau^2(\omega)}. \tag{5.14}$$

The harmonic power is written in terms of the power $P(\omega)$ in Eqs 5.9 and 5.11–5.13 as

$$P(2\omega, t) \propto P^2(\omega, t) = P(2\omega) \exp\left[-\frac{4t^2}{\tau^2(\omega)}\right] \rightarrow \tau(2\omega) = \frac{\tau(\omega)}{\sqrt{2}}, \tag{5.15}$$

and again compression of the second harmonic signal occurs, but in this case, in time. One then concludes that nonlinear optical interactions *in general* will lead to *compression of beams in both space and time*. The higher the order of the nonlinear susceptibility involved, the stronger the compression.

As stated above, most measurements involve measuring pulse energies, i.e., $\Delta E(\omega_i) = \int_{-\infty}^{\infty} P(\omega_i, t)\, dt$, which gives

$$\Delta E(2\omega) = \sqrt{\frac{\pi}{4}} \tau(\omega) P(2\omega, 0). \tag{5.16}$$

Combining the results for space and time in the thin sample approximation gives

$$\frac{\Delta E(2\omega)}{\Delta E^2(\omega)} = \frac{1}{\sqrt{\pi}\tau(\omega)} \left\{ \frac{P(2\omega, 0)}{P^2(\omega, 0)} \right\} = \frac{2\omega^2 L^2 \left| \hat{d}_{\text{eff}}^{(2)} \right|^2}{\pi^{3/2}\tau(\omega) w_0^2(\omega) c^3 \varepsilon_0 n(2\omega) n^2(\omega)}. \tag{5.17}$$

5.1.4 Beam Walk-Off in Space

To this point it has been assumed that the fundamental and the harmonic beam energy travel along the z-axis. In the most common method of wave-vector matching, i.e., "critical" wave-vector matching, there is always a walk-off between the fundamental and the second harmonic. (This is discussed in Chapter 3, where an expression for the walk-off angle δ is given in Eqs 3.10 and 3.11.) The problem is illustrated in Fig. 5.4. The walk-off is assumed, for simplicity, to be limited to a single axis, which is the usual case in uniaxial crystals. To simplify the analysis, a thin sample will be assumed so that $w(z) \cong w_0$, i.e., the diffraction can be neglected (2).

The fundamental field is written as

$$\mathcal{E}(z', \omega, x', y') \cong \mathcal{E}_0^2(\omega) \exp\left[-\frac{(x'^2 + y'^2)}{w_0^2(\omega)}\right], \tag{5.18}$$

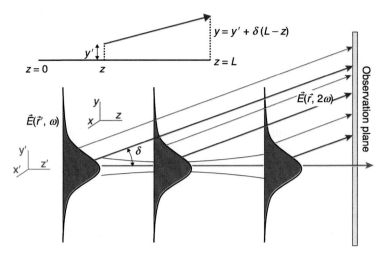

FIGURE 5.4   Walk-off between the fundamental and the harmonic for *eoo* critical wavevector matching. The primed coordinate system is for the fundamental and the unprimed one is for the second harmonic. The upper inset defines the relation between y and y'.

and the harmonic field as

$$\mathcal{E}(L, 2\omega, x, y) \cong i\frac{\omega \hat{d}_{eff}^{(2)}}{n(2\omega)c}\mathcal{E}_0^2(\omega)\int_0^L \exp\left[-\frac{2(x^2 + [y - \delta(L-z)]^2)}{w_0^2(\omega)}\right]dz. \quad (5.19)$$

Defining new variables and the function $F(u, \zeta)$ as (2)

$$u = \sqrt{2}\frac{y - \delta L}{w_0(\omega)}, \quad \xi = \sqrt{2}\frac{z\delta}{w_0(\omega)}, \quad \zeta = \sqrt{2}\frac{\delta L}{w_0(\omega)}, \quad F(u, \zeta) = \frac{1}{\zeta}\int_0^\zeta e^{-(u+\xi)^2}d\xi$$

$$(5.20)$$

gives

$$\mathcal{E}(L, 2\omega) \cong i\frac{\omega \hat{d}_{eff}^{(2)}}{n(2\omega)c}L\mathcal{E}_0^2(\omega)\exp\left[-\frac{2x^2}{w_0^2(\omega)}\right]F(u, \zeta). \quad (5.21)$$

The shape of the harmonic field in the observation plane for various values of ζ is shown in Fig. 5.5. The parameter ζ is essentially the ratio of the amount of walk-off to the harmonic beam radius.

The larger the walk-off relative to the beam radius, the smaller the peak conversion, the smaller the overall SHG, and the more diffuse the SHG is along the y-axis—all undesirable characteristics. The reduction in the total second harmonic generated can be quantified by Ref. 2

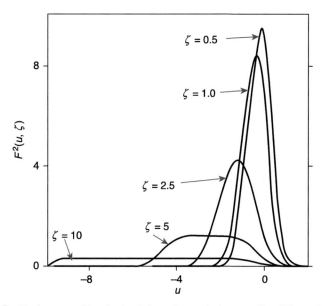

FIGURE 5.5   The beam profiles obtained along the y-axis due to walk-off for various values of ζ. Reproduced with permission from American Physical Society (2).

$$G(\zeta) = \sqrt{\frac{2}{\pi}} \int_{-\infty}^{\infty} F^2(u, \zeta)\, du \;\rightarrow\; \frac{P(2\omega)}{P(\omega)} = \frac{2\omega^2 L^2 \left|\hat{d}_{eff}^{(2)}\right|^2}{\pi w_0^2(\omega)\varepsilon_0 c^3 n(2\omega) n^2(\omega)} G(\zeta) P(\omega).$$

(5.22)

This function is plotted in Fig. 5.6 versus $\zeta' = \sqrt{2\pi} L/\ell_{wo}$, with $\ell_{wo} = \sqrt{\pi}[w_0(\omega)/\delta]$. This basically shows that to keep loss in conversion efficiency to less than 50%, $\zeta' \leq 3$ is needed.

Clearly, the walk-off length ℓ_{wo} is also a very important characteristic length and for optimum conversion $\ell_{wo} \geq L$ is also needed in addition to $\ell_{coh} > L > \ell_{pg}$

5.1.5   Examples of Finite Beam SHG

Example 5.1.5.1

KH_2PO_4, Type 1, SHG (*eoo*) for the continuous-wave, 1.06-μm fundamental input, 1-cm-long crystal.

(a) Calculate what intensity $I(\omega)$ is needed for a conversion of 58% for a plane-wave input? What are the angular bandwidth and the walk-off angle?

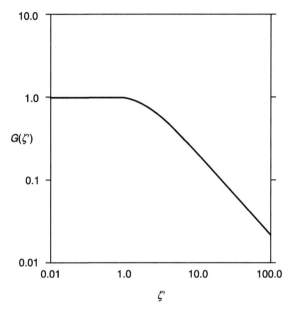

FIGURE 5.6   The decrease in the total second harmonic radiated versus the ratio of the beam walk-off to the beam propagation distance of the fundamental. Reproduced with permission from American Physical Society (2).

(b) For a low conversion efficiency of 10%, what is the fundamental power required for $L = 2z_0$? What is w_0 at the center of the crystal? What is w at the crystal input?

(c) What is the approximate harmonic loss due to walk-off for (b)?

Solution:

To solve (a) the wave-vector-match angle, the optimum orientation for SHG, and the actual $\hat{d}_{eff}^{(2)}$ need to be calculated first. Furthermore, optimum conversion requires operating on wave-vector match (which must be calculated from the refractive index data) and using the solutions for large conversion. The angular bandwidth and the walk-off angle are standard calculations obtained using Eqs 3.10 and 3.15. Thus,

$$n_e(\omega) = 1.4603, \quad n_o(\omega) = 1.4942, \quad n_e(2\omega) = 1.4709, \quad n_o(2\omega) = 1.5129$$

$$n_o(\omega) = n_e(2\omega, \theta)$$

$$\rightarrow \quad \sin(\theta_{PM}) = \frac{n_e(2\omega)}{n_o(\omega)} \sqrt{\frac{n_o^2(2\omega) - n_o^2(\omega)}{n_o^2(2\omega) - n_e^2(2\omega)}}$$

$$= 0.65942$$

$$\rightarrow \quad \theta_{PM} = 41.26°,$$

$$KH_2PO_4: \quad \bar{4}2m \quad \rightarrow \quad \hat{d}_{eff}^{(2)} = \hat{d}_{14}^{(2)} \sin(\theta_{PM}) \sin(2\phi) \rightarrow \phi_{opt} = \frac{\pi}{4},$$

and

$$\hat{d}_{14}^{(2)} = 0.43 \, \text{pm/V} \rightarrow \hat{d}_{eff}^{(2)} = 0.43 \sin(41.26°) = 0.28 \, \text{pm/V}.$$

For conversion efficiencies of $>10\%$ with plane waves, it is necessary to first calculate the parametric gain length and then use $I(2\omega)/I(\omega) = \tanh^2\left(L/\ell_{pg}\right)$ to calculate the efficiency.

For $L = \ell_{pg}$, $I(2\omega)/I(\omega) = \tanh^2(1) = 58\%$, and ℓ_{pg} expressed in terms of $I(\omega)$:

$$\ell_{pg}^{-1}(\text{in cm}^{-1}) = \frac{2\sqrt{2}\pi|\hat{d}_{eff}^{(2)}|(\text{in m/V})}{n(\omega)\sqrt{n(\omega)c\varepsilon_0}\lambda_{vac}(\text{in m})}\sqrt{I(\omega)(\text{in W/m}^2)},$$

$$\ell_{pg}^{-1}(\text{in cm}^{-1}) = 0.172\frac{\hat{d}_{eff}^{(2)}(\text{in pm/V})}{\lambda_{vac}(\text{in μm})n^{3/2}}\sqrt{I(\omega)(\text{in MW/cm}^2)}\ (\text{general result}),$$

$$\text{(5.23)}$$

and

$$\ell_{pg}^{-1}(\text{in cm}^{-1}) = 1 = \frac{0.172 \times 0.28}{1.06 \times (1.49)^{3/2}}\sqrt{I(\text{MW/cm}^2)} \rightarrow 1.6 \, \text{GW/cm}^2.$$

The angular bandwidth and the walk-off angle are given as

$$\Delta\theta_{PM} = \frac{\lambda_{vac}(\omega)}{4L}\left[|\{n_o(2\omega)-n_e(2\omega)\}|\sin(2\theta_{PM})\right]^{-1}$$

$$= \frac{1.06 \times 10^{-4}}{4 \times 0.042 \sin(82.52°)} = 0.64 \, \text{mrad},$$

$$\delta = \frac{n_e(2\omega)-n_o(2\omega)}{n_o(2\omega)}\sin(2\theta_{PM}) \cong 28 \, \text{mrad}.$$

(b) For $L = 2z_0$,

$$\frac{P(2\omega)}{P(\omega)} \cong 3.2\frac{\omega^2|\hat{d}_{eff}^{(2)}|^2}{n^2(\omega)n_{eff}(2\omega)c^3\varepsilon_0\,\lambda_{vac}(\omega)}LP(\omega)$$

$$\text{(5.24)}$$

$$= 0.475 \times 10^{-3}\frac{\left[|\hat{d}_{eff}^{(2)}|(\text{in pm, /V, })\right]^2}{n^2(\omega)n_{eff}(2\omega)\lambda_{vac}^3(\text{in μm})}L(\text{in cm})P(\omega)(\text{in W}).$$

(Note that Eq. 5.24 is *a general formula* that can be used for any case.)

$$\frac{P(2\omega)}{P(\omega)} = 0.1 = 0.475 \times 10^{-3} \frac{[0.28]^2}{1.49^3 \times 1.06^3} P(\omega)(\text{in W}) \rightarrow P(\omega) \cong 10.6\,\text{kW},$$

$$\frac{\pi w_0^2 n}{\lambda_{\text{vac}}} = \frac{L}{2} \rightarrow w_0 = 33.6\,\mu\text{m},$$

$$\text{At input facet}: \quad w\left(\frac{L}{2}\right) = w_0\left[1 + \frac{z^2}{z_0^2}\right] = \sqrt{1+1} = 47.6\,\mu\text{m}.$$

(c) The walk-off loss can be calculated approximately from $\zeta = \sqrt{2}L\delta/w_0 = 1.416 \times 0.028/33.6 \times 10^{-4} = 12$. From Fig. 5.6, only $\approx 20\%$ of the harmonic calculated without walk-off will actually be obtained because of the walk-off.

Example 5.1.5.2

LiNbO$_3$ for 1.06-μm input, birefringent and QPM "noncritical" wave-vector match ($L = 1$ cm). LiNbO$_3$ is a very popular material and is used in two forms: birefringent wave-vector match and quasi-phase-match. Wave-vector matching with eoo occurs approximately at 1.06 μm, and for simplicity, it will be assumed that minor temperature tuning can bring it onto a wave-vector match.

(a) For a birefringent Type 1 wave-vector match, calculate the input intensity required for the plane-wave input for a conversion of 58% into the harmonic. Assuming a beam input with $L = 2z_0$, find the input power required for a conversion efficiency of 10%.

(b) What is the angular bandwidth?

Solution:

Figure 5.7 shows the geometry for Type 1 eoo wave-vector match. Thus,

$$n_o(\omega) \cong n_e(2\omega) \cong 2.24, \qquad n_o(2\omega) = 2.32, \qquad d_{31} = 5.95\,\text{pm/V},$$

$$\ell_{\text{pg}}^{-1}(\text{in cm}^{-1}) = 0.172 \frac{\hat{d}_{\text{eff}}^{(2)}(\text{in pm/V})}{\lambda_{\text{vac}}(\text{in }\mu\text{m})n^{3/2}} \sqrt{I(\omega)(\text{in MW/cm}^2)} \rightarrow I(\omega) = 12.1\,\text{MW/cm}^2,$$

$$\frac{P(2\omega)}{P(\omega)} \cong 0.475 \times 10^{-3} \frac{\left[\left|\hat{d}_{\text{eff}}^{(2)}\right|(\text{in pm/V})\right]^2}{n^2(\omega)n(2\omega)\lambda_{\text{vac}}^3(\text{in }\mu\text{m})} L(\text{in cm})P(\omega)(\text{in W}),$$

$$0.1 \cong 0.475 \times 10^{-3} \frac{[5.95]^2}{2.24^3 \times 1.06^3} P(\omega)(\text{in W}) \rightarrow 80\,\text{W}.$$

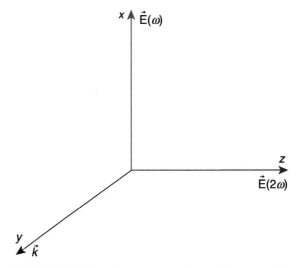

FIGURE 5.7   Type 1 *eoo* wave-vector-matching geometry.

Since the wave vector $\vec{k}$ lies along one of the crystal axes, this is noncritical wave-vector matching:

$$\Delta\theta_{\mathrm{PM}} \cong \left\{ \frac{\lambda_{\mathrm{vac}}(\omega)}{4L[n_o(2\omega)-n_e(2\omega)]} \right\}^{1/2}$$

$$\cong \left\{ \frac{1.06}{4\times 10^4 \times 0.08} \right\}^{1/2}$$

$$\approx 1^{\circ}.$$

(b) Now repeat the above steps for QPM LiNbO$_3$ and use the general result in Eq. 5.24. Thus,
For $p=1$, $\hat{d}_{\mathrm{eff}}^{(2)} = \frac{2}{\pi}\,\hat{d}_{33}^{(2)}$ and $\hat{d}_{33}^{(2)} = 28\,\mathrm{pm/V} \rightarrow \hat{d}_{\mathrm{eff}}^{(2)} = 18\,\mathrm{pm/V}$.

$$\ell_{\mathrm{pg}}^{-1}(\mathrm{in\,cm}^{-1}) = 0.172\frac{\hat{d}_{\mathrm{eff}}^{(2)}(\mathrm{in\,pm/V})}{\lambda_{\mathrm{vac}}(\mathrm{in\,\mu m})n^{3/2}}\sqrt{I(\omega)(\mathrm{in\,MW/cm}^2)} \rightarrow I(\omega) \cong 1.3\,\mathrm{MW/cm}^2,$$

$$\frac{P(2\omega)}{P(\omega)} \cong 0.475\times 10^{-3}\frac{\left[\left|\hat{d}_{\mathrm{eff}}^{(2)}\right|(\mathrm{in\,pm/V})\right]^2}{n^2(\omega)n(2\omega)\lambda_{\mathrm{vac}}^3(\mathrm{in\,\mu m})}L(\mathrm{in\,cm})P(\omega)(\mathrm{in\,W})$$

$$0.1 \cong 0.475\times 10^{-3}\frac{[18]^2}{2.24^2\times 2.32\times 1.06^3}P(\omega)(\mathrm{in\,W}) \rightarrow 9.0\,\mathrm{W}.$$

The formula for the angular bandwidth has already been derived in Eq. 3.26 (in Chapter 3):

$$\Delta\theta_{\mathrm{PM}} \cong \sqrt{\frac{\lambda_{\mathrm{vac}}(\omega)}{4L}} \left| \left\{ [n_o(\omega)-n_o(2\omega)] - [n_e(\omega)-n_e(2\omega)] - \frac{\lambda_{\mathrm{vac}}(\omega)}{4\Lambda} \right\}^{-1} \right|.$$

Note that more accurate values for the refractive index are needed to calculate the birefringence:

$$n_o(\omega)-n_e(\omega)=0.090, \quad n_o(2\omega)-n_e(2\omega)=0.107, \quad n_e(2\omega)-n_e(\omega)=0.083.$$

The poling contribution is given by $\frac{\lambda_{\mathrm{vac}}(\omega)}{4\Lambda}=\frac{1}{2}[n_e(2\omega)-n_e(\omega)]=0.042$. Therefore,

$$\Delta\theta_{\mathrm{PM}} \cong \sqrt{\frac{1.06\times10^{-4}}{4}}|\{0.090-0.107-0.042\}^{-1}|=\sqrt{\frac{1.06\times10^{-4}}{4\times0.059}}=0.021\,\mathrm{rad}\cong1.2°.$$

The advantage of using QPM is obvious.

5.2 UNIQUE AND PERFORMANCE-ENHANCED APPLICATIONS OF PERIODICALLY POLED LiNbO₃ (PPLN)

QPM has made possible a number of interesting applications because both the period of the poling and the width of the poling region relative to the input beam width can be engineered. Martin Fejer and his colleagues at Stanford have pioneered many of these ideas (3,4).

5.2.1 Tunable Frequency Conversion

Periodically poled LiNbO₃ (PPLN) has been shown to provide a solution to a common problem encountered with angle tuning by using birefringent phase matching. As angle is tuned in a birefringent wave-vector match, the spatial walk-off between the fundamental and the harmonic typically requires adjustment of the crystal geometry or complicated additional optics. With PPLN this is simply achieved with samples fabricated with a poling period that varies linearly along the x-axis, i.e., $\Lambda(x)$ in the sample shown in Fig. 5.8 (5). Therefore, as the input beam moves along the y-axis, the wave-vector-matching condition changes and different frequencies are efficiently doubled. This also alleviates the need for precise fabrication of the poling period. The drawback is that the effective length available for doubling any frequency is dramatically reduced. Fortunately, PPLN has been fabricated in lengths up to 8 cm (6).

5.2.2 Enhanced Frequency Bandwidth at Doubled Frequencies

The concept of enhanced frequency bandwidth at doubled frequencies can best be understood by interpreting the generated SHG field in the weak conversion limit

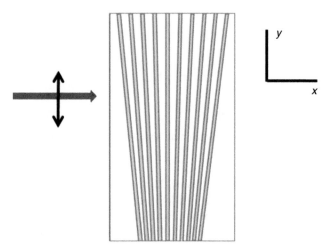

FIGURE 5.8 "Fan-"shaped poling period $\Lambda(x) = 2\pi/K(x)$ such that the wave-vector-matching condition changes along the x-axis.

(see Fig. 5.9). In the weak conversion limit, the harmonic field for poling, optimized for the first grating order, is given by

$$\mathcal{E}_3(L, 2\omega) = i\frac{\omega}{n(2\omega)c}\mathcal{E}_3^2(\omega)\int_0^L e^{i[\Delta k + K]x}\hat{d}_{1,\text{eff}}^{(2)}(x)\,dx, \tag{5.25}$$

which can be interpreted as a Fourier transform of the nonlinearity.

Consider now a poled sample with two overlapping periodicities, K and K_g, as shown in Fig. 5.10a. The Fourier spectrum of the wave-vector-matching condition now contains two peaks. This results in two second harmonic peaks, as illustrated in Fig. 5.10b.

If the poling periodicity is now random along the x-axis (see Fig. 5.11a), then there is a quasi-continuum of wave-vector-matching conditions. For samples of finite length, there are oscillations in the response and a smooth flat response requires an infinitely long sample (see Fig. 5.11b). As a result, the Fourier spectrum of the wave-vector-matching condition, and therefore the SHG bandwidth, is enhanced. However, each Fourier component for a given sample length is smaller and therefore the second harmonic power is reduced relative to the single-period case. It was predicted

FIGURE 5.9 (a) Quasi-phase matching with the uniform period. (b) Wave-vector-matching condition due to the uniform quasi-phase-matching period.

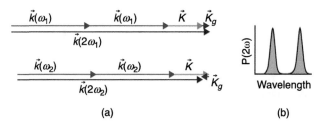

(a)   (b)

FIGURE 5.10   (a) The wave-vector-matching conditions for both periodicities. (b) Second harmonic response.

theoretically that the product of the bandwidth and the peak conversion efficiency is a constant for long enough samples, a well-known result in filter theory (7). This trade-off was verified numerically as shown in Fig. 5.12a, as well as the oscillations which occur for finite samples. As shown in Fig. 5.12b, the bandwidth increases by a factor of ≈10 and the SHG peak reduces by the same factor. This result was found to be in good agreement with that of the experiment (Fig. 5.12b).

5.2.3 Doubling of Ultrashort Pulses

PPLN offers some advantages for efficient doubling of ultrafast pulses. Pulse walk-off in time and space between the fundamental and the harmonic pulses is a serious issue with fs pulses since it limits the usable length of a frequency doubler and hence its efficiency. Furthermore, the wide bandwidths of fs pulses also require short sample lengths so that the frequency bandwidth of the doubler can accommodate the pulse bandwidth (8). Because of such complications, usually the doubling efficiency with *fs* pulses scales with pulse energy instead of the intensity. A useful figure of merit (FOM) is the percent conversion per nanojoule (*%/nJ*) of fundamental energy (8).

Because of dispersion in the refractive index with frequency, the group velocity (i.e., the velocity at which the energy of a pulse is propagated) depends on the frequency. The higher frequency components of the signal usually travel slower than the lower frequency components. In frequency doublers, there is difference of a factor of 2 between the frequencies of the fundamental and the harmonic, which makes pulse walk-off a serious problem. An example of pulse walk-off is illustrated in Fig. 5.13 in the time domain.

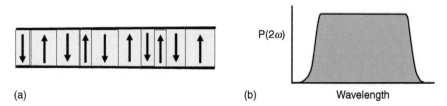

(a)   (b)   Wavelength

FIGURE 5.11   (a) An example of randomly sized domain reversal in a sample. (b) The theoretical harmonic spectrum for an infinitely long sample.

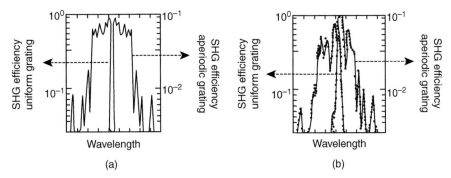

FIGURE 5.12 (a) Theoretical calculation of the SHG efficiency for a uniform grating and an aperiodic grating. (b) Experimental measurement of the SHG efficiency for a uniform grating and an aperiodic grating. Reproduced with permission from the Institution of Engineering and Technology (7).

For pulses of width τ, the temporal separation between the two pulses should be kept smaller than τ. For longer times (and crystal lengths), the conversion efficiency decreases dramatically. The requirement for useful conversion can be summarized as follows:

$$\tau \geq \frac{L}{v_g(2\omega)} - \frac{L}{v_g(\omega)} = \frac{L}{c}\left\{n_g(2\omega) - n_g(\omega)\right\} = \frac{L\Delta n_g}{c} \Rightarrow L_{wo} \leq \frac{c\tau}{\Delta n_g}, \quad (5.26)$$

where L_{wo} is the maximum usable length for walk-off, the group velocity $v_g = c/n_g$, and n_g is the group index.

A second limiting length arises from bandwidth considerations; i.e., the SHG spectral bandwidth $\Delta\lambda_{\text{phase match}}$ should be larger than the pulse bandwidth ($\Delta\lambda_{\text{pulse}}$) so that the full spectrum of a pulse is doubled (see Fig. 5.14).

The SHG bandwidth, given by $\Delta kL = \pi \rightarrow \Delta\lambda_{\text{pm}} = \lambda^2/L$, must be larger than $\Delta\lambda_{\text{pulse}}$. Therefore, the maximum crystal length L_{bw} for efficient SHG must be less

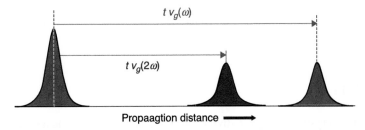

FIGURE 5.13 The pulse separation between a fundamental pulse (red) and a harmonic pulse (blue) with different group velocities typical (blue slower than red) in the "normal group velocity dispersion" region of a material.

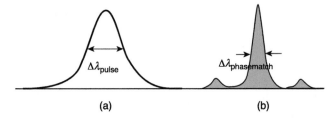

FIGURE 5.14 Bandwidth of (a) an ultrashort pulse and (b) the SHG process.

TABLE 5.1 Comparison of the Parameters of Three Commonly Used Materials for Doubling Ultrafast Pulses[a]

Material[b]	Wave-Vector Match	Figure of Merit of Ultrafast Pulses (pm^2/V^2)	Efficiency (%/nJ)	L_{max} (mm)
Periodically poled LiNbO$_3$	Noncritical	710	95	0.4
LBO	Noncritical	42	6	5.0
KTP	Type 2 critical	20	1.5	1.5

[a] Ref. 8.
[b] The properties of LBO and KTP are discussed in Chapter 7.

than $\lambda^2/\Delta\lambda_{pulse}$. To avoid a walk-off and to maintain optimum conversion, L_{max} must be smaller than L_{wo} and L_{bw}. It is then useful to define the FOM for SHG of ultrafast pulses as

$$\text{FOM}_{uf} = \text{FOM}_{cw}L_{max} = \left[\frac{\hat{d}_{eff}^{(2)}}{n}\right]^2 L_{max}. \tag{5.27}$$

As compared with LBO and KTP, PPLN has superior performance. Despite the short length of its crystals for subpicosecond pulses, PPLN still has the maximum efficiency (see Table 5.1).

PROBLEMS

1. Consider a crystal of length L divided into N small elements of length $\Delta L = N/L$, where ΔL is the walk-off length. Using the figure below and assuming a wave-vector-matched SHG in each segment, show that the ratio of the total second harmonic power generated for the walk-off to the no walk-off case for the condition

that the second harmonic wave-vector matched to a fundamental beam (FW) is given approximately by $\Delta L/L$.

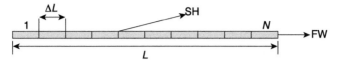

2. A fundamental beam with a Gaussian shape in space of the form $\frac{1}{2}\mathcal{E}_0\,e^{-\rho'^2/a^2}\,\hat{e}_f\,e^{i[k(\omega)z'-\omega t]}+\text{c.c.}$ is incident onto a $\chi^{(2)}$ medium. A nonlinear polarization with a polarization direction $\hat{e}_f$ is formed via $\chi^{(2)}$. Due to normal dispersion in the refractive index, the harmonic wave vector is $k(2\omega)>2k(\omega)$. This problem can be solved by using "Green's function" techniques to find the generated harmonic field in the far-field limit. The nonlinear polarization $\vec{P}^{\text{NL}}(\vec{r}')$ induced in a small volume element $\Delta V'$ radiates an electric field in all directions (see figure below). These fields are integrated over the volume dV' of the induced polarization to give a field at the point $\vec{r}$ given by

$$\mathcal{E}(2\omega,\vec{r})=\frac{1}{4\pi}\int\frac{e^{ik(2\omega)|\vec{r}-\vec{r}'|}}{|\vec{r}-\vec{r}'|}\left\{\mu_0\frac{d^2\mathcal{P}^{\text{NL}}(\vec{r}',t')}{dt'^2}\right\}dV'.$$

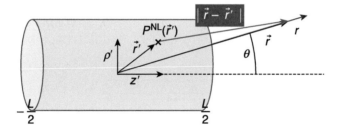

(a) Show that in the far field ($|r|\gg r'$), this result can be written as

$$\mathcal{E}(2\omega,\vec{r})=A\frac{e^{ik(2\omega)r}}{4\pi}\int e^{-ik(2\omega)\hat{r}\cdot\vec{r}'+2ik(\omega)z'}\,e^{-2\rho'^2/a^2}\,dV',\quad A=-2\omega^2\varepsilon_0\mu_0 d^{(2)}_{\text{eff}}\mathcal{E}_0^2(\omega).$$

[Hint: Expand $|\vec{r}-\vec{r}'|=\sqrt{(\vec{r}-\vec{r}')\cdot(\vec{r}-\vec{r}')}$ for $r\gg r'$ and keep only the leading terms.]

(b) Assuming the obvious cylindrical symmetry, show that the harmonic field is given by

$$\mathcal{E}(2\omega)=A\frac{e^{ik(2\omega)r}}{4\pi r}L\,\text{sinc}\left[\{2k(\omega)-k(2\omega)\cos\theta\}\frac{L}{2}\right]f[a,\sin\theta],$$

where

$$f[a, \sin\theta] = \int_0^\infty e^{-(2\rho'^2/a^2)-ik(2\omega)\sin\theta\rho'} 2\pi\rho' \, d\rho'.$$

Find the intensity pattern in space and give a simple interpretation for the harmonic radiation pattern.

(c) Using the mathematical identity $\sin^2(Lx)/\pi L x^2 \xrightarrow{L\to\infty} \delta(x)$, show that the radiated harmonic intensity grows linearly with L. Also show that for small L the intensity grows quadratically with L, i.e., L^2. Explain the physical origin of this difference.

3. There is a relatively rare form of wave-vector matching that is based on achieving phase matching in a single segment that is small relative to a coherence length and then cascading N segments (see figure below). Phase matching in a segment is obtained by dividing each segment into two non-phase-matched parts, a and b, arranged such that the phase mismatches $\Delta\phi_a$ and $\Delta\phi_b$ for the second harmonic in the two subsegments cancel each other. The propagation wave vectors are k_{1a} and k_{2a} for the fundamental and the harmonic, respectively, in subsegment a and are k_{1b} and k_{2b} for the fundamental and the harmonic, respectively, in subsegment b. Assume that there is no loss and the depletion of the fundamental can be neglected.

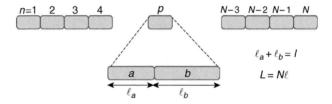

The fundamental propagating in the a and b regions of the pth segment are written, respectively, as

$$\frac{1}{2}\mathcal{E}_{1,p-1}(z) \, e^{ik_{1,a}(z-[p-1]\ell)} + \text{c.c.} \quad \text{and} \quad \frac{1}{2}\mathcal{E}_{1,p-1}(z) \, e^{ik_{1,b}(z-[p-1]\ell+k_{1,a}\ell_a)} + \text{c.c.}$$

(a) Show that the second harmonic generated by the end of the pth segment is

$$E_{2,p}(p\ell) = \frac{i}{2}\mathcal{E}_{1,p-1}^2([p-1]\ell) \, e^{2i\bar{k}_1(p-1)\ell}$$

$$\times \left[\kappa_a \ell_a \, e^{i\{\Delta\phi_a + k_{2a}\ell_a + k_{2b}\ell_b\}} \sin(\Delta\phi_a) + \kappa_b \ell_b \, e^{i\{\Delta\phi_b + 2k_{1a}\ell_a + k_{2b}\ell_b\}} \sin(\Delta\phi_b) \right] + \text{c.c.},$$

where

$$\kappa_a = \frac{\omega \hat{d}^{(2)}_{a,\text{eff}}}{c n_a(2\omega)}, \qquad \kappa_b = \frac{\omega \hat{d}^{(2)}_{b,\text{eff}}}{c n_b(2\omega)}, \qquad \Delta\phi_a = (2k_{1a}-k_{2a})\frac{\ell_a}{2},$$

$$\Delta\phi_b = (2k_{1b}-k_{2b})\frac{\ell_b}{2},$$

$$E_{2,p}(N\ell) = \frac{i}{2}\mathcal{E}_1^2(0)K\, e^{2i\bar{k}_1(p-1)\ell + (N-p)\ell\bar{k}_2} + \text{c.c.},$$

$$\bar{k}_1 = \frac{\ell_a k_{1a}+\ell_b k_{1b}}{\ell_a+\ell_b}, \qquad \bar{k}_2 = \frac{\ell_a k_{2a}+\ell_b k_{2b}}{\ell_a+\ell_b},$$

and

$$K = \kappa_a\ell_a\, e^{i\{\Delta\phi_a+k_{2a}\ell_a+k_{2b}\ell_b\}}\sin(\Delta\phi_a) + \kappa_b\ell_b\, e^{i\{\Delta\phi_b+2k_{1a}\ell_a+k_{2b}\ell_b\}}\sin(\Delta\phi_b).$$

(b) Show that the ratio of the total (summed overall p) second harmonic intensity to the incident intensity is given by

$$\frac{I(2\omega)}{I(0)} = |K|^2\frac{\sin^2[N(\Delta\phi_a+\Delta\phi_b)]}{\sin^2(\Delta\phi_a+\Delta\phi_b)}.$$

(c) For the perfect wave-vector match $I_{\text{opt}}(2\omega)=|\kappa NP|^2$, show that

$$\frac{I_{\text{seg}}(2\omega,L)}{I_{\text{opt}}(2\omega,L)} = \frac{\sin c^2[N(\Delta\phi_a+\Delta\phi_b)]}{\sin c^2(\Delta\phi_a+\Delta\phi_b)}\left\{\left[\frac{\kappa_a\ell_a}{\kappa\ell}\sin c(\Delta\phi_a)\right]^2\right.$$

$$\left.+\left[\frac{\kappa_b\ell_b}{\kappa\ell}\sin c(\Delta\phi_b)\right]^2+2\frac{\kappa_a\kappa_b\ell_a\ell_b}{(\kappa\ell)^2}\sin c(\Delta\phi_a)\sin c(\Delta\phi_b)\cos(\Delta\phi_a+\Delta\phi_b)\right\}.$$

(d) Show that optimum conversion occurs for $\Delta\phi_a = -\Delta\phi_b$.

(e) Assuming further that $\hat{d}^{(2)}_{\text{eff},a} = \hat{d}^{(2)}_{\text{eff},b}$ and $\kappa_a = \kappa_b$, show that

$$\frac{I_{\text{seg}}(2\omega,L)}{I_{\text{opt}}(2\omega,L)} = \sin c^2(\Delta\phi_a) \xrightarrow{\Delta\phi_a\ll\pi/2} 1.$$

4. A bulk medium with $\hat{d}^{(2)}_{\text{eff}}\neq 0$ has its nonlinearity and linear refractive index modulated with the same periodicity as shown below. Either the index grating or the nonlinearity grating, or both can be used to wave-vector match second harmonic generation.

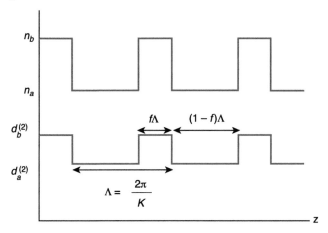

(a) Assuming both modulations as small perturbations, show that the coupled mode equations for second harmonic generation are given by

$$\frac{da_1(z)}{dz} - i \sum_{q=-\infty}^{\infty} \kappa_q^L(\omega)\, e^{iqKz} a_1(z) = i \sum_{p=-\infty}^{\infty} \tilde{\kappa}_p^{NL}\, e^{i[k_2 - 2k_1 - pK]z} a_2(z) a_1^*(z)$$

and

$$\frac{da_2(z)}{dz} - i \sum_{q=-\infty}^{\infty} \kappa_q^L(2\omega)\, e^{iqKz} a_2(z) = i \sum_{p=-\infty}^{\infty} \tilde{\kappa}_p^{NL}\, e^{i[2k_1 - k_2 + pK]z} a_1^2(z),$$

where the fundamental and the harmonic fields, respectively, are written as

$$\frac{1}{2} a_1(z)\, e^{i[k_1 z - \omega t]} + \text{c.c.} \quad \text{and} \quad \frac{1}{2} a_2(z)\, e^{i[k_2 z - 2\omega t]} + \text{c.c.}$$

(b) Find a substitution $b_i(z) = a_i(z)f(z)$, which, combined with keeping only the $q = \pm 1$ terms, will reduce the equations with $\phi_L = \dfrac{4\kappa_1^L(\omega) - 2\kappa_1^L(2\omega)}{K}$ to

$$\frac{db_1(z)}{dz} = i \sum_{p=-\infty}^{\infty} \tilde{\kappa}_p^{NL}\, e^{i[k_2 - 2k_1 - pK]z - i\phi_L \sin(Kz)} b_2(z) b_1^*(z)$$

and

$$\frac{db_2(z)}{dz} = i \sum_{p=-\infty}^{\infty} \tilde{\kappa}_p^{NL}\, e^{-i[k_2 - 2k_1 - pK]z + i\phi_L \sin(Kz)} b_1^2(z).$$

(c) Using the standard expansion for $e^{i\phi_L \sin(Kz)}$ in terms of Bessel functions, show that the equations for first-order grating wave-vector match reduce further to

$$\frac{db_1(z)}{dz} = i\tilde{\kappa}_{\text{eff}}^{\text{NL}} \, e^{i[k_2 - 2k_1 - Kz]} b_2(z) b_1^*(z),$$

$$\frac{db_2(z)}{dz} = i\tilde{\kappa}_{\text{eff}}^{\text{NL}} \, e^{-i[k_2 - 2k_1 - Kz]} b_1^2(z),$$

$$\text{with } \tilde{\kappa}_{\text{eff}}^{\text{NL}} = \tilde{\kappa}_1 + \tilde{\kappa}_0 \left[\frac{4\kappa_1^L(\omega) - 2\kappa_1^L(2\omega)}{K} \right].$$

Note that these two familiar coupled mode equations can now be solved in the usual way. Furthermore, in quasi-phase matching lithium niobate there is no index modulation, just nonlinearity.

REFERENCES

1. G. D. Boyd and D. A. Kleinman, "Parametric interaction of focused Gaussian beams," J. Appl. Phys., **39**, 3597–3639 (1968).
2. G. D. Boyd, A. Ashkin, J. M. Dziedzic, and D. A. Kleinman, "Second-harmonic generation of light with double refraction," Phys. Rev., **137**, A1305–A1320 (1965).
2. Fejer group publications: www.stanford.edu/group/fejer/cgi-bin/publications.php.
4. C. Langrock, S. Kumar, J. E. McGeehan, A. E. Willner, and M. M. Fejer, "All-optical signal processing using $\chi^{(2)}$ nonlinearities in guided-wave devices," J. Lightwave Technol., **24**, 2579–2592 (2006).
5. Y. Ishigame, T. Suhara, and H. Nishihara, "LiNbO$_3$ waveguide second-harmonic-generation device phase matched with a fan-out domain-inverted grating," Opt. Lett., **16**, 375–377 (1991).
6. D. Hofmann, G. Schreiber, C. Haase, H. Herrmann, W. Grundkötter, R. Ricken, and W. Sohler, "Quasi-phase-matched difference-frequency generation in periodically poled Ti: LiNbO$_3$ channel waveguides," Opt. Lett., **24**, 896–898 (1999).
7. M. L. Bortz, M. Fujimura, and M.M. Fejer, "Increased acceptance bandwidth for quasi-phase-matched second harmonic generation in LiNbO$_3$ waveguides," Electron. Lett., **30**, 34–35 (1994).
8. M. A. Arbore, M. M. Fejer, M. E. Fermann, A. Hariharan, A. Galvanauskas, and D. Harter, "Frequency doubling of femtosecond erbium-fiber soliton lasers in periodically poled lithium niobate," Opt. Lett., **22**, 13–15 (1997).

SUGGESTED FURTHER READING

R. W. Boyd, Nonlinear Optics, 3rd Edition (Academic Press, Burlington, MA, 2008).

Y. R. Shen, Principles of Nonlinear Optics (John Wiley & Sons, New York, 1984).

CHAPTER 6

Three-Wave Mixing, Optical Amplifiers, and Generators

In previous chapters, discussions were focused on second harmonic doublers, which are by far the most common nonlinear optics devices. Almost as popular, especially for obtaining tunable radiation, are optical parametric oscillators. Sum frequency generation (SFG), difference frequency generation (DFG), and optical parametric amplifiers (OPAs) are intermediate effects and devices in which a signal is amplified at the expense of a higher frequency pump beam. When an OPA is placed in a cavity with feedback at some of the frequencies present, it is called an optical parametric oscillator (OPO). The devices discussed in this chapter operate in the Kleinman limit, i.e., far from any material resonances, to minimize loss and optimize efficiency.

6.1 THREE-WAVE MIXING PROCESSES

Three different waves exist in a $\chi^{(2)}$-active medium with frequencies $\omega_c > \omega_b > \omega_a$, which are common to SFG, DFG, OPA and OPO (see Fig. 6.1). These waves interact strongly via $\chi^{(2)}$ when they satisfy the conditions $\omega_c = \omega_a + \omega_b$ and $k_a + k_b - k_c \cong 0$. The coupled wave equations satisfied by the three fields in the Kleinman limit are given in Chapter 2 as

$$\frac{d\mathcal{E}_c}{dz} = i\frac{\omega_c}{n_c c}\tilde{d}_{\text{eff}}^{(2)}\mathcal{E}_a\mathcal{E}_b e^{i\Delta kz}, \quad \frac{d\mathcal{E}_a}{dz} = i\frac{\omega_a}{n_a c}\tilde{d}_{\text{eff}}^{(2)}\mathcal{E}_c\mathcal{E}_b^* e^{-i\Delta kz}, \quad \frac{d\mathcal{E}_b}{dz} = i\frac{\omega_b}{n_b c}\tilde{d}_{\text{eff}}^{(2)}\mathcal{E}_c\mathcal{E}_a^* e^{-i\Delta kz}.$$

(6.1)

These different nonlinear processes are frequently summarized graphically. In Fig. 6.2, the input waves in the interaction are shown to the left of the dashed line and the relevant output waves to the right. In SFG, two lower energy photons combine to produce a higher energy photon, and in DFG, a higher energy photon (ω_c) mixes with a low frequency photon (ω_b) to produce another ω_b photon and a

Nonlinear Optics: Phenomena, Materials, and Devices, George I. Stegeman and Robert A. Stegeman.
© 2012 John Wiley & Sons, Inc. Published 2012 by John Wiley & Sons, Inc.

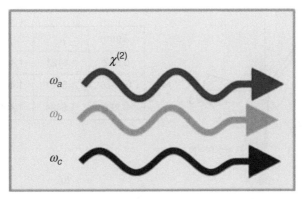

FIGURE 6.1   Three waves of different frequencies interacting via $\chi^{(2)}$.

different low frequency photon (ω_a). In an OPA, the high energy photon breaks up into two lower frequency photons. In an OPO, the interaction occurs in an optical cavity tuned for cavity resonance on one, two, or even all three of the interacting waves.

The wave-vector mismatch is defined by

$$\Delta k = k_a + k_b - k_c = \frac{n_a \omega_a}{c} + \frac{n_b \omega_b}{c} - \frac{n_c \omega_c}{c}, \quad \Delta k = 0 \;\Rightarrow\; n_c \omega_c = n_a \omega_a + n_b \omega_b. \tag{6.2}$$

Finding the right condition for wave-vector matching is more complicated than finding it for harmonic generation (SHG) and is usually done numerically. Consider a specific example of finding the wave-vector-matching condition for SFG in

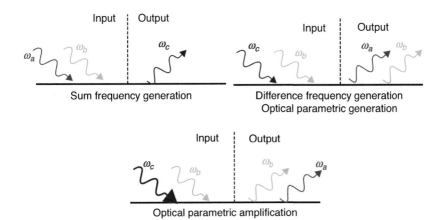

FIGURE 6.2   The basic three-wave interactions examined in this chapter.

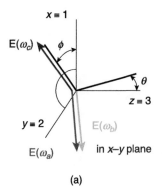

λ (μm)	n_o	n_e
0.532	1.5121	1.4705
0.6328	1.5071	1.4649
0.289	1.5494	1.4990

(a)   (b)

FIGURE 6.3 (a) KH_2PO_4 *eoo* geometry for sum frequency generation. (b) Refractive indices at the interacting wavelengths.

KH_2PO_4 *eoo* (Fig. 6.3a), with $\lambda_a = 0.6328$ μm and $\lambda_b = 0.532$ μm. The Sellmeier equations for dispersion in the refractive indices are as follows:

$$\frac{1}{\lambda_a} + \frac{1}{\lambda_b} = \frac{1}{\lambda_c} \quad \rightarrow \quad \frac{1}{0.532} + \frac{1}{0.6328} = \frac{1}{\lambda_c} \quad \rightarrow \quad \lambda_c = 0.289 \text{ μm},$$

$$n_e^2 = 2.1295 + \frac{0.0097}{\lambda^2 - 0.0014} + 0.7580 \frac{\lambda^2}{\lambda^2 - 127.0535} \quad \text{(with } \lambda \text{ in μm)},$$

$$n_o^2 = 2.2576 + \frac{0.0101}{\lambda^2 - 0.0142} + 1.7623 \frac{\lambda^2}{\lambda^2 - 57.8984}.$$

The wave-vector-matching condition gives the phase-match angle and $\hat{d}_{\text{eff}}^{(2)}$ as follows:

$$\frac{n_o(\omega_a)}{\lambda_a} + \frac{n_o(\omega_b)}{\lambda_b} = \frac{n_e(\omega_c, \theta)}{\lambda_c} \quad \rightarrow \quad \frac{n_e(\omega_c, \theta)}{0.289} = 5.27407 \quad \rightarrow \quad n_e(\omega_c, \theta) = 1.5224,$$

$$n_e(\omega_c, \theta) = \frac{n_o(\omega_c) n_e(\omega_c)}{\sqrt{n_o^2(\omega_c)\sin^2\theta + n_e^2(\omega_c)\cos^2\theta}} \quad \rightarrow \quad \theta = 61.96°,$$

$$\hat{d}_{\text{eff}}^{(2)} = \hat{d}_{14}\sin\theta\sin(2\phi) \quad \rightarrow \quad \phi_{\text{opt}} = \pi/4, \quad \hat{d}_{14}^{(2)} = 0.39 \text{ pm/V} \quad \rightarrow \quad \hat{d}_{\text{eff}}^{(2)} = 0.34 \text{ pm/V}.$$

6.2 MANLEY–ROWE RELATIONS

Energy is conserved in all the processes discussed in this chapter, as explained in Chapter 2. The analysis here follows that established for SHG in Chapter 2. Multiplying Eq. 6.1 for ω_a by $\frac{1}{2}n_a\varepsilon_0 c \mathcal{E}_a^*$ and adding the complex conjugate gives

$$\frac{1}{\omega_a}\frac{dI_a}{dz} = \frac{1}{2}i\varepsilon_0 \tilde{d}_{\text{eff}}^{(2)} \mathcal{E}_c \mathcal{E}_b^* \mathcal{E}_a^* e^{-i\Delta kz} + \text{c.c.} \tag{6.3a}$$

Similarly, for the frequencies ω_b and ω_c, we obtain

$$\frac{1}{\omega_b}\frac{dI_b}{dz} = \frac{1}{2}i\varepsilon_0\tilde{d}_{\text{eff}}^{(2)}\mathcal{E}_c\mathcal{E}_b^*\mathcal{E}_a^* e^{-i\Delta kz} + \text{c.c.} \tag{6.3b}$$

and

$$\frac{1}{\omega_c}\frac{dI_c}{dz} = -\frac{1}{2}i\varepsilon_0\tilde{d}_{\text{eff}}^{(2)}\mathcal{E}_c\mathcal{E}_b^*\mathcal{E}_a^* e^{-i\Delta kz} + \text{c.c.} \tag{6.3c}$$

Since the right-hand sides of all the three equations are identical (except for a minus sign), we obtain

$$-\frac{1}{\hbar\omega_a}\frac{dI_a}{dz} = -\frac{1}{\hbar\omega_b}\frac{dI_b}{dz} = \frac{1}{\hbar\omega_c}\frac{dI_c}{dz}. \tag{6.4}$$

Defining the photon flux in the usual way by $N_i = I_i/\hbar\omega_i$ gives

$$\frac{dN_c}{dz} = -\frac{dN_b}{dz} = -\frac{dN_a}{dz} \quad \rightarrow \quad \frac{d}{dz}\left[N_c + \frac{1}{2}(N_a + N_b)\right] = 0. \tag{6.5}$$

For sum frequency case, when a ω_c photon is created, it is at the expense of both a ω_a and a ω_b photon, which are annihilated in the process. Furthermore, from Eq. 6.4, we obtain

$$\frac{d\{I_c + I_a + I_b\}}{dz} = \frac{-\omega_c + \omega_a + \omega_b}{\omega_a}\frac{dI_a}{dz} = 0 \quad \rightarrow \quad \frac{d\{I(z,\omega_c) + I(z,\omega_a) + I(z,\omega_b)\}}{dz} = 0, \tag{6.6}$$

which shows that the intensity, and thus power, is conserved in these three-wave-mixing interactions.

6.3 SUM FREQUENCY GENERATION

6.3.1 Low Depletion Limit with Wave-Vector Mismatch

One of the early applications of SFG was to translate a signal from the infrared region into the visible region, where more sensitive and less noisy detectors are available (see Fig. 6.4).

Although the general solutions for $\Delta k = 0$ are given in terms of the Jacobi elliptic function solutions (see Chapter 4), these solutions are not transparent for negligible pump depletion or the inclusion of wave-vector mismatch. Hence, it is important to

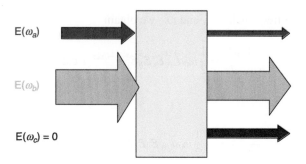

$E(\omega_a)$

$E(\omega_b)$

$E(\omega_c) = 0$

FIGURE 6.4 Up-conversion of a weak signal at ω_a to a higher frequency signal at ω_c.

consider this case separately. For negligible pump depletion at ω_b and $\Delta k \neq 0$ with $\tilde{\kappa}_c = \omega_c \tilde{d}_{\text{eff}}^{(2)}/cn_c$ and $\tilde{\kappa}_a = \omega_a \tilde{d}_{\text{eff}}^{(2)}/cn_a$, the coupled wave equations reduce to

$$\frac{d\mathcal{E}_c(z)}{dz} = i\tilde{\kappa}_c \mathcal{E}_a(z)\mathcal{E}_b(0)\,e^{i\Delta kz}, \qquad \frac{d\mathcal{E}_a(z)}{dz} = i\tilde{\kappa}_c \mathcal{E}_c(z)\mathcal{E}_b^*(0)\,e^{-i\Delta kz}. \tag{6.7}$$

Taking the derivative d/dz of both equations in 6.7, reducing each to a single variable, and finally solving gives

$$\mathcal{E}_c(z) = A\sin(\Gamma z) + B\cos(\Gamma z) \quad \text{and} \quad \mathcal{E}_a(z) = C\sin(\Gamma z) + D\cos(\Gamma z),$$

with $\Gamma = \sqrt{\tilde{\kappa}_a \tilde{\kappa}_c |\mathcal{E}_b(0)|^2}$. $\tag{6.8}$

Applying the boundary conditions gives the fields and intensities as

$$\mathcal{E}(z,\omega_a) = \mathcal{E}(0,\omega_a)\cos(\Gamma z), \qquad \mathcal{E}(z,\omega_c) = i\sqrt{\frac{\tilde{\kappa}_c}{\tilde{\kappa}_a}}\mathcal{E}(0,\omega_a)\sin(\Gamma z),$$

$$I(z,\omega_a) = I(0,\omega_a)\cos^2(\Gamma z), \qquad I(z,\omega_c) = \frac{n_c\tilde{\kappa}_c}{n_a\tilde{\kappa}_a}I(0,\omega_a)\sin^2(\Gamma z). \tag{6.9}$$

For low conversion efficiency of the signal, $\frac{\pi}{2} \gg \Gamma z \to \sin(\Gamma z) \approx \Gamma z$,

$$\frac{I(z,\omega_c)}{I(0,\omega_a)} = \frac{n_c\tilde{\kappa}_c}{n_a\tilde{\kappa}_a}\Gamma^2 z^2 \propto I(0,\omega_b). \tag{6.10}$$

6.3.2 Strong Interaction Limit with Depletion

Equations for this case have already been derived in Chapter 4 (Eq. 4.36) for $\Delta k = 0$. A sample calculation for the normalized photon fluxes is shown in Fig. 6.5, which is

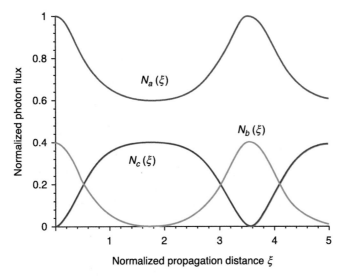

FIGURE 6.5 Sample calculation of the evolution of the normalized photon fluxes for sum frequency generation with initial conditions $N_a = 1.0$ and $N_b = 0.4$.

similar to the one seen previously for strong harmonic generation. The conversion into ω_c photons takes place up to the point where the ω_b wave is fully depleted, and at that point the process is reversed and energy converts back to the two input waves. Note that in this case the photon flux is *not* conserved.

6.4 OPTICAL PARAMETRIC AMPLIFIERS

The beam geometry and inputs for an OPA are shown in Fig. 6.6a. In OPA, the idler is generated along with the amplified signal; i.e., the idler is a by-product of the breakup of the pump photons.

6.4.1 Undepleted Pump Approximation $(\mathcal{E}(z, \omega_c) = \text{constant})$

In this case it is assumed that $I(\omega_c) \gg I(\omega_a)$, $I(\omega_b)$ and that $I(\omega_c)$ is undepleted. The weak "signal" beam at frequency ω_a is amplified, an "idler" beam at ω_b is generated, and energy $\hbar\omega_c = \hbar\omega_a + \hbar\omega_b$ is conserved. Furthermore, $k_c \cong k_a + k_b$ and so the wave-vector mismatch is small. With these approximations, the pertinent equations can be written as

$$\frac{d\mathcal{E}(z, \omega_a)}{dz} = i\frac{\omega_a}{n(\omega_a)c}\tilde{d}_{\text{eff}}^{(2)}\mathcal{E}(0, \omega_c)\mathcal{E}^*(z, \omega_b)\,e^{-i\Delta kz},$$

$$\frac{d\mathcal{E}(z, \omega_b)}{dz} = i\frac{\omega_b}{n(\omega_b)c}\tilde{d}_{\text{eff}}^{(2)}\mathcal{E}(0, \omega_c)\mathcal{E}^*(z, \omega_a)\,e^{-i\Delta kz}.$$

$$(6.11)$$

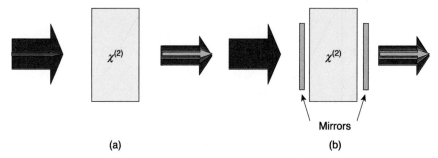

FIGURE 6.6 (a) An optical parametric amplifier in which the input signal frequency is amplified and an idler wave is generated by the breakup of pump beam photons. (b) An optical parametric oscillator in which only the pump beam is incident and both a signal and a idler wave is generated. The process is initiated by noise.

Defining $\tilde{\kappa}_a = \omega_a \tilde{d}_{\text{eff}}^{(2)}/n_a c \mathcal{E}(0, \omega_c)$ and $\tilde{\kappa}_b = \omega_b \tilde{d}_{\text{eff}}^{(2)}/n_b c \mathcal{E}(0, \omega_c)$ and substituting $\mathcal{E}(z, \omega_i) = \bar{\mathcal{E}}(z, \omega_i) e^{-i(\Delta k z)/2}$, Eqs 6.11 become

$$\frac{d\bar{\mathcal{E}}(z, \omega_a)}{dz} = i\frac{\Delta k}{2}\bar{\mathcal{E}}(z, \omega_a) + i\tilde{\kappa}_a \bar{\mathcal{E}}^*(z, \omega_b), \tag{6.12a}$$

$$\frac{d\bar{\mathcal{E}}(z, \omega_b)}{dz} = i\frac{\Delta k}{2}\bar{\mathcal{E}}(z, \omega_b) + i\tilde{\kappa}_b \bar{\mathcal{E}}^*(z, \omega_a). \tag{6.12b}$$

Taking the derivative d/dz of both equations, simplifying, and finally substituting for $d\bar{\mathcal{E}}(z, \omega_a)/dz$ and $d\bar{\mathcal{E}}^*(z, \omega_b)/dz$ from Eqs 6.12 gives the simple differential equation for $\bar{\mathcal{E}}(z, \omega_a)$:

$$\frac{d^2\bar{\mathcal{E}}(z, \omega_a)}{dz^2} = \left\{ \tilde{\kappa}_a \tilde{\kappa}_b - \frac{\Delta k^2}{4} \right\}\bar{\mathcal{E}}(z, \omega_a). \tag{6.13}$$

The solutions to this equation are well known, namely, $e^{\pm \Gamma z}$ with $\Gamma^2 = \tilde{\kappa}_a \tilde{\kappa}_b - \Delta k^2/4$. Defining

$$\frac{1}{\ell_{\text{OPA}}} = \sqrt{\tilde{\kappa}_a \tilde{\kappa}_b} = \sqrt{\frac{\omega_a \omega_b}{n_a n_b c^2}}\tilde{d}_{\text{eff}}^{(2)}\mathcal{E}(0, \omega_c) = \sqrt{\frac{4\pi^2}{n_a n_b \lambda_{\text{vac}}(\omega_a)\lambda_{\text{vac}}(\omega_b)}}\tilde{d}_{\text{eff}}^{(2)}\mathcal{E}(0, \omega_c)$$

$$\tag{6.14}$$

gives

$$\Gamma(\text{in cm}^{-1}) = \frac{1}{\ell_{\text{OPA}}(\text{in cm})} = 0.1726\frac{\tilde{d}_{\text{eff}}^{(2)}(\text{in pm/V})I_p^{1/2}(\text{in MW/cm}^2)}{\sqrt{n_a n_b n_p}\lambda_a(\text{in }\mu\text{m})\lambda_b(\text{in }\mu\text{m})}. \tag{6.15}$$

Clearly, the functional behavior of the solutions will depend on the sign of Γ^2. The behavior near and on the wave-vector match with $\Gamma^2 > 0$ shows an exponential growth. Therefore,

$$\frac{1}{\ell_{\text{OPA}}^2} - \frac{\Delta k^2}{4} > 0 \quad \rightarrow \quad \Gamma = \pm\sqrt{\frac{1}{\ell_{\text{OPA}}^2} - \frac{\Delta k^2}{4}} \tag{6.16}$$

$$\rightarrow \quad \mathcal{E}(z, \omega_a) = \left[A_+ \, e^{\Gamma z} + A_- \, e^{-\Gamma z} \right] e^{-i(\Delta k z)/2}.$$

Using the input boundary condition $\mathcal{E}(0, \omega_a) = A_+ + A_-$ and the boundary condition $\mathcal{E}(0, \omega_b) = 0$ for the idler, we obtain

$$\mathcal{E}(z, \omega_a) = \mathcal{E}(0, \omega_a) \left[\cosh(\Gamma z) + i\frac{\Delta k}{2\Gamma} \sinh(\Gamma z) \right] e^{-i(\Delta k z)/2}, \tag{6.17a}$$

$$\mathcal{E}(z, \omega_b) = i\tilde{\kappa}_b \mathcal{E}^*(0, \omega_a) \frac{\sinh(\Gamma z)}{\Gamma} e^{-i(\Delta k z)/2}. \tag{6.17b}$$

The gain is the largest on the wave-vector match. The field evolution (see Fig. 6.7) is given by

$$\mathcal{E}(z, \omega_a) = \mathcal{E}(0, \omega_a) \cosh\left(\frac{z}{\ell_{\text{OPA}}}\right), \qquad \mathcal{E}(z, \omega_b) = i\sqrt{\frac{n_a \omega_b}{n_b \omega_a}} \mathcal{E}^*(0, \omega_a) \sinh\left(\frac{z}{\ell_{\text{OPA}}}\right).$$
$$\tag{6.18}$$

Both fields grow exponentially (in the nondepletion pump regime) when $z \gg \ell_{\text{OPA}}$, with field and intensity gain coefficients given by ℓ_{OPA}^{-1} and $2\ell_{\text{OPA}}^{-1}$, respectively.

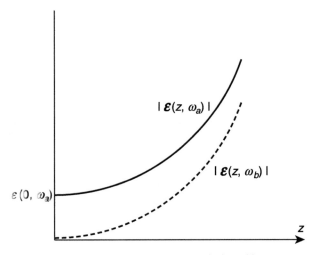

FIGURE 6.7 The evolution of the amplified signal $|\mathcal{E}(z, \omega_a)|$ and the generated idler $|\mathcal{E}(z, \omega_b)|$.

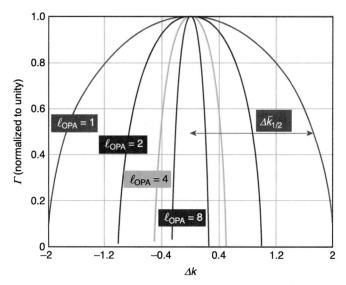

FIGURE 6.8 The parametric gain coefficient normalized to unity at $\Delta k = 0$ versus detuning from the $\Delta k = 0$ condition for different values of ℓ_{OPA}. $2\Delta\bar{k}_{1/2}$ is the gain coefficient bandwidth.

The gain coefficient versus wave-vector detuning when $\Delta k = 0$ is shown in Fig. 6.8. Clearly, the larger the ℓ_{OPA}, the larger the range of wave-vector detuning for which gain is achieved, i.e., the broader the gain curve. The detuning at which the gain coefficient (not the gain) drops to half of its peak value is defined as $\Delta\bar{k}_{1/2} = \sqrt{3}/\ell_{\text{OPA}}$. For $L \gg \ell_{\text{OPA}} \rightarrow \Delta\bar{k}_{1/2}L = \sqrt{3}L/\ell_{\text{OPA}} \gg 1$, the signal grows so fast over a coherence length that the usual destructive interference over a coherence length is ineffective.

The signal gain is defined as

$$G(L, \omega_a) = \frac{I(L, \omega_a) - I(0, \omega_a)}{I(0, \omega_a)} = \left(1 + \frac{\Delta k^2}{4\Gamma^2}\right) \sinh^2(\Gamma L). \qquad (6.19)$$

Its variation with detuning will be shown in Fig. 6.10. The full-signal bandwidth $2\Delta\bar{\bar{k}}_{1/2}$, i.e., the detuning at which the signal *gain* falls to half of its peak value, can be estimated only approximately without resorting to numerical calculations.

The solutions for $\Delta k^2 > 4\tilde{\kappa}_a\tilde{\kappa}_b$ with Γ^2 redefined as $\Gamma^2 = \Delta k^2/4 - \tilde{\kappa}_a\tilde{\kappa}_b$ are oscillatory, with fields given by

$$\mathcal{E}(z, \omega_a) = \mathcal{E}(0, \omega_a)\left[\cos(\Gamma z) + i\frac{\Delta k}{2\Gamma}\sin(\Gamma z)\right]e^{-i(\Delta kz)/2},$$

$$\mathcal{E}(z, \omega_b) = i\tilde{\kappa}_b\mathcal{E}^*(0, \omega_a)\frac{\sin(\Gamma z)}{\Gamma}e^{-i(\Delta kz)/2}. \qquad (6.20)$$

The field evolution for this case is shown in Fig. 6.9.

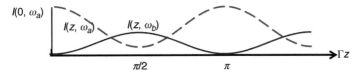

FIGURE 6.9 The evolution of the signal and idler intensities in the limit $\Delta k^2 > 4\tilde{\kappa}_a\tilde{\kappa}_b$ and $\omega_a > \omega_b$.

The actual signal gain for this regime is defined as

$$G(L, \omega_a) = \frac{I(L, \omega_a) - I(0, \omega_a)}{I(0, \omega_a)} = \left(1 + \frac{\Delta k^2}{4\Gamma^2}\right)\sin^2(\Gamma L). \qquad (6.21)$$

This is plotted in Fig. 6.10 for both regimes, i.e., $\Delta k^2 > 4\tilde{\kappa}_a\tilde{\kappa}_b$ and $4\tilde{\kappa}_a\tilde{\kappa}_b > \Delta k^2$. For large L/ℓ_{OPA}, the oscillations still exist but are too small to be seen on this scale. The "negative gain" oscillations occur because for $\tilde{\kappa}_a\tilde{\kappa}_b < \Delta k^2/4$ the signal solution is a sine function and the signal input energy is shared with the idler that is generated. The signal bandwidth increases with decreasing ℓ_{OPA}, as expected from the behavior of the gain coefficient versus detuning.

A numerical sample is used to illustrate an approximately degenerate OPA based on LiNbO$_3$. It is almost birefringent wave-vector matched for SHG at $\lambda = 1.06\,\mu\mathrm{m}$, and so the pump will be at $\lambda = 0.53\,\mu\mathrm{m}$ and the signal and the idler will have almost equal wavelengths detuned slightly from $\lambda = 1.06\,\mu\mathrm{m}$, i.e., at $\lambda = (1.06 \pm \Delta\lambda)\,\mu\mathrm{m}$.

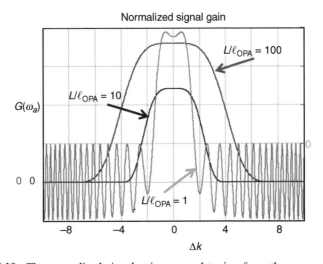

FIGURE 6.10 The normalized signal gain versus detuning from the wave-vector match ($\Delta k = 0$) for three different values of L/ℓ_{OPA}. Note that the (green) zero gain point on the vertical axis for the $L/\ell_{\mathrm{OPA}} = 1$ case (right-hand side) is different from that for the $L/\ell_{\mathrm{OPA}} = 10$ and 100 cases (left-hand side).

The pertinent parameters are $n_e(\omega_c) = 2.227$, $n_o(\omega_a) \cong n_o(\omega_b) = 2.234$, $\tilde{d}_{31}^{(2)} = 5.95 \text{ pm/V}$, $L = 1 \text{ cm}$, and $I(0, \omega_c) = 45 \text{ MW/cm}^2$. Thus,

$$\frac{1}{\ell_{OPA}(\text{in cm})} = \frac{0.172 \tilde{d}_{eff}^{(2)}(\text{in pm/V})}{\sqrt{n_a n_b n_c \lambda_a (\text{in } \mu\text{m}) \lambda_b (\text{in } \mu\text{m})}} \sqrt{I(0, \omega_c)(\text{in MW/cm}^2)} = 1.92 \text{ cm}^{-1},$$

$$\Delta k = 0 \quad \rightarrow \quad G(L, \omega_a) = \sinh^2(\Gamma L) = \frac{1}{4}(e^{1.92} - e^{-1.92})^2 = 1.7.$$

6.4.2 Strong Interaction Limit

The form of the exact solutions for sum frequency in Eqs 4.36 is generic to three-wave interactions and is also valid for OPAs (1). The boundary conditions are different for OPAs since here the inputs are $N_c(0) \neq 0$, $N_a(0) \neq 0$, and $N_b(0) = 0$. Normally $N_c(0) \gg N_a(0)$, which is not necessary for the general solution but is obviously useful for an effective amplifier. With these boundary conditions, the exact OPA solution for $N_c(\xi)$, $N_b(\xi)$, and $N_a(\xi)$ follow from the relations that describe the physics of the process, i.e., for every ω_c photon annihilated, one photon at ω_a and one at ω_b are created. This translates into the equations

$$\begin{aligned}
N_a(\xi) &= N_a(0) + [N_c(0) - N_c(\xi)], \\
N_b(\xi) &= N_c(0) - N_c(\xi),
\end{aligned} \tag{6.22}$$

which indicate that the key term is $N_c(\xi)$. The solution after the usual tedious mathematics is (1)

$$N_c(\xi) = N_c(0)\left[1 - \text{sn}^2\left(N_c(0)[\xi - \xi_0], \sqrt{\frac{N_c(0)}{N_c(0) + N_a(0)}}\right)\right], \tag{6.23a}$$

$$N_a(\xi) = N_a(0) + N_c(0)\left[1 - \text{sn}^2\left(N_c(0)[\xi - \xi_0], \sqrt{\frac{N_c(0)}{N_c(0) + N_a(0)}}\right)\right], \tag{6.23b}$$

$$N_b(\xi) = N_c(0)\left[1 - \text{sn}^2\left(N_c(0)[\xi - \xi_0], \sqrt{\frac{N_c(0)}{N_c(0) + N_a(0)}}\right)\right]. \tag{6.23c}$$

In this case, the constant of integration ξ_0 is needed to provide the offset (half the period of the sn^2 function) required to ensure that the solution for the pump beam starts at unity.

An example of the evolution of the photon fluxes is shown in Fig. 6.11 for the initial conditions $N_c(0) = 1$, $N_a(0) = 0.001$, and $\tilde{\kappa} = 1$. This amplifier response is periodic in distance and pump power and there is no limiting saturation as with electronic amplifiers. The idler growth is exponential until a significant depletion of the pump wave occurs at which time it peaks and then decreases again. For the

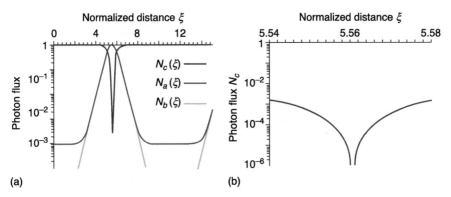

FIGURE 6.11   (a) The evolution of the photon fluxes with propagation distance. (b) The enhancement of the pump photon flux in the vicinity of complete pump depletion.

signal wave, the initial growth with the propagation distance is subexponential until its photon flux is approximately equal to that of the idler photon flux. Essentially, complete pump depletion can occur (limited in this graph by the finite step size used for ξ).

6.5 OPTICAL PARAMETRIC OSCILLATOR

OPOs are very common in most laboratories that rely on tunable sources (1). They can be either pulsed wave or continuous wave and are commercially available worldwide. Their schematic geometries are shown in Fig. 6.6b. The enormous flexibility of OPOs is shown in Fig. 6.12. A single OPO material can span the wavelength range from 270 to 2400 nm, replacing the need for multiple lasers.

For an OPO, the gain medium ($\chi^{(2)}$ active) is usually placed in a cavity whose mirrors have high reflectivity at the signal frequency (singly resonant cavity) or at both the signal and idler frequencies (doubly resonant cavity). This multipass approach allows better pump depletion. Normally, the cavity is not resonant at the pump frequency because this requires sophisticated stabilization techniques (which have been developed for low power continuous-wave pump beams). The resonant geometry is shown in Fig. 6.13.

Initially in an OPO, "noise" at the frequencies ω_a and ω_b provides the "seed" to Eqs 6.1, which is amplified by interaction with the beam at ω_c. A complete analysis of the noise and the initial evolution is beyond the scope of this textbook. Instead, the *steady state* after all the cavity "turn-on" oscillations have died down will be discussed. It will be assumed that the medium in the cavity is lossless due to absorption, and the only loss will be leakage through the mirrors (necessary to get light into and out of the cavity). Furthermore, the analysis will be in the negligible pump depletion limit. (The more general solutions on wave-vector match simply require Eqs 6.23, already discussed for the OPA. However, for $\Delta k \neq 0$, the analysis requires detailed numerical calculations.)

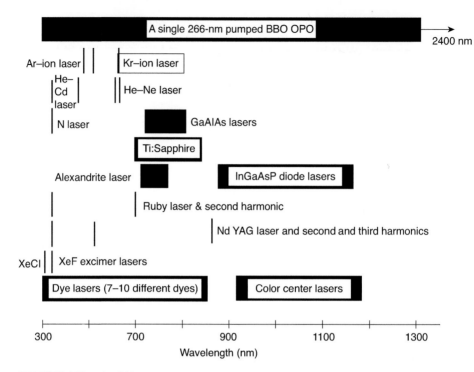

FIGURE 6.12   The different lasers whose wavelengths are spanned by a single barium borate optical parametric oscillator. Reproduced with permission from Harwood Publishers (2).

6.5.1   Doubly Resonant Cavity

In this case, mirrors are coated for high reflectivity for the fields at ω_a (input R_a and output R'_a, typically 90–99%) and ω_b (input R_b and output R'_b, again typically 90–99%). With such high reflectivity, the cavity modes will play an important role at these frequencies. Note that as indicated in Fig. 6.14a, the idler (ω_b) and signal (ω_a) beams experience gain in only one direction because only in the forward direction the total wave vector can be matched; i.e., $\Delta \vec{k} = \vec{k}_c - \vec{k}_a - \vec{k}_b \cong 0$. However, in the backward direction, $\Delta \vec{k} = \vec{k}_c - \vec{k}_a - \vec{k}_b \approx 2\vec{k}_c$.

After one forward pass in the cavity (Fig. 6.14b), the growth of field a due to the interaction of the "incident" field a with the pump field c and due to the interaction of the incident b field with the pump field c results in the fields at the end of the cavity given by

$$\mathcal{E}(L, \omega_a) = \mathcal{E}(0, \omega_a) \left[\cosh(\Gamma L) + i \frac{\Delta k}{2\Gamma} \sinh(\Gamma L) \right] e^{-i(\Delta k L/2)},$$

$$\mathcal{E}(L, \omega_a) = i\tilde{\kappa}_a \mathcal{E}^*(0, \omega_b) \frac{\sinh(\Gamma L)}{\Gamma} e^{-i(\Delta k L/2)},$$

(6.24)

FIGURE 6.13   The cavity geometry for a singly or doubly resonant cavity.

respectively. Similarly, the growth of field b due to the interaction of the incident field b with the pump field c and due to the interaction of the incident a field with the pump field c results in

$$\mathcal{E}(L, \omega_b) = \mathcal{E}(0, \omega_b) \left[\cosh(\Gamma L) + i \frac{\Delta k}{2\Gamma} \sinh(\Gamma L) \right] e^{-i(\Delta k L/2)},$$

$$\mathcal{E}(L, \omega_b) = i\tilde{\kappa}_b \mathcal{E}^*(0, \omega_a) \frac{\sinh(\Gamma L)}{\Gamma} e^{-i(\Delta k L/2)},$$

(6.25)

respectively. Summing the terms for both $\mathcal{E}(L, \omega_a)$ and $\mathcal{E}(L, \omega_b)$ gives

$$\mathcal{E}(L, \omega_a) = \left\{ \left[\cosh \Gamma L + i \frac{\Delta k}{2\Gamma} \sinh \Gamma L \right] \mathcal{E}(0, \omega_a) + i \frac{\tilde{\kappa}_a}{\Gamma} \sinh(\Gamma L) \mathcal{E}^*(0, \omega_b) \right\} e^{-i(\Delta k L/2)},$$

(6.26a)

$$\mathcal{E}^*(L, \omega_b) = \left\{ -i \frac{\tilde{\kappa}_b}{\Gamma} \sinh(\Gamma L) \mathcal{E}(0, \omega_a) + \left[\cosh \Gamma L - i \frac{\Delta k}{2\Gamma} \sinh \Gamma L \right] \mathcal{E}^*(0, \omega_b) \right\} e^{i(\Delta k L/2)}.$$

(6.26b)

For a round trip, each beam accumulates a linear phase on both forward and backward propagation. In the absence of the interactions, the linear phase shift terms are $2ik_a L$ for $\mathcal{E}(2L, \omega_a)$ and $2ik_b L$ for $\mathcal{E}(2L, \omega_b)$. Interactions introduce additional phase shifts given by $\pm i \Delta k L/2$. Both the signal and the idler waves also experience transmission losses at the mirrors given by $1 - \sqrt{R_a R_a'}$ and $1 - \sqrt{R_b R_b'}$. Therefore, in the steady state after one round trip,

$$\mathcal{E}(2L \equiv z = 0, \omega_a) = \mathcal{E}(L, \omega_a) e^{2ik_a L} \sqrt{R_a R_a'},$$
$$\mathcal{E}(2L \equiv z = 0, \omega_b) = \mathcal{E}(L, \omega_b) e^{2ik_b L} \sqrt{R_b R_b'}.$$

(6.27)

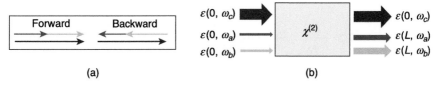

FIGURE 6.14   (a) Wave-vector matching in the forward direction and the huge wave-vector mismatch in the backward direction. (b) Fields assumed incident on the crystal and output from the crystal in the steady state.

This can be written in the matrix form as

$$
\begin{bmatrix} \mathcal{E}(2L,\omega_a) \\ \mathcal{E}^*(2L,\omega_b) \end{bmatrix} = \vec{\vec{M}} \begin{bmatrix} \mathcal{E}(0,\omega_a) \\ \mathcal{E}^*(0,\omega_b) \end{bmatrix} \xrightarrow{\text{steady state}} \begin{bmatrix} \mathcal{E}(0,\omega_a) \\ \mathcal{E}^*(0,\omega_b) \end{bmatrix} = \vec{\vec{M}} \begin{bmatrix} \mathcal{E}(0,\omega_a) \\ \mathcal{E}^*(0,\omega_b) \end{bmatrix}. \quad (6.28)
$$

The transfer matrix $\vec{\vec{M}}$ on wave vector is given by

$$
\vec{\vec{M}} = \begin{bmatrix} \sqrt{R_a R_a'} \cosh\dfrac{L}{\ell_{\text{OPA}}} e^{2ik_a L} & i\sqrt{R_a R_a'} \tilde{\kappa}_a \ell_{\text{OPA}} \sinh\dfrac{L}{\ell_{\text{OPA}}} e^{2ik_a L} \\[4mm] -i\sqrt{R_b R_b'} \tilde{\kappa}_b \ell_{\text{OPA}} \sinh\dfrac{L}{\ell_{\text{OPA}}} e^{-2ik_b L} & \sqrt{R_b R_b'} \cosh\dfrac{L}{\ell_{\text{OPA}}} e^{-2ik_b L} \end{bmatrix}. \quad (6.29)
$$

The steady-state case given in Eq. 6.28 requires that $\det\left|\vec{\vec{M}} - \vec{\vec{I}}\right| = 0$, with $\vec{\vec{I}} = \begin{bmatrix} 1 & 0 \\ 0 & 1 \end{bmatrix}$.
Assuming for simplicity $R_a = R_a'$ and $R_b = R_b'$ and using $\tilde{\kappa}_a \tilde{\kappa}_b = \ell_{\text{OPA}}^{-2}$, $\det\left|\vec{\vec{M}} - \vec{\vec{I}}\right| = 0$ gives

$$
\left(R_a \cosh\left[\frac{L}{\ell_{\text{OPA}}}\right] e^{2ik_a L} - 1 \right)\left(R_b \cosh\left[\frac{L}{\ell_{\text{OPA}}}\right] e^{-2ik_b L} - 1 \right)
$$
$$
- R_a R_b \sinh^2\left[\frac{L}{\ell_{\text{OPA}}}\right] e^{2i(k_a - k_b)L} = 0. \quad (6.30)
$$

For minimum gain threshold, all terms must be real; i.e., $2k_a L = 2m_a\pi$ and $2k_b L = 2m_b\pi$, with m_a and m_b being integers. These two conditions require that the signal and idler beams are both standing waves of the cavity. This reduces Eq. 6.30 to

$$
(R_a + R_b)\cosh\left(\frac{L}{\ell_{\text{OPA}}}\right) - 1 = R_a R_b \quad \rightarrow \quad \cosh\left(\frac{L}{\ell_{\text{OPA}}}\right) = \frac{1 + R_a R_b}{R_a + R_b}, \quad (6.31)
$$

which is now the minimum threshold condition. A detuned cavity will have a larger threshold. For the case that the gain is small per pass,

$$
1 \gg \frac{L}{\ell_{\text{OPA}}} \quad \rightarrow \quad \cosh\left(\frac{L}{\ell_{\text{OPA}}}\right) \cong 1 + \frac{1}{2}\left(\frac{L}{\ell_{\text{OPA}}}\right)^2, \quad (6.32)
$$

$$
\rightarrow \quad \frac{L}{\ell_{\text{OPA}}} \cong \sqrt{\frac{2(1 - R_a)(1 - R_b)}{R_a + R_b}} \quad \text{for } R_a \cong R_b \cong 1 \quad \rightarrow \quad \frac{L}{\ell_{\text{OPA}}} \cong \sqrt{(1 - R_a)(1 - R_b)}. \quad (6.33)
$$

Substituting for ℓ_{OPA} from Eq. 6.14 gives

$$\frac{8\pi^2\left[\tilde{d}_{eff}^{(2)}L\right]^2}{n_a n_b n_c \lambda_a \lambda_b c \varepsilon_0}I(0,\omega_c)=(1-R_a)(1-R_b)$$

$$\rightarrow I_{th}(0,\omega_c)=\frac{n_a n_b n_c \lambda_a \lambda_b \varepsilon_0 c}{8\pi^2 L^2\left|\tilde{d}_{eff}^{(2)}\right|^2}(1-R_a)(1-R_b)$$

(6.34)

for the threshold intensity. A higher input intensity gives a positive gain for the signal and the idler. A simple formula for $I_{th}(0,\omega_c)$ in the usual units used is

$$I_{th}(0,\omega_c)(\text{in MW/cm}^2)=33.6\frac{n_a n_b n_c \lambda_a(\text{in }\mu\text{m})\lambda_b(\text{in }\mu\text{m})}{[L(\text{in cm})]^2\left[\tilde{d}_{eff}^{(2)}(\text{in pm/V})\right]^2}(1-R_a)(1-R_b).$$

(6.35)

For gain per pass, which is not small, it is necessary to solve Eq. 6.31 to obtain ℓ_{OPA}.

In Section 6.4.1, Type 1 birefringent wave-vector-matched LiNbO$_3$ with $\tilde{d}_{31}^{(2)}=$ 5.95 pm/V, $L=1$ cm, $\lambda_c=0.53\,\mu\text{m}$, $\lambda_a=\lambda_b=1.06\,\mu\text{m}$ (near degeneracy), and $n_a\cong n_b\cong n_c\cong 2.24$ was used as an example for an OPA calculation. The same parameters are used to calculate the threshold intensity for a 1-cm-long crystal in a cavity with mirror reflectivities $R_a=R_b=0.98$. Substituting these values into Eq. 6.35 gives $I_{th}(0,\omega_c)=4.8\,\text{kW/cm}^2$—quite a modest intensity.

An important issue in OPOs is the stability of the output frequency. Mechanical instabilities, such as vibrations, optical mount creep, and relaxation and thermal drift, cause cavity length changes and hence changes in the cavity mode frequencies and in turn changes in the OPO output frequency. The cavity resonant frequencies are

$$\omega_a = m_a\frac{c\pi}{n_a L} \quad \text{and} \quad \omega_b = m_b\frac{c\pi}{n_b L}.$$

(6.36)

A change in m_a and/or m_b causes changes in the mode frequencies by $\delta\omega_a$ and $\delta\omega_b$ given by

$$\Delta m_a = 1 \quad \rightarrow \quad \delta\omega_a = \frac{\pi c}{L n_a} \quad \text{and} \quad \Delta m_b = 1 \quad \rightarrow \quad \delta\omega_b = \frac{\pi c}{L n_b}.$$

(6.37)

The key question is: "How many of these cavity modes exist within the gain bandwidth?" The pump frequency does not "see" a cavity since the mirror reflectivity is small at that frequency. Since $\omega_c = \omega_a + \omega_b = $ constant, any change in $\omega_a \rightarrow \omega_a + \Delta\omega_a \Rightarrow \omega_b \rightarrow \omega_b - \Delta\omega_a$. Assuming that the pump frequency is a constant, i.e., it should be stabilized for demanding applications,

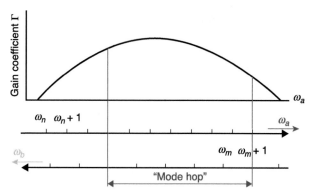

FIGURE 6.15   A typical gain curve for an optical parametric oscillator and the comb of cavity modes at frequencies ω_a and ω_b that exist under the gain curve. The arrows show the direction in which the combs move relative to the gain curve due to a change in frequency $\Delta\omega_a$. Reproduced with permission from John Wiley & Sons (3).

a wave-vector mismatch Δk of the order π/L due to a change in the cavity causes a change in frequency $\Delta\omega_a$ given by

$$\Delta k = \Delta(k_a + k_b - k_c) = \Delta\omega_a\left\{\frac{n_a}{c} - \frac{n_b}{c}\right\} = \frac{2\pi}{L} \quad\rightarrow\quad \Delta\omega_a = \frac{2c\pi}{L(n_a - n_b)}.$$

$$(6.38)$$

Since $n_a \gg n_a - n_b \rightarrow \Delta\omega_a \gg \delta\omega_a = \frac{\pi c}{Ln_a}$, normally many cavity modes exist within the gain curve (see Fig. 6.15).

The "operating point" for the OPO occurs at the frequency when the two sets of cavity modes coincide (at least approximately) at a larger gain than all the other coincidences. Note that when ω_a drifts up in frequency, ω_b drifts down in frequency—a consequence of ω_c being decoupled from the cavity. If the mirror separation (cavity length) or ω_c changes, the next operating point when the cavity modes coincide can cause a large shift (called a "mode hop") in the output frequency. Note that if the pump frequency ω_c changes, this moves the gain curve relative to the cavity frequencies and again a mode hop can occur.

6.5.2   Singly Resonant Cavity

For this case, the cavity is resonant at only one frequency, usually the desired signal (ω_a). This corresponds to mirror reflectivities $R_a \cong 1$ and $R_b \rightarrow 0$. The threshold condition is easily obtained by setting $R_b \rightarrow 0$ in Eq. 6.31 and expanding $\cosh(L/\ell_{OPA}) \cong 1 + (1/2)(L/\ell_{OPA})^2$ for the small gain per pass case. Thus,

$$\frac{L}{\ell_{OPA}} \cong \sqrt{\frac{2(1 - R_a)}{R_a}}.$$

$$(6.39)$$

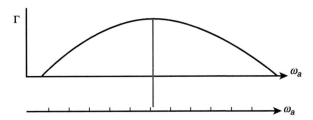

FIGURE 6.16 The comb of the longitudinal cavity modes and the gain curve of the optical parametric oscillator.

Substituting for ℓ_{OPA} from Eq. 6.14 gives

$$I_{th}(0, \omega_c) = \frac{n_a n_b n_c \varepsilon_0 c \lambda_a \lambda_b (1 - R_a)}{4\pi^2 L^2 \left(\tilde{d}_{eff}^{(2)}\right)^2 R_a}. \tag{6.40}$$

Comparing the thresholds for the singly resonant OPO (SRO) and doubly resonant OPO (DRO) gives

$$\frac{I_{th}(SRO)}{I_{th}(DRO)} = \frac{2}{R_a(1 - R_b)} \quad \text{for } R_b = 98\% \quad \rightarrow \quad \frac{I_{th}(SRO)}{I_{th}(DRO)} \cong 100. \tag{6.41}$$

As expected, the threshold for the singly resonant case is much larger than for the doubly resonant case.

On the other hand, the singly resonant OPO is inherently more stable against mechanical cavity changes than the doubly resonant one. This is clear from Fig. 6.16. The idler is now also decoupled from the cavity so that there is only one "comb" of cavity modes. Again, the OPO operates on the cavity mode, which experiences the largest gain. If the frequency ω_a changes, the OPO will jump only to the nearest cavity mode and the mode hop is much smaller than for the doubly resonant OPO.

6.5.3 OPO Power Output

At the threshold intensity, gain equals loss. The OPO output depends on the mirror transmission coefficients as just discussed. If $I(\omega_c) > I_{th}(\omega_c)$, the input photons in excess of the threshold are converted into signal and idler photons and the signal and idler outputs above this threshold are linear in the input intensity (see Fig. 6.17); i.e.,

$$\frac{I(\omega_c) - I_{th}(\omega_c)}{\hbar\omega_c} = \frac{I(\omega_a)}{\hbar\omega_a} = \frac{I(\omega_b)}{\hbar\omega_b} \quad \rightarrow \quad I(\omega_a) = \frac{\omega_a}{\omega_c}[I(\omega_c) - I_{th}(\omega_c)]. \tag{6.42}$$

The ratio ω_a/ω_c is called the "slope efficiency" and is always less than unity since $\omega_a < \omega_c$.

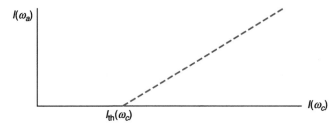

FIGURE 6.17   The signal output from the optical parametric oscillator versus the input pump beam intensity.

6.5.4 Frequency Tuning of OPOs

Just as in the frequency-doubling case, the wave-vector mismatch and hence the signal output frequency can be tuned by changing the angle of the input beam relative to the crystal or by changing the temperature or both. Of these two options, angle tuning usually produces a much larger range of output frequencies.

Consider the angle tuning of a uniaxial crystal in the geometry shown in Fig. 6.18. As discussed previously, both wave-vector matching and energy conservation must be satisfied in the OPO; i.e.,

$$\omega_c = \omega_a + \omega_b, \quad n_e(\omega_c, \theta)\omega_c = n_o(\omega_a)\omega_a + n_o(\omega_b)\omega_b, \tag{6.43}$$

with the pump frequency fixed. Solving over the full tuning range is a computer analysis problem of Eq. 6.43 (see, e.g., Fig. 6.19). When the angle θ is changed by a *small* amount $\Delta\theta$,

$$
\begin{aligned}
&\Delta(\omega_a + \omega_b) = 0 \quad \rightarrow \quad \Delta\omega_b = -\Delta\omega_a, \\
&n_c(\theta + \Delta\theta, \omega_c) \quad \rightarrow \quad n_c(\theta, \omega_c) + \Delta n_c(\theta), \\
&n_o(\omega_{a,b}) \quad \rightarrow \quad n_o(\omega_{a,b}) + \Delta n_{a,b} \\
&\rightarrow \quad \omega_c[n_c(\theta, \omega_c) + \Delta n_c(\theta)] = (\omega_a + \Delta\omega_a)(n_a + \Delta n_a) + (\omega_b - \Delta\omega_a)(n_b + \Delta n_b).
\end{aligned}
\tag{6.44}
$$

Neglecting the small $\Delta n \Delta \omega$ terms gives

$$\omega_c\Delta n_c(\theta) - \omega_a\Delta n_a - \omega_b\Delta n_b = (n_a - n_b)\Delta\omega_a \quad \rightarrow \quad \Delta\omega_a = \frac{\omega_c\Delta n_c(\theta) - \omega_a\Delta n_a - \omega_b\Delta n_b}{n_a - n_b}. \tag{6.45}$$

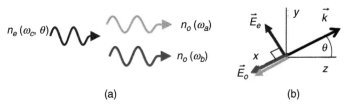

(a)   (b)

FIGURE 6.18   (a) Interacting waves for the optical parametric oscillator example. (b) Sample and field geometry.

Writing

$$\Delta n_a = \frac{\partial n_a}{\partial \omega}\bigg|_{\omega_a} \Delta \omega_a, \quad \Delta n_b = -\frac{\partial n_b}{\partial \omega}\bigg|_{\omega_b} \Delta \omega_a, \quad \Delta n_c = \frac{\partial n_c(\theta)}{\partial \theta}\bigg|_{\theta_{PM}} \Delta \theta$$

$$\rightarrow \quad \frac{\partial \omega_a}{\partial \theta} = \frac{\omega_c \dfrac{\partial n_c(\theta)}{\partial \theta}\bigg|_{\theta_{PM}}}{(n_a - n_b) + \omega_a \dfrac{\partial n_a}{\partial \omega_a} - \omega_b \dfrac{\partial n_b}{\partial \omega_b}}. \tag{6.46}$$

Substituting for $\dfrac{\partial n_c(\omega, \theta)}{\partial \theta}\bigg|_{\theta_{PM}}$ from Eq. 3.22 gives

$$\frac{\partial \omega_a}{\partial \theta}\bigg|_{\theta_{PM}} = -\frac{1}{2}\omega_c \sin(2\theta_{PM}) \frac{n_0(\omega_c)}{n_e^2(\omega_c, \theta_{PM})} \frac{n_0^2(\omega_c) - n_e^2(\omega_c, \theta_{PM})}{(n_a - n_b) + \omega_a(\partial n_a/\partial \omega_a) - \omega_b(\partial n_b/\partial \omega_b)}. \tag{6.47}$$

An example of an angle-tuned OPO is shown in Fig. 6.19.

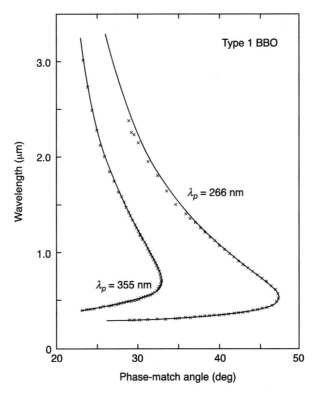

FIGURE 6.19 Response of an angle-tuned β-BaB$_2$O$_4$ parametric oscillator pumped at two different wavelengths λ_p: wavelength versus crystal angle. The wavelength at which the response curve is a maximum corresponds to the degeneracy wavelength at which the signal and idler wavelengths (and therefore frequencies) are equal. Reproduced with permission from the American Institute of Physics (4).

6.6 MID-INFRARED QUASI-PHASE MATCHING PARAMETRIC DEVICES

6.6.1 Multifrequency Outputs

Obtaining two or even three outputs simultaneously at different frequencies has a number of applications, most recently for frequency shifting into specific mid-infrared regions of the spectrum. This is a very important application of $LiNbO_3$, quasi-phase matching (QPM), and OPO devices up to 5 μm, although losses do start to become an issue beyond 4 μm (5).

A great deal of research and development has recently led to the availability of compact laser pump sources, such as Nd-based solid-state lasers (and their harmonics) or rare-earth-doped fiber lasers, which provide high peak powers when operated in pulsed mode or high average power when operated in continuous-wave or quasi-continuous-wave mode. Coupled with novel frequency conversion devices based on specially engineered, periodically poled, QPM crystals, such as $LiNbO_3$ or GaAs, tunable multiwavelength sources with useful powers in the mid-infrared region have now become a reality (5–8). To make practical multiwavelength devices, it is common either to use multiple nonlinear crystals inside a common resonator or to engineer the QPM gratings on a single crystal such that multiple wavelengths can be generated. These two cases are illustrated in Figs 6.20 and 6.21. The optical engineering required pushes the state of the art, and an example of each is discussed below in sufficient detail to highlight the optical engineering issues that must be considered (9).

Consider first the case of a multizone, periodically poled, single $LiNbO_3$ crystal (see Fig. 6.20). A fixed-frequency laser, such as a Nd-YAG or a Yb-doped fiber laser emitting near 1 μm (pump frequency ω_p), can be used as the pump source. Various applications require a wavelength converting device that must emit two discrete frequencies in the infrared region such that $\omega_p/2 > \omega_1 > \omega_2$ and maximum conversion efficiency such that $P_{\omega_2} > P_{\omega_1} \gg P_{\omega_p}$ *out* of the device; i.e., most of the laser power should be converted to ω_1 and ω_2. To efficiently create two frequencies that are lower in frequency than the degeneracy frequency $\omega_p/2$ of the 1-μm laser pump, it is desirable to first shift the 1-μm pump energy to a lower frequency. A convenient frequency of the "shifted" pump energy at ω_p' would be $\omega_{p'} > \omega_1 > \omega_{p'}/2 > \omega_2$. These conditions on the frequencies and the relative powers desired at each frequency can be implemented in a single $LiNbO_3$ crystal by poling two successive different regions with different periods (see the crystal in Fig. 6.20). Consider the first QPM

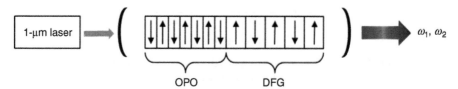

FIGURE 6.20 Laser-resonator-crystal geometry required for a two-wavelength mid-infrared source using one nonlinear crystal containing separate quasi-phase-matched zones.

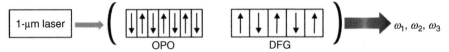

FIGURE 6.21   Laser-resonator-crystal geometry required for a three-wavelength mid-infrared source using two separate quasi-phase-matched nonlinear crystals.

zone whose primary function is to produce radiation at ω_p', which is designed to satisfy the wave-vector-matching criteria $k_p = k_p' + k_2 \pm K_{OPO}$, with the poling period ($\Lambda_{OPO} = 2\pi/K_{OPO}$) chosen so that the wave-vector-matching equation is satisfied. By convention, the first QPM zone is the OPO stage due to the resonating beam created at ω_p'. Since $P_{\omega_2} > P_{\omega_1}$ is required, some power is generated at ω_2 and used subsequently as a seed for the second nonlinear process. To create power at ω_1 and additional power at ω_2, the second QPM zone of the crystal can be designed such that $k_p' = k_1 + k_2 \pm K_{DFG}$, with the poling period ($\Lambda_{DFG} = 2\pi/K_{DFG}$) chosen so that the second wave-vector-matching equation is satisfied. Again by convention, the second zone is labeled the DFG zone due to the nonresonant beams created at ω_1 and ω_2.

To enhance the efficiency by using both forward and backward propagating beams, a cavity is designed and optimized for maximum energy extraction at ω_1 and ω_2. To maximize pump depletion, the strong beam at ω_p' should resonate inside the device cavity by using cavity mirror reflectivities $R(\omega_p') \cong 100\%$. To maximize energy extraction from the laser cavity at ω_1 and ω_2, the reflectivity of the output coupler of the resonator cavity need to be set such that $R(\omega_1, \omega_2) \cong 0\%$. To further take advantage of the backward propagating pump, the mirror reflectivities at the input of the resonator are chosen to be $R(\omega_1, \omega_2) \cong 100\%$. This design is attractive because of the inherent mechanical stability of using a single crystal.

For more than two simultaneous output frequencies, using multiple crystals is desirable due to the limited crystal lengths and lower damage thresholds available in

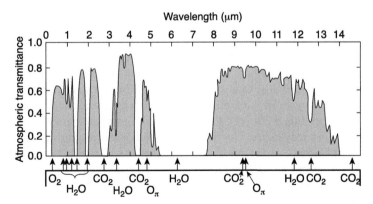

FIGURE 6.22   Atmospheric transmittance in the spectral range of 0.5–15 μm. The absorption signatures of various molecules can be seen. Reproduced from Wikipedia (11).

some materials. Consider again a fixed-frequency laser, such as a Nd-YAG or Yb-doped fiber laser emitting near 1 μm as the pump source. The wavelength-converting resonator should be designed to emit three discrete frequencies (see Fig. 6.21) such that $\omega_p/2 > \omega_1 > \omega_2 > \omega_3$ and maximize conversion efficiency such that $P_{\omega_3}, P_{\omega_1}, P_{\omega_2} \gg P_{\omega_p}$ at the output; i.e., the device needs to be efficient. In this scenario, it is desirable to use two separate nonlinear crystals with QPM gratings, although, in principle, an extension of the previous example can be implemented onto a single crystal with multiple QPM zones. To efficiently create three frequencies that are lower in frequency than the degeneracy frequency $\omega_p/2$ of the 1-μm pump, it is again useful to shift the 1-μm pump energy to a lower frequency. A convenient frequency of the shifted pump energy ω_p' would be $\omega_{p'} > \omega_1 > \omega_{p'}/2 > \omega_2$ or ω_3. The first QPM crystal is designed to satisfy the wave-vector-matching criteria $k_p = k_p' + k_2 \pm \mathrm{K_{OPO}}$ to generate some power at ω_2 and a resonant beam at ω_p'. To create power at ω_1 and ω_3, the second QPM zone of the crystal can be designed such that $k_p' = k_1 + k_3 \pm \mathrm{K_{DFG}}$. The power generation at ω_1, ω_2, and ω_3 and pump depletion are maximized by carefully choosing a resonator design.

The basic physics of these devices is relatively simple, but to determine specific values for the resonator design, the multilayer film coatings for specific reflectivities require rigorous modeling. Relevant input parameters such as pump beam characteristics (spatial and temporal shapes and pulse energy), crystal characteristics (nonlinear coefficient, absorption coefficients at particular frequencies, length of each QPM zone), and mirror reflectivities are required to accurately predict the desired output of an ideal device. To converge to more realistic solutions of performance, the random variations of the periodic poling quality in the QPM sections of real crystals must be accounted for in the design. This is due to the phase mismatch introduced by nonperfect domains in the crystal, which affect the energy transfer between beams of different frequencies. The design and modeling of a practical frequency conversion device is an art. There have already been reports of similar multielement devices operating in the mid-infrared region (9,10).

6.6.2 Comb Generation

Another OPO example using QPM to generate multiple frequencies in the mid-infrared region is comb generation. The goal is to create, over a specified bandwidth, a series of frequencies that are exactly equally spaced from one another. Many organic and inorganic molecules in the atmosphere exhibit vibrational modes in the mid-infrared region (see Fig. 6.20) (11). Rotational mode absorption features, which can be quite narrow in bandwidth, exist within large vibrational mode absorption bands. A frequency comb source in the mid-infrared region would enable the detection of narrow, closely spaced absorption features. Sources using frequency combs can be useful for a number of applications. For example, a molecular "fingerprint" spectroscopy system for determining the chemical composition of a substance can be a powerful tool to determine the chemical composition and concentration of an unknown substance, such as waste being discarded by a chemical manufacturing plant or the remote detection of the chemical composition and concentration of aerosols and aromatics from a gas plume or cloud.

The recent development of periodically poled GaAs, known as OP-GaAs, with its large second-order nonlinearity $d_{14}^{(2)} = 94$ pm/V used with beam propagation along the $\langle 011 \rangle$ axis and its excellent transparency up to 17 μm has opened up the possibility of parametric generation up to that wavelength (12). OP-GaAs crystals have been produced with thicknesses of 1 mm and lengths up to 50 mm (13). A thin layer of GaAs is first grown on a $0\bar{1}1$-oriented GaAs substrate upon which a thin layer of Ge is subsequently deposited, usually via molecular beam epitaxy. The Ge layer "buffers" the substrate layer of GaAs, and a second layer of GaAs is grown on the Ge layer. The second GaAs layer has the opposite orientation relative to the GaAs substrate layer. Chemical etching is then used to expose opposite orientations of the GaAs, creating a template of uniform domain reversed areas. Hydride vapor-phase epitaxy regrowth of the template pattern creates OP-GaAs crystals, which have the appropriate thickness for bulk nonlinear optics experiments. An OP-GaAs crystal is shown in the inset of Fig. 6.23 (12).

As shown in Fig. 6.23, a mode-locked laser generates the frequency combs near 2.45 μm. The high peak power pulses are then injected into an OPO cavity, which contains an OP-GaAs crystal oriented at the Brewster angle. When the nonlinear crystal is oriented/engineered to phase match near the degeneracy frequency, i.e., $\omega_{pump} \approx 2\omega_{OPO}$, the OPO cavity can then be adjusted to operate as a doubly degenerate broadband frequency comb source. The result is a mid-infrared frequency comb source extending from 4.4 to 5.4 μm, whose output spectrum after transmission through air is shown in Fig. 6.24 along with calibrated air transmission spectra (12). Note that this particular OPO cavity does not require dispersion compensation due to the short interaction length within the OP-GaAs crystal.

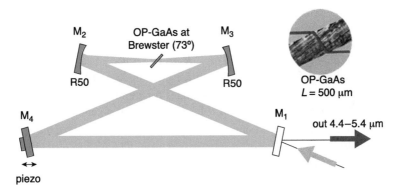

FIGURE 6.23 An OP-GaAs optical parametric oscillator cavity pumped by mode-locked 2.45-μm laser (not shown), which produces frequency combs from 4.4 to 5.4 μm. The inset shows a photograph of an OP-GaAs crystal and the beam direction. Reproduced with permission from the Optical Society of America (12).

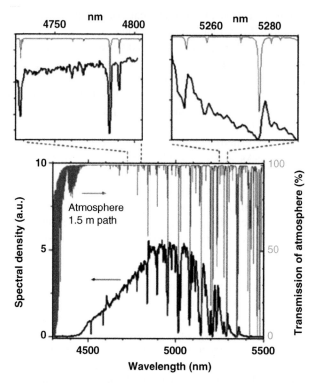

FIGURE 6.24 Envelope of the OPO output comb spectrum measured by a grating mono-chromator. The atmospheric transmission spectrum of air in the same spectral region is superimposed (gray curves) from above. The insets show the comb transmission spectrum and air-calibrated air transmission spectrum in expanded wavelength ranges. Reproduced with permission from the Optical Society of America (12).

PROBLEMS

1. Consider a sum frequency (ω_3) generation geometry in which there are two input beams of frequency $\omega_1 = R_1\omega_3$ and $\omega_2 = R_2\omega_3$ ($R_1 + R_2 = 1$) and unequal power. Using the notation from Chapter 4 for ρ_1, ρ_2, ρ_3, θ, Δs and $S = \pm$.

   (a) show that there are nonlinear eigenmodes for sum frequency generation and that two values are possible for θ for the nonlinear eigenmodes;

   (b) show that the normalized sum frequency field is given in terms of the two input fields by

$$\rho_3 = -\frac{\Delta s \rho_1 \rho_2}{2S(R_2\rho_1^2 + R_1\rho_2^2)} \pm \frac{1}{2S(R_2\rho_1^2 + R_1\rho_2^2)}\sqrt{\Delta s^2\rho_1^2\rho_2^2 + 4S^2\rho_1^2\rho_2^2(R_2\rho_1^2 + R_1\rho_2^2)};$$

   (c) show that for the case $\rho_1^2 = \rho_2^2$, there are two values possible for the funda-mental field for each ρ_3, provided $1 \geq \rho_3 > |\Delta s|$.

2. Consider generation of radiation at 4.5 μm by difference frequency generation in a LiNbO$_3$ quasi-phase-matched crystal by mixing strong pump sources at 1.064 μm (Nd:YAG laser) and 2.05 μm (holmium laser). The dispersion in the refractive index for extraordinarily polarized light in periodically poled LiNbO$_3$ is given by

$$n_e^2 = 4.5820 + \frac{0.099169}{\lambda^2 - 0.04443} - 0.021950\lambda^2.$$

   (a) Find the quasi-phase-matching grating period required.
   (b) For a 3-cm-long periodically poled LiNbO$_3$ crystal and equal intensities in the plane-wave input beams, find what input intensity is required for a conversion efficiency of 5% for [I(4.5 μm) × 100/{I(1.064 μm) + I(2.05 μm)}].
   (c) Now repeat (b) assuming that the inputs are beams with Gaussian profiles in space characterized by $w_0 = 0.5$ mm for the Nd:YAG laser and $w_0 = 2$ mm for the holmium laser. The two beams have equal peak intensities, the same as the values found in (b).
   (d) Calculate the output beam w_0 and the power conversion efficiency [P(4.5 μm) × 100/{P(1.064 μm) + P(2.05 μm)}].

3. One way to reduce feature size for nanoelectronics photolithography is to use radiation with as short a wavelength as possible. Li$_2$B$_4$O$_7$ is a negative uniaxial nonlinear crystal with 4mm tetragonal crystal symmetry and has a "transparency window" from 160 to 3600 nm. It has a nonlinearity $\hat{d}_{31}^{(2)} = 0.12$ pm/V, and wavelength-dependent refractive indices are given by

$$n_0^2(\lambda) = 2.56431 + \frac{0.012337}{\lambda^2 - 0.013103} - 0.019075\lambda^2,$$

$$n_e^2(\lambda) = 2.38651 + \frac{0.010664}{\lambda^2 - 0.012878} - 0.012813\lambda^2.$$

   (a) For noncritical phase matching, plot the sum frequency wavelengths with all the possible combinations of the two input wavelengths, limiting the range of input wavelengths dictated by the absorption edge at 160 nm.
   (b) For critical phase matching, plot the sum frequency phase-match angle θ versus the sum frequency wavelength for one input beam with a wavelength of 190 or 205 nm derived from the fourth harmonic (189–210 nm) of a Ti: sapphire laser and the second tunable one from 1.6 to 2.5 μm from a near-infrared optical parametric oscillator whose output wavelength is tuned to map out the possible critical phase-matching angles.

4. The smallest wavelength generated by nonlinear optics techniques in the deep ultraviolet region has used the KBe$_2$BO$_3$F$_2$ crystal.. KBe$_2$BO$_3$F$_2$ is a negative uniaxial nonlinear crystal with 32 rhombohedral crystal symmetry. It has a transparency window from 155 to 3600 nm. Its refractive indices, as known at

the time, are given by

$$n_0^2(\lambda) = 1 + \frac{1.169725\lambda^2}{\lambda^2 - 0.0062400} - 0.009904\lambda^2,$$

$$n_e^2(\lambda) = 1 + \frac{0.956611\lambda^2}{\lambda^2 - 0.0061926} - 0.027849\lambda^2.$$

By sum frequency mixing of the fourth harmonic of a tunable Ti:sapphire laser (780–800 nm) with the output of the laser, it has been possible to obtain wavelengths in the 156–160-nm range.

(a) For Type 1 wave-vector match, calculate the phase-match angle over this wavelength range, the acceptance angle at 156 nm for an effective 5.6-mm optical path, and the walk-off angle.

(b) The most recent measurements for index dispersion are

$$n_0^2(\lambda) = 1 + \frac{1.168705\lambda^2}{\lambda^2 - 0.0062782} - 0.0096676\lambda^2,$$

$$n_e^2(\lambda) = 1 + \frac{0.957724\lambda^2}{\lambda^2 - 0.0059816} - 0.028510\lambda^2.$$

Does this give a better fit to the data: 60.7° at 156 nm?

5. A case encountered occasionally in sum frequency generation is that a material is effectively lossless at one input frequency, but the second input has absorption with an amplitude absorption coefficient α_1.

(a) Show that in the negligible fundamental depletion limit, on wave-vector match, saturation of the conversion efficiency with propagation distance results. Also, derive a formula for the sum frequency field.

(b) Calculate in the negligible depletion limit the intensity conversion efficiency with propagation distance in the case of wave-vector mismatch in terms of $\ell_{coh}\Delta k/\pi = 0, 0.5, 1.0, 2.0$. Find the asymptotic conversion efficiency for long samples.

6. The periodicity with distance in three-wave mixing depends strongly on $\delta = [\rho_a^2(0) - \rho_b^2(0)]/[\rho_a^2(0) + \rho_b^2(0)]$. Show this numerically by plotting $\rho_a^2(\xi)$ as a function of ξ for a few values of δ. [Hint: Try $\delta = 0.6, 5 \times 10^{-4}, 5 \times 10^{-7}$.]

7. Barium borate is frequently used for broad bandwidth optical parametric oscillators because of its large transparency range. It belongs to the crystal class $3m$, with refractive indices given by

$$n_0^2(\lambda) = 2.7359 + \frac{0.01878}{\lambda^2 - 0.01822} - 0.01354\lambda^2,$$

$$n_e^2(\lambda) = 2.3753 + \frac{0.01224}{\lambda^2 - 0.01667} - 0.01516\lambda^2.$$

Its most important nonlinearity is $\hat{d}_{22}^{(2)} = 2.5\,\text{pm/V}$ ($\hat{d}_{15}^{(2)} = 0.14\,\text{pm/V}$). For pumping at 266 nm (doubled 532 nm), calculate

(a) the optimum crystal orientation at the degeneracy point and, for that crystal orientation, the angle tuning curve of an optical parametric oscillator for the signal and the idler using Type 1 wave-vector matching;

(b) the single pass gain for an injected signal over the complete tuning range, assuming a $1.0\,\text{GW/cm}^2$ input; and a 1 mm cubic crystal;

(c) the threshold pump intensity over the complete tuning range, assuming that mirrors with intensity reflection coefficients of 0.99 are used over this complete tuning range for a doubly resonant optical parametric oscillator; and

(d) the output intensity of the optical parametric oscillator for the lower frequency beam for a pump intensity of $20\,\text{GW/cm}^2$.

8. Consider quasi-phase-matched LiNbO_3 used in the first-grating order. Here $n_c^2 = 4.54535 + [0.0938326/(\lambda^2 - 0.0488982)] + 0.0228750\lambda^2$ and $\hat{d}_{33}^{(2)} = 28\,\text{pm/V}$.

(a) For sum harmonic generation with a fundamental wavelength of 1550 nm,
  (i) what is the grating period required?
  (ii) what is the optimum conversion efficiency for a 10-mW input in a 5-cm-long waveguide with an effective cross-sectional area of $5\,\mu\text{m}^2$?
  (iii) calculate the low signal bandwidth in nanometers.

(b) For the same device used as an optical parametric amplifier,
  (i) what is the intensity required at the degeneracy point for a gain coefficient of $1\,\text{cm}^{-1}$?
  (ii) what is the corresponding gain bandwidth?

(c) For a singly resonant optical parametric oscillator with $R = 97\%$ for both mirrors,
  (i) what is the threshold intensity at degeneracy?
  (ii) what is the threshold power for the waveguide?

9. Lithium triborate is frequently used for broad bandwidth optical parametric oscillators because of its large transparency range. It is biaxial and belongs to the crystal class $mm2$, with refractive indices given by

$$n_x^2 = 2.454140 + \frac{0.011249}{\lambda^2 - 0.011350} - 0.014591\lambda^2 - 6.60 \times 10^{-5}\lambda^4,$$

$$n_y^2 = 2.539070 + \frac{0.012711}{\lambda^2 - 0.012523} - 0.018540\lambda^2 - 2.00 \times 10^{-5}\lambda^4,$$

$$n_z^2 = 2.586179 + \frac{0.013099}{\lambda^2 - 0.011893} - 0.017968\lambda^2 - 2.26 \times 10^{-4}\lambda^4.$$

The nonlinear coefficients are $\hat{d}_{31}^{(2)} = 1.05\,\text{pm/V}$, $\hat{d}_{32}^{(2)} = -0.98\,\text{pm/V}$, and $\hat{d}_{33}^{(2)} = 0.05\,\text{pm/V}$. For pumping at 355 nm (doubled 532 nm), calculate

(a) the optimum crystal orientation at the degeneracy point and, for that crystal orientation, the angle tuning curve of an optical parametric oscillator for the signal and the idler using Type 1 wave-vector matching in the y–z plane;

(b) the single pass gain for an injected signal over the complete tuning range, assuming a 100 MW/cm^2 input;

(c) the threshold pump intensity over the complete tuning range, assuming that mirrors with intensity reflection coefficients of 0.99 are used over this complete tuning range for a doubly resonant optical parametric oscillator.

(d) the output intensity of the optical parametric oscillator for the lower frequency beam for a pump intensity of 20 MW/cm^2.

10. Mid-infrared optical parametric oscillators have been demonstrated in AgGaSe$_2$ nonlinear crystals using a Nd:YAG (1.34 μm) pump laser. AgGaSe$_2$ is a tetragonal $\bar{4}2m$ negative uniaxial crystal, which is transparent from 0.7 to 18 μm. Its nonlinear coefficient is $\hat{d}_{36}^{(2)} = 39$ pm/V. The Sellmeier equations (with λ in micrometers) are given by

$$n_o^2 = 4.6453 + \frac{2.2057}{1 - (0.1879/\lambda^2)} + \frac{1.8577}{1 - (1600/\lambda^2)},$$

$$n_e^2 = 5.2912 + \frac{1.3970}{1 - (0.2845/\lambda^2)} + \frac{1.9282}{1 - (1600/\lambda^2)}.$$

(a) Calculate the Type 1 tuning curves, the walk-off angles, and the angular bandwidth for Nd:YAG excitation as a function of phase-match angle.

(b) For the ND:YAG case, the optical parametric oscillator cavity's input mirror reflectivity was ~98% and the output mirror reflectivity was 52% for 1.5–1.7 μm. Assuming lossless 10-mm-long AgGaSe$_2$ crystal, calculate the approximate threshold intensity over 1.5–1.7 μm. At 15× threshold intensity, evaluate, on wave-vector match, the gain coefficient, single pass gain, and the output intensity in both the signal and the idler. Assume that the mirrors are lossless at all other wavelengths.

11. There has been interest in producing radiation at ~2.4, ~3.4, and ~4.0 μm. One approach has been to pump a KTiOPO$_4$ optical parametric oscillator (Type 2 noncritical phase matching, propagation along the x-axis) with a Nd:YLF laser operating at 1.047 μm to produce a 1.54-μm signal source which pumps a periodically poled LiNbO$_3$ optical parametric oscillator to give simultaneously (signal + idler) outputs at ~2.5 and ~4.0 μm. The KTiOPO$_4$ optical parametric oscillator idler output was at 3.5 μm. The temperature dependent Sellmeier equation for periodically poled LiNbO$_3$ is

$$f(T) = (T - 24.5)(T - 570.82)\ \text{Crystal Technology website}$$

$$n_e^2(\lambda) = 5.35583 - 4.629 \times 10^{-7} f(T) - \frac{0.100473}{\lambda^2 - (0.20692 - 3.862 \times 10^{-8} \times f(T))^2}$$

$$- \frac{100}{\lambda^2 - 11.34927^2} - 0.015334 \times \lambda^2,$$

and for KTiOPO$_4$ the equations are

$$n_x^2 = 3.0065 + \frac{0.03901}{\lambda^2 - 0.04251} - 0.01327\lambda^2,$$

$$n_y^2 = 3.0333 + \frac{0.04154}{\lambda^2 - 0.04547} - 0.01408\lambda^2,$$

$$n_z^2 = 3.3134 + \frac{0.05694}{\lambda^2 - 0.05658} - 0.01682\lambda^2.$$

The actual system developed had both the KTiOPO$_4$ optical parametric oscillator and the periodically poled LiNbO$_3$ optical parametric oscillator inside the laser cavity (intracavity), which results in complex cavity and mirror design beyond the scope of this book. Here we will look at each component separately and independently, i.e., not intracavity.

(a) Calculate and plot the wavelengths generated for noncritical phase matching versus pump wavelength. What are the exact wavelengths generated by the KTiOPO$_4$ optical parametric oscillator from the Nd:YLF pump laser?

(b) Assuming that the input and output KTiOPO$_4$ optical parametric oscillator mirrors have reflectivities of 98%, respectively, calculate the threshold intensity for a 1-cm-long crystal. The nonlinear coefficients for KTiOPO$_4$ are $\hat{d}_{31}^{(2)} = 2.54$ pm/V, $\hat{d}_{32}^{(2)} = 4.35$ pm/V, $\hat{d}_{33}^{(2)} = 16.9$ pm/V, $\hat{d}_{24}^{(2)} = 3.64$ pm/V, and $\hat{d}_{15}^{(2)} = 1.91$ pm/V. The periodically poled LiNbO$_3$ crystal has $\hat{d}_{33}^{(2)} = 27$ pm/V; it was 1.9 cm long and had anti-reflection-coated end faces. The periodically poled LiNbO$_3$ optical parametric oscillator cavity has an output reflectivity of 90% and an input reflectivity of 70% over the 2.5–4.0-µm wavelength range.

(c) For a grating period of 32.3 µm used in first order at room temperature, calculate the signal and idler wavelengths. LiNbO$_3$ is usually used at elevated temperatures to avoid photorefractive damage. The thermal expansion coefficient along the x-axis is $2 \times 10^{-6}°C^{-1}$.

(d) Calculate the idler wavelength over the temperature range 50–150°C.

(e) Calculate the threshold intensity of the periodically poled LiNbO$_3$ optical parametric oscillator at room temperature.

(f) Assuming that the ∼4-µm output intensity required is twice the threshold intensity for the periodically poled LiNbO$_3$, what is the intensity needed for the KTiOPO$_4$ crystal?

12. There has been interest in producing radiation at ∼2.4, ∼3.4, and ∼4.0 µm. One approach has been to pump a noncritically phase-matched KTiOPO$_4$ optical parametric oscillator (Type 2 noncritical phase matching, propagation along the x-axis) with a Nd:YLF laser operating at 1.047 µm to produce a ∼1.54 µm beam

which pumps an AgGaSe$_2$ optical parametric oscillator to give simultaneously (signal + idler) outputs at ~2.5 and ~4.0 μm. The KTiOPO$_4$ optical parametric oscillator idler output was at ~3.5 μm. The Sellmeier equations for AgGaSe$_2$ are

$$n_e^2 = 5.2912 + \frac{1.3970}{1 - (0.2845/\lambda^2)} + \frac{1.9282}{1 - (1600/\lambda^2)},$$

$$n_o^2 = 4.6453 + \frac{2.2057}{1 - (0.1879/\lambda^2)} + \frac{1.8577}{1 - (1600/\lambda^2)},$$

and for KTiOPO$_4$ the equations are

$$n_x^2 = 3.0065 + \frac{0.03901}{\lambda^2 - 0.04251} - 0.01327\lambda^2,$$

$$n_y^2 = 3.0333 + \frac{0.04154}{\lambda^2 - 0.04547} - 0.01408\lambda^2,$$

$$n_z^2 = 3.3134 + \frac{0.05694}{\lambda^2 - 0.05658} - 0.01682\lambda^2.$$

The actual system developed had both the KTiOPO$_4$ optical parametric oscillator and the AgGaSe$_2$ optical parametric oscillator inside the laser cavity (intracavity), which results in complex cavity and mirror design beyond the scope of this textbook. Here we will look at each component separately and independently, i.e., not intracavity.

(a) Calculate the exact wavelengths generated by the KTiOPO$_4$ optical parametric oscillator from the pump laser.

(b) Assuming that the input and output KTiOPO$_4$ optical parametric oscillator mirrors have reflectivities of 98%, calculate the threshold intensity for a 1-cm-long crystal. The nonlinear coefficients for KTiOPO$_4$ are $\hat{d}_{31}^{(2)} = 2.54$ pm/V, $\hat{d}_{32}^{(2)} = 4.35$ pm/V, $\hat{d}_{33}^{(2)} = 16.9$ pm/V, $\hat{d}_{24}^{(2)} = 3.64$ pm/V, and $\hat{d}_{15}^{(2)} = 1.91$ pm/V. AgGaSe$_2$ is a tetragonal $\bar{4}2m$ uniaxial crystal, which is transparent from 0.7 to 18 μm. Its relevant nonlinear coefficient is $\hat{d}_{36}^{(2)} = 39$ pm/V. The crystal was 3 cm long with a Type 1 (*eoo*) phase-matching angle of $\theta = 74°$ and had AR-coated end faces. The cavity has an input mirror with a transmission >90% at 1.52 μm and a reflection coefficient of 95% between 2.5 and 4 μm. The output coupler had 90% at 2.4 μm and 98% transmission at 4 μm. The cavity was singly resonant at 2.5 μm.

(c) Calculate the exact output wavelengths from the AgGaSe$_2$ optical parametric oscillator.

(d) Find $\hat{d}_{eff}^{(2)}$ for this crystal and find the optimum ϕ. What is the optimum $\hat{d}_{eff}^{(2)}$?

(e) Calculate the threshold intensity of the optical parametric oscillator.

(f) Assuming that the ~4-μm output intensity required is twice the threshold intensity for the periodically poled LiNbO$_3$, what is the intensity needed for the KTiOPO$_4$ crystal?

13. Mid-infrared optical parametric oscillators have been demonstrated in $AgGaSe_2$ nonlinear crystals using a Ho:YLF (2.05 µm) pump laser. $AgGaSe_2$ is a tetragonal $\bar{4}2m$ negative uniaxial crystal, which is transparent from 0.7 to 18 µm. Its relevant nonlinear coefficient is $\hat{d}^{(2)}_{36} = 39 \, pm/V$. The Sellmeier equations (with λ in micrometers) are given by

$$n_o^2 = 4.6453 + \frac{2.2057}{1 - (0.1879/\lambda^2)} + \frac{1.8577}{1 - (1600/\lambda^2)},$$

$$n_e^2 = 5.2912 + \frac{1.3970}{1 - (0.2845/\lambda^2)} + \frac{1.9282}{1 - (1600/\lambda^2)}.$$

(a) Calculate the Type 1 tuning curves, the angular acceptance angle, and the walk-off angle as a function of wavelength for excitation at 2.05 µm.

(b) For $AgGaSe_2$, the optical parametric oscillator cavity's input mirror had a reflectivity of ∼98% and the output mirror had a reflectivity of 65% for 2.9–4.1 µm. Assuming a lossless 10-mm-long $AgGaSe_2$ crystal, calculate the approximate threshold intensity at wave-vector match and 15× threshold, the gain coefficient, single pass gain, and the signal and idler output intensities. Consider the mirror reflectivity to be zero at other wavelengths.

14. The exact solutions for the optical amplifier show additional (to those discussed in the textbook) features that are different from the undepleted approximation. Show that both the regions of exponential growth and the distance to maximum conversion depend on the ratio of the input signal to pump fluxes. [Hint: A simple way to do this is to evaluate the photon fluxes with normalized distance for $\rho_a^2(0) = 0.0001$ and 0.001 and compare the results.]

REFERENCES

1. J. A. Armstrong, N. Bloembergen, J. Ducuing, and P. S. Pershan, "Interaction between light waves in a nonlinear dielectric," Phys. Rev., **127**, 1918–1939 (1662).

2. C. L. Tang and L. K. Cheng, Fundamentals of Optical Parametric Processes and Oscillators (Harwood, Amsterdam, 1995).

3. F. A. Hopf and G. I. Stegeman, Applied Classical Electrodynamics, Volume 2: Nonlinear Optics (John Wiley & Sons, New York, 1985).

4. L. K. Cheng, W. R. Bosenberg, and C. L. Tang, "Broadly tunable optical parametric oscillation in β-BaB₂O₄," Appl. Phys. Lett., **53**, 175–177 (1988).

5. L. E. Myers, W. R. Bosenberg, and C. L. Tang, "Periodically poled lithium niobate and quasi-phase-matched optical parametric oscillators," IEEE J. Quantum Electron., **QE-33**, 1663–1672 (1997).

6. P. S. Kuo, K. L. Vodopyanov, M. M. Fejer, D. M. Simanovskii, X. Yu, J. S. Harris, D. Bliss, and D. Weyburne, "Optical parametric generation of a mid-infrared continuum in orientation-patterned GaAs," Opt. Lett., **31**, 71–73 (2006).

7. P. S. Kuo and M. M. Fejer, "Microstructured semiconductors for mid-infrared nonlinear optics," in Mid-Infrared Coherent Sources and Applications, NATO Advanced Research Workshop on Middle Infrared Coherent Sources, edited by M. Ebrahim-Zadeh and I. T. Sorokina, pp. 149–170. (Springer, Dordrecht, 2007).

8. M. Ebrahim-Zadeh, "Continuous-wave optical parameteric oscillators," in Mid-Infrared Coherent Sources and Applications, NATO Advanced Research Workshop on Middle Infrared Coherent Sources, edited by M. Ebrahim-Zadeh and I. T. Sorokina, pp. 347–376. (Springer, Dordrecht, 2007).

9. H. C. Guo, S. H. Tang, Z. D. Gao, Y. Q. Qin, S. N. Zhu, and Y. Y. Zhu, "Multiple-channel mid-infrared optical parametric oscillator in periodically poled MgO:LiNbO$_3$," J. Appl. Phys., **101**, 113112 (2007).

10. N. Dixit, R. Mahendra, O. P. Naraniya, A. N. Kaul, and A. K. Gupta, "High repetition rate mid-infrared generation with singly resonant optical parametric oscillator using multi-grating periodically poled MgO:LiNbO$_3$," Opt. Laser Technol., **42**, 18–22 (2010).

11. http://en.wikipedia.org/wiki/Infrared_window.

12. K. L. Vodopyanov, E. Sorokin, I. T. Sorokina, and P. G. Schunemann, "Mid-IR frequency comb source spanning 4:4–5:4 μm based on subharmonic GaAs optical parametric oscillator," Opt. Lett., **36**, 2275–2277 (2011).

13. L. A. Eyres, P. J. Tourreau, T. J. Pinguet, C. B. Ebert, J. S. Harris, M. M. Fejer, L. Becouarn, B. Gerard, and E. Lallier, "All-epitaxial fabrication of thick, orientation-patterned GaAs films for nonlinear optical frequency conversion," Opt. Lett., **36**, 2275–2277 (2001).

SELECTED FURTHER READING

R. W. Boyd, Nonlinear Optics, 3rd Edition (Academic Press, Burlington, MA, 2008).

M. Mahric, Fiber Optical Parametric Amplifiers, Oscillators and Related Devices (Cambridge University Press, New York, 2007).

Y. R. Shen, Principles of Nonlinear Optics (John Wiley & Sons, New York, 1984).

C. L. Tang and L. K. Cheng, Fundamentals of Optical Parametric Processes and Oscillators (Harwood, Amsterdam, 1995).

$\chi^{(2)}$ Materials and Their Characterization

Since the early days of nonlinear optics when $\hat{d}_{eff}^{(2)} \approx 1\,\mathrm{pm/V}$ was considered normal, there have been great advances in the nonlinearities available in new materials. For example, in organic materials, $\hat{d}_{eff}^{(2)}$ of about hundreds to thousands of picometers per volt have now been reported (1). Nevertheless, oxide-based crystals with small to moderate nonlinearities are still the only ones used in commercial devices. This chapter discusses the reasons for this and highlights a number of materials and their important properties. Two common techniques for measuring nonlinear coefficients are described in Section 7.4.

7.1 SURVEY OF MATERIALS

Three "classes" of materials have been investigated for $\chi^{(2)}$ activity. Extensive tabulations of nonlinearity can be found in Ref. 2. Dielectric crystals, almost exclusively containing oxygen bonding to various atoms, exhibit nonlinearities up to tens of picometers per volt. Semiconductors typically have $d^{(2)}$'s of hundreds of picometers per volt. Although these are large nonlinearities, it is difficult to achieve wave-vector matching in the common semiconductors whose refractive index is isotropic, and they are not available commercially as doublers, except occasionally in the infrared region where their index dispersion is very small. There are exceptions such as patterned GaAs, which is optically isotropic but can be quasi-phase matched (3). Finally, organic materials have been grown as crystals with reported values as high as 1000 pm/V. Approximately 20–25% of the organic crystals that have noncentrosymmetric molecules are noncentrosymmetric and therefore exhibit $\chi^{(2)}$ activity.

The material properties needed for second harmonic generation (SHG) go beyond just wave-vector matching and large nonlinear coefficients. For a start, all the expressions for SHG in previous chapters had a common term, namely, $|\hat{d}_{eff}^{(2)}|^2/n^3$, which is now accepted universally as a useful figure of merit for comparing second-order materials. The following list gives this and other desirable characteristics:

Nonlinear Optics: Phenomena, Materials, and Devices, George I. Stegeman and Robert A. Stegeman.
© 2012 John Wiley & Sons, Inc. Published 2012 by John Wiley & Sons, Inc.

1. Large figure of merit is $|\hat{d}_{eff}^{(2)}|^2/n^2$ or $\dfrac{|\hat{d}_{eff}^{(2)}|^2}{n^2\alpha_1^2}$. In some cases, the linear absorption coefficient (α_1) limits the usable length of a material.

2. The possibility to wave-vector match is critical; i.e., $\Delta k = 0$ must be possible. For birefringent wave-vector matching, this implies small birefringence for useful tuning bandwidths.

3. The possibility to minimize the beam cross-sectional area is important. The material can be fabricated into waveguides, and this is a classic way of maintaining wavelength-sized transverse dimensions over centimeter-long propagation distances, usually limited only by waveguide loss or absorption.

4. The linear and nonlinear absorption should be small since it leads to thermal detuning and limits effective interaction lengths.

5. Environmental and chemical long-term stability.

6. Good mechanical properties (for polishing of surfaces).

7. Material homogeneity.

8. Easy to grow in large single crystals.

Table 7.1 lists some second-order active media with their important properties. The chemical formulas of these organic materials, which correspond to the common names used for them in the chemical literature, are given in the table footnote. The table reveals some striking features. The first grouping consists of oxide crystals that require birefringent wave-vector matching, except $LiNbO_3$, which can also be quasi-phase matched. These materials have very wide transparency windows. However, a great deal of progress has been made in quasi-phase matching. In organic crystals, the larger the nonlinearity, the smaller is the transparency window.

TABLE 7.1 Some $d^{(2)}$-Active Materials, Their Largest On- (il) and Off-Diagonal (iJ) Nonlinear Coefficients, and Their Approximate Transparency Regions

Material	d_{iJ} (pm/V)	d_{il} (pm/V)	Transparency (nm)
BaB_2O_4		1.6	190 to 2500
LiB_3O_5	1.1	0.06	180 to > 2800
$LiNbO_3$	5.8	30	400 to > 2500
$KTiOPO_4$	4.4	18.5	350 to 4500
$KNbO_3$	−18	−27	400 to > 3500
GaAs	125	0	900 to > 10,000
NPP	84	30	480 to 1800
DMNP	90	30	450 to 1700
DAST	30	600	750 to 1700

NPP, N-(4-nitrophenyl)-L-propinol; DMNP, 3,5-dimethyl-1-(4-nitrophenyl) pyrazole; DAST, dimethyl-amino-4-N-methylstilbazolium tosylate.

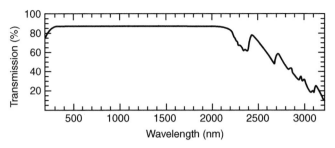

FIGURE 7.1   The transparency region of β-BaB$_2$O$_4$. Reproduced with permission from the American Institute of Physics (5).

7.2   OXIDE-BASED DIELECTRIC CRYSTALS

Although many different crystals have been "designed" for niche applications, assessed for SHG activity, and grown as single crystals, only a few are available commercially and used in commercial devices. The most common ones are β-BaB$_2$O$_4$ and KTiOPO$_4$. A very useful source of data on the material properties and tuning curves of these crystals is the catalog of CASTECH (4).

7.2.1   β-BaB$_2$O$_4$ (Trigonal 3*m* Crystal)

The transparency region of a β-BaB$_2$O$_4$ crystal is shown in Fig. 7.1 (5). This crystal has the following properties:

1. Broad wave-vector-matchable SHG from 410 to 3500 nm.
2. Wide temperature bandwidth of 55°C (for Type 1 SHG 1064 nm).
3. High optical homogeneity with $\Delta n \approx 10^{-6}\,\mathrm{cm}^{-1}$.
4. Useful for optical parametric amplifiers (OPAs) and optical parametric oscillators (OPOs).
5. Useful for second, third, fourth, and fifth harmonic generation of Nd:YAG lasers.

7.2.2   KTiOPO$_4$ (Orthorhombic *mm*2 Crystal)

The transparency region of a KTiOPO$_4$ crystal is shown in Fig. 7.2 (6). This crystal has the following properties:

1. Wide angular, spectral, and temperature bandwidth with a small walk-off angle.
2. Parametric sources (OPG, OPA, and OPO) for 0.6–4.5-μm tunable output.
3. Frequency doubling of Nd:YAG and Nd:YLF lasers.

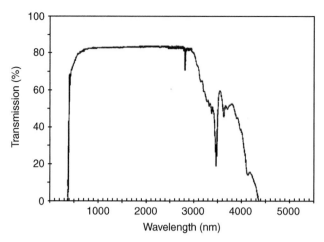

FIGURE 7.2 The transparency region of KTiOPO$_4$. Reproduced with permission from the Optical Society of America (6).

7.3 ORGANIC MATERIALS

Single-crystal organic materials exhibit *huge* second-order nonlinear coefficients, exceeding 100 pm/V (see Table 7.1). High optical quality crystals have been grown successfully. The largest nonlinearities occur through a mechanism called charge transfer, but unfortunately the stronger the nonlinearity, the narrower the transmission window. Examples of organic crystals will be discussed in Section 7.3.2, which highlights the reasons why they are rarely used because of this adverse trade-off. The following problems hamper progress in the commercialization of organic crystals for nonlinear optics:

1. Very limited transparency window relative to oxide dielectric crystals (7). In these materials, absorption due to the overtones of the C−H vibrations is responsible for the limited transmission at long wavelengths, in the >1.5-μm spectral region, typically $\alpha_1 \geq 0.1$ cm^{-1}. At wavelengths >1.5 μm, losses due to vibrations make the materials unusable for efficient doubling. The short wavelength absorption edge moves to longer wavelengths as the strength of the charge transfer mechanism responsible for the nonlinearity increases. This is a common characteristic of the charge transfer mechanism in molecules. The transmission window is usually limited to about a factor of 2–3 in wavelength.

2. The charge transfer mechanism also increases the refractive index along the charge transfer axis and leads to a large birefringence. This birefringence can lead to wave-vector matching, but with very small acceptance and large walk-off angles.

However, there are niche applications. For example, the very large nonlinearities coupled with short interaction lengths are useful for characterizing femtosecond pulses (8).

FIGURE 7.3 The prototype charge transfer molecule with acceptor and donor groups separated by the π-electron bridge.

7.3.1 Charge Transfer Molecules

The charge transfer mechanism is the basic nonlinear process in organic molecules that leads to large permanent dipole moments and large second-order nonlinearities (7, 9–11). The prototype molecule is shown in Fig. 7.3. Groups with different electron affinities are attached to opposite ends of a "bridge" whose function is to facilitate transfer of electrons between the end groups. This is usually accomplished with linear chains of carbon atoms whose p_z orbitals overlap to form new delocalized π orbitals that allow the electrons to move more easily between end groups. (Electron delocalization is discussed more extensively in Chapter 10.) Another example of a bridge group is single or multiple benzene groups, which contain carbon atoms with overlapping orbitals in a hexagonal structure. The end groups are either donors of electrons (D) or acceptors of electrons (A), and there occurs a net charge transfer between the two groups. This is illustrated in Fig. 7.4.

The net result is that the molecule has a dipole moment and is nonsymmetric, i.e., $\bar{\beta}_{ijk}^{(2)}(-[\omega_1 + \omega_2]; \omega_1, \omega_2) \neq 0$ which is the molecular hyperpolarizability in the molecule's frame of reference, and $\chi_{ijk}^{(2)}(-[\omega_1 + \omega_2]; \omega_1, \omega_2) = Nf^{(2)}\langle \bar{\beta}_{ijk}^{(2)}(-[\omega_1 + \omega_2]; \omega_1, \omega_2)\rangle$, where N is the molecular density, $f^{(2)}$ is the appropriate nonlinear local field factor (discussed in detail in Chapter 8), and $\langle \cdots \rangle$ is an average over the molecular orientation.

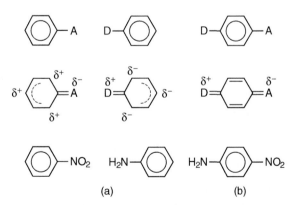

(a)   (b)

FIGURE 7.4 (a) The charge distributions induced by attaching an acceptor or donor group to a benzene bridge group. The bottom row shows a simple example of such molecules. (b) The charge distributions induced by attaching both an acceptor and a donor group to a benzene bridge group. The bottom row shows nitroaniline as an example of a charge transfer molecule. Reproduced with permission from Academic Press (9).

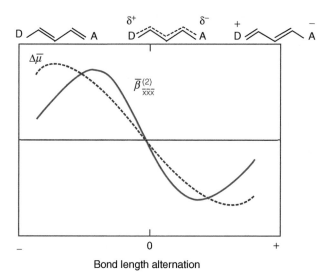

FIGURE 7.5 The calculated values of the charge-transfer-induced dipole moment ($\Delta\bar{\mu}$) and first-order hyperpolarizability ($\bar{\beta}^{(2)}_{\bar{x}\bar{x}\bar{x}}$) versus the change in the C—C bond lengths in a charge transfer molecule. Reproduced with permission from John Wiley & Sons (10).

The nonlinearity produced by this mechanism can be analyzed in terms of a simple two level; quantum mechanical model and chemical intuition, which yield excellent predictive power for optimizing $\bar{\beta}^{(2)}$ (10,11). This model includes contributions *only* from the difference in permanent dipole moments and does not include contributions from one photon transitions. Here, the principal ideas and approximations are listed briefly (with details in Appendix 7.1):

1. The actual molecular structure is a linear combination of the two extreme states (see the upper part of Fig. 7.5).
2. "Electrostatic interaction" occurs due to transferred charge, which leads to new ground and excited states.
3. The C—C and C=C bond lengths change (called bond length alternation) along the electron bridge (10,12).
4. The resultant change in elastic energy along the bonds leads to a new minimum energy equilibrium configuration.
5. An applied field $\mathcal{E}$ modulates the charge transfer and hence the induced linear and nonlinear polarization.

The final result is that the polarization induced in the new ground state $\bar{E}_{gr}$ by an applied field is $\bar{p}_{\bar{x}} = -d\bar{E}_{gr}(\mathcal{E})/d\mathcal{E}$ for a single molecule and the $\bar{\beta}^{(2)}$ along the charge transfer axis is given by

$$\bar{\beta}^{(2)}_{\bar{x}\bar{x}\bar{x}} = \frac{1}{2}\frac{d^2\bar{p}_x}{d\mathcal{E}^2}\bigg|_{\mathcal{E}=0}. \tag{7.1}$$

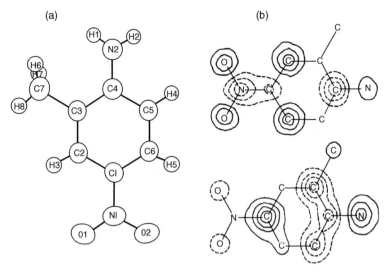

FIGURE 7.6   (a) The equilibrium for MNA molecules. (b) The electron distribution in the ground state (lower) and first excited state (upper) of MNA. The dashed lines indicate positively charged regions, and the solid lines indicate negatively charged regions. Reproduced with permission from Academic Press (13).

The key point is that this model, combined with values of the electron affinity of different groups, can be used to predict the molecular nonlinearity of single molecules and how it can be optimized (see Fig. 7.4) (12).

Figure 7.6a shows the unit cell of an MNA molecule, which consists of carbon, hydrogen, oxygen, and nitrogen atoms (13). The bridge group is benzene, and the donor and acceptor groups are the NH_2 and NO_2 groups, respectively. The detailed charge redistribution that leads to a change in the permanent dipole moment of both the ground state and the first excited state is shown in Fig. 7.6b. This molecule was the first "designer" molecule crystallized for its $\chi^{(2)}$ activity.

The presence of a noncentrosymmetric molecule does not guarantee a nonzero $\chi^{(2)}$ response after crystallization. The relative orientation of the individual molecules can lead to the cancellation of the permanent dipole moments over a unit cell and hence a zero second-order susceptibility over an optical wavelength. As an approximate number, only 20–25% of anisotropic molecules lead to $\chi^{(2)}$ activity, although crystals with charge transfer molecules and no $\chi^{(2)}$ activity tend to have large quadrupole moments. The science of which molecules will have second-order susceptibilities in the crystal form is not well established. Although quite a few organic crystals have been synthesized and crystallized to date, usually their transparency regions are rather limited, especially if they have large nonlinearities. They are in fact more suitable for electrooptical applications since they are reasonably transparent between 800 and 1500 nm.

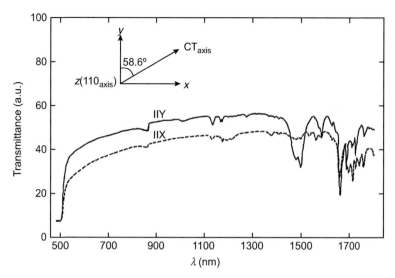

FIGURE 7.7 The transparency window for *N*-4-nitrophenyl-L-prolinol. Reproduced with permission from the Optical Society of America (14).

7.3.2 Examples of Single Crystal Organics

7.3.2.1 N-4-Nitrophenyl-L-prolinol (Monoclinic 2 Crystal). This crystal was also custom "designed." Its transparency window is shown in Fig. 7.7. This crystal was used to demonstrate an OPO, tuning curve shown in Fig. 7.8.

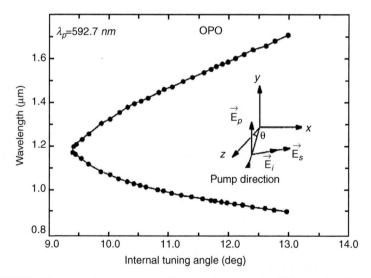

FIGURE 7.8 The optical parametric oscillator tuning curve for *N*-4-nitrophenyl-L-prolinol. Reproduced with permission from the Optical Society of America (14).

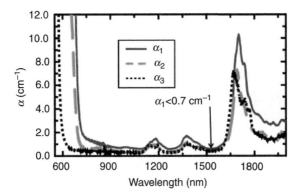

FIGURE 7.9   The absorption coefficient of 4-*N,N*-dimethylamino-4-*N*-methyl-stilbazolium 2,4,6-trimethylbenzenesulfonate along the three optical axes. Reproduced with permission from the Optical Society of America (15).

7.3.2.2   4-N,N-Dimethylamino-4-N-methyl-stilbazolium 2,4,6-Trimethyl-benzenesulfonate (Monoclinic m Crystal). This material has $d^{(2)}_{xxx} > 200$ pm/V, and its properties are typical of organic crystals with large nonlinear coefficients (12). Its absorption coefficient is shown in Fig. 7.9. Note the minimum absorption coefficient of 0.7 cm^{-1}, which is typical of highly nonlinear organic crystals.

7.3.2.3   Poled Polymers. There are various methods to align charge transfer molecules to produce a net second-order coefficient $\chi^{(2)}$. These methods rely on the large permanent electric dipole moment of such molecules in their ground state. The molecules can either be "dissolved" in a suitable polymer or attached as side chains (like pendants) to a suitable polymer or chemically bonded into the structure of a suitable polymer chain. Normally, the orientation of the molecules is random and there is no net $\chi^{(2)}$. However, when strong electric fields are applied, some net orientation of the molecules can be induced via the permanent dipole moment at elevated temperatures (see Fig. 7.10) (parallel plate poling). The field is applied above the glass transition temperature for the host polymer, at which the host polymer is soft and partial reorientation of molecules can occur. Once the molecules are reoriented to some degree, the structure is cooled down to below the glass transition temperature, effectively freezing in the partial orientation. This is usually done in thin films to guide light so as to optimize the beam cross section. Although various "tricks" have been used to achieve phase matching, overall the efficiency was never large enough to be practical (16).

7.4   MEASUREMENT TECHNIQUES

The most common materials used for SHG are single crystals. Growing large single crystals is a time-consuming art. Fortunately, a simple method has been developed for

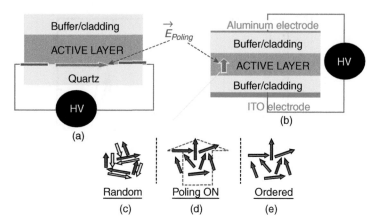

FIGURE 7.10   (a) In-plane poling and (b) parallel poling of the active layer. (c) Random orientation of molecules prior to poling. (d) Partial alignment of molecules by field. (e) "Frozen"-in structure.

assessing whether the material can be wave-vector matched by using *small* crystallites. This is known as the Kurtz method (17). However, this method is not suitable for measuring individual nonlinear coefficients, and the Maker fringe method is used instead.

7.4.1 The Kurtz Method

The first few attempts at growing new crystals typically yield small crystallites of random shape, orientation, and size. To assess their potential for SHG, first the crystallites are passed through a series of sieves (filters) to separate out crystallites of different average sizes. Next, the second harmonic light generated by each size range is measured as a function of the average crystallite size $\langle r \rangle$. The key to the measurement is to use a parabolic reflector to gather as much of the generated light as possible (see Fig. 7.11). The dependence of the total harmonic power on the average size $\langle r \rangle$ is then used to identify wave-vector-matchable materials.

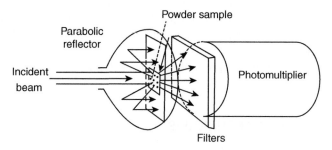

FIGURE 7.11   Second harmonic collection from crystallites and detection system for the Kurtz method. Reproduced with permission from the American Institute of Physics (17).

There is a specific connection between the SHG signal and the crystallite size. Define a coherence length averaged over all crystal orientations as $\langle \ell_{coh} \rangle = \pi/\overline{\Delta k} = \pi c/2\omega \overline{\left[n(2\omega) - n(\omega) \right]}$ and consider the second harmonic intensity radiated in two limits.

Case I: $\langle \ell_{coh} \rangle \gg \langle r \rangle$ (the smallest crystallites have some directions in which SHG can grow). The second harmonic signal is proportional to the product of the number of crystallites and the length over which the harmonic signal grows; i.e.,

$$I(2\omega) \propto \frac{1}{\langle r \rangle^3} \times \langle r \rangle^2 (\text{cross-sectional area})$$

$$\times \langle r \rangle^2 (L^2 = \langle r^2 \rangle) \times \left| \hat{d}_{eff}^{(2)} \right|^2 \rightarrow I(2\omega) \propto \langle r \rangle. \tag{7.2}$$

Case II: $\langle r \rangle \gg \ell_{coh}$ Two possibilities exist in this case:

(a) The crystal has no wave-vector-matching directions, i.e., small coherence lengths:

$$I(2\omega) \propto \frac{1}{\langle r \rangle^3} \times \hat{r}^2 (\text{cross-sectional area})$$

$$\times \langle \ell_{coh}^2 \rangle (L^2 = \langle \ell_{coh}^2 \rangle) \left| \hat{d}_{eff}^{(2)} \right|^2 \rightarrow I(2\omega) \propto \langle r \rangle^{-1}. \tag{7.3}$$

(b) The crystal can be wave-vector matched over some directions. These directions give $I(2\omega) \propto \langle r \rangle$ and these crystals should be developed further. There are also crystal directions that cannot be wave-vector matched; i.e., $I(2\omega) \propto \langle r \rangle^{-1}$. Detailed averaging gives $I(2\omega) = \text{constant}$. Figure 7.12 shows the typical dependence of the harmonic signal on $\langle r \rangle$ for these two cases.

7.4.2 The Maker Fringe Method

The Maker fringe method uses thin oriented crystals with light incidence approximately parallel to the wave-vector-matching direction, normal to the plane of the thin crystal (18). When the crystal is rotated away from the wave-vector-match direction, a phase difference is introduced between the fundamental and the harmonic signal, which changes with angle. The incident light–crystal geometry is shown in Fig. 7.13a. The wave equation at 2ω in the medium is

$$\nabla^2 \vec{E}(2\omega, \vec{r}) - \frac{n^2(2\omega, \theta)}{c^2} \frac{\partial^2 \vec{E}(2\omega, \vec{r})}{\partial t^2} = \mu_0 \frac{\partial^2 \vec{P}^{NL}(2\omega, \vec{r})}{\partial t^2}. \tag{7.4}$$

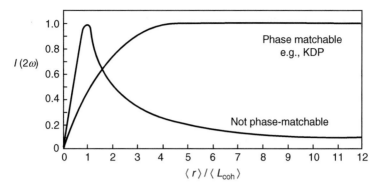

FIGURE 7.12   Variation of the second harmonic power with the crystallite size using the Kurtz method for both a wave-vector-matchable and a non-wave-vector-matchable crystal. Reproduced with permission from the American Institute of Physics (17).

For incident beams many wavelengths wide the wave-vector component of the fundamental parallel to the interface is conserved such that the nonlinear polarization $P(2\omega)$ and the generated harmonic field have the same component $2k_{\parallel}(\omega)$ parallel to the interface. The solution field is written as $\vec{E}_{\text{hom}}(2\omega,\vec{r})+\vec{E}_{\text{inhom}}(2\omega,\vec{r})$, where $\vec{E}_{\text{hom}}(2\omega,\vec{r})$ is the solution to Eq. 7.4, with $\vec{P}^{\text{NL}}(2\omega,\vec{r})=0$, and $\vec{E}_{\text{inhom}}(2\omega,\vec{r})$ is the particular solution for $\vec{P}^{\text{NL}}(2\omega,\vec{r})\neq 0$. Since the harmonic grows from the boundary at $z=0$, $E_{\text{hom}}(2\omega,0)+E_{\text{inhom}}(2\omega,0)=0$ and the solution for the two field components is

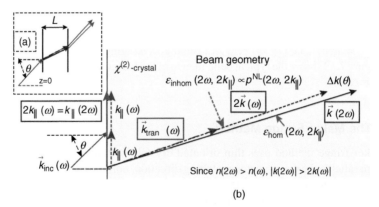

FIGURE 7.13   (a) Overview of the sample geometry and the incident beam. The red line identifies the wave-vector direction of the incident fundamental field; the dashed blue line is the wave-vector direction of the nonlinear polarization field; and the solid blue line is the wave-vector direction for a freely propagating second harmonic field, i.e., the homogeneous solution. (b) Wave vectors of the homogeneous and inhomogeneous field solutions and the resulting Δk between them due to dispersion in the refractive index.

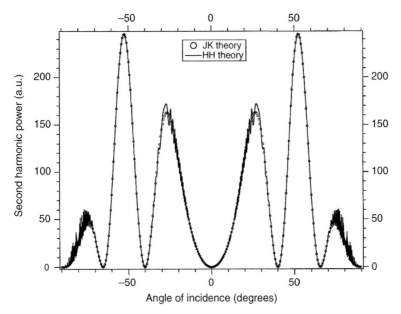

FIGURE 7.14 Second harmonic Maker fringes obtained by rotating the angle of incidence of the fundamental beam for a 300-mm-thick piece of X-cut quartz. Reproduced with permission from the American Institute of Physics (18).

$$\mathcal{E}_{\text{inhom}}(2\omega) = \frac{\hat{d}^{(2)}_{\text{eff}}}{c^2\left[4k^2_\perp(\omega) - k^2_\perp(2\omega)\right]}, \qquad \mathcal{E}_{\text{hom}}(2\omega) = -\frac{\hat{d}^{(2)}_{\text{eff}}}{c^2\left[4k^2_\perp(\omega) - k^2_\perp(2\omega)\right]}.$$

$$(7.5)$$

Note that $k(2\omega) > 2k(\omega)$ because $n(2\omega) > n(\omega)$ due to dispersion in the refractive index in the visible and infrared regions. Furthermore, since both wave vectors have the same $2k_\parallel(\omega)$ parallel to the interface, this results in a $\Delta k(\theta)$ parallel to the x-axis, as shown in Fig. 7.13b. Therefore the two fields will interfere with one another as the angle θ varies. This results in a fringe pattern (see Fig. 7.14), called the Maker fringes. The nonlinearity $\hat{d}^{(2)}_{\text{eff}}$ is obtained from the magnitude of the fringes.

APPENDIX 7.1: QUANTUM MECHANICAL MODEL FOR CHARGE TRANSFER MOLECULAR NONLINEARITIES

This model was first put forward by Lu et al. (11). It is assumed that molecules with charge transfer properties can be described by a linear combination of two states: a *valence* bond state characterized by no charge transfer (Fig. A.7.1.1a) and a *charge transfer* state in which the maximum possible charge transfer occurs (Fig. A.7.1.1b). The wave functions of the molecules are Ψ_{VB} and Ψ_{CT} with energy levels $\bar{E}_{\text{VB}}$ and $\bar{E}_{\text{CT}}$, respectively. As discussed in Section 7.3.1, the bridge in Fig. A.7.1.1 is a chain of

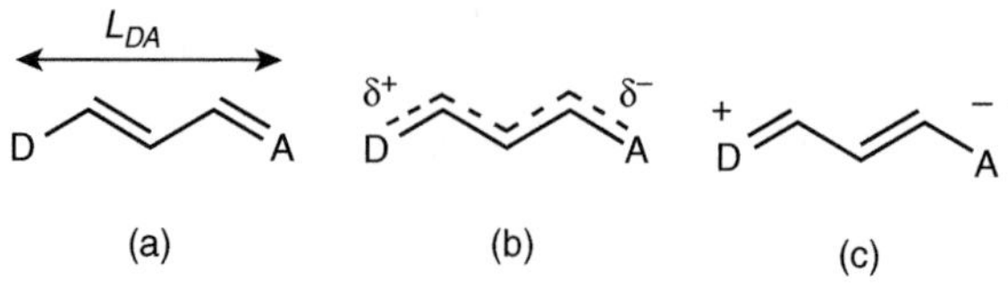

FIGURE A.7.1.1   A prototype charge transfer molecule showing net charges and the bridge bonding structure of (a) the valence-band state, (b) the equilibrium charge transfer state for a molecule, and (c) the charge transfer state, maximum charge transfer.

carbon atoms in which D and A refer to the electron donor and acceptor groups. L_{DA} is the initial acceptor–donor separation, and δ^+ and δ^- are the partial charges transferred in the equilibrium state of the charge transfer molecule (Fig. A.7.1.1c).

The mixing of the valence bond and charge transfer states via the change in the charge distribution is described by the Hamiltonian H_0:

$$\bar{H}_0 = \begin{pmatrix} 0 & -\bar{t} \\ -\bar{t} & \bar{V} \end{pmatrix}, \qquad \bar{V} = \bar{E}_{CT} - \bar{E}_{VB}, \qquad -\bar{t} = \langle \bar{\Psi}_{VB} | \mathcal{H} | \bar{\Psi}_{CT} \rangle, \qquad (A7.1)$$

where the energy of the valence-band state is taken as zero since it is the energy difference between states that is important and the off-diagonal elements $\bar{t}$ quantify the Coulomb interaction between the charged donor and acceptor groups. The solution to Schrödinger's equation gives the ground- and excited-state wave functions $\bar{\Psi}_{gr}$ and $\bar{\Psi}_{ex}$ and energies $\bar{E}_{gr}$ and $\bar{E}_{ex}$, respectively, for this two-level model. Defining f as the actual fraction of charge transferred gives $\bar{\Psi}_{gr} = \sqrt{1-\bar{f}^2}\,\bar{\Psi}_{VB} + \bar{f}\,\bar{\Psi}_{CT}$ and

$$\bar{E}_{gr} = \frac{1}{2}\left[\bar{V} - \sqrt{\bar{V}^2 + 4\bar{t}^2} \right], \qquad \bar{E}_{ex} = \frac{1}{2}\left[\bar{V} + \sqrt{\bar{V}^2 + 4\bar{t}^2} \right],$$

$$\bar{f} = \frac{\bar{E}_{gr}^2}{\bar{t}^2 + \bar{E}_{gr}^2} = \frac{1}{2} - \frac{\bar{V}}{2\sqrt{\bar{V}^2 + 4\bar{t}^2}} \qquad\qquad (A.7.1.2)$$

$$\Rightarrow \quad \bar{f} = \frac{d\bar{E}_{gr}}{d\bar{V}}.$$

The charge transfer creates a change in the C–C bond lengths. The relevant force constant $\bar{k}$ ("spring constant" for the equivalent simple harmonic oscillator) between the carbons is the same one that provides the restoring force for vibrations (optical phonons) polarized along the charge transfer axis ($\bar{x}$). This displacement contributes the elastic energy $\frac{1}{2}\bar{k}(\bar{q}-\bar{q}_{VB}^0)^2$ to both states, where $\bar{q}$ is the molecule's actual bond length. The parameters $\bar{q}_{CT}^0$ and q_{CT}^0 are the bond lengths for the original valence bond and charge transfer states. The new minimum of energy for the ground state is

$$\bar{E}_{gr} = \frac{1}{2}\left[\bar{V}_0 + \frac{1}{2}\bar{k}\left\{ (\bar{q}-\bar{q}_{VB}^0)^2 + (\bar{q}-\bar{q}_{CT}^0)^2 \right\} - \sqrt{\bar{V}^2 + 4\bar{t}^2} \right], \qquad (A.7.1.3)$$

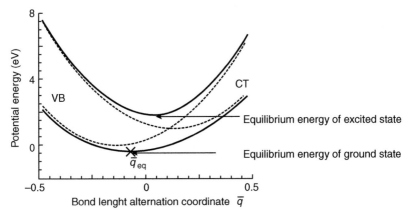

FIGURE A.7.1.2   The potential energy of the original valence-band (VB) and charge-transfer (CT) states (dashed lines) and of the ground and excited states, including charge transfer versus the bond change parameter $\bar{q}$. The equilibrium ground-state bond change, which minimizes the ground-state energy, is identified as $\bar{q}_{\mathrm{eq}}$. Reproduced with permission from American Chemical Society (11).

which is a function of $\bar{q}$. $\bar{V}_0$ is the adiabatic energy difference between the ground and excited states, i.e., at their *minimum* energy value. Hence, the minimum energy for the ground state can be calculated by finding $d\bar{E}_{\mathrm{gr}}(\bar{q})/d\bar{q} = 0$. The calculation is shown in Fig. A.7.1.2. The parameters used were $\bar{V} = 1\,\mathrm{eV}$, $\bar{t} = 1.1\,\mathrm{eV}$, and $\bar{k} = 33.55\,\mathrm{eV/\mathring{A}^2}$. The equilibrium coordinates for valence bond and charge transfer are shifted by -0.12 and $0.12\,\mathring{A}$, respectively. The symbol ✗ marks the equilibrium coordinate $\bar{q}_{\mathrm{eq}}$ on the ground-state potential surface.

The fraction of charge transfer can now be defined as

$$\bar{f} = \frac{\bar{q}_{\mathrm{VB}}^0 - \bar{q}_{\mathrm{eq}}}{\bar{q}_{\mathrm{VB}}^0 - \bar{q}_{\mathrm{CT}}^0}. \tag{A.7.1.4}$$

The maximum possible dipole moment due to the charge transfer state is $\bar{\mu}_{\mathrm{CT}}^0 = QL_{\mathrm{DA}}$, and therefore the resulting equilibrium molecule's value is $\Delta\bar{\mu} = \bar{f}\bar{\mu}_{\mathrm{CT}}^0$. For an applied electric field $\mathcal{E}$, an additional term must be added to the *original* Hamiltonian; namely,

$$\bar{H} = \begin{pmatrix} \bar{E}_{\mathrm{VB}}' & -\bar{t} \\ -\bar{t} & \bar{E}_{\mathrm{CT}}' - \bar{\mu}_{\mathrm{CT}}^0 \mathcal{E} \end{pmatrix}$$

$$\rightarrow \quad \bar{V}_{\mathcal{E}} = \bar{E}_{\mathrm{CT}}' - \bar{E}_{\mathrm{VB}}' - \bar{\mu}_{\mathrm{CT}}^0 \mathcal{E} \tag{A.7.1.5}$$

$$\rightarrow \quad \frac{d\bar{f}}{d\mathcal{E}} = \frac{d\bar{f}}{d\bar{V}}\frac{d\bar{V}_{\mathcal{E}}}{d\mathcal{E}} = \frac{2\bar{t}^2 \bar{\mu}_{\mathrm{CT}}^0}{\bar{E}_{\mathrm{gr}}^3}.$$

The polarization induced in the ground state by the applied field $\mathcal{E}$ is

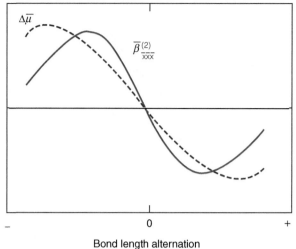

FIGURE A.7.1.3   The dipole moment $\Delta\bar{\mu}$ and the molecular first-order hyperpolarizability $\bar{\beta}^{(2)}_{\bar{x}\bar{x}\bar{x}}$ versus $\bar{f}$. Reproduced with permission from John Wiley & Sons (10).

$$\bar{p}_{\bar{x}} = -\frac{d\bar{E}_{\text{gr}}(\mathcal{E})}{d\mathcal{E}} = \mu^0_{\text{CT}} \frac{d\bar{E}_{\text{gr}}}{d\bar{V}_\mathcal{E}}\bigg|_{\mathcal{E}=0} = \bar{f}\mu^0_{\text{CT}}, \qquad (A.7.1.6)$$

which gives the molecular polarizability

$$\bar{\alpha}_{\bar{x}\bar{x}} = \frac{d\bar{p}_x}{d\mathcal{E}}\bigg|_{\mathcal{E}=0} = -(\bar{\mu}^0_{\text{CT}})^2 \frac{d\bar{f}}{d\bar{V}_\mathcal{E}}\bigg|_{\mathcal{E}=0} = \frac{2\bar{t}^2(\bar{\mu}^0_{\text{CT}})^2}{\bar{E}^3_{\text{gr}}}. \qquad (A.7.1.7)$$

The molecular first-order hyperpolarizability $\bar{\beta}^{(2)}_{xxx}$ is

$$\bar{\beta}^{(2)}_{\bar{x}\bar{x}\bar{x}} = \frac{1}{2}\frac{d^2 p\bar{x}}{d\mathcal{E}^2}\bigg|_{\mathcal{E}=0} = -\mu^3_{\text{CT}} \frac{d^2\bar{f}}{d\bar{V}^2_\mathcal{E}}\bigg|_{\mathcal{E}=0} = \frac{3\bar{t}^2(\bar{\mu}^0_{\text{CT}})^3\bar{V}}{\bar{E}^5_{\text{gr}}}. \qquad (A.7.1.8)$$

This is plotted in Fig. A.7.1.3. This curve leads to the design of molecules so as to maximize $\bar{\beta}^{(2)}_{\bar{x}\bar{x}\bar{x}}$.

In the same way, the second-order hyperpolarizability $\bar{\gamma}^{(3)}_{\bar{x}\bar{x}\bar{x}\bar{x}}$ (which leads to $\bar{\chi}^{(3)}_{xxxx}$) can be calculated:

$$\bar{\gamma}^{(3)}_{\bar{x}\bar{x}\bar{x}\bar{x}} = \frac{1}{6}\frac{d^3\bar{p}_x}{d\mathcal{E}^3}\bigg|_{\mathcal{E}=0} = -(\bar{\mu}^0_{\text{CT}})^4 \frac{d^3\bar{f}}{d\bar{V}^3_\mathcal{E}}\bigg|_{\mathcal{E}=0} = \frac{4\bar{t}^2(\mu^0_{\text{CT}})^4[\bar{V}^2-\bar{t}^2]}{\bar{E}^7_{\text{gr}}}. \qquad (A.7.1.9)$$

The dependence of $\bar{\gamma}^{(3)}_{\bar{x}\bar{x}\bar{x}\bar{x}}$ on the charge transfer fraction is illustrated in Fig. A.7.1.4 (11, 19).

Note that these expressions do not include the local field correction, which will be discussed in detail in Chapter 8.

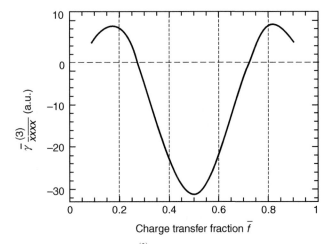

FIGURE A.7.1.4   The dependence of $\bar{\gamma}^{(3)}_{xxxx}$ on the charge transfer fraction. Reproduced with permission from American Chemical Society (11).

REFERENCES

1. U. Meier, M. Bösch, Ch. Bosshard, and P. Günter, "DAST a high optical nonlinearity organic crystal," Synth. Met., **109**, 19–22 (2000).

2. S. Singh, "Nonlinear optical materials: organic and inorganic materials," in Handbook of Laser Science and Technology, Supplement 2: Optical Materials, edited by M. J. Weber (CRC Press, Bocda Raton, FL, 1995), pp. 147–249.

3. X. Yu, L. Scaccabarozzi, O. Levi, T. J. Pinguet, M. M. Fejer, and J. S. Harris Jr., "Template design and fabrication for low-loss orientation-patterned nonlinear AlGaAs waveguides pumped at 1.55 µm," J. Cryst. Growth, **251**, 794–799 (2003).

4. CASTECH catalog of crystals: http://www.castech.com.

5. D. Eimerl, L. Davis, S. Velsko, E. K. Graham, and A. Zalkin, "Optical, mechanical and thermal properties of barium borate," J. Appl. Phys., **62**, 1668–1983 (1967).

6. G. Hansson, H. Carlsson, S. Wang, and F. Laurell, "Transmission measurements in KTP and isomorphic compounds," Appl. Opt., **39**, 5058–5069 (2000).

7. I. Ledoux, "New advances in molecular engineering for quadratic nonlinear optics," Synth. Metals, **54**, 123–137 (1993).

8. N. A. van Dantzig, P. C. M. Planken, and H. J. Bakker, "Far-infrared second-harmonic generation and pulse characterization with the organic salt DAST," Opt. Lett., **23**, 466–468 (1998).

9. J. F. Nicoud and R. J. Twieg, "Design and synthesis of organic molecular compounds for efficient second-harmonic generation," in Nonlinear Optical Properties of Organic Molecules and Crystals: Volume 1, edited by D. S. Chemla and J. Zyss (Academic Press, Orlando, FL, 1987), pp. 227–296.

10. S. Barlow and S. R. Marder, "Nonlinear optical properties of organic materials," in Functional Organic Molecules, edited by T. J. J. Müller and U. H. F. Bunz (John Wiley and Sons, New York, 2007), pp. 393–437.

11. D. Lu, G. Chen, J. W. Perry, and W. A. GoddardIII, "Valence-bond charge-transfer model for nonlinear optical properties of charge-transfer organic molecules," J. Chem. Soc. Am., **116**, 10679–10685 (1994).

12. Ch. B. Gorman and S. R. Marder, "An investigation of the interrelationships between linear and nonlinear polarizabilities and bond-length alternation in conjugated organic molecules," Proc. Natl. Acad. Sci. U.S.A., **90**, 11297–11301 (1993).

13. K. D. Singer, S. L. Lalama, J. E. Sohn, and R. D. Small, "Electro-optic organic materials," in Nonlinear Optical Properties of Organic Molecules and Crystals: Volume 1, edited by D. S. Chemla and J. Zyss (Academic Press, Orlando, FL, 1987), 437–468.

14. D. Josse, S. X. Dou, and J. Zyss, "Near-infrared pulsed optical parametric oscillation in N-(4-nitrophenyl)-L-prolinol at the 1-ns time scale," J. Opt. Soc. Am. B, **10**, 1708–1715 (1993).

15. L. Mutter, F. D. Brunner, Z. Yang, M. Jazbinšek, and P. Günter, "Linear and nonlinear optical properties of the organic crystal DSTMS," J. Opt. Soc. Am. B, **24**, 2556–2561 (2007).

16. T. Pliska, W.-R. Cho, J. Meier, A.-C. Le Duff, V. Ricci, A. Otomo, M. Canva, G. I. Stegeman, P. Raymond, and F. Kajzar, "Comparative study of nonlinear-optical polymers for guided-wave second-harmonic generation at telecommunication wavelengths," J. Opt. Soc. Am B, **17**, 1554–1564 (2000).

17. S. K. Kurtz and T. T. Perry, "A powder technique for the evaluation of nonlinear optical materials," J. Appl. Phys., **39**, 3798–3813 (1968).

18. W. N. Herman and L. M. Hayden, "Maker fringes revisited: second-harmonic generation from birefringent or absorbing materials," J. Opt. Soc. Am., **12**, 416–427 (1995).

19. J. M. Hales, J. W. Perry, "Organic and polymeric 3rd-order nonlinear optical materials and device applications," in Introduction to Organic Electronic and Optoelectronic Materials and Devices, edited by S.-S. Sun and L. Dalton (CRC Press, Orlando, FL, 2008), Chap. 17, pp. 521–579.

SUGGESTED FURTHER READING

D. S. Chemla and J. Zyss, eds, Nonlinear Optical Properties of Organic Molecules and Crystals: Volume 1 (Academic Press, Orlando, FL, 1987).

V. G. Dmitriev, G. G. Gurzadyan, and D. N. Nikogosyan, Handbook of Nonlinear Optical Crystals, 3rd Edition (Springer, Berlin 1999).

S. Miyata and H. Sasabe, Poled Polymers and Their Applications to SHG and EO Devices (CRC Press, Boca Raton, FL, 1997).

H. S. Nalwa and S. Miyata, eds, Nonlinear Optics of Organic Molecules and Polymers (CRC Press, Boca Raton, FL, 1997).

NONLINEAR SUSCEPTIBILITIES

CHAPTER 8

Second- and Third-Order Susceptibilities: Quantum Mechanical Formulation

Second-order nonlinear optics was developed in the preceding chapters on the basis of an anharmonic oscillator model with phenomenological nonlinear "force" constants that cannot be calculated from theories of the interaction of light with matter. This approach produced useful results for two reasons: (1) the interactions of interest are usually performed far enough from resonances so that the nonlinear susceptibility is approximately a constant and (2) the magnitudes of the nonlinear susceptibilities were determined experimentally.

Using first-order perturbation theory for allowed electric dipole transitions, formulas will be derived in this chapter for the second- and third-order nonlinear susceptibilities of a single isolated molecule with a given set of energy levels. The results, called the some over states (SOS), will be expressed in terms of the energy separations between the excited-state energy levels m and the ground state g, $\hbar\bar{\omega}_{mg}$, between excited states n and m, $\hbar\bar{\omega}_{nm}$, the photon energy of the incident light, $\hbar\omega_p$, and the transition electric dipole moments $\vec{\mu}_{mg}$ and $\vec{\mu}_{nm}$ between the states. In contrast to the anharmonic oscillator approach, all these parameters either can be calculated from first principles or can be obtained from linear and nonlinear spectroscopies.

Although the electrons can exist in any of the discrete states, they are predominantly in the lowest energy ground state. Electrons in the discrete excited states can be excited by some mechanism from the ground state. Here the discussion is limited to excitation due to light, normally accompanied by the absorption of energy from the light fields. Electrons in an excited odd symmetry state m will return to the ground state accompanied by a spontaneous emission of radiation of frequency $\bar{\omega}_{mg}$. The average electron lifetime in the excited state, $\bar{\tau}_{mg}$, will be introduced phenomenologically. Electrons in the ground state are assumed to have an "infinite" lifetime in the absence of interaction with a field. In a gas, there are collisions that can contribute to electron excitation into higher lying energy levels and also that reduce the natural lifetime associated with the excited states of molecules. In a solid there are

Nonlinear Optics: Phenomena, Materials, and Devices, George I. Stegeman and Robert A. Stegeman.
© 2012 John Wiley & Sons, Inc. Published 2012 by John Wiley & Sons, Inc.

intermolecular forces that alter the lifetimes, energy levels, and so on. However, since the binding of the electrons to the nucleus is normally the dominant effect, these other effects only decrease the lifetimes of the excited states and broaden their frequency spectra.

The analysis here deals with just one excited electron per molecule. The radiation interaction with the electromagnetic field is dealt with exclusively in the electric dipole approximation. Higher order effects, such as the electric quadrupole interaction, are typically much weaker and are ignored.

The philosophy of the derivations here is as follows:

1. The Schröedinger equation and first-order perturbation theory are used to find the probability of an electron being excited to state m by an interaction with the total electromagnetic field present.

2. The permanent and induced (by the interaction) dipole moments in a molecule are calculated.

3. The linear and nonlinear susceptibilities are deduced from the induced dipole moments.

4. The assumed infinite lifetime of the ground state will lead to some unphysical divergences. This problem has been resolved in the literature by modifying the procedure outlined here (beyond the scope of this text) and those solutions will be adopted.

5. Examples of the relevant susceptibilities will be derived for different light wave mixing interactions for the simple case of a "two-level" molecule and a three-level symmetric molecule.

8.1 PERTURBATION THEORY OF FIELD INTERACTION WITH MOLECULES

Before interaction with an electromagnetic field, it is assumed that the electrons are in the ground state g. $\hat{\bar{\Psi}}(\vec{r}, t)$ is the electron wave function, and $\left|\hat{\bar{\Psi}}(\vec{r}, t)\right|^2 d\vec{r}$ is the probability of finding an electron in volume $d\vec{r} = dx\, dy\, dz$ at time t. Since the electron must be somewhere, $\int_{-\infty}^{\infty}\int_{-\infty}^{\infty}\int_{-\infty}^{\infty} \left|\hat{\bar{\Psi}}(\vec{r}, t)\right|^2 d\vec{r} = 1$. The stationary discrete states are solutions of the Schrödinger equation $i\hbar(\partial\hat{\bar{\Psi}}/\partial t) = \bar{H}_0\hat{\bar{\Psi}}$. The wave function for the mth eigenstate is written as $\hat{\bar{\psi}}_m = \bar{u}_m(\vec{r})\, e^{-i\hat{\bar{\omega}}_m t}$, where $\bar{u}_m(\vec{r})$ is the spatial distribution of the wave function and $\hat{\bar{\omega}}_{mg} = \bar{\omega}_{mg} - i/\bar{\tau}_{mg}$ is a complex quantity with usually $\bar{\omega}_{mg}\bar{\tau}_{mg} \gg 1$, which reduces to $\hat{\bar{\omega}}_g = \bar{\omega}_g$ for the ground state, which does not decay. The parameter $\bar{\tau}_{mg}$ is the decay time from the excited state to the ground state. This is the physical significance of the imaginary part of the complex transition frequencies. The eigenstates are "orthogonal" in the sense that $\int_{-\infty}^{\infty}\int_{-\infty}^{\infty}\int_{-\infty}^{\infty} \bar{u}_m^*(\vec{r})\, \bar{u}_n(\vec{r})\, dx\, dy\, dz = \delta_{mn}$. The ground state wave function is $\hat{\bar{\Psi}}^{(s=0)}(\vec{r}, t) = \hat{a}_0\bar{u}_g(\vec{r})\, e^{-i\bar{\omega}_g t}$. The superscript $s = 0$ identifies the case that no interaction has yet occurred,

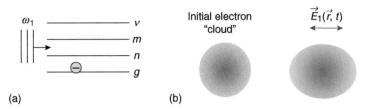

FIGURE 8.1 (a) A spherically symmetric molecule with *the* electron in the ground state prior to incidence of the field. (b) The initial electron cloud is distorted by the field that mixes the discrete eigenstates.

and $s > 0$ identifies the number of interactions between the electron and an electromagnetic field. The initial total wave function is given by

$$\hat{\bar{\Psi}}(\vec{r}, t) = \sum_m \hat{\bar{a}}_n^{(0)} \bar{u}_m(\vec{r}) e^{-i\hat{\omega}_m t} \quad \text{with } \hat{\bar{a}}_m^{(0)}(t) = 0 \text{ for } m \geq 1. \tag{8.1}$$

The summation over m is over all the states of the molecule that satisfy the Schrödinger equation, including the ground state; i.e.,

$$i\hbar \frac{\partial \hat{\bar{\psi}}_m^{(0)}}{\partial t} = \hbar\hat{\omega}_m \hat{\bar{\psi}}_m^{(0)} = \bar{H}_o \hat{\bar{\psi}}_m^{(0)} \quad \rightarrow \quad \bar{E}_m = \hbar\hat{\omega}_m. \tag{8.2}$$

8.1.1 Field–Molecule Interaction

Consider now an electromagnetic field incident on a symmetric molecule, as indicated in Fig. 8.1.

The electromagnetic field "shakes" the electrons, mixing the eigenstates via the electric dipole interaction and the electron cloud is distorted from its ground-state shape. Assuming that eigenstates and eigenfunctions are not significantly changed and that only small changes in the *populations* of the states occur, the probability of the electron being in the mth excited state due to one interaction is proportional to $\left| \hat{\bar{a}}_m^{(1)} \right|^2$ and the total wave function now becomes

$$\hat{\bar{\Psi}}(\vec{r}, t) = \hat{\bar{\Psi}}^{(0)}(\vec{r}, t) + \hat{\bar{\Psi}}^{(1)}(\vec{r}, t),$$

$$\hat{\bar{\Psi}}^{(1)}(\vec{r}, t) = \sum_m \hat{\bar{a}}_m^{(1)} \bar{u}_m(\vec{r}) e^{-i\hat{\omega}_m t} \quad \text{with } \hat{\bar{a}}_m^{(1)}(t) \neq 0 \text{ for some } m. \tag{8.3}$$

The electron cloud interacts again with either the same field or a different field. The electron cloud is distorted further, and the eigenstates are mixed further.

The probability of the electron in state m, $\left|\hat{\bar{a}}_m^{(2)}\right|^2$, changes, and the new wave function is

$$\hat{\bar{\Psi}}(\vec{r}, t) = \hat{\bar{\Psi}}^{(0)}(\vec{r}, t) + \hat{\bar{\Psi}}^{(1)}(\vec{r}, t) + \hat{\bar{\Psi}}^{(2)}(\vec{r}, t),$$

$$\hat{\bar{\Psi}}^{(2)}(\vec{r}, t) = \sum_n \hat{\bar{a}}_n^{(2)} \bar{u}_n(\vec{r})\, e^{-i\bar{\omega}_n t} \quad \text{with } \hat{\bar{a}}_n^{(2)}(t) \neq 0 \text{ for some } n. \tag{8.4}$$

Noting that this derivation is valid for $\left|\hat{\bar{a}}_0^{(2)}\right|^2 \gg \sum_{m>0} \left|\hat{\bar{a}}_m^{(2)}\right|^2$, i.e., weak excitation of the excited states, an illustration of the eigenstate probabilities is shown in Fig. 8.2. After a third interaction with an electromagnetic field (which is sufficient for calculating $\chi^{(3)}$),

$$\hat{\bar{\Psi}}^{(3)}(\vec{r}, t) = \sum_v \hat{\bar{a}}_v^{(3)} \bar{u}_v(\vec{r})\, e^{-i\bar{\omega}_v t} \quad \text{with } \hat{\bar{a}}_v^{(3)}(t) \neq 0 \text{ for some } v \text{ and so on.} \tag{8.5}$$

Note that the summations over m, n, and v are each over all the states.

In quantum mechanics, the interactions are governed by the interaction potential $\bar{V}(\vec{r}, t)$, which is inserted into the Schrödinger equation as follows:

$$i\hbar \frac{\partial}{\partial t}\Psi = (\bar{H}_0 + \varsigma\bar{V}(\vec{r}, t))\Psi, \tag{8.6}$$

and $\zeta = 1$ indicates that the interaction is on and $\zeta = 0$ indicates that it is off. In the electric dipole approximation,

$$\bar{V}(\vec{r}, t) = \vec{\bar{\mu}}(\vec{r}) \cdot \vec{E}_{\text{loc}}(\vec{r}, t) = -e\vec{r} \cdot \vec{E}_{\text{loc}}(\vec{r}, t), \tag{8.7}$$

where $\vec{E}_{\text{loc}}(\vec{r}, t)$ is the "local field" at the molecule and $\vec{\bar{\mu}}$ is the induced or permanent dipole moment. Thus the total wave function can be written in terms of the number of interactions as follows:

$$\hat{\bar{\Psi}} = \hat{\bar{\Psi}}^{(0)} + \varsigma\hat{\bar{\Psi}}^{(1)} + \varsigma^2\hat{\bar{\Psi}}^{(2)} + \varsigma^3\hat{\bar{\Psi}}^{(3)} + \cdots. \tag{8.8}$$

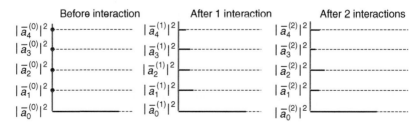

FIGURE 8.2 Illustration of the evolution of the probabilities for the first four states (three excited states) after two interactions with electromagnetic fields.

Defining

$\langle \vec{p}(t) \rangle =$ induced polarization, averaged over the electron distribution,

$$= \langle \hat{\Psi} | \hat{\vec{\mu}} | \hat{\Psi} \rangle = -\bar{e} \int_{-\infty}^{\infty} \hat{\Psi}^*(\vec{r}, t) \vec{r} \hat{\Psi}(\vec{r}, t) \, d\vec{r}$$

$$\equiv -\bar{e} \int_{-\infty}^{\infty} \int_{-\infty}^{\infty} \int_{-\infty}^{\infty} \hat{\Psi}^*(\vec{r}, t) \vec{r} \hat{\Psi}(\vec{r}, t) \, dx \, dy \, dz$$

$$\rightarrow \quad \langle \vec{p}(t) \rangle = \langle \vec{p}^{(0)}(t) \rangle + \varsigma \langle \vec{p}^{(1)}(t) \rangle + \varsigma^2 \langle \vec{p}^{(2)}(t) \rangle + \varsigma^3 \langle \vec{p}^{(3)}(t) \rangle + \cdots . \quad (8.9)$$

The term $\langle \vec{p}^{(0)}(t) \rangle$ is the permanent dipole moment (which is independent of time), $\langle \vec{p}^{(1)}(t) \rangle$ is the linear polarization induced via the molecular linear polarizability $\bar{\alpha}_{ij}$ and $\bar{\alpha}_{ij} \xrightarrow{\text{leads to}} \chi_{ij}^{(1)}$, $\langle \vec{p}^{(2)}(t) \rangle$ is the nonlinear polarization induced via the second-order molecular susceptibility $\bar{\beta}_{ijk}^{(2)} \xrightarrow{\text{leads to}} \chi_{ijk}^{(2)}$, $\langle \vec{p}^{(3)}(t) \rangle$ is the nonlinear polarization induced via the third-order molecular susceptibility $\bar{\gamma}_{ijk\ell}^{(3)} \xrightarrow{\text{leads to}} \chi_{ijk\ell}^{(3)}$, $\langle \vec{p}^{(4)}(t) \rangle$ to $\chi_{ijk\ell mn}^{(4)}$, and so on.

Successive applications of first-order perturbation theory can now be used to calculate these susceptibilities. Inserting the expansion of the total wave function into terms of the number of interactions,

$$i\hbar \frac{\partial \left[\hat{\Psi}^{(0)} + \varsigma \hat{\Psi}^{(1)} + \varsigma^2 \hat{\Psi}^{(2)} + \varsigma^3 \hat{\Psi}^{(3)} + \cdots \right]}{\partial t} = \bar{H}_0 \left[\hat{\Psi}^{(0)} + \varsigma \hat{\Psi}^{(1)} + \varsigma^2 \hat{\Psi}^{(2)} + \varsigma^3 \hat{\Psi}^{(3)} + \cdots \right]$$

$$+ \varsigma \bar{V}(t) \left[\hat{\Psi}^{(0)} + \varsigma \hat{\Psi}^{(1)} + \varsigma^2 \hat{\Psi}^{(3)} + \cdots \right],$$

$$(8.10)$$

and equating terms with the same power of ζ gives

$$i\hbar \frac{\partial \hat{\Psi}^{(0)}}{\partial t} = \bar{H}_0 \hat{\Psi}^{(0)}, \qquad i\hbar \frac{\partial \hat{\Psi}^{(1)}}{\partial t} = \bar{H}_0 \hat{\Psi}^{(1)} + \bar{V}(t) \hat{\Psi}^{(0)},$$

$$(8.11)$$

$$i\hbar \frac{\partial \hat{\Psi}^{(2)}}{\partial t} = \bar{H}_0 \hat{\Psi}^{(2)} + \bar{V}(t) \hat{\Psi}^{(1)}, \qquad i\hbar \frac{\partial \hat{\Psi}^{(3)}}{\partial t} = \bar{H}_0 \hat{\Psi}^{(3)} + \bar{V}(t) \hat{\Psi}^{(2)}.$$

Substituting the expansions for the wave functions and the electric dipole interaction into the Schrödinger equation, the amplitude function $\hat{\bar{a}}_\ell^{(N)}$ after N interactions is given by the probability amplitude distribution of the states $\hat{\bar{a}}_{\ell'}^{(N-1)}$ after $N-1$ interactions by

$$i\hbar[\hat{\bar{a}}_{\ell}^{(N)}(t) - i\hat{\bar{\omega}}_{\ell}\hat{\bar{a}}_{\ell}^{(N)}(t)]\bar{u}_{\ell}(\vec{r})\,e^{-i\hat{\bar{\omega}}_{\ell}t} = \hbar\hat{\bar{\omega}}_{\ell}\hat{\bar{a}}_{\ell}^{(N)}(t)\bar{u}_{\ell}(\vec{r})e^{-i\hat{\bar{\omega}}_{\ell}t}$$

$$+ \sum_{\ell'}\hat{\bar{a}}_{\ell'}^{(N-1)}(t)\bar{V}(\vec{r},t)\bar{u}_{\ell'}(\vec{r})\,e^{-i\hat{\bar{\omega}}_{\ell'}t}. \tag{8.12}$$

Equation 8.12 is simplified by multiplying it by $\bar{u}_{\ell}^{*}(\vec{r})\,e^{i\hat{\bar{\omega}}_{\ell}t}$ and integrating over all space to give

$$i\hbar\int_{-\infty}^{\infty}\bar{u}_{\ell}^{*}(\vec{r})\bar{u}_{\ell}(\vec{r})\,d\vec{r}[\hat{\bar{a}}_{\ell}^{(N)}(t) - i\hat{\bar{\omega}}_{\ell}\hat{\bar{a}}_{\ell}^{(N)}(t)] = \hbar\hat{\bar{\omega}}_{\ell}\hat{\bar{a}}_{\ell}^{(N)}(t)\int_{-\infty}^{\infty}\bar{u}_{\ell}^{*}(\vec{r})\bar{u}_{\ell}(\vec{r})\,d\vec{r}$$

$$+ \sum_{\ell'}\hat{\bar{a}}_{\ell'}^{N-1}(t)\int_{-\infty}^{\infty}\bar{u}_{\ell}^{*}(\vec{r})\bar{V}(\vec{r},t)\bar{u}_{\ell'}(\vec{r})\,d\vec{r}\}\,e^{i[\hat{\bar{\omega}}_{\ell}-\hat{\bar{\omega}}_{\ell'}]t}. \tag{8.13}$$

Applying the normalization and orthogonality relations gives

$$\hat{\bar{a}}_{\ell}^{(N)}(t) = \frac{1}{i\hbar}\sum_{\ell'}\hat{\bar{a}}_{\ell'}^{(N-1)}\bar{V}_{\ell\ell'}(t)\,e^{i(\hat{\bar{\omega}}_{\ell}-\hat{\bar{\omega}}_{\ell'})t}(t)$$

$$\text{with } \bar{V}_{\ell\ell'}(t) = \int_{-\infty}^{\infty}\bar{u}_{\ell}^{*}(\vec{r})\bar{V}(\vec{r},t)\bar{u}_{\ell'}(\vec{r})\,d\vec{r}. \tag{8.14}$$

Defining $\hat{\bar{\omega}}_{\ell\ell'} = \hat{\bar{\omega}}_{\ell} - \hat{\bar{\omega}}_{\ell'}$ and integrating from $t' = -\infty$ to t gives

$$\hat{\bar{a}}_{\ell}^{(N)}(t) = \frac{1}{i\hbar}\sum_{\ell'}\int_{-\infty}^{t}\bar{V}_{\ell\ell'}(t')\hat{\bar{a}}_{\ell'}^{(N-1)}(t')\,e^{i\hat{\bar{\omega}}_{\ell\ell'}t'}\,dt', \tag{8.15}$$

with $|\hat{\bar{a}}_{\ell}^{(N)}|^2$ being the probability that the electron is in the state ℓ, given the probability that it was in the ℓ' state after $N-1$ interactions. This is a general quantum mechanics result obtained by applying first-order perturbation theory N times.

8.1.2 Electric Dipole Interaction

The total electromagnetic field present at the site of a molecule, $\vec{E}_{\text{loc}}(\vec{r},t)$, can consist of a number of different frequencies $\omega_{p'}$ with $p' > 0$ so that

$$\vec{E}_{\text{loc}}(\vec{r},t) = \frac{1}{2}\sum_{p'}\vec{E}_{\text{loc}}(\omega_{p'})\,e^{-i\omega_{p'}t} + \text{c.c.} = \frac{1}{2}\sum_{p'}[\vec{E}_{\text{loc}}(\omega_{p'})\,e^{-i\omega_{p'}t} + \vec{E}_{\text{loc}}^{*}(\omega_{p'})\,e^{i\omega_{p'}t}]. \tag{8.16}$$

Note that $\vec{E}_{loc}^*(\omega_{p'}) = \vec{E}_{loc}(-\omega_{p'})$ and that in nonlinear optics, $\vec{E}_{loc}(\omega_{p'})\, e^{-i\omega_{p'}t}$ and $\vec{E}_{loc}^*(\omega_{p'})\, e^{i\omega_{p'}t}$ can be considered to be separate input modes. Although this is strictly incorrect in linear optics, it does facilitate the analysis of wave mixing interactions. Therefore, substituting Eq. 8.16 into Eq. 8.14 gives

$$\bar{V}_{\ell\ell'}(t) = \int_{-\infty}^{\infty} \bar{u}_{\ell}^*(\vec{r})\vec{\bar{\mu}}(\vec{r}) \cdot \vec{E}_{loc}(\vec{r},t)\bar{u}_{\ell'}(\vec{r})\, d\vec{r}$$

$$= \sum_{p'} \vec{\bar{\mu}}_{\ell\ell'} \cdot \frac{1}{2}\{\vec{E}_{loc}(\omega_{p'})\, e^{-i\omega_{p'}t} + \vec{E}_{loc}^*(\omega_{p'})\, e^{+i\omega_{p'}t}\}, \tag{8.17a}$$

$$\vec{\bar{\mu}}_{\ell\ell'} = \int_{-\infty}^{\infty} \bar{u}_{\ell}^*(\vec{r})\vec{\bar{\mu}}(\vec{r})\bar{u}_{\ell'}(\vec{r})\, d\vec{r}. \tag{8.17b}$$

8.1.2.1 *First Interaction of the Molecules with the Field.* Starting from

the ground state $\hat{\bar{\Psi}}^{(0)}(\vec{r},t) = \hat{a}_0\bar{u}_g(\vec{r})\, e^{-i\hat{\bar{\omega}}_g t}\, (g \equiv \ell')$ and integrating Eq. 8.15 from $t' = -\infty$ to t, the probability amplitude for $m \equiv \ell$ is

$$\hat{\bar{a}}_m^{(1)}(t) = \frac{1}{2\hbar}\sum_{p'}\left\{\frac{\vec{\bar{\mu}}_{mg} \cdot \vec{E}_{loc}(\omega_{p'})}{\hat{\bar{\omega}}_{mg} - \omega_{p'}}e^{i(\hat{\bar{\omega}}_{mg}-\omega_{p'})t} + \frac{\vec{\bar{\mu}}_{mg} \cdot \vec{E}_{loc}^*(\omega_{p'})}{\hat{\bar{\omega}}_{mg} + \omega_{p'}}e^{i(\hat{\bar{\omega}}_{mg}+\omega_{p'})t}\right\}. \tag{8.18}$$

It is useful to redefine the summation over p' to a summation over p, with p going from $-p_{max}$ to p_{max} (where p_{max} is the total number of fields present), and for negative p, $\omega_{-p} = -\omega_p$ and $\vec{E}_{loc}(\omega_{-p}) = \vec{E}_{loc}^*(\omega_p)$ and $p=0$ gives no contribution. For example, the sum of the $p=2$ and $p=-2$ terms is

$$\frac{1}{2\hbar}\left\{\frac{\vec{\bar{\mu}}_{mg} \cdot \vec{E}_{loc}(\omega_2)}{\hat{\bar{\omega}}_{mg} - \omega_2}e^{i(\hat{\bar{\omega}}_{mg}-\omega_2)t} + \frac{\vec{\bar{\mu}}_{mg} \cdot \vec{E}_{loc}^*(\omega_2)}{\hat{\bar{\omega}}_{mg} + \omega_2}e^{i(\hat{\bar{\omega}}_{mg}+\omega_2)t}\right\}. \tag{8.19}$$

Equation 8.18 becomes

$$\bar{a}_m^{(1)}(t) = \frac{1}{2\hbar}\sum_{p}\frac{\vec{\bar{\mu}}_{mg} \cdot \vec{E}_{loc}(\omega_p)}{\hat{\bar{\omega}}_{mg} - \omega_p}e^{i(\hat{\bar{\omega}}_{mg}-\omega_p)t}. \tag{8.20}$$

8.1.2.2 Second Interaction of the Molecules with the Field. For the second interaction of the total field with the molecule, from Eq. 8.15 we obtain

$$
\hat{a}_n^{(2)}(t) = \frac{1}{i\hbar} \sum_m \int_{-\infty}^t \bar{V}_{nm}(t') \hat{a}_m^{(1)}(t') \, e^{i\hat{\omega}_{nm}t'} \, dt'
$$

$$
= -\frac{1}{2i\hbar^2} \sum_m \sum_q \int_{-\infty}^t \hat{a}_m^{(1)}(t') \vec{\mu}_{mn} \cdot \vec{E}_{\text{loc}}(\omega_q) \, e^{i(\hat{\omega}_{nm}-\omega_q)t'} \, dt' \tag{8.21}
$$

and substituting Eq. 8.20 for $\hat{a}_m^{(1)}(t')$ gives

$$
\hat{a}_n^{(2)}(t) = \frac{1}{4\hbar^2} \sum_q \sum_p \sum_m \frac{[\vec{\mu}_{nm} \cdot \vec{E}_{\text{loc}}(\omega_q)][\vec{\mu}_{mg} \cdot \vec{E}_{\text{loc}}(\omega_p)]}{(\hat{\omega}_{ng} - \omega_q - \omega_p)(\hat{\omega}_{mg} - \omega_p)} e^{i(\hat{\omega}_{ng}-\omega_q-\omega_p)t}. \tag{8.22}
$$

The case $n = g$ will ultimately require a refinement of this theory. For example, if $n = g$ and $\pm\omega_q = \mp\omega_p$, then $\hat{\omega}_{gg} = 0$ leads to a divergence, which will be discussed in sections 8.2.2.4 and 8.2.2.5. Note that with the simplified notation, Eq. 8.22 actually contains four terms at the frequencies $\hat{\omega}_{ng} \pm \omega_q \pm \omega_p$. This approach can easily be extended to three interactions with the fields:

$$
\hat{a}_v^{(3)}(t) = \frac{1}{8\hbar^3} \sum_{n,m} \sum_{r,q,p} \frac{[\vec{\mu}_{vm} \cdot \vec{E}_{\text{loc}}(\omega_r)][\vec{\mu}_{mn} \cdot \vec{E}_{\text{loc}}(\omega_q)][\vec{\mu}_{ng} \cdot \vec{E}_{\text{loc}}(\omega_p)]}{(\hat{\omega}_{vg} - \omega_r - \omega_q - \omega_p)(\hat{\omega}_{ng} - \omega_q - \omega_p)(\hat{\omega}_{mg} - \omega_p)} e^{i(\hat{\omega}_{vg}-\omega_r-\omega_q-\omega_p)t}.
$$

$$
\tag{8.23}
$$

The summations over n and m are *both* over *all* the states. Also summations over p, q, and r are *each* over *all* the fields present. Note that states m and n can be the same state, states m and v can be the same state, and so on. Finally, note that there appears to be a time sequence for the interactions with fields which is p, q, r. However, since each of p, q, r is over the total field, all the possible permutations of p, q, r approximate an "instantaneous interaction." For example, assume that there are two *optical* fields present: $\vec{E}_{\text{loc}}(\omega_1)$ and $\vec{E}_{\text{loc}}(\omega_2)$. Therefore for $a_v^{(2)}$, p, and q each run from -2 to $+2$, excluding 0, and there are $4 \times 4 = 16$ different contributing field combinations, each defining a time sequence.

For *each* field combination of initial and final states, there are multiple possible "intermediate" states (pathways to state v), denoted by m and n, which can be identical, different, and so on. For example, if there is the ground state g and three excited states, one of which is the state $v = 2$, then the "pathways" to $v = 2$ could be $g \rightarrow 2 \rightarrow 1 \rightarrow 2$, $g \rightarrow 3 \rightarrow 1 \rightarrow 2$, $g \rightarrow 2 \rightarrow g \rightarrow 2$, and so on. The probability for each step in the pathway, e.g., state m to state n is given by the transition dipole matrix element $|\vec{\mu}_{nm}|^2$. Normally, there are only a few states linked by strong transition moments in a given molecule that simplifies the SOS calculation. The probability of the exciting state m also depends, via the resonant denominators, on how close the energy difference is between the ground state (initial electronic state before any interaction) and the state m, i.e., whether it matches the energy obtained from the electromagnetic fields in reaching state m via state n and the other states in that particular pathway.

8.2 OPTICAL SUSCEPTIBILITIES

It is now possible to calculate the "expectation value" of the linear and nonlinear polarization terms, which will ultimately lead to the linear and nonlinear susceptibilities. From Eq. 8.9, we obtain

$$\langle \vec{p}(t) \rangle = \int \sum_{\ell} \hat{\bar{a}}_{\ell}^{*}(t) \bar{u}_{\ell}^{*}(\vec{r}) \vec{\mu}(\vec{r}) \sum_{\ell'} \hat{\bar{a}}_{\ell'}(t) u_{\ell'}(\vec{r}) \, d\vec{r} \, e^{-i(\hat{\bar{\omega}}_{\ell'} - \hat{\bar{\omega}}_{\ell})t}. \tag{8.24}$$

There is a well-defined value (could be zero) for $\vec{p}(t)$ in the ℓth state when $\ell = \ell'$. $\vec{p}_{\ell}$ is the permanent dipole moment in state ℓ, independent of time; i.e.,

$$\vec{p}_{\ell} \xrightarrow[\text{for molecule in excited state}]{\text{"permanent" dipole moment}} \int \bar{u}_{\ell}^{*}(\vec{r}) \vec{\mu}(\vec{r}) \bar{u}_{\ell}(\vec{r}) \, d\vec{r} = \vec{\mu}_{\ell\ell} \text{ (real quantity).} \tag{8.25}$$

Using the general expansion for $\langle \vec{p}(t) \rangle$ in powers of ζ in Eq. 8.9, we obtain

$$\langle \vec{p}(t) \rangle = \left[\left\langle \hat{\bar{\Psi}}^{(0)} | \vec{\mu} | \hat{\bar{\Psi}}^{(0)} \right\rangle + \varsigma \left\{ \left\langle \hat{\bar{\Psi}}^{(0)} | \vec{\mu} | \hat{\bar{\Psi}}^{(1)} \right\rangle + \left\langle \hat{\bar{\Psi}}^{(1)} | \vec{\mu} | \hat{\bar{\Psi}}^{(0)} \right\rangle \right\} \right.$$

$$+ \varsigma^2 \left\{ \left\langle \hat{\bar{\Psi}}^{(0)} | \vec{\mu} | \hat{\bar{\Psi}}^{(2)} \right\rangle + \left\langle \hat{\bar{\Psi}}^{(1)} | \vec{\mu} | \hat{\bar{\Psi}}^{(1)} \right\rangle + \left\langle \hat{\bar{\Psi}}^{(2)} | \vec{\mu} | \hat{\bar{\Psi}}^{(0)} \right\rangle \right\}$$

$$+ \varsigma^3 \left\{ \left\langle \hat{\bar{\Psi}}^{(0)} | \vec{\mu} | \hat{\bar{\Psi}}^{(3)} \right\rangle + \left\langle \hat{\bar{\Psi}}^{(1)} | \vec{\mu} | \hat{\bar{\Psi}}^{(2)} \right\rangle \right.$$

$$\left. \left. + \left\langle \hat{\bar{\Psi}}^{(2)} | \vec{\mu} | \hat{\bar{\Psi}}^{(1)} \right\rangle + \left\langle \hat{\bar{\Psi}}^{(3)} | \vec{\mu} | \hat{\bar{\Psi}}^{(0)} \right\rangle \right\} \right], \tag{8.26}$$

where integration over space is implicit in $\langle || \rangle$ and the wave functions are all functions of space and time, i.e., $(\vec{r}, t)$. The terms in this equation have a physical interpretation as illustrated by the two examples shown in Fig. 8.3.

8.2.1 Linear Susceptibility

The terms proportional to ς^1 ultimately give rise to $\chi^{(1)}$; i.e.,

$$\varsigma^1 \rightarrow \left\langle \vec{p}^{(1)} \right\rangle = \left\langle \hat{\vec{\Psi}}^{(0)} | \vec{\hat{\mu}} | \hat{\vec{\Psi}}^{(1)} \right\rangle + \left\langle \hat{\vec{\Psi}}^{(1)} | \vec{\hat{\mu}} | \hat{\vec{\Psi}}^{(0)} \right\rangle \rightarrow \chi^{(1)}. \tag{8.27}$$

Here $\left\langle \hat{\vec{\Psi}}^{(0)} | \vec{\hat{\mu}} | \hat{\vec{\Psi}}^{(1)} \right\rangle = \int \hat{\vec{\Psi}}^{*(0)} \vec{\hat{\mu}} \hat{\vec{\Psi}}^{(1)} \, d\vec{r}$ and $\left\langle \hat{\vec{\Psi}}^{(1)} | \vec{\hat{\mu}} | \hat{\vec{\Psi}}^{(0)} \right\rangle = \int \hat{\vec{\Psi}}^{*(1)} \vec{\hat{\mu}} \hat{\vec{\Psi}}^{(0)} \, d\vec{r}$, with

$$\hat{\vec{\Psi}}^{(0)}(\vec{r}, t) = \bar{u}_g(\vec{r}) \, e^{-i\bar{\omega}_g t},$$

$$\hat{\vec{\Psi}}^{(1)}(\vec{r}, t) = \frac{1}{2\hbar} \sum_m \sum_p \frac{\vec{\bar{\mu}}_{mg} \cdot \vec{E}_{\text{loc}}(\omega_p)}{\hat{\bar{\omega}}_{mg} - \omega_p} e^{i(\hat{\bar{\omega}}_{mg} - \omega_p)t} \bar{u}_m(\vec{r}) \, e^{-i\hat{\bar{\omega}}_m t} \tag{8.28}$$

$$\rightarrow \left\langle \hat{\vec{\Psi}}^{(1)} | \vec{\hat{\mu}} | \hat{\vec{\Psi}}^{(0)} \right\rangle = \frac{1}{2\hbar} \sum_m \sum_{p>0} \frac{[\vec{\bar{\mu}}^*_{mg} \cdot \vec{E}^*_{\text{loc}}(\omega_p)]}{\hat{\bar{\omega}}^*_{mg} - \omega_p} e^{-i\bar{\omega}_g t + i(\omega_p + \bar{\omega}_g)t} \int \bar{u}^*_m(\vec{r}) \vec{\bar{\mu}}(\vec{r}) \bar{u}_g(\vec{r}) \, d\vec{r} \quad (p > 0)$$

$$+ \frac{1}{2\hbar} \sum_m \sum_{p<0} \frac{[\vec{\bar{\mu}}^*_{mg} \cdot \vec{E}_{\text{loc}}(\omega_p)]}{\hat{\bar{\omega}}^*_{mg} + \omega_p} e^{i\bar{\omega}_g t - i(\omega_p + \bar{\omega}_g)t} \int \bar{u}^*_m(\vec{r}) \vec{\bar{\mu}}(\vec{r}) \bar{u}_g(\vec{r}) \, d\vec{r} \quad (p < 0)$$

$$= \frac{1}{2\hbar} \Bigg\{ \sum_m \sum_{p>0} \frac{[\vec{\bar{\mu}}^*_{mg} \cdot \vec{E}^*_{\text{loc}}(\omega_p)]\vec{\bar{\mu}}_{mg}}{\hat{\bar{\omega}}^*_{mg} - \omega_p} e^{+i\omega_p t}$$

$$+ \sum_m \sum_{p<0} \frac{[\vec{\bar{\mu}}^*_{mg} \cdot \vec{E}_{\text{loc}}(\omega_p)]\vec{\bar{\mu}}_{mg}}{\hat{\bar{\omega}}^*_{mg} + \omega_p} e^{-i\omega_p t} \tag{8.29a}$$

$$\rightarrow \int \hat{\vec{\Psi}}^{*(0)} \vec{\hat{\mu}} \hat{\vec{\Psi}}^{(1)} \, d\vec{r} = \frac{1}{2\hbar} \sum_m \sum_{p>0} \frac{\vec{\bar{\mu}}_{mg} \cdot \vec{E}_{\text{loc}}(\omega_p)}{\hat{\bar{\omega}}_{mg} - \omega_p} e^{i(\bar{\omega}_g - \omega_p - \bar{\omega}_g)t} \int \bar{u}^*_g(\vec{r}) \vec{\bar{\mu}}(\vec{r}) \bar{u}_m(\vec{r}) \, d\vec{r}$$

$$+ \frac{1}{2\hbar} \sum_m \sum_{p<0} \frac{\vec{\bar{\mu}}_{mg} \cdot \vec{E}^*_{\text{loc}}(\omega_p)}{\hat{\bar{\omega}}_{mg} + \omega_p} e^{-i(\bar{\omega}_g - \omega_p - \bar{\omega}_g)t} \int \bar{u}^*_g(\vec{r}) \vec{\bar{\mu}}(\vec{r}) u_m(\vec{r}) \, d\vec{r}$$

$$= \frac{1}{2\hbar} \sum_m \sum_{p>0} \frac{\vec{\bar{\mu}}_{gm}[\vec{\bar{\mu}}_{mg} \cdot \vec{E}_{\text{loc}}(\omega_p)]}{\hat{\bar{\omega}}_{mg} - \omega_p} e^{-i\omega_p t}$$

$$+ \sum_m \sum_{p<0} \frac{[\vec{\bar{\mu}}_{mg} \cdot \vec{E}^*_{\text{loc}}(\omega_p)]\vec{\bar{\mu}}_{gm}}{\hat{\bar{\omega}}_{mg} + \omega_p} e^{+i\omega_p t}. \tag{8.29b}$$

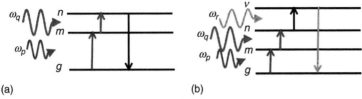

FIGURE 8.3 (a) Illustration of the pathway $g|\vec{\mu}_{gn}|n \xleftarrow{\vec{\mu}_{nm}} m \xleftarrow{\vec{\mu}_{mg}} g$ for $\left\langle \Psi^{(0)}|\vec{\mu}|\Psi^{(2)} \right\rangle$.
(b) Illustration of the pathway $g \xrightarrow{\vec{\mu}_{gv}} v|\vec{\mu}_{vn}|n \xleftarrow{\vec{\mu}_{nm}} m \xleftarrow{\vec{\mu}_{mg}} g$ for $\left\langle \Psi^{(1)}|\vec{\mu}|\Psi^{(2)} \right\rangle$.

Since $\vec{\mu}_{gm}^{*} = \vec{\mu}_{mg}$, $\left\langle \hat{\Psi}^{(0)}|\hat{\vec{\mu}}|\hat{\Psi}^{(1)} \right\rangle + \left\langle \hat{\Psi}^{(1)}|\hat{\vec{\mu}}|\hat{\Psi}^{(0)} \right\rangle$ from Eq. 8.29 is

$$
\frac{1}{2\hbar} \sum_{m} \left\{ \sum_{p>0} \frac{\vec{\mu}_{gm}[\vec{\mu}_{mg} \cdot \vec{E}_{\text{loc}}(\omega_p)]}{\hat{\omega}_{mg} - \omega_p} + \sum_{p<0} \frac{\vec{\mu}_{mg}[\vec{\mu}_{gm} \cdot \vec{E}_{\text{loc}}(\omega_p)]}{\hat{\omega}_{mg}^{*} + \omega_p} \right\} e^{-i\omega_p t}
$$

$$
+ \sum_{m} \left\{ \sum_{p>0} \frac{[\vec{\mu}_{gm} \cdot \vec{E}_{\text{loc}}^{*}(\omega_p)]\vec{\mu}_{mg}}{\hat{\omega}_{mg}^{*} - \omega_p} + \sum_{p<0} \frac{\vec{\mu}_{gm}[\vec{\mu}_{mg} \cdot \vec{E}_{\text{loc}}^{*}(\omega_p)]}{\hat{\omega}_{mg} + \omega_p} \right\} e^{+i\omega_p t}. \tag{8.30}
$$

This finally gives for the induced linear polarization

$$
\left\langle \vec{p}^{(1)}(t) \right\rangle = \frac{1}{2\hbar} \sum_{m} \sum_{p} \left\{ \frac{\vec{\mu}_{gm}[\vec{\mu}_{mg} \cdot \vec{E}_{\text{loc}}(\omega_p)]}{\hat{\omega}_{mg} - \omega_p} + \frac{\vec{\mu}_{mg}[\vec{\mu}_{gm} \cdot \vec{E}_{\text{loc}}(\omega_p)]}{\hat{\omega}_{ng}^{*} + \omega_p} \right\} e^{-i\omega_p t} + \text{c.c.}
$$

$$
\tag{8.31}
$$

The two denominator terms in Eq. 8.31 are referred to as the "resonant" and "antiresonant" terms. The former has the form $[\bar{\omega}_{mg} - i/\bar{\tau}_{mg} - \omega_p]^{-1}$, and $\langle \vec{p}^{(1)} \rangle$ is enhanced when $\bar{\omega}_{mg} \cong \omega_p$, and hence the name "resonant." For the $[\bar{\omega}_{mg} + i/\bar{\tau}_{mg} + \omega_p]^{-1}$ term, the denominator remains large and hence the name "antiresonant" is appropriate. Note that although the resonant contribution is dominant when the photon energy is comparable to $\hbar\bar{\omega}_{mg}$, in the zero-frequency limit ($\hbar\bar{\omega}_{mg} \gg \hbar\omega_p$) the two terms are comparable.

Perhaps a more physical interpretation can be given in terms of the time that the field interacts with the molecule as interpreted by the uncertainty principle. When an electromagnetic field interacts with the electron cloud, there can be energy exchange between the molecule and the field. The uncertainty principle $\Delta E \Delta t \geq \hbar$ can be interpreted in terms of ΔE being the allowed "uncertainty" in energy and Δt being the maximum time over which it can occur. Within this constraint, a photon can be absorbed and reemitted *or* emitted and then reabsorbed. These two processes are shown in Fig. 8.4.

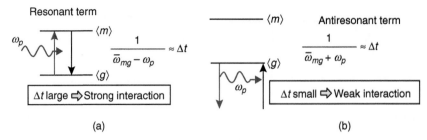

FIGURE 8.4 (a) The resonant term for which Δt is interpreted as a "long" interaction time, resulting in a strong net interaction. (b) The nonresonant term gives a much shorter interaction time, leading to a much weaker net interaction.

8.2.1.1 *Local Field Correction.*

The basic concept of the local field was discussed in Chapter 1, where it was applied to the linear susceptibility. In Chapter 1, it was shown that the macroscopic linear polarization is given by

$$\vec{P}^{(1)}(\vec{r},t) = \frac{\varepsilon_r + 2}{3} N \vec{\tilde{\alpha}} \cdot \vec{E}(\vec{r},t), \tag{8.32}$$

and so the local field and the local field correction $f^{(1)}$ are

$$\vec{E}_{\mathrm{loc}}(\vec{r},t) = \frac{\varepsilon_r(\omega)+2}{3}\vec{E}(\vec{r},t), \qquad f^{(1)} = \frac{\varepsilon_r(\omega)+2}{3}, \tag{8.33}$$

respectively. Hence in the present context, Eq. 8.31 can be rewritten as

$$\left\langle \vec{p}^{(1)}(t) \right\rangle = \frac{1}{2\hbar} f^{(1)} \sum_m \sum_p \left\{ \frac{\vec{\mu}_{gm}[\vec{\mu}_{mg} \cdot \vec{E}(\omega_p)]}{\hat{\tilde{\omega}}_{mg} - \omega_p} + \frac{\vec{\mu}_{mg}[\vec{\mu}_{gm} \cdot \vec{E}(\omega_p)]}{\hat{\tilde{\omega}}^*_{mg} + \omega_p} \right\} e^{-i\omega_p t} + \mathrm{c.c.} \tag{8.34}$$

Furthermore, since the polarization is usually expressed in terms of the Maxwell field $\vec{E}(\omega_p)$,

$$\vec{P}^{(1)}(\vec{r},t) = N\left\langle \vec{p}^{(1)}(\vec{r},t) \right\rangle = \frac{1}{2}\varepsilon_0 \sum_p \vec{\tilde{\chi}}^{(1)}(-\omega_p;\omega_p) \cdot \vec{E}(\omega_p) e^{-i\omega_p t} + \mathrm{c.c.}$$

$$\Rightarrow \vec{\tilde{\chi}}^{(1)}(-\omega_p;\omega_p) = \frac{N}{\hbar\varepsilon_0} f^{(1)} \sum_m \left\{ \frac{\vec{\mu}_{gm}\vec{\mu}_{mg}}{\tilde{\omega}_{mg} - \omega_p - i\bar{\tau}_{mg}^{-1}} + \frac{\vec{\mu}_{mg}\vec{\mu}_{gm}}{\tilde{\omega}_{mg} + \omega_p + i\bar{\tau}_{mg}^{-1}} \right\}$$

$$\Rightarrow \vec{\tilde{\chi}}^{(1)}(-\omega_p;\omega_p) = \frac{N}{\hbar\varepsilon_0} f^{(1)} \sum_m |\vec{\mu}_{mg}|^2\, 2\tilde{\omega}_{mg} \left\{ \frac{\tilde{\omega}_{mg}^2 - \tilde{\omega}_p^2 + \tau_{mg}^{-2} + 2i\bar{\tau}_{mg}^{-1}\omega}{[(\tilde{\omega}_{mg} - \omega_p)^2 + \bar{\tau}_{mg}^{-2}][(\tilde{\omega}_{mg} + \omega_p)^2 + \bar{\tau}_{mg}^{-2}]} \right\}. \tag{8.35}$$

It is interesting to compare this dispersion with the dispersion from the electron on a spring model (Eq. 1.26):

$$
\hat{\chi}_{ii}^{(1)}(-\omega_p;\omega_p) = \frac{N\bar{e}^2 f^{(1)}(\omega_p)}{\bar{m}_e \varepsilon_0} \sum_m \frac{\bar{\omega}_{mg}^2 - \omega_p^2 + 2i\omega_p \bar{\tau}_{mg}^{-1}}{(\bar{\omega}_{mg}^2 - \omega_p^2)^2 + 4\omega_p^2 \bar{\tau}_{mg}^{-2}},
\tag{8.36a}
$$

and neglecting the shift in the resonance frequency due to overdamping and rewriting Eq. 8.35 as

$$
\hat{\chi}_{ii}^{(1)}(-\omega_p;\omega_p) = \frac{2\bar{\omega}_{mg}N}{\hbar\varepsilon_0} f^{(1)}(\omega_p) \sum_m |\vec{\mu}_{mg}|^2 \left\{ \frac{\bar{\omega}_{mg}^2 - \omega_p^2 + 2i\omega_p \bar{\tau}_{mg}^{-1}}{(\bar{\omega}_{mg}^2 - \omega_p^2)^2 + 2\bar{\tau}_{mg}^{-2}(\bar{\omega}_{mg}^2 + \omega_p^2)} \right\}
\tag{8.36b}
$$

indicates that the equations are identical near and on resonance where $\omega_{mg} \cong \omega_p$, which allows the transition dipole moment to be identified in terms of the harmonic oscillator parameters as $|\vec{\mu}_{mg}|^2 = \bar{e}^2 \hbar / 2\bar{m}_e \bar{\omega}_{mg}$.

The local field correction is necessary *only* when a macroscopic material response is calculated from single molecule properties. Always keep in mind that there are many other complications in condensed matter that make this approach approximate. For example, lifetimes of states and so on are all usually altered in the condensed matter state. Most linear and nonlinear optical properties are determined experimentally, e.g., the second-order coefficients $\hat{\chi}_{ijk}^{(2)}$ and $\hat{d}_{ijk}^{(2)}$. Hence the local field effects are already included in the values of the experimental coefficients.

8.2.2 Second-Order Susceptibility (Three ζ^2 Terms)

There are three terms that contribute to the induced second-order nonlinear polarization in a single molecule; namely,

$$
\left\langle \vec{\hat{p}}^{(2)}(t) \right\rangle = \left\langle \hat{\vec{\Psi}}^{(0)} | \hat{\vec{\mu}} | \hat{\vec{\Psi}}^{(2)} \right\rangle + \left\langle \hat{\vec{\Psi}}^{(1)} | \hat{\vec{\mu}} | \hat{\vec{\Psi}}^{(1)} \right\rangle + \left\langle \hat{\vec{\Psi}}^{(2)} | \hat{\vec{\mu}} | \hat{\vec{\Psi}}^{(0)} \right\rangle.
\tag{8.37}
$$

The wave functions $\hat{\vec{\Psi}}^{(0)}(\vec{r}, t)$ and $\hat{\vec{\Psi}}^{(1)}(\vec{r}, t)$ are already given in Eq. 8.28. $\hat{\vec{\Psi}}^{(2)}(\vec{r}, t)$ is given as

$$
\hat{\vec{\Psi}}^{(2)}(\vec{r}, t) = \frac{1}{4\hbar^2} \sum_{n,m} \sum_{q,p} \frac{[\vec{\mu}_{nm} \cdot \vec{E}_{\text{loc}}(\omega_q)][\vec{\mu}_{mg} \cdot \vec{E}_{\text{loc}}(\omega_p)]}{(\hat{\omega}_{ng} - \omega_q - \omega_p)(\hat{\omega}_{mg} - \omega_p)} e^{-i(\bar{\omega}_g + \omega_q + \omega_p)t} \bar{u}_n(\vec{r}).
\tag{8.38}
$$

Following the same procedure and keeping in mind that the summations over n and m and summations over p and q are over all the states and all the fields, respectively, for the sum frequency case $\omega_p + \omega_q$,

$$\left\langle \vec{p}^{(2)}(t) \right\rangle = \frac{1}{4\hbar^2} \sum_{n,m} \sum_{q,p} \left\{ \frac{\vec{\mu}_{gn}[\vec{\mu}_{nm} \cdot \vec{E}_{\text{loc}}(\omega_q)][\vec{\mu}_{mg} \cdot \vec{E}_{\text{loc}}(\omega_p)]}{(\hat{\omega}_{ng} - \omega_q - \omega_p)(\hat{\omega}_{mg} - \omega_p)} \right.$$

$$+ \frac{[\vec{\mu}_{gn} \cdot \vec{E}_{\text{loc}}(\omega_q)]\vec{\mu}_{nm}[\vec{\mu}_{mg} \cdot \vec{E}_{\text{loc}}(\omega_p)]}{(\hat{\omega}_{ng}^* + \omega_q)(\hat{\omega}_{mg} - \omega_p)}$$

$$\left. + \frac{[\vec{\mu}_{nm} \cdot \vec{E}_{\text{loc}}(\omega_p)][\vec{\mu}_{gn} \cdot \vec{E}_{\text{loc}}(\omega_q)]\vec{\mu}_{mg}}{(\hat{\omega}_{mg}^* + \omega_q + \omega_p)(\hat{\omega}_{ng}^* + \omega_q)} \right\} e^{-i(\omega_p + \omega_q)t} + \text{c.c.} \quad (8.39)$$

For the difference frequency case $\omega_p - \omega_q$,

$$\left\langle \vec{p}^{(2)}(t) \right\rangle = \frac{1}{4\hbar^2} \sum_{n,m} \sum_{q,p} \left\{ \frac{\vec{\mu}_{gn}[\vec{\mu}_{nm} \cdot \vec{E}_{\text{loc}}^*(\omega_q)][\vec{\mu}_{mg} \cdot \vec{E}_{\text{loc}}(\omega_p)]}{(\hat{\omega}_{ng} + \omega_q - \omega_p)(\hat{\omega}_{mg} - \omega_p)} \right.$$

$$+ \frac{[\vec{\mu}_{gn} \cdot \vec{E}_{\text{loc}}^*(\omega_q)]\vec{\mu}_{nm}[\vec{\mu}_{mg} \cdot \vec{E}_{\text{loc}}(\omega_p)]}{(\hat{\omega}_{ng}^* - \omega_q)(\hat{\omega}_{mg} - \omega_p)}$$

$$\left. + \frac{[\vec{\mu}_{nm} \cdot \vec{E}_{\text{loc}}(\omega_p)][\vec{\mu}_{gn} \cdot \vec{E}_{\text{loc}}^*(\omega_q)]\vec{\mu}_{mg}}{(\hat{\omega}_{ng}^* - \omega_q)(\hat{\omega}_{mg}^* + \omega_p - \omega_q)} \right\} e^{-i(\omega_p - \omega_q)t}. \quad (8.40)$$

8.2.2.1 Local Field Correction for Second-Order Interactions.

The local field correction $f^{(2)}$ is *not* simply given by $f^{(1)}(\omega_p)f^{(1)}(\omega_q)$ because three frequencies ω_p, ω_q, and $\omega_p \pm \omega_q$ and fields are present. Therefore this case and also those for higher order susceptibilities must be reformulated carefully. Consider the local field problem for which a second-order, nonlinear, Maxwell polarization $\vec{P}(\omega')$ exists throughout the medium at the nonlinearly generated frequency $\omega' = \omega_p \pm \omega_q$ due to the nonlinear interaction of the Maxwell field with the medium. According to Eq. 8.32,

$$\vec{E}_{\text{loc}}(\omega') = \vec{E}(\omega') + \frac{1}{3\varepsilon_0}\vec{P}(\omega'). \quad (8.41)$$

Including now the nonlinear polarization field $\vec{p}^{(2)}(\omega')$ induced at the molecule by the mixing of fields at the molecule and the "cavity" field at the molecule at frequency ω' due to contribution from all the other molecules outside the cavity, the total dipole moment induced at the molecule is

$$\vec{p}(\omega') = \overset{\leftrightarrow}{\alpha} \cdot \left[\vec{E}(\omega') + \frac{1}{3\varepsilon_0}\vec{P}(\omega') \right] + \vec{p}^{(2)}(\omega'). \quad (8.42)$$

Since $\vec{P}(\omega') = N\vec{p}(\omega')$ and from Eq. 1.23, we obtain

$$\vec{P}(\omega') = \left[\frac{\varepsilon_r(\omega')+2}{3}\right]\left[N\vec{\alpha}\cdot\vec{E}(\omega') + N\vec{p}^{(2)}(\omega')\right] \;\rightarrow\; \vec{P}^{(2)}(\omega') = N\left[\frac{\varepsilon_r(\omega')+2}{3}\right]\vec{p}^{(2)}(\omega').$$

(8.43)

Now defining $\vec{P}^{(2)}(\vec{r},t)$ in the usual way as

$$\vec{P}^{(2)}(\vec{r},t) = \frac{1}{2}\vec{P}^{(2)}(\omega_p \pm \omega_q)\,e^{-i(\omega_p \pm \omega_q)t} + \text{c.c.}$$

$$= \frac{1}{4}\varepsilon_0 \overset{\overset{\Rightarrow}{\approx}}{\chi}^{(2)}(-[\omega_p \pm \omega_q];\omega_p,\pm\omega_q):\vec{E}(\omega_p)$$

$$\times \vec{E}(+\omega_q)\,e^{-i(\omega_p \pm \omega_q)t} + \text{c.c.},$$

(8.44)

with

$$\vec{P}^{(2)}(\omega_p \pm \omega_q) = \frac{N}{2\hbar^2}\left[\frac{\varepsilon_r(\omega_p \pm \omega_q)+2}{3}\right]\sum_{n,m}\sum_{q,p}\frac{\vec{\mu}_{gn}[\vec{\mu}_{nm}\cdot\vec{E}_{\text{loc}}(\pm\omega_q)][\vec{\mu}_{mg}\cdot\vec{E}_{\text{loc}}(\omega_p)]}{(\hat{\tilde{\omega}}_{ng}\mp\omega_q-\omega_p)(\hat{\tilde{\omega}}_{mg}-\omega_p)}$$

$$+\frac{[\vec{\mu}_{gn}\cdot\vec{E}_{\text{loc}}(\pm\omega_q)]\vec{\mu}_{nm}[\vec{\mu}_{mg}\cdot\vec{E}_{\text{loc}}(\omega_p)]}{(\hat{\tilde{\omega}}_{ng}^{*}\pm\omega_q)(\hat{\tilde{\omega}}_{mg}-\omega_p)}$$

$$+\frac{[\vec{\mu}_{gn}\cdot\vec{E}_{\text{loc}}(\pm\omega_q)][\vec{\mu}_{nm}\cdot\vec{E}_{\text{loc}}(\omega_p)]\vec{\mu}_{mg}}{(\hat{\tilde{\omega}}_{mg}^{*}\pm\omega_q+\omega_p)(\hat{\tilde{\omega}}_{ng}^{*}\pm\omega_q)}$$

(8.45)

$$\Rightarrow\quad \hat{\chi}_{ijk}^{(2)}(-[\omega_p \pm \omega_q];\omega_p,\omega_q) = \frac{N}{\hbar^2\varepsilon_0}\frac{\varepsilon_r(\omega_p \pm \omega_q)+2}{3}\frac{\varepsilon_r(\omega_p)+2}{3}\frac{\varepsilon_r(\omega_q)+2}{3}$$

$$\sum_{n,m}\left\{\frac{\bar{\mu}_{gn,i}\bar{\mu}_{nm,k}\bar{\mu}_{mg,j}}{(\hat{\tilde{\omega}}_{ng}\mp\omega_q-\omega_p)(\hat{\tilde{\omega}}_{mg}-\omega_p)} + \frac{\bar{\mu}_{gn,k}\bar{\mu}_{nm,i}\bar{\mu}_{mg,j}}{(\hat{\tilde{\omega}}_{ng}^{*}\pm\omega_q)(\hat{\tilde{\omega}}_{mg}-\omega_p)}\right.$$

$$\left.+\frac{\bar{\mu}_{nm,j}\bar{\mu}_{gn,k}\bar{\mu}_{mg,i}}{(\hat{\tilde{\omega}}_{mg}^{*}\pm\omega_q+\omega_p)(\hat{\tilde{\omega}}_{ng}^{*}+\omega_q)}\right\}.$$

(8.46a)

Note that the nonlinear local field correction is

$$f^{(2)} = \frac{\varepsilon_r(\omega_p \pm \omega_q)+2}{3}\frac{\varepsilon_r(\omega_p)+2}{3}\frac{\varepsilon_r(\omega_q)+2}{3}.$$

(8.46b)

That is, it contains an extra term relative to the linear case at the generated frequency $\omega_p \pm \omega_q$.

8.2.2.2 Example of Type 1 Second Harmonic Generation.

Consider the simple case of *eoo* in a uniaxial crystal. Since only one input frequency is present, the nonlinear polarization is

$$P_i^{(2)}(2\omega_p) = \frac{1}{2}\varepsilon_0\hat{\chi}_{ijj}^{(2)}(-2\omega;\ \omega,\omega)E_j(\omega)E_j(\omega). \tag{8.47}$$

8.2.2.3 Example of Type 2 Sum Frequency Generation.

An example with two orthogonally polarized inputs $E_x(\omega_1)$ and $E_z(\omega_2)$ can produce an x-polarized sum frequency nonlinear polarization. Therefore there are two possible combinations of (ω_p, ω_q), i.e., (ω_1, ω_2) and (ω_2, ω_1), so that

$$
\begin{aligned}
P_x^{(2)}(\omega_1 + \omega_2) = \frac{1}{2}\varepsilon_0 \Big[& \hat{\chi}_{xzx}^{(2)}(-(\omega_1+\omega_2);\ \omega_1,\omega_2) \\
& + \hat{\chi}_{xxz}^{(2)}(-(\omega_1+\omega_2);\ \omega_2,\omega_1) \Big] E_x(\omega_1)E_z(\omega_2).
\end{aligned}
\tag{8.48}
$$

Note that when the frequencies are interchanged, the polarizations of their respective fields are also interchanged.

8.2.2.4 Example of Divergence.

Consider the case of two orthogonally polarized fundamental eigenmodes $E_x(\omega)$ and $E_z^*(\omega) \equiv E_z(-\omega)$, considered as two separate input modes with frequencies $\pm\omega$. Therefore there are two possible combinations of (ω_p, ω_q), i.e., $(\omega, -\omega)$ and $(-\omega, \omega)$, so that for a x-polarized nonlinear polarization at zero frequency

$$P_x^{(2)}(0) = \frac{1}{2}\varepsilon_0\Big[\hat{\chi}_{xxz}^{(2)}(-0;\ \omega,-\omega) + \hat{\chi}_{xzx}^{(2)}(-0;-\omega,\omega)\Big]E_x(\omega)E_z(-\omega), \tag{8.49}$$

with

$$
\therefore \hat{\chi}_{ijk}^{(2)}(0;-\omega,\omega) + \chi_{ijk}^{(2)}(0;\omega,-\omega) = \frac{N}{\varepsilon_0\hbar^2}\Big[\frac{\varepsilon_r(\omega)+2}{3}\Big]\Big[\frac{\varepsilon_r(\omega)+2}{3}\Big]\Big[\frac{\varepsilon_r(0)+2}{3}\Big]
$$

$$
\sum_{n,m}\Big\{ \frac{\bar{\mu}_{gn,i}\bar{\mu}_{mg,j}\bar{\mu}_{nm,k}}{\hat{\bar{\omega}}_{ng}(\hat{\bar{\omega}}_{mg}+\omega)} + \frac{\bar{\mu}_{nm,i}\bar{\mu}_{mg,j}\bar{\mu}_{gn,k}}{(\hat{\bar{\omega}}_{ng}^*+\omega)(\hat{\bar{\omega}}_{mg}+\omega)} + \frac{\bar{\mu}_{mg,i}\bar{\mu}_{nm,j}\bar{\mu}_{gn,k}}{(\hat{\bar{\omega}}_{ng}^*+\omega)\hat{\bar{\omega}}_{mg}^*}
$$

$$
+\frac{\bar{\mu}_{gn,i}\bar{\mu}_{mg,j}\bar{\mu}_{nm,k}}{\hat{\bar{\omega}}_{ng}\hat{\bar{\omega}}_{mg}-\omega)} + \frac{\bar{\mu}_{nm,i}\bar{\mu}_{mg,j}\bar{\mu}_{gn,k}}{(\hat{\bar{\omega}}_{ng}^*-\omega)(\hat{\bar{\omega}}_{mg}-\omega)} + \frac{\bar{\mu}_{mg,i}\bar{\mu}_{nm,j}\bar{\mu}_{gn,k}}{(\hat{\bar{\omega}}_{ng}^*-\omega)\hat{\bar{\omega}}_{mg}^*}\Big\} \tag{8.50}
$$

The current summation over n and m includes $n = g$ and $m = g$ and $n = m = g$, i.e., the ground state. By definition, the ground state does not decay; i.e., $\bar{\omega}_{gg}^* = \bar{\omega}_{gg} = 0$. As a result, some of the terms (circled in red) diverge and must be treated very carefully.

8.2.2.5 Divergence Removed.
The corrected procedure that deals with the divergences is given in Ref. 1. Unfortunately, the derivations given in Ref. 1 are beyond the scope of this textbook. The correct general formula taken from that reference is given below

$$\hat{\chi}_{ijk}^{(2)}\left(-\left[\omega_p + \omega_q\right]; \omega_p, \omega_q\right) = \frac{N}{\varepsilon_0 \hbar^2} f^{(2)} \sum_{n,m}' \left\{ \frac{\bar{\mu}_{gn,i}\left(\bar{\mu}_{nm,k} - \bar{\mu}_{gg,k}\right)\bar{\mu}_{mg,j}}{(\hat{\omega}_{ng} - \omega_p - \omega_q)(\hat{\omega}_{mg} - \omega_p)} \right.$$

$$+ \frac{\bar{\mu}_{gn,k}\left(\bar{\mu}_{nm,j} - \bar{\mu}_{gg,j}\right)\bar{\mu}_{mg,i}}{(\hat{\omega}_{ng}^* + \omega_q)(\hat{\omega}_{mg}^* + \omega_p + \omega_q)}$$

$$\left. + \frac{\bar{\mu}_{gn,k}\left(\bar{\mu}_{nm,i} - \bar{\mu}_{gg,i}\right)\bar{\mu}_{mg,j}}{(\hat{\omega}_{ng}^* + \omega_q)(\hat{\omega}_{mg} - \omega_p)} \right\} \tag{8.51}$$

and the prime in $\sum_{n,m}'$ means that the ground state is excluded from the summation over the states, i.e., the *summation is taken over only the excited states*. Note that the summation includes contributions from permanent dipole moments in the ground state and excited states (case $n = m$).

8.2.2.6 Centrosymmetric versus Noncentrosymmetric Molecules and Crystals.
The key question is whether a medium exhibits complete inversion symmetry, i.e., along all crystal axes. Crystals are composed of identically aligned unit cells, each of which can consist of one or more molecules, as shown in Fig. 8.5a. The molecules themselves may or may not exhibit inversion symmetry

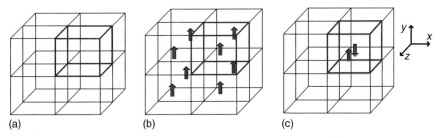

(a)   (b)   (c)

FIGURE 8.5 (a) Example of unit cell arrangement in a crystal. A single unit cell is shown by the heavier lines. (b) The case when an average over a unit cell yields a net parallel $\chi_{yyy}^{(2)} \neq 0$ (denoted by red arrows) in each unit cell. (c) Antiparallel arrangement of $\overline{\chi_{yyy}^{(2)}} \neq 0$ in adjacent unit cells so that $\chi_{ijk}^{(2)} = 0$ over a small fraction of an optical wavelength.

along some, but not all, molecular axes (see Fig. 8.5b and c). If the molecules exhibit complete inversion symmetry when the average over all unit cells is taken, the medium is centrosymmetric and $\hat{\chi}^{(2)} = 0$. If the molecules are noncentrosymmetric and if the alignment of the molecules in the unit cell results in a noncentrosymmetric unit cell, then some elements of $\hat{\chi}_{ijk}^{(2)}$ are not equal to zero (see Fig. 8.5b). However, the average over the unit cell may still be centrosymmetric if the individual molecules inside the unit cell are aligned in a manner that cancels out the net inversion symmetry of the unit cell, by, e.g., counteralignment of molecules. This last case is illustrated in Fig. 8.5c.

Whether a molecule is centrosymmetric is determined by its symmetry properties. If all the spatial wave functions of the electronic states have no symmetry, i.e., are asymmetric, they can be considered as having both a symmetric and an asymmetric part. A nonzero electric dipole transition moment between states ℓ and ℓ' defined by Eq. 8.17b, i.e., $\vec{\bar{\mu}}_{\ell\ell'} = \int_{-\infty}^{\infty} \bar{u}_\ell^*(\vec{r})\vec{\bar{\mu}}(\vec{r})\bar{u}_{\ell'}(\vec{r})\,d\vec{r}$, requires a change in the symmetry of the spatial wave functions $\bar{u}_\ell(\vec{r})$ and $\bar{u}_{\ell'}(\vec{r})$. When the electronic states exhibit both even and odd symmetries, $\vec{\bar{\mu}}_{\ell\ell'} \neq 0$ for some states. The case $\ell = \ell'$ gives the permanent dipole moments. Therefore the terms in the numerator of Eq. 8.51, namely, $\bar{\mu}_{gn}, \bar{\mu}_{nm}, \bar{\mu}_{gg}$, and $\bar{\mu}_{mg}$, are all nonzero and hence some elements of $\hat{\chi}_{ijk}^{(2)}$ are not equal to zero.

On the other hand, for centrosymmetric molecules the molecular spatial wave functions exhibit either even (gerade, subscript g) or odd (ungerade, subscript u) symmetry, with the ground state having even symmetry. Therefore the electric dipole transition moments are nonzero only if there is a change in wave function symmetry. Substituting for the symmetry of the wave functions as either even (g) or odd (u) into the numerators of Eq. 8.51 gives $\bar{\mu}_{gu}(\bar{\mu}_{uu} - \bar{\mu}_{gg})\bar{\mu}_{ug}$. The terms in the parentheses are zero and hence $\chi^{(2)} = 0$. This is a proof that centrosymmetric molecular media do not exhibit second-order nonlinearities.

8.2.2.7   *Nonresonant Limit ($\omega \to 0$).*   Since in the usual case $\bar{\omega}_{\ell g} \gg \tau_{\ell g}^{-1}$ for all the excited states ℓ,

$$\hat{\chi}_{ijk}^{(2)}(0;0,0) \simeq \frac{N}{\varepsilon_0\hbar^2}f^{(2)}\sum_{n,m}' \frac{1}{\bar{\omega}_{ng}\bar{\omega}_{mg}}\left\{\bar{\mu}_{gn,i}(\bar{\mu}_{nm,k} - \bar{\mu}_{gg,k})\bar{\mu}_{mg,j}\right.$$

$$\left. + \bar{\mu}_{gn,k}(\bar{\mu}_{nm,j} - \bar{\mu}_{gg,j})\bar{\mu}_{mg,i} + \bar{\mu}_{gn,k}(\bar{\mu}_{nm,i} - \bar{\mu}_{gg,i})\bar{\mu}_{mg,j}\right\},$$

$$\text{with } f^{(2)} = \left[\frac{\varepsilon_r(0)+2}{3}\right]\left[\frac{\varepsilon_r(0)+2}{3}\right]^2. \tag{8.52}$$

The essential result is that the same susceptibility is obtained for second harmonic generation, sum frequency generation, and difference frequency generation, as expected for Kleinman symmetry.

8.2.3 Third-Order Susceptibility

It is of course possible to continue the derivations along the lines used for the first- and second-order susceptibilities. However, this leads to a formulation for $\chi^{(3)}$, which also exhibits divergences in certain (important) cases. Instead quoted here is the result with the divergences removed (1):

$$\hat{\chi}_{ijkl}^{(3)}\left(-[\omega_p + \omega_q + \omega_r]; \omega_p, \omega_q, \omega_r\right)$$

$$= \frac{N}{\varepsilon_0 \hbar^3} f^{(3)} \sum_{v,n,m}' \left\{ \frac{\bar{\mu}_{gv,i}(\bar{\mu}_{vn,l} - \bar{\mu}_{gg,l})(\bar{\mu}_{nm,k} - \bar{\mu}_{gg,k})\bar{\mu}_{mg,j}}{(\hat{\bar{\omega}}_{vg} - \omega_p - \omega_q - \omega_r)(\hat{\bar{\omega}}_{ng} - \omega_q - \omega_p)(\hat{\bar{\omega}}_{mg} - \omega_p)} \right.$$

$$+ \frac{\bar{\mu}_{gv,j}(\bar{\mu}_{vn,k} - \bar{\mu}_{gg,k})(\bar{\mu}_{nm,i} - \bar{\mu}_{gg,i})\bar{\mu}_{mg,l}}{(\hat{\bar{\omega}}_{vg}^* + \omega_p)(\hat{\bar{\omega}}_{ng}^* + \omega_q + \omega_p)(\hat{\bar{\omega}}_{mg} - \omega_r)}$$

$$+ \frac{\bar{\mu}_{gv,l}(\bar{\mu}_{vn,i} - \bar{\mu}_{gg,i})(\bar{\mu}_{nm,k} - \bar{\mu}_{gg,k})\bar{\mu}_{mg,j}}{(\hat{\bar{\omega}}_{vg}^* + \omega_r)(\hat{\bar{\omega}}_{ng} - \omega_q - \omega_p)(\hat{\bar{\omega}}_{mg} - \omega_p)}$$

$$+ \left. \frac{\bar{\mu}_{gv,j}(\bar{\mu}_{vn,k} - \bar{\mu}_{gg,k})(\bar{\mu}_{nm,l} - \bar{\mu}_{gg,l})\bar{\mu}_{mg,i}}{(\hat{\bar{\omega}}_{vg}^* + \omega_p)(\hat{\bar{\omega}}_{ng}^* + \omega_q + \omega_p)(\hat{\bar{\omega}}_{mg}^* + \omega_p + \omega_q + \omega_r)} \right\}$$

$$- \frac{N}{\varepsilon_0 \hbar^3} f^{(3)} \sum_{n,m}' \left\{ \frac{\bar{\mu}_{gn,i}\bar{\mu}_{ng,l}\bar{\mu}_{gm,k}\bar{\mu}_{mg,j}}{(\hat{\bar{\omega}}_{ng} - \omega_p - \omega_q - \omega_r)(\hat{\bar{\omega}}_{ng} - \omega_r)(\hat{\bar{\omega}}_{mg} - \omega_p)} \right.$$

$$+ \frac{\bar{\mu}_{gn,i}\bar{\mu}_{ng,l}\bar{\mu}_{gm,k}\bar{\mu}_{mg,j}}{(\hat{\bar{\omega}}_{mg}^* + \omega_q)(\hat{\bar{\omega}}_{ng} - \omega_r)(\hat{\bar{\omega}}_{mg} - \omega_p)}$$

$$+ \frac{\bar{\mu}_{gn,l}\bar{\mu}_{ng,i}\bar{\mu}_{gm,j}\bar{\mu}_{mg,k}}{(\hat{\bar{\omega}}_{ng}^* + \omega_r)(\hat{\bar{\omega}}_{mg}^* + \omega_p)(\hat{\bar{\omega}}_{mg} - \omega_q)}$$

$$+ \left. \frac{\bar{\mu}_{gn,l}\bar{\mu}_{ng,i}\bar{\mu}_{gm,j}\bar{\mu}_{mg,k}}{(\hat{\bar{\omega}}_{ng}^* + \omega_r)(\hat{\bar{\omega}}_{mg}^* + \omega_p)(\hat{\bar{\omega}}_{ng}^* + \omega_p + \omega_q + \omega_r)} \right\}, \quad (8.53)$$

where the prime in $\sum_{n,m}'$ and $\sum_{v,n,m}'$ means that the ground state is excluded from the summation over the states; i.e., the *summation is taken over only the excited states*. Note that the first summation includes contributions from permanent dipole moments in the ground state and excited states whereas the second summation does not. There are a total of 48 terms possible for the nonlinear polarization before the summation over excited states. Finally,

$$f^{(3)} = \left[\frac{\varepsilon_r(\omega_p) + 2}{3}\right]\left[\frac{\varepsilon_r(\omega_q) + 2}{3}\right]\left[\frac{\varepsilon_r(\omega_r) + 2}{3}\right]\left[\frac{\varepsilon_r(\omega_p + \omega_q + \omega_r) + 2}{3}\right]$$

is the local field correction.

8.2.3.1 Symmetry Properties of Third-Order Susceptibilities.

Since the third-order susceptibility is of the form $\hat{\chi}^{(3)}_{ijk\ell}$, with the subscripts each being either x, y, or z, there is a maximum of $3^4 = 81$ independent elements. However, as shown in Chapter 3 in the case of $\hat{\chi}^{(2)}_{ijk}$, the symmetry properties of a material can reduce this number drastically. The number of independent $\hat{\chi}^{(3)}_{ijk\ell}$ coefficients was given previously in Table 3.1, and the relations between the different nonlinear coefficients are given in Appendix 8.1. Because the number of nonzero third-order coefficients increases rapidly with decreasing material symmetry, the complexity of the analysis of third-order phenomena can quickly become overwhelming unless the propagation direction and field polarization lie along crystal axes. Nonisotropic media are typically characterized by their symmetry rotation axes and mirror reflection planes. Furthermore, the numerical values of all the independent coefficients for materials of lower symmetry than for those of cubic symmetry are simply not known. The net result is that frequently isotropic media are used in calculations as a first approximation to more general cases.

It is very important to point out that that there are no symmetry restrictions that cause any material to have a zero third-order susceptibility. Although it is highly unlikely that the third-order susceptibility will be zero for some given propagation direction and polarization, this cannot be ruled out from symmetry considerations. This is in contrast to the case of the second-order (and higher even orders) susceptibility for which the existence of a nonzero $\hat{\chi}^{(2)}_{ijk}$ for a specific propagation direction and polarization is determined by crystal symmetry.

8.2.3.1.1 Isotropic Media.

In an isotropic medium, all coordinate systems are equivalent. Therefore in terms of the subscripts $ijk\ell$, for the nonlinear coefficients we have

$$
\begin{aligned}
&xxxx \equiv yyyy \equiv zzzz : \text{in general}, \delta_{ij}\,\delta_{jk}\,\delta_{k\ell}; \\
&yyzz \equiv yyxx \equiv xxzz \equiv xxyy \equiv zzxx \equiv zzyy : \text{in general}, \delta_{ij}\,\delta_{k\ell}; \\
&xyyx \equiv xzzx \equiv yxxy \equiv yzzy \equiv zxxz \equiv zyyz : \text{in general}, \delta_{i\ell}\,\delta_{jk}; \\
&xyxy \equiv xzxz \equiv yxyx \equiv yzyz \equiv zxzx \equiv zyzy : \text{in general}, \delta_{ik}\,\delta_{j\ell}.
\end{aligned}
\tag{8.54}
$$

In addition, any arbitrary rotation of a coordinate system must lead to the same nonlinear polarization for the given field directions specified by the inputs. This leads to a simple relation between the coefficients in Eq. 8.54. Assume the general case of three, parallel, copolarized (along, e.g., the x-axis) input fields E_1, E_2, and E_3 with different frequencies $\omega_1, \omega_2, \omega_3$ producing the field ω_4 via $\hat{\chi}^{(3)}_{xxxx}(-\omega_4; \omega_1, \omega_2, \omega_3)$. This produces the nonlinear polarization (along the x-axis)

$$P^{(3)}_x(\omega_4) = \frac{1}{4} \varepsilon_0 \hat{\chi}^{(3)}_{xxxx}(-\omega_4; \omega_3, \omega_2, \omega_1) E_1 E_2 E_3. \tag{8.55}$$

Now consider the axis system (x', y') rotated $45°$ from the original x-axis in the x–y plane. The three fields have the following components along the x'-axis and the y'-axis:

$$E_{1x'} = \frac{1}{\sqrt{2}} E_1, \quad E_{2x'} = \frac{1}{\sqrt{2}} E_2, \quad E_{3x'} = \frac{1}{\sqrt{2}} E_3,$$

$$E_{1y'} = \frac{1}{\sqrt{2}} E_1, \quad E_{2y'} = \frac{1}{\sqrt{2}} E_2, \quad E_{3y'} = \frac{1}{\sqrt{2}} E_3. \tag{8.56}$$

Noting that the arbitrary choice of axes in isotropic materials means that

$$\hat{\chi}^{(3)}_{x'x'x'x'}(-\omega_4; \omega_3, \omega_2, \omega_1) = \hat{\chi}^{(3)}_{xxxx}(-\omega_4; \omega_3, \omega_2, \omega_1),$$

$$\hat{\chi}^{(3)}_{xxyy}(-\omega_4; \omega_3, \omega_2, \omega_1) = \hat{\chi}^{(3)}_{x'x'y'y'}(-\omega_4; \omega_3, \omega_2, \omega_1), \tag{8.57}$$

and so on, the nonlinear third-order polarization induced along the x'-axis is given by

$$\begin{aligned}
P^{(3)}_{x'}(\omega_4) = \frac{1}{4} \varepsilon_0 \Big[&\hat{\chi}^{(3)}_{xxxx}(-\omega_4; \omega_3, \omega_2, \omega_1) E_{1x'} E_{2x'} E_{3x'} \\
&+ \hat{\chi}^{(3)}_{xxyy}(-\omega_4; \omega_3, \omega_2, \omega_1) E_{1y'} E_{2y'} E_{3x'} \\
&+ \hat{\chi}^{(3)}_{xyyx}(-\omega_4; \omega_3, \omega_2, \omega_1) E_{1x'} E_{2y'} E_{3y'} \\
&+ \hat{\chi}^{(3)}_{xyxy}(-\omega_4; \omega_3, \omega_2, \omega_1) E_{1y'} E_{2x'} E_{3y'} \Big]
\end{aligned}$$

$$\begin{aligned}
\rightarrow \quad P^{(3)}_{x'}(\omega_4) = \frac{1}{4} \varepsilon_0 \frac{1}{2\sqrt{2}} \Big[&\hat{\chi}^{(3)}_{xxxx}(-\omega_4; \omega_3, \omega_2, \omega_1) \\
&+ \hat{\chi}^{(3)}_{xxyy}(-\omega_4; \omega_3, \omega_2, \omega_1) \\
&+ \hat{\chi}^{(3)}_{xyyx}(-\omega_4; \omega_3, \omega_2, \omega_1) + \hat{\chi}^{(3)}_{xyxy}(-\omega_4; \omega_3, \omega_2, \omega_1) \Big] E_1 E_2 E_3.
\end{aligned} \tag{8.58}$$

The nonlinear polarization $P_{x'}^{(3)}(\omega_4)$ in Eq. 8.58 can also be obtained by projecting the nonlinear polarization given by Eq. 8.55 onto the x'-axis to give

$$P_{x'}^{NL}(\omega_4) = \frac{1}{4}\varepsilon_0 \frac{1}{\sqrt{2}} \hat{\chi}_{xxxx}^{(3)}(-\omega_4, \omega_3, \omega_2, \omega_1)E_1E_2E_3. \qquad (8.59)$$

Equating Eqs 8.58 and 8.59 gives the following relation between the third-order susceptibilities:

$$\hat{\chi}_{xxxx}^{(3)}(-\omega_4; \omega_3, \omega_2, \omega_1) = \hat{\chi}_{xxyy}^{(3)}(-\omega_4; \omega_3, \omega_2, \omega_1) + \hat{\chi}_{xyyx}^{(3)}(-\omega_4; \omega_3, \omega_2, \omega_1)$$
$$+ \hat{\chi}_{xyxy}^{(3)}(-\omega_4; \omega_3, \omega_2, \omega_1). \qquad (8.60)$$

Therefore, these coupled with the relations in Eq. 8.54 valid, in general, for *isotropic* media give

$$\hat{\chi}_{ijkl}^{(3)}(-\omega; \omega_p, \omega_q, \omega_r) = \hat{\chi}_{1122}^{(3)}(-\omega; \omega_p, \omega_q, \omega_r)\delta_{i\ell}\delta_{jk}$$
$$+ \hat{\chi}_{1221}^{(3)}(-\omega; \omega_p, \omega_q, \omega_r)\delta_{i\ell}\delta_{jk}$$
$$+ \hat{\chi}_{1212}^{(3)}(-\omega; \omega_p, \omega_q, \omega_r)\delta_{ik}\delta_{j\ell}. \qquad (8.61)$$

Note that this is true for any set of input frequencies, on or off resonance. This relation not only reduces the number of independent elements to 3 but is also a very powerful restriction on the nonlinear coefficients for *all* isotropic media, including liquids (but not liquid crystals exhibiting long-range orientation correlations) and gases of molecules or electrons. In the case of Kleinman symmetry (only), i.e., all $\bar{\omega}_{mg} \gg \omega$, permutation symmetry holds and

$$\frac{1}{3}\tilde{\chi}_{xxxx}^{(3)}(-\omega_4; \omega_3, \omega_2, \omega_1) = \tilde{\chi}_{xxyy}^{(3)}(-\omega_4; \omega_3, \omega_2, \omega_1) = \tilde{\chi}_{xyyx}^{(3)}(-\omega_4; \omega_3, \omega_2, \omega_1)$$
$$= \tilde{\chi}_{xyxy}^{(3)}(-\omega_4; \omega_3, \omega_2, \omega_1), \qquad (8.62)$$

and there is only *one* independent nonlinear coefficient.

Similar arguments can be applied to obtain relations between the higher order odd susceptibilities. Unfortunately the number of terms describing the mixed x and y polarization terms increases rapidly with m in $\chi^{(2m+1)}$; e.g., there are 15 terms for $m = 2$ with 5 permutations of the subscripts $xxyyyy$ and 10 for the subscripts $xxxxyy$, with the position of only the first x fixed in each case by the output polarization. The total increases to 63 for $m = 3$, 255 for $m = 4$, and so on. However, in the Kleinman limit ($\omega \to 0$, with tilde above) all the cross-polarization terms are equal, which simplifies the situation considerably.

8.2.3.2 Formalism for General Third-Order Polarization. The link between the general formula for third-order nonlinear polarization given in Chapter 1 and the susceptibilities derived in this chapter will now be established. Based on Eq. 1.4, the general expression for the local third-order polarization induced in a material is

$$\vec{P}_i^{(3)}(t) = \varepsilon_0 \iiint \chi_{ijk\ell}^{(3)}(t - t', t - t'', t - t''') E_j(t') E_k(t'') E_\ell(t''') \, dt' \, dt'' \, dt. \quad (8.63)$$

Following the same procedure as given in Chapter 1 for the linear susceptibility and after some manipulation with Fourier transforms

$$P_i^{(3)}(\vec{r}, \omega) = \varepsilon_0 \sum_p \sum_q \sum_r \iiint \hat{\chi}_{ijkl}^{(3)}(-\omega; \omega_p, \omega_q, \omega_r) E_j(\vec{r}, \omega_p) E_k(\vec{r}, \omega_q) E_\ell(\vec{r}, \omega_r)$$

$$\times \delta(\omega - \omega_p - \omega_q - \omega_r) \, d\omega_p \, d\omega_q \, d\omega_r, \quad (8.64)$$

in which all the quantities such as $P_i^{(3)}(\vec{r}, \omega)$ and $E_k(\vec{r}, \omega_q)$, are the Fourier transforms in frequency of $P_i^{(3)}(\vec{r}, t)$ and $E_k(\vec{r}, t)$, respectively, the summations over p, q, and r are each over all the incident fields and

$$\hat{\chi}_{ijkl}^{(3)}(-\omega; \omega_p, \omega_q, \omega_r) = \int_{-\infty}^{\infty} \int_{-\infty}^{\infty} \int_{-\infty}^{\infty} \hat{\chi}_{ijk\ell}^{(3)}(t - t', t - t'', t - t''')$$

$$e^{i\omega_p(t-t') + i\omega_q(t-t'') + i\omega_r(t-t''')} \, d(t - t') \, d(t - t'') \, d(t - t'''). \quad (8.65)$$

Note that $\hat{\chi}_{ijk\ell}^{(3)}(-\omega; \omega_p, \omega_q, \omega_r) = \hat{\chi}_{ijk\ell}^{(3)*}(\omega; -\omega_p, -\omega_q, -\omega_r)$.

8.2.3.3 Examples of Third-Order Processes. Consider the most general case of three frequency inputs: ω_a, ω_b, ω_c. Therefore there are $6 \times 6 \times 6 = 216$ possible combinations of $E_j(\vec{r}, \omega_a) E_k(\vec{r}, \omega_b) E_\ell(\vec{r}, \omega_c)$, which include the complex conjugates and give frequency outputs at $\pm\omega_a \pm \omega_b \pm \omega_c$. Of these, there are only a limited number that are of general interest in third-order nonlinear optics and they will be discussed in the next section.

8.2.3.3.1 Single-Beam (Eigenmode) Input and Polarization Along Crystal Axis. The Fourier transform of the input for insertion into Eq. 8.64 is given by Eq. 1.3 as

$$\vec{E}(\omega_p) = \frac{1}{2}\hat{e}_x \mathrm{E}(\omega_a)\delta(\omega_p - \omega_a) + \frac{1}{2}\hat{e}_x \mathrm{E}^*(\omega_a)\delta(\omega_p + \omega_a)$$

$$\rightarrow \quad \vec{E}(\omega_p)\vec{E}(\omega_q)\vec{E}(\omega_r) = \frac{1}{8}\left\{ \left[\hat{e}_x \mathrm{E}(\omega_a)\delta(\omega_p - \omega_a) + \hat{e}_x \mathrm{E}^*(\omega_a)\delta(\omega_p + \omega_a)\right] \right.$$

$$\times \left[\hat{e}_x \mathrm{E}(\omega_a)\delta(\omega_q - \omega_a) + \hat{e}_x \mathrm{E}^*(\omega_a)\delta(\omega_q + \omega_a)\right]$$

$$\left. \times \left[\hat{e}_x \mathrm{E}(\omega_a)\delta(\omega_r - \omega_a) + \hat{e}_x \mathrm{E}^*(\omega_a)\delta(\omega_r + \omega_a)\right] \right\},$$

$$(8.66)$$

where, as discussed in Chapter 1, $\mathrm{E}(\omega_a)$ is the field in the expansion

$$E(\vec{r}, t) = \frac{1}{2}\mathrm{E}(\omega_a)\,e^{-i\omega_a t} + \text{c.c.} \tag{8.67}$$

Inserting this into Eq. 8.64 and integrating over frequency gives the nonlinear polarization (see Eq. 1.2 for the notation) as

$$P_x^{(3)}(\omega) = \frac{1}{8}\varepsilon_0 \hat{\chi}_{xxxx}^{(3)}(-3\omega_a; \omega_a, \omega_a, \omega_a)\mathrm{E}_x^3(\omega)\delta(\omega - 3\omega_a)$$

$$+ \frac{1}{8}\varepsilon_0 \left[\hat{\chi}_{xxxx}^{(3)}(-\omega_a; -\omega_a, \omega_a, \omega_a) + \hat{\chi}_{xxxx}^{(3)}(-\omega_a; \omega_a, -\omega_a, \omega_a)\right.$$

$$\left. + \hat{\chi}_{xxxx}^{(3)}(-\omega_a; \omega_a, \omega_a, -\omega_a)\right]$$

$$\mathrm{E}_x(\omega_a)|\mathrm{E}_x(\omega_a)|^2 \delta(\omega - \omega_a) + \text{c.c.} \tag{8.68}$$

The first term gives third harmonic generation and the next three give a nonlinear correction to the linear polarization at frequency ω, normally linked to intensity-dependent refraction and absorption (discussed in Chapter 9).

8.2.3.3.1.1   THIRD HARMONIC GENERATION.   For this case, $\omega_p = \omega_q = \omega_r = \omega$. Fourier transforming Eq. 8.68 back into the time domain gives

$$P_x^{(3)}(\vec{r}, t) = \frac{1}{8}\varepsilon_0 \hat{\chi}_{xxxx}^{(3)}(-3\omega; \omega, \omega, \omega)\mathrm{E}_x^3(\omega)\,e^{-3i\omega t} + \text{c.c.}$$

$$(8.69)$$

$$\rightarrow \quad \boldsymbol{P}_x^{(3)}(3\omega) = \frac{1}{4}\varepsilon_0 \hat{\chi}_{xxxx}^{(3)}(-3\omega; \omega, \omega, \omega)\boldsymbol{\mathcal{E}}_x^3(\omega),$$

with

$$\hat{\chi}_{xxxx}^{(3)}(-3\omega; \omega, \omega, \omega) = \frac{N}{\varepsilon_0 \hbar^3} f^{(3)} \sum_{v,n,m}' \left\{ \frac{\bar{\mu}_{gv,x}(\bar{\mu}_{vn,x} - \bar{\mu}_{gg,x})(\bar{\mu}_{nm,x} - \bar{\mu}_{gg,x})\bar{\mu}_{mg,x}}{(\hat{\bar{\omega}}_{vg} - 3\omega)(\hat{\bar{\omega}}_{ng} - 2\omega)(\hat{\bar{\omega}}_{mg} - \omega)} \right.$$

$$+ \frac{\bar{\mu}_{gv,x}(\bar{\mu}_{vn,x} - \bar{\mu}_{gg,x})(\bar{\mu}_{nm,x} - \bar{\mu}_{gg,x})\bar{\mu}_{mg,x}}{(\hat{\bar{\omega}}_{vg}^* + \omega)(\hat{\bar{\omega}}_{ng}^* + 2\omega)(\hat{\bar{\omega}}_{mg} - \omega)}$$

$$+ \frac{\bar{\mu}_{gv,x}(\bar{\mu}_{vn,x} - \bar{\mu}_{gg,x})(\bar{\mu}_{nm,x} - \bar{\mu}_{gg,x})\bar{\mu}_{mg,x}}{(\hat{\bar{\omega}}_{vg}^* + \omega)(\hat{\bar{\omega}}_{ng} - 2\omega)(\hat{\bar{\omega}}_{mg} - \omega)}$$

$$+ \left. \frac{\bar{\mu}_{gv,x}(\bar{\mu}_{vn,x} - \bar{\mu}_{gg,x})(\bar{\mu}_{nm,x} - \bar{\mu}_{gg,x})\bar{\mu}_{mg,x}}{(\hat{\bar{\omega}}_{vg}^* + \omega)(\hat{\bar{\omega}}_{ng}^* + 2\omega)(\hat{\bar{\omega}}_{mg}^* + 3\omega)} \right\}$$

$$- \frac{N}{\varepsilon_0 \hbar^3} f^{(3)} \sum_{n,m}' \left\{ \frac{\bar{\mu}_{gn,x}\,\bar{\mu}_{ng,x}\,\bar{\mu}_{gm,x}\,\bar{\mu}_{mg,x}}{(\hat{\bar{\omega}}_{ng} - 3\omega)(\hat{\bar{\omega}}_{ng} - \omega)(\hat{\bar{\omega}}_{mg} - \omega)} \right.$$

$$+ \frac{\bar{\mu}_{gn,x}\,\bar{\mu}_{ng,x}\,\bar{\mu}_{gm,x}\,\bar{\mu}_{mg,x}}{(\hat{\bar{\omega}}_{mg}^* + \omega)(\hat{\bar{\omega}}_{ng} - \omega)(\hat{\bar{\omega}}_{mg} - \omega)}$$

$$+ \frac{\bar{\mu}_{gn,x}\,\bar{\mu}_{ng,x}\,\bar{\mu}_{gm,x}\,\bar{\mu}_{mg,x}}{(\hat{\bar{\omega}}_{ng}^* + \omega)(\hat{\bar{\omega}}_{mg}^* + \omega)(\hat{\bar{\omega}}_{mg} - \omega)}$$

$$+ \left. \frac{\bar{\mu}_{gn,x}\,\bar{\mu}_{ng,x}\,\bar{\mu}_{gm,x}\,\bar{\mu}_{mg,x}}{(\hat{\bar{\omega}}_{ng}^* + \omega)(\hat{\bar{\omega}}_{mg}^* + \omega)(\hat{\bar{\omega}}_{ng}^* + 3\omega_r)} \right\},$$

with $f^{(3)} = \left[\dfrac{\varepsilon_r(\omega) + 2}{3}\right]^3 \left[\dfrac{\varepsilon_r(3\omega) + 2}{3}\right].$ (8.70)

There are eight terms before the summation over excited states is taken. Note that there are resonant enhancements whenever some resonant frequency $\bar{\omega}_{vg} \cong 3\omega$, $\bar{\omega}_{ng} \cong 2\omega$, and/or $\bar{\omega}_{mg} \cong \omega$ occurs.

8.2.3.3.1.2 INTENSITY-DEPENDENT REFRACTION AND ABSORPTION. For this case there are three possibilities in which $\omega_p + \omega_q + \omega_r = \omega$: $\omega_p = -\omega$, $\omega_q = \omega_r = \omega$; $\omega_p = \omega_r = \omega$, $\omega_q = -\omega$; and $\omega_p = \omega_q = \omega$, $\omega_r = -\omega$.

$$P_x^{(3)}(\vec{r}, t) = \frac{1}{8}\varepsilon_0 \left[\hat{\chi}_{xxxx}^{(3)}(-\omega; -\omega, \omega, \omega) + \hat{\chi}_{xxxx}^{(3)}(-\omega; \omega, -\omega, \omega)\right.$$

$$\left. + \hat{\chi}_{xxxx}^{(3)}(-\omega; \omega, \omega, -\omega)\right]|E_x(\omega)|^2 E_x(\omega)\, e^{-i\omega t} + \text{c.c.} \qquad (8.71)$$

$$\rightarrow \quad \boldsymbol{\mathcal{P}}_x^{(3)}(\omega) = \frac{1}{4}\varepsilon_0 \left[\hat{\chi}_{xxxx}^{(3)}(-\omega; -\omega, \omega, \omega) + \hat{\chi}_{xxxx}^{(3)}(-\omega; \omega, -\omega, \omega)\right.$$

$$\left. + \hat{\chi}_{xxxx}^{(3)}(-\omega; \omega, \omega, -\omega)\right]\boldsymbol{\mathcal{E}}_x(\omega)|\boldsymbol{\mathcal{E}}_x(\omega)|^2. \qquad (8.72)$$

Defining

$$\hat{\tilde{\chi}}_{xxxx}^{(3)}(-\omega; \omega, -\omega, \omega) = \frac{1}{3}\left[\hat{\chi}_{xxxx}^{(3)}(-\omega; -\omega, \omega, \omega) + \hat{\chi}_{xxxx}^{(3)}(-\omega; \omega, -\omega, \omega)\right.$$

$$\left. + \hat{\chi}_{xxxx}^{(3)}(-\omega; \omega, \omega, -\omega)\right]$$

$$\rightarrow \quad \boldsymbol{\mathcal{P}}_x^{(3)}(\omega) = \frac{3}{4}\varepsilon_0 \hat{\tilde{\chi}}_{xxxx}^{(3)}(-\omega; \omega, -\omega, \omega)\boldsymbol{\mathcal{E}}_x(\omega)|\boldsymbol{\mathcal{E}}_x(\omega)|^2. \qquad (8.73)$$

Note that the superscripts $\hat{A}$ and $\widehat{A}$ are meant to identify a complex quantity and a simplified form for susceptibilities, respectively.

It is common notation in the literature to represent the sum of the three susceptibilities by $3\chi_{xxxx}^{(3)}(-\omega; \omega, -\omega, \omega)$, which is written here as $3\hat{\tilde{\chi}}_{xxxx}^{(3)}(-\omega; \omega, -\omega, \omega)$ to avoid confusion. In the Kleinman limit ($\omega \rightarrow 0$),

$$\tilde{\chi}_{xxxx}^{(3)}(-\omega; -\omega, \omega, \omega) = \tilde{\chi}_{xxxx}^{(3)}(-\omega; \omega, -\omega, \omega) = \tilde{\chi}_{xxxx}^{(3)}(-\omega; \omega, \omega, -\omega)$$

$$\rightarrow \quad \boldsymbol{\mathcal{P}}_x^{(3)}(\omega) = \frac{3}{4}\varepsilon_0 \tilde{\chi}_{xxxx}^{(3)}(-\omega; \omega, -\omega, \omega)\boldsymbol{\mathcal{E}}_x(\omega)|\boldsymbol{\mathcal{E}}_x(\omega)|^2. \qquad (8.74)$$

The factor of 3 in Eqs 8.73 and 8.74 is called the permutation parameter, sometimes written as **P**. In this notation the second part of Eq. 8.74 would be written as

$$\boldsymbol{\mathcal{P}}_x^{(3)}(\omega_a) = \mathbf{P}\frac{1}{4}\varepsilon_0 \tilde{\chi}_{xxxx}^{(3)}(-\omega_a; \omega_a, -\omega_a, \omega_a)\boldsymbol{\mathcal{E}}_x(\omega_a)|\boldsymbol{\mathcal{E}}_x(\omega_a)|^2. \qquad (8.75)$$

In this general case, there are 24 terms before the summation over the excited states. The induced polarization is exactly at the same frequency as the input frequency. Since the linear polarization $P_x^{(1)}(\vec{r}, t, \omega)$ directly leads to the complex refractive index, the nonlinear polarization proportional to $|E_x|^2$ results in an intensity-dependent contribution to the refractive index $\Delta n = n_2(-\omega; \omega)I$ and/or the field absorption coefficient $\Delta\alpha = \alpha_2(-\omega; \omega)I$, which are the most important consequences of the third-order nonlinearity, with

$$n_2(-\omega; \omega) = \frac{1}{4n_0^2\varepsilon_0 c}\, \mathfrak{Real}\Big\{\hat{\chi}_{xxxx}^{(3)}(-\omega; \omega, -\omega, \omega) + \hat{\chi}_{xxxx}^{(3)}(-\omega; \omega, \omega, -\omega)$$

$$+ \hat{\chi}_{xxxx}^{(3)}(-\omega; -\omega, \omega, \omega)\Big\},$$

$$\alpha_2(-\omega; \omega) = \frac{\omega}{2n_0^2\varepsilon_0 c^2}\, \mathfrak{Imag}\Big\{\hat{\chi}_{xxxx}^{(3)}(-\omega; \omega, -\omega, \omega) + \hat{\chi}_{xxxx}^{(3)}(-\omega; \omega, \omega, -\omega)$$

$$+ \hat{\chi}_{xxxx}^{(3)}(-\omega; -\omega, \omega, \omega)\Big\}. \tag{8.76}$$

Here in the notation $(-\omega; \omega)$, the first $-\omega$ identifies the beam in which the changes occur and the second ω identifies the beam responsible for the change, in analogy to the notation used for $\chi^{(2)}$. The response is effectively instantaneous since it is associated with the distortion of the electron cloud. Many other physical mechanisms also lead to intensity-dependent changes in index and absorption and will be discussed later in Chapter 11.

8.2.3.3.2 Two-Beam (Eigenmodes) Input and Polarizations Along Crystal Axes. A number of common interactions are considered here.

8.2.3.3.2.1   CASE I: COHERENT BEAMS, EQUAL FREQUENCIES, AND ORTHOGONAL POLARIZATION.   This is a case commonly encountered in fibers. The Fourier components of the inputs are

$$\vec{E}(\omega_p) = \frac{1}{2}\Big\{ \big[\hat{e}_x\big[E_x(\omega_a)\delta(\omega_p - \omega_a) + E_x^*(\omega_a)\delta(\omega_p + \omega_a)\big]$$

$$+ \hat{e}_y[E_y(\omega_a)\delta(\omega_p - \omega_a) + E_y^*(\omega_a)\delta(\omega_p + \omega_a)]\Big\}. \tag{8.77}$$

The equations for $\vec{E}(\omega_q)$ and $\vec{E}(\omega_r)$ are the same with just different subscripts for the polarizations. The products of the fields lead to a number of phenomena in addition to third harmonic generation and intensity-dependent refraction and absorption for each beam separately. They are the two-beam effects of cross-polarized third harmonic generation via $|E_x(\omega)|^2 E_y(\omega)$ and $|E_y(\omega)|^2 E_x(\omega)$ and nonlinear refraction and absorption of one polarization induced onto another, labeled here as cross-nonlinear refraction (typically called cross-phase modulation) via $|E_x(\omega)|^2 E_y(\omega)$ and $|E_y(\omega)|^2 E_x(\omega)$. In addition, there is four-wave mixing via $E_x^2(\omega)E_y^*(\omega)$ and $E_y^2(\omega)E_x^*(\omega)$ by which energy is exchanged between the two polarizations.

The two new important effects are the cross-nonlinear refraction and the four-wave mixing.

8.2.3.3.2.1.1   Cross-Nonlinear Refraction   For this case there are six different nonlinear susceptibilities that contribute to each nonlinear polarization; namely,

$$P_x^{(3)}(\omega) = \frac{\varepsilon_0}{4}|E_y(\omega)|^2 E_x(\omega)\left[\left\{\hat{\chi}_{xxyy}^{(3)}(-\omega;\,\omega,-\omega,\omega) + \hat{\chi}_{xxyy}^{(3)}(-\omega;\,\omega,\omega,-\omega)\right\}\right.$$
$$+\left\{\hat{\chi}_{xyyx}^{(3)}(-\omega;\,\omega,-\omega,\omega) + \hat{\chi}_{xyyx}^{(3)}(-\omega;-\omega,\omega,\omega)\right\}$$
$$\left.+\left\{\hat{\chi}_{xyxy}^{(3)}(-\omega;-\omega,\omega,\omega) + \hat{\chi}_{xyxy}^{(3)}(-\omega;\,\omega,\omega,-\omega)\right\}\right], \qquad (8.78a)$$

$$P_y^{(3)}(\omega) = \frac{\varepsilon_0}{4}|E_x(\omega)|^2 E_y(\omega)\left[\left\{\hat{\chi}_{yyxx}^{(3)}(-\omega;\,\omega,-\omega,\omega) + \hat{\chi}_{yyxx}^{(3)}(-\omega;\,\omega,\omega,-\omega)\right\}\right.$$
$$+\left\{\hat{\chi}_{yxxy}^{(3)}(-\omega;\,\omega,-\omega,\omega) + \hat{\chi}_{yxxy}^{(3)}(-\omega;-\omega,\omega,\omega)\right\}$$
$$\left.+\left\{\hat{\chi}_{yxyx}^{(3)}(-\omega;-\omega,\omega,\omega) + \hat{\chi}_{yxyx}^{(3)}(-\omega;\,\omega,\omega,-\omega)\right\}\right]. \qquad (8.78b)$$

In the Kleinman limit all the cross-polarization terms are equal *in isotropic media* and so

$$P_x^{(3)}(\omega) = \frac{2\varepsilon_0}{4}|E_y(\omega)|^2 E_x(\omega)\tilde{\chi}_{xxxx}^{(3)}(-\omega;\,\omega,-\omega,\omega), \qquad (8.79a)$$

$$P_y^{(3)}(\omega) = \frac{2\varepsilon_0}{4}|E_x(\omega)|^2 E_y(\omega)\tilde{\chi}_{xxxx}^{(3)}(-\omega;\,\omega,-\omega,\omega). \qquad (8.79b)$$

Comparing them with Eq. 8.74, the cross-phase modulation term is 2/3 of the self-phase modulation.

For *nonisotropic media*, the relations between the different terms can be found in Appendix 8.1.

8.2.3.3.2.1.2   Four-Wave Mixing   The four-wave mixing nonlinear polarizations are

$$P_x^{(3)}(\omega) = \frac{\varepsilon_0}{4}E_y^2(\omega)E_x^*(\omega)\left\{\hat{\chi}_{xxyy}^{(3)}(-\omega;-\omega,\omega,\omega)\right.$$
$$\left.+\hat{\chi}_{xyyx}^{(3)}(-\omega;\,\omega,\omega,-\omega) + \hat{\chi}_{xyxy}^{(3)}(-\omega;\,\omega,-\omega,\omega)\right\}, \qquad (8.80a)$$

$$P_y^{(3)}(\omega) = \frac{\varepsilon_0}{4}E_x^2(\omega)E_y^*(\omega)\left\{\hat{\chi}_{yyxx}^{(3)}(-\omega;-\omega,\omega,\omega)\right.$$
$$\left.+\hat{\chi}_{yxxy}^{(3)}(-\omega;\,\omega,\omega,-\omega) + \hat{\chi}_{yxyx}^{(3)}(-\omega;\,\omega,-\omega,\omega)\right\}, \qquad (8.80b)$$

which in the Kleinman limit are

$$P_x^{(3)}(\omega) = \frac{\varepsilon_0}{4} E_y^2(\omega) E_x^*(\omega) \tilde{\chi}_{xxxx}^{(3)}(-\omega; -\omega, \omega, \omega), \tag{8.81a}$$

$$P_y^{(3)}(\omega) = \frac{\varepsilon_0}{4} E_x^2(\omega) E_y^*(\omega) \tilde{\chi}_{xxxx}^{(3)}(-\omega; -\omega, \omega, \omega). \tag{8.82b}$$

Adding the three effects—self-nonlinear refraction, cross-nonlinear refraction, and four-wave mixing—together in the Kleinman limit for isotropic media gives

$$\mathcal{P}_x^{(3)}(\omega) = \frac{3\varepsilon_0}{4} \tilde{\chi}_{xxxx}^{(3)}(-\omega; \omega, -\omega, \omega) \left[\left\{ |\mathcal{E}_x(\omega)|^2 + \frac{2}{3} |\mathcal{E}_y(\omega)|^2 \right\} \mathcal{E}_x(\omega) \right.$$

$$\left. + \frac{1}{3} |\mathcal{E}_y^2(\omega) \mathcal{E}_x^*(\omega)| \right] e^{i\{[2k_y(\omega)-k_x(\omega)]z + 2\phi_y - \phi_x\}}, \tag{8.83}$$

$$\mathcal{P}_y^{(3)}(\omega) = \frac{3\varepsilon_0}{4} \tilde{\chi}_{xxxx}^{(3)}(-\omega; \omega, -\omega, \omega) \left[\left\{ |\mathcal{E}_y(\omega)|^2 + \frac{2}{3} |\mathcal{E}_x(\omega)|^2 \right\} \mathcal{E}_y(\omega) \right.$$

$$\left. + \frac{1}{3} |\mathcal{E}_x^2(\omega) \mathcal{E}_y^*(\omega)| \right] e^{-i\{[2k_x(\omega)-k_y(\omega)] + 2\phi_x - \phi_y\}},$$

in which ϕ_x and ϕ_y are phases of two polarizations at the input. Normally for an isotropic medium, $k_y = k_x$ and the transfer of energy can be complete between the polarizations with the initial phase difference determining which beam initially gains energy. However, in fibers, e.g., there may be a small stress- or geometry-induced index birefringence that introduces a $k_y \neq k_x$ and results in a periodic exchange of energy between the two polarizations with the propagation distance.

8.2.3.3.2.2   CASE II: COHERENT BEAMS, UNEQUAL FREQUENCIES, AND PARALLEL POLARIZA-
TION.   For this case,

$$\vec{E}(\omega_p) = \frac{1}{2} \left\{ \hat{e}_x \left[E_x(\omega_a) \delta(\omega_p - \omega_a) + E_x^*(\omega_a) \delta(\omega_p + \omega_a) \right] \right.$$

$$\left. + \hat{e}_x \left[E_x(\omega_b) \delta(\omega_p - \omega_b) + E_x^*(\omega_b) \delta(\omega_p + \omega_b) \right] \right\}. \tag{8.84}$$

In addition to the previously discussed cases at each frequency, there are two different effects here that are commonly encountered in nonlinear optics. The cross-nonlinear refraction polarization is given by

$$\boldsymbol{P}_x^{(3)}(\omega_a) = \frac{1}{4}\varepsilon_0 \Big\{ \hat{\chi}_{xxxx}^{(3)}(-\omega_a;\ \omega_a,\omega_b,-\omega_b) + \hat{\chi}_{xxxx}^{(3)}(-\omega_a;\ \omega_a,-\omega_b,\omega_b)$$

$$+\ \hat{\chi}_{xxxx}^{(3)}(-\omega_a;\ \omega_b,\omega_a,-\omega_b) + \hat{\chi}_{xxxx}^{(3)}(-\omega_a;-\omega_b,\omega_a,\omega_b)$$

$$+\ \hat{\chi}_{xxxx}^{(3)}(-\omega_a;-\omega_b,\omega_b,\omega_a) + \hat{\chi}_{xxxx}^{(3)}(-\omega_a;\ \omega_b,-\omega_b,\omega_a)\Big\}$$

$$|\boldsymbol{\mathcal{E}}_x(\omega_b)|^2 \boldsymbol{\mathcal{E}}_x(\omega_a), \tag{8.85}$$

with a similar equation for $\boldsymbol{P}_x^{(3)}(\omega_b)$. If there is frequency dispersion in suscepti-bilities, it becomes difficult to link this result to the single frequency cases. In the absence of dispersion, e.g., in the Kleinman limit,

$$\boldsymbol{P}_x^{(3)}(\omega_a) \xrightarrow{\omega_a,\,\omega_b\to 0} \frac{6}{4}\varepsilon_0\tilde{\chi}_{xxxx}^{(3)}(-\omega_a;\ \omega_a,-\omega_b,\omega_b)|\boldsymbol{\mathcal{E}}_x(\omega_b)|^2\boldsymbol{\mathcal{E}}_x(\omega_a), \tag{8.86a}$$

$$\boldsymbol{P}_x^{(3)}(\omega_b) \xrightarrow{\omega_a,\,\omega_b\to 0} \frac{6}{4}\varepsilon_0\tilde{\chi}_{xxxx}^{(3)}(-\omega_b;\ \omega_b,-\omega_a,\omega_a)|\boldsymbol{\mathcal{E}}_x(\omega_a)|^2\boldsymbol{\mathcal{E}}_x(\omega_b). \tag{8.86b}$$

which, when compared to Eq. 8.74, shows that cross-nonlinear refraction for different frequencies is twice as effective as for the single-beam case.

8.2.3.3.2.3 NONLINEAR RAMAN SPECTROSCOPY.   Two unequal frequency beams have also been used for nonlinear spectroscopy. For coherent anti-Stokes Raman scattering ($\omega_b > \omega_a$) and an output signal at $2\omega_b - \omega_a$,

$$P_x^{(3)}(2\omega_b - \omega_a) = \frac{1}{4}\varepsilon_0\Big\{\hat{\chi}_{xxxx}^{(3)}(-[2\omega_b-\omega_a];\ \omega_b,\omega_b,-\omega_a)$$

$$+\ \hat{\chi}_{xxxx}^{(3)}(-[2\omega_b-\omega_a];\ \omega_b,-\omega_a,\omega_b)$$

$$+\ \hat{\chi}_{xxxx}^{(3)}(-[2\omega_b-\omega_a];-\omega_a,\omega_b,\omega_b)]\Big\}$$

$$\boldsymbol{\mathcal{E}}_x^2(\omega_b)\boldsymbol{\mathcal{E}}_x^*(\omega_a)\ e^{i[2k(\omega_b)-k(\omega_a)-k(2\omega_b-\omega_a)]z}$$

$$\xrightarrow[\text{symmetry}]{\text{Kleinman}} \frac{3}{4}\varepsilon_0\tilde{\chi}_{xxxx}^{(3)}(-[2\omega_b-\omega_a];\ \omega_b,-\omega_a,\omega_b)\boldsymbol{\mathcal{E}}_x^2(\omega_b)$$

$$\boldsymbol{\mathcal{E}}_x^*(\omega_a)\ e^{i[2k(\omega_b)-k(\omega_a)-k(2\omega_b-\omega_a)]z}. \tag{8.87}$$

For a signal at $2\omega_a - \omega_b$, known as coherent Stokes Raman scattering,

$$\mathcal{P}_x^{(3)}(2\omega_a - \omega_b) = \frac{1}{4}\varepsilon_0 \Big[\hat{\chi}_{xxxx}^{(3)}(-[2\omega_a - \omega_b]; \omega_a, \omega_a, -\omega_b)$$

$$+ \hat{\chi}_{xxxx}^{(3)}(-[2\omega_a - \omega_b]; \omega_a, -\omega_b, \omega_a)$$

$$+ \hat{\chi}_{xxxx}^{(3)}(-[2\omega_a - \omega_b]; -\omega_b, \omega_a, \omega_a) \Big] \mathcal{E}_x^2(\omega_a) \mathcal{E}_x^*(\omega_b)$$

$$\xrightarrow[\text{symmetry}]{\text{Kleinman}} \frac{3}{4}\varepsilon_0 \hat{\chi}_{xxxx}^{(3)}(-[2\omega_a - \omega_b]; \omega_a, -\omega_b, \omega_a) \mathcal{E}_x^2(\omega_a) \mathcal{E}_x^*(\omega_b). \qquad (8.88)$$

These processes are resonantly enhanced when $\omega_a - \omega_b$ is close to a vibration frequency in the medium. However, due to dispersion in index with frequency, these processes are not automatically wave-vector matched. Nonlinear spectroscopy will be discussed in more detail in section 15.3.

8.2.3.3.3 Two Incoherent Beams, Arbitrary Frequency, and Polarization.
Noting that the interactions in which relative phase does not occur are self- and cross-nonlinear refraction, these effects occur only between incoherent beams.

8.2.3.4 Susceptibilities in the Nonresonant Limit ($\omega \to 0$, $\bar{\omega}_{ng}\bar{\tau}_{ng} \gg 1$).

$$\chi_{ijkl}^{(3)}(-[\omega_p + \omega_q + \omega_r]; \omega_p, \omega_q, \omega_r) \xrightarrow{\omega_{p,q,r} \to 0}$$

$$\frac{N}{\varepsilon_0 \hbar^3} \Bigg[f^{(3)} {\sum_{v,n,m}}' \frac{1}{\bar{\omega}_{vg}\bar{\omega}_{mg}\bar{\omega}_{ng}} \times \Big\{ \bar{\mu}_{gv,i}(\bar{\mu}_{vn,l} - \bar{\mu}_{gg,l})(\bar{\mu}_{nm,k} - \bar{\mu}_{gg,k})\bar{\mu}_{mg,j}$$

$$+ \bar{\mu}_{gv,l}(\bar{\mu}_{vn,i} - \bar{\mu}_{gg,i})(\bar{\mu}_{nm,k} - \bar{\mu}_{gg,k})\bar{\mu}_{mg,j}$$

$$+ \bar{\mu}_{gv,j}(\bar{\mu}_{vn,k} - \bar{\mu}_{gg,k})(\bar{\mu}_{nm,i} - \bar{\mu}_{gg,i})\bar{\mu}_{mg,l}$$

$$+ \bar{\mu}_{gv,j}(\bar{\mu}_{vn,k} - \bar{\mu}_{gg,k})(\bar{\mu}_{nm,l} - \bar{\mu}_{gg,l})\bar{\mu}_{mg,i} \Big\}$$

$$- {\sum_{n,m}}' \left(\frac{1}{\bar{\omega}_{ng}} + \frac{1}{\bar{\omega}_{mg}} \right) \frac{\bar{\mu}_{gn,i}\bar{\mu}_{ng,l}\bar{\mu}_{gm,k}\bar{\mu}_{mg,j} + \bar{\mu}_{gn,l}\bar{\mu}_{ng,i}\bar{\mu}_{gm,j}\bar{\mu}_{mg,k}}{\bar{\omega}_{mg}\bar{\omega}_{ng}}.$$

$$(8.89)$$

Again, just as in the second-order susceptibility case, all the third-order susceptibilities are identical in this limit.

APPENDIX 8.1: $\chi_{ijk\ell}^{(3)}$ SYMMETRY PROPERTIES FOR DIFFERENT CRYSTAL CLASSES

A.8.1.1 Triclinic

For both classes (1 and $\bar{1}$) there are 81 independent nonzero elements.

A.8.1.2 Monoclinic

For all three classes (2, m, and $2/m$) there are 41 independent nonzero elements consisting of

3 elements with suffixes all equal;
18 elements with suffixes equal in pairs;
12 elements with suffixes having two y's, one x, and one z;
4 elements with suffixes having three x's and one z;
4 elements with suffixes having three z's and one x.

A.8.1.3 Orthorhombic

For all three classes (222, $mm2$, and mmm) there are 21 independent nonzero elements, consisting of

3 elements with all suffixes equal;
18 elements with suffixes equal in pairs.

A.8.1.4 Tetragonal

For the three classes 4, $\bar{4}$, and $4/m$, there are 41 nonzero elements of which only 21 are independent. They are as follows:

$$xxxx = yyyy \quad zzzz$$

$$zzxx = zzyy \quad xyzz = -yxzz \quad xxyy = yyxx \quad xxxy = -yyyx$$

$$xxzz = yyzz \quad zzxy = -zzyx \quad xyxy = yxyx \quad xxyx = -yyxy$$

$$zxzx = zyzy \quad xzyz = -yzxz \quad xyyx = yxxy \quad xyxx = -yxyy$$

$$xzxz = yzyz \quad zxzy = -zyzx \quad yxxx = -xyyy$$

$$zxxz = zyyz \quad zxyz = -zyxz$$

$$xzzx = yzzy \quad xzzy = -yzzx$$

For the four classes 422, 4mm, 4/mmm, and $\bar{4}2m$, there are 21 nonzero elements of which only 11 are independent. They are as follows:

$$xxxx = yyyy \quad zzzz$$
$$yyzz = xxzz \quad yzzy = xzzx \quad xxyy = yyxx \quad zzyy = zzxx \quad yzyz = xzxz$$
$$xyxy = yxyx \quad zyyz = zxxz \quad zyzy = zxzx \quad xyyx = yxxy$$

A.8.1.5 Cubic

For the two classes 23 and m3, there are 21 nonzero elements of which only 7 are independent. They are as follows:

$$xxxx = yyyy = zzzz \quad yyzz = zzxx = xxyy \quad zzyy = xxzz = yyxx \quad yzyz = zxzx = xyxy$$
$$zyzy = xzxz = yxyx \quad yzzy = zxxz = xyyx \quad zyyz = xzzx = yxxy$$

For the three classes 432, $\bar{4}3m$, and $m3m$, there are 21 nonzero elements of which only 4 are independent. They are as follows:

$$xxxx = yyyy = zzzz \quad yyzz = zzxx = xxyy = zzyy = xxzz = yyxx$$
$$yzyz = zxzx = xyxy = zyzy = xzxz = yxyx \quad yzzy = zxxz = xyyx = zyyz = xzzx = yxxy$$

A.8.1.6 Trigonal

For the two classes 3 and $\bar{3}$, there are 73 nonzero elements of which only 27 are independent. They are as follows:

$$zzzz$$
$$xxxx = yyyy = xxyy + xyyx + xyxy \quad xxyy = yyxx \quad xyyx = yxxy \quad xyxy = yxyx$$
$$yyzz = xxzz \quad xyzz = -yxzz \quad zzyy = zzxx \quad zzxy = -zzyx \quad yzyz = xzxz \quad xzyz = -yzxz$$
$$zyyz = zxxz \quad zxyz = -zyxz \quad yzzy = xzzx \quad xzzy = -yzzx \quad zyzy = zxzx \quad zyzx = -zyzx$$
$$xxxy = -yyyx = yyxy + yxyy + xyyy \quad yyxy = -xxyx \quad yxyy = -xyxx \quad xyyy = -yxxx$$
$$yyyz = -yxxz = -xyxz = -xxyz \quad yyzy = -yxzx = -xyzx = -xxzy$$
$$yzyy = -yzxx = -xzyx = -xzxy$$
$$zyyy = -zyxx = -zxyx = -zxxy \quad xxxz = -xyyz = -yxyz = -yyxz$$
$$xxzx = -xyzy = -yxzy = -yyzx$$
$$xzxx = -yzxy = -yzyx = -xzyy \quad zxxx = -zxyy = -zyxy = -zyyx$$

For the three classes $3m$, $\bar{3}m$, and 32, there are 37 nonzero elements of which only 14 are independent. They are as follows:

zzzz

$$xxxx = yyyy = xxyy + xyyx + xyxy \quad xxyy = yyxx \quad xyyx = yxxy \quad xyxy = yxyx$$

$$yyzz = xxzz \quad zzyy = zzxx \quad zyyz = zxxz \quad yzzy = xzzx \quad yzyz = xzxz \quad zyzy = zxzx$$

$$xxxz = -xyyz = -yxyz = -yyxz \quad xxzx = -xyzx = -yxzy = -yyzx$$

$$zxxx = -zxyy = -zyxy = -zyyx \quad xzxx = -xzyy = -yzxy = -yzyx$$

A.8.1.7   Hexagonal

For the three classes 6, $\bar{6}$, and $6/m$, there are 41 nonzero elements of which only 19 are independent. They are as follows:

zzzz

$$xxxx = yyyy = xxyy + xyyx + xyxy \quad xxyy = yyxx \quad xyyx = yxxy \quad xyxy = yxyx$$

$$yyzz = xxzz \quad xyzz = -yxzz \quad zzyy = zzxx \quad zzxy = -zzyx$$

$$zyyz = zxxz \quad zxyz = -zyxz \quad yzzy = xzzx \quad xzzy = -yzzx$$

$$yzyz = xzxz \quad xzyz = -yzxz \quad zyzy = zxzx \quad zxzy = -zyzx$$

$$xxxy = -yyyx = yyxy + yxyy + xyyy \quad yyxy = -xxyx \quad xyxy = -xyxx$$

$$xyyy = -yxxx$$

For the four classes 622, $6\,mm$, $6/mmm$, and $\bar{6}m2$, there are 21 nonzero elements of which only 10 are independent. They are as follows:

zzzz

$$xxxx = yyyy = xxyy + xyyx + xyxy \quad xxyy = yyxx \quad xyyx = yxxy \quad xyxy = yxyx$$

$$yyzz = xxzz \quad zzyy = zzxx \quad zyyz = zxxz \quad yzzy = xzzx \quad yzyz = xzxz \quad zyzy = zxzx$$

PROBLEMS

1.

   (a) Show that the original and "corrected" expressions for Type I second harmonic generation give identical results for the second-order susceptibility calculated by using first-order perturbation theory.

(b) Show that the simple final (corrected) expression for $\chi^{(2)}_{ijk}(0; \omega, -\omega)$ in the off-resonance limit by using the first-order perturbation theory is given by

$$\hat{\chi}^{(2)}_{ijk}(0; \omega, -\omega) + \hat{\chi}^{(2)}_{ijk}(0; -\omega, \omega) = 2\frac{N}{\varepsilon_0\hbar^2}|\bar{\mu}_{10}|^2(\bar{\mu}_{11} - \bar{\mu}_{00})\frac{3\bar{\omega}^2_{10} - \omega^2}{(\bar{\omega}^2_{10} - \omega^2)^2}.$$

2. Consider $\hat{\chi}^{(5)}$ with a single frequency and polarization input.

   (a) What nonlinear processes are possible for this case? Write down the nonlinear polarization $P^{(5)}$ for each case. Assume Kleinman symmetry and be sure to identify the degeneracy factors for each case.

   (b) What new terms are introduced when an orthogonally polarized input of the same frequency appears only once in the mixing process?

3. Consider $\hat{\chi}^{(4)}$ with a single frequency and polarization input.

   (a) What nonlinear processes are possible for this case? Write down the nonlinear polarization $P^{(4)}$ for each case. Assume Kleinman symmetry and be sure to identify the degeneracy factors for each case.

   (b) What new terms are introduced when an orthogonally polarized input of the same frequency appears only once in the mixing process?

4. Prove that for an isotropic material, $\hat{\chi}^{(3)}_{xxxx} = \hat{\chi}^{(3)}_{xxyy} + \hat{\chi}^{(3)}_{xyxy} + \hat{\chi}^{(3)}_{xyyx}$, given that

$$\hat{\chi}^{(3)}_{xxxx} = \hat{\chi}^{(3)}_{yyyy} = \hat{\chi}^{(3)}_{zzzz},$$

$$\hat{\chi}^{(3)}_{xxyy} = \hat{\chi}^{(3)}_{yyxx} = \hat{\chi}^{(3)}_{xxzz} = \hat{\chi}^{(3)}_{zzxx} = \hat{\chi}^{(3)}_{zzyy} = \hat{\chi}^{(3)}_{yyzz},$$

$$\hat{\chi}^{(3)}_{xyxy} = \hat{\chi}^{(3)}_{yxyx} = \hat{\chi}^{(3)}_{xzxz} = \hat{\chi}^{(3)}_{zxzx} = \hat{\chi}^{(3)}_{zyzy} = \hat{\chi}^{(3)}_{yzyz},$$

$$\hat{\chi}^{(3)}_{xyyx} = \hat{\chi}^{(3)}_{yxxy} = \hat{\chi}^{(3)}_{xzzx} = \hat{\chi}^{(3)}_{zxxz} = \hat{\chi}^{(3)}_{zyyz} = \hat{\chi}^{(3)}_{yzzy}$$

by taking a field with an *arbitrary* direction in space defined by θ and ϕ and by applying the condition that the nonlinear polarization induced must be independent of the choice of coordinates.

5. Consider nonlinear optics at the molecular level for two-dimesional disk-shaped molecules with a third-order molecular nonlinearity that is isotropic in the plane of the disk: $\bar{\gamma}^{(3)}_{1111} = \bar{\gamma}^{(3)}_{2222} \neq 0$. Show that $\bar{\gamma}^{(3)}_{\bar{x}\bar{x}\bar{x}\bar{x}} = \bar{\gamma}^{(3)}_{\bar{x}\bar{x}\bar{y}\bar{y}} + \bar{\gamma}^{(3)}_{\bar{x}\bar{y}\bar{y}\bar{x}} + \bar{\gamma}^{(3)}_{\bar{x}\bar{y}\bar{x}\bar{y}}$ for *all* possible field directions in the plane.

6. Consider inducing the largest possible index change in a beam of frequency ω_a with an intense beam of frequency ω_b. Assuming that the material is isotropic and Kerr and that the two frequencies are far from any resonance, would it be more efficient if the two beams were copolarized or orthogonally polarized? By what factor is it better?

7. Find the nonlinear polarizations that occur for two coherent, cross-polarized, and unequal frequency beams.

8. Show that in the nonresonant limit, for arbitrary input frequencies $\omega_1, \omega_2, \omega_3, \omega_4, \omega_5$

$$\tilde{\chi}^{(5)}_{xxxxxx}(-\omega_6; \omega_5, \omega_4, \omega_3, \omega_2, \omega_1) = 5\tilde{\chi}^{(5)}_{yyxxxx}(-\omega_6; \omega_5, \omega_4, \omega_3, \omega_2, \omega_1).$$

REFERENCE

1. B. J. Orr and J. F. Ward, "Perturbation theory of the nonlinear optical polarization of an isolated system," Mol. Phys., **20**(3), 513–26 (1971).

SUGGESTED FURTHER READING

N. Bloembergen, Nonlinear Optics, A Lecture Note and Reprint Volume (W. A. Benjamin, New York, 1965).

M. Born and E. Wolf, Principles of Optics, 7th Edition (Cambridge University Press, Cambridge, UK, 1999).

R. W. Boyd, Nonlinear Optics, 3rd Edition (Academic Press, Burlington MA, 2008).

F. A. Hopf and G. I. Stegeman, Applied Classical Electrodynamics, Volume 2: Nonlinear Optics (John Wiley & Sons, New York, 1985).

E. Merzbacher, Quantum Mechanics (John Wiley & Sons, New York, 1972).

D. L. Mills, Nonlinear Optics: Basic Concepts, 2nd Edition (Springer, New York, 1998).

Y. R. Shen, Principles of Nonlinear Optics (John Wiley & Sons, New York, 1984).

CHAPTER 9

Molecular Nonlinear Optics

In Chapter 8 the general sum-over-states (SOS) model for the nonlinear suscepti-bilities of molecules was discussed. In general, the application of this model to a specific molecule requires that the energies of the discrete electronic states above the ground state and the transition dipole moments between them be known. Although, in principle, this can be done and the physical chemistry community has developed computer routines for this purpose, it is a formidable task and simpler approaches have been developed and tested that give useful approximations to the full SOS case (1). In this chapter the predictions of such simple models are discussed.

As discussed previously for $\chi^{(2)}$, the key question is whether the unit cell in a crystal exhibits inversion symmetry. For centrosymmetric molecules, $\chi^{(2)} = 0$. As can be easily seen from Eq. 8.51 and the discussion in Section 8.2.2.6, for asymmetric unit cells composed of asymmetric molecules only one excited state is sufficient to produce a nonzero $\chi^{(2)}$. Hence the minimum model for $\chi^{(2)}$ will involve only two states, i.e., a two-level model. SOS again gives two contributions (see Eq. 8.53) to the third-order susceptibility. Since the first contribution can produce two-photon effects with just one ground state and one excited state, again a two-level model is the minimum sufficient to describe the nonlinearity $\chi^{(3)}$.

On the other hand, for $\chi^{(3)}$, in centrosymmetric molecules the spatial wave functions exhibit either even (gerade, subscript g) or odd (ungerade, subscript u) symmetry, with the ground state having even symmetry. Furthermore, these mole-cules have zero permanent dipole moments. SOS gives two sets of contributions as shown in Eq. 8.53. The first set has products of the form $\bar{\mu}_{gv}\bar{\mu}_{vn}\bar{\mu}_{nm}\bar{\mu}_{mg}$ in the numerator. Since there must be a change in the parity of the wave functions between the initial and final states for each nonzero electric dipole transition moment, this product must have the form $\bar{\mu}_{gu'}\bar{\mu}_{u'g'}\bar{\mu}_{g'u}\bar{\mu}_{ug}$ (in terms of the symmetry of the states), in which u and u' may refer to the same or different odd symmetry states and g and g' refer to different even symmetry states. ($g' = g$ is not allowed since the summation *excludes* the ground state.) Therefore a minimum of three states are required to have two photon states with the pathway $g \rightarrow u \rightarrow g' \rightarrow u \rightarrow g$. Furthermore, numerators of the second set of terms in Eq. 8.53, $\bar{\mu}_{gn}\bar{\mu}_{ng}\bar{\mu}_{gm}\bar{\mu}_{mg}$ require only a minimum of two levels since $n = m$ is allowed. Hence for symmetric molecules, a three-level model is

Nonlinear Optics: Phenomena, Materials, and Devices, George I. Stegeman and Robert A. Stegeman.
© 2012 John Wiley & Sons, Inc. Published 2012 by John Wiley & Sons, Inc.

adopted for the first set of terms and a two-level model for the second set of terms in Eq. 8.53 and the combination is called a three-level model.

9.1 TWO-LEVEL MODEL

The simplest possible description of a molecule is that it has one ground state and one excited state. Despite its simplicity, this model has proven very valuable for understanding general trends of susceptibilities with frequency for noncentrosymmetric molecules. Also, it provides a good representation of charge transfer molecules (2).

The relevant states for the SOS calculation are shown in Fig. 9.1.

9.1.1 Second-Order Susceptibilities

It is immediately clear from Eq. 8.51 that one requirement for $\chi^{(2)}$-active molecules in this model is that they must exhibit a permanent dipole moment in either the ground or the excited state. In this limit, Eq. 8.51 simplifies considerably to

$$\hat{\chi}_{ijk}^{(2)}(-[\omega_p + \omega_q]; \omega_p, \omega_q) = \frac{N}{\varepsilon_0 \hbar^2} f^{(2)} \left\{ \frac{\bar{\mu}_{01,i}(\bar{\mu}_{11,k} - \bar{\mu}_{00,k})\bar{\mu}_{10,j}}{(\hat{\bar{\omega}}_{10} - \omega_p - \omega_q)(\hat{\bar{\omega}}_{10} - \omega_p)} \right.$$

$$\left. + \frac{\bar{\mu}_{01,k}(\bar{\mu}_{11,j} - \bar{\mu}_{00,j})\bar{\mu}_{10,i}}{(\hat{\bar{\omega}}_{10}^* + \omega_q)(\hat{\bar{\omega}}_{10}^* + \omega_p + \omega_q)} + \frac{\bar{\mu}_{01,k}(\bar{\mu}_{11,i} - \bar{\mu}_{00,i})\bar{\mu}_{10,j}}{(\hat{\bar{\omega}}_{10}^* + \omega_q)(\hat{\bar{\omega}}_{10} - \omega_p)} \right\}. \quad (9.1)$$

The permanent dipole moments in the ground and excited states are written as $\vec{\mu}_{00}$ and $\vec{\mu}_{11}$, respectively, and the transition dipole moment is $\vec{\mu}_{10} = \vec{\mu}_{01}$. Normally frequency conversion via $\hat{\chi}^{(2)}$ is performed in the off-resonance or nonresonant regime. Therefore it is assumed in the following examples that all the optical frequencies involved are far enough from $\bar{\omega}_{10}$, and so the $\bar{\tau}_{10}^{-1}$ part of $\hat{\bar{\omega}}_{10}$ in the denominators can be neglected; i.e., $\hat{\bar{\omega}}_{10}$ is real. This is reasonable since parametric mixing is performed at frequencies for which the loss is small.

9.1.1.1 Example of Sum Harmonic Generation in a Periodically Poled Crystal. In periodically poled lithium niobate, e.g., the dominant second-order nonlinearity lies along the z-axis. For this case, the off-resonance result in the

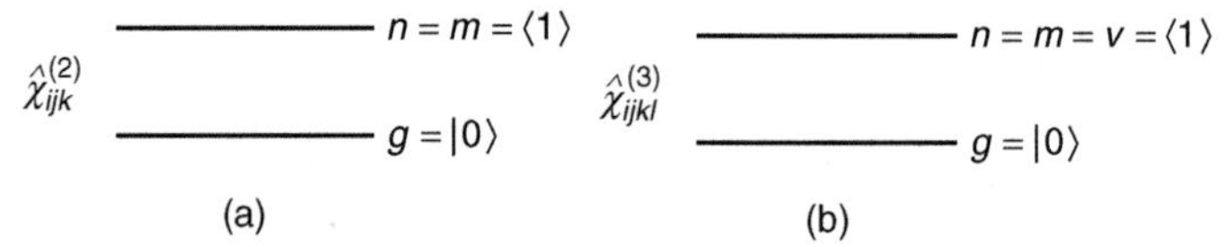

$$\hat{\chi}_{ijk}^{(2)} \quad \frac{}{} n = m = \langle 1 | \qquad \hat{\chi}_{ijkl}^{(3)} \quad \frac{}{} n = m = v = \langle 1 |$$

$$\frac{}{} g = |0\rangle \qquad \qquad \frac{}{} g = |0\rangle$$

(a) \qquad\qquad\qquad (b)

FIGURE 9.1 The two-level model for calculating (a) $\hat{\chi}^{(2)}$ and (b) $\hat{\chi}^{(3)}$ for noncentrosymmetric molecules.

two-level model is

$$\Re e a l\left\{\hat{\chi}_{zzz}^{(2)}(-2\omega;\,\omega,\omega)\right\} = \frac{N}{\varepsilon_0\hbar^2}f^{(2)}|\bar{\mu}_{10}|^2(\bar{\mu}_{11}-\bar{\mu}_{00})$$

$$\times\left\{\frac{1}{(\bar{\omega}_{10}-2\omega)(\bar{\omega}_{10}-\omega)}+\frac{1}{(\bar{\omega}_{10}+\omega)(\bar{\omega}_{10}+2\omega)}+\frac{1}{(\bar{\omega}_{10}+\omega)(\bar{\omega}_{10}-\omega)}\right\}$$

$$\rightarrow\Re e a l\left\{\hat{\chi}_{zzz}^{(2)}(-2\omega;\,\omega,\omega)\right\} = \frac{N}{\varepsilon_0\hbar^2}f^{(2)}|\bar{\mu}_{10}|^2(\bar{\mu}_{11}-\bar{\mu}_{00})\frac{3\bar{\omega}_{10}^2}{(\bar{\omega}_{10}^2-\omega^2)(\bar{\omega}_{10}^2-4\omega^2)}.$$
$$(9.2)$$

This result has *similar but not identical resonances* to those predicted by the electron on a spring model discussed in Chapter 2, e.g., see Eq. 2.8 in the limit $\tau_{mg}^{-1}\rightarrow 0$, which has $(\bar{\omega}_{10}^2-\omega^2)^2(\bar{\omega}_{10}^2-4\omega^2)$ in the denominator; i.e., the resonance is "stronger" in the anharmonic oscillator model. And in SOS the transition and permanent dipole moments are measurable quantities.

9.1.1.2 *Example of Sum Frequency Generation in Periodically Poled Lithium Niobate.*

In this case there are two input frequencies, namely, ω_1 and ω_2. In the two-level limit, far from resonance

$$P_z^{(3)}(\omega_1+\omega_2) = \frac{1}{2}\varepsilon_0\left[\hat{\chi}_{zzz}^{(2)}(-[\omega_2+\omega_1];\,\omega_1,\omega_2)+\hat{\chi}_{zzz}^{(2)}(-[\omega_2+\omega_1];\,\omega_2,\omega_1)\right]$$
$$\times E_z(\omega_1)E_z(\omega_2),$$
$$(9.3)$$

$$\Re e a l\left\{\hat{\chi}_{zzz}^{(2)}(-[\omega_2+\omega_1];\,\omega_2,\omega_1)+\hat{\chi}_{zzz}^{(2)}(-[\omega_2+\omega_1];\,\omega_1,\omega_2)\right\}=f^{(2)}\frac{N}{\varepsilon_0\hbar^2}|\bar{\mu}_{01}|^2(\bar{\mu}_{11}-\bar{\mu}_{00})$$

$$\left\{\frac{1}{(\bar{\omega}_{10}-\omega_2-\omega_1)(\bar{\omega}_{10}-\omega_1)}+\frac{1}{(\bar{\omega}_{10}-\omega_2-\omega_1)(\bar{\omega}_{10}-\omega_2)}\right.$$

$$+\frac{1}{(\bar{\omega}_{10}+\omega_1)(\bar{\omega}_{10}+\omega_2+\omega_1)}+\frac{1}{(\bar{\omega}_{10}+\omega_2)(\bar{\omega}_{10}+\omega_2+\omega_1)}$$

$$\left.+\frac{1}{(\bar{\omega}_{10}+\omega_1)(\bar{\omega}_{10}-\omega_2)}+\frac{1}{(\bar{\omega}_{10}-\omega_1)(\bar{\omega}_{10}+\omega_2)}\right\},$$
$$(9.4)$$

$$P_z^{(3)}(\omega_1+\omega_2) = \frac{\bar{\omega}_{10}^2 N}{\hbar^2}f^{(2)}|\bar{\mu}_{01}|^2(\bar{\mu}_{11}-\bar{\mu}_{00})$$
$$\times\frac{\bar{\omega}_{10}^2[3\bar{\omega}_{10}^2-(\omega_1+\omega_2)^2+\omega_1\omega_2]E_z(\omega_1)E_z(\omega_2)}{[\bar{\omega}_{10}^2-(\omega_1+\omega_2)^2][\bar{\omega}_{10}^2-\omega_1^2][\bar{\omega}_{10}^2-\omega_2^2]}.$$
$$(9.5)$$

Again, the frequency dispersion is different from the anharmonic oscillator result, which has no frequency dependence in the numerator.

9.1.2 Third-Order Susceptibilities

The third-order susceptibility for the two-level model is

$$
\hat{\chi}_{ijkl}^{(3)}\left(-\left[\omega_p+\omega_q+\omega_r\right];\,\omega_p,\omega_q,\omega_r\right)
$$

$$
=\frac{N}{\varepsilon_0\hbar^3}f^{(3)}\left\{\frac{\bar{\mu}_{01,i}(\bar{\mu}_{11,l}-\bar{\mu}_{00,l})(\bar{\mu}_{11,k}-\bar{\mu}_{00,k})\bar{\mu}_{10,j}}{(\hat{\bar{\omega}}_{10}-\omega_p-\omega_q-\omega_r)(\hat{\bar{\omega}}_{10}-\omega_q-\omega_p)(\hat{\bar{\omega}}_{10}-\omega_p)}\right.
$$

$$
+\frac{\bar{\mu}_{01,j}(\bar{\mu}_{11,k}-\bar{\mu}_{00,k})(\bar{\mu}_{11,i}-\bar{\mu}_{00,i})\bar{\mu}_{10,l}}{(\hat{\bar{\omega}}_{10}^{*}+\omega_p)(\hat{\bar{\omega}}_{10}^{*}+\omega_q+\omega_p)(\hat{\bar{\omega}}_{10}-\omega_r)}
$$

$$
+\frac{\bar{\mu}_{01,l}(\bar{\mu}_{11,i}-\bar{\mu}_{00,i})(\bar{\mu}_{11,k}-\bar{\mu}_{00,k})\bar{\mu}_{10,j}}{(\hat{\bar{\omega}}_{10}^{*}+\omega_r)(\hat{\bar{\omega}}_{10}-\omega_q-\omega_p)(\hat{\bar{\omega}}_{10}-\omega_p)}
$$

$$
+\left.\frac{\bar{\mu}_{01,j}(\bar{\mu}_{11,k}-\bar{\mu}_{00,k})(\bar{\mu}_{11,l}-\bar{\mu}_{00,l})\bar{\mu}_{10,i}}{(\hat{\bar{\omega}}_{10}^{*}+\omega_p)(\hat{\bar{\omega}}_{10}^{*}+\omega_q+\omega_p)(\hat{\bar{\omega}}_{10}^{*}+\omega_p+\omega_q+\omega_r)}\right\}
$$

$$
-\frac{N}{\varepsilon_0\hbar^3}f^{(3)}\left\{\frac{\bar{\mu}_{01,i}\,\bar{\mu}_{10,l}\,\bar{\mu}_{01,k}\,\bar{\mu}_{10,j}}{(\hat{\bar{\omega}}_{10}-\omega_p-\omega_q-\omega_r)(\hat{\bar{\omega}}_{10}-\omega_r)(\hat{\bar{\omega}}_{10}-\omega_p)}\right.
$$

$$
+\frac{\bar{\mu}_{01,i}\,\bar{\mu}_{10,l}\,\bar{\mu}_{01,k}\,\bar{\mu}_{10,j}}{(\hat{\bar{\omega}}_{10}^{*}+\omega_q)(\hat{\bar{\omega}}_{10}-\omega_r)(\hat{\bar{\omega}}_{10}-\omega_p)}+\frac{\bar{\mu}_{01,l}\,\bar{\mu}_{10,i}\,\bar{\mu}_{01,j}\,\bar{\mu}_{10,k}}{(\hat{\bar{\omega}}_{10}^{*}+\omega_r)(\hat{\bar{\omega}}_{10}^{*}+\omega_p)(\hat{\bar{\omega}}_{10}-\omega_q)}
$$

$$
+\left.\frac{\bar{\mu}_{01,l}\,\bar{\mu}_{10,i}\,\bar{\mu}_{01,j}\,\bar{\mu}_{10,k}}{(\hat{\bar{\omega}}_{10}^{*}+\omega_r)(\hat{\bar{\omega}}_{10}^{*}+\omega_p)(\hat{\bar{\omega}}_{10}^{*}+\omega_p+\omega_q+\omega_r)}\right\}. \tag{9.6}
$$

9.1.2.1 *Third Harmonic Generation.* With a single z-polarized input beam,

$$
\hat{\chi}_{zzzz}^{(3)}(-3\omega;\,\omega,\omega,\omega)=\frac{N}{\varepsilon_0\hbar^3}f^{(3)}\left\{\frac{(\bar{\mu}_{11,z}-\bar{\mu}_{00,z})^2|\bar{\mu}_{10,z}|^2}{(\hat{\bar{\omega}}_{10}-3\omega)(\hat{\bar{\omega}}_{10}-2\omega)(\hat{\bar{\omega}}_{10}-\omega)}\right.
$$

$$
+\frac{(\bar{\mu}_{11,z}-\bar{\mu}_{00,z})^2|\bar{\mu}_{10,z}|^2}{(\hat{\bar{\omega}}_{10}^{*}+\omega)(\hat{\bar{\omega}}_{10}^{*}+2\omega)(\hat{\bar{\omega}}_{10}-\omega)}+\frac{(\bar{\mu}_{11,z}-\bar{\mu}_{00,z})^2|\bar{\mu}_{10,z}|^2}{(\hat{\bar{\omega}}_{10}^{*}+\omega)(\hat{\bar{\omega}}_{10}-2\omega)(\hat{\bar{\omega}}_{10}-\omega)}
$$

$$
+\left.\frac{(\bar{\mu}_{11,z}-\bar{\mu}_{00,z})^2|\bar{\mu}_{10,z}|^2}{(\hat{\bar{\omega}}_{10}^{*}+\omega)(\hat{\bar{\omega}}_{10}^{*}+2\omega)(\hat{\bar{\omega}}_{10}^{*}+3\omega)}\right\}-\frac{N}{\varepsilon_0\hbar^3}f^{(3)}\left\{\frac{|\bar{\mu}_{10,z}|^4}{(\hat{\bar{\omega}}_{10}-3\omega)(\hat{\bar{\omega}}_{10}-\omega)(\hat{\bar{\omega}}_{10}-\omega)}\right.
$$

$$+ \frac{|\bar{\mu}_{10,z}|^4}{(\hat{\bar{\omega}}_{10}^*+\omega)(\hat{\bar{\omega}}_{10}-\omega)(\hat{\bar{\omega}}_{10}-\omega)} + \frac{|\bar{\mu}_{10,z}|^4}{(\hat{\bar{\omega}}_{10}^*+\omega)(\hat{\bar{\omega}}_{10}^*+\omega)(\hat{\bar{\omega}}_{10}-\omega)}$$

$$+ \left. \frac{|\bar{\mu}_{10,z}|^4}{(\hat{\bar{\omega}}_{10}^*+\omega)(\hat{\bar{\omega}}_{10}^*+\omega)(\hat{\bar{\omega}}_{10}^*+3\omega_r)} \right\}. \tag{9.7}$$

For materials in which a two-level system would be valid, third harmonic resonance peaks occur for $3\omega = \bar{\omega}_{10}$ and $2\omega = \bar{\omega}_{10}$.

9.1.2.2 Nonlinear Refraction and Absorption.

The starting point is Eq. 8.53. As shown in Fig. 9.2, there are three cases of $\hat{\chi}^{(3)}$, each corresponding to a different ordering of the frequencies $(\omega, \omega, -\omega)$. Note, however, that there can be strong interferences between the contributing terms and so the only physically relevant value is the sum of the terms.

Cases I and II go through the ground state with a zero frequency response in an intermediate step, whereas Case III goes through a second harmonic response.

Case I: $\hat{\chi}^{(3)}_{xxxx}(-\omega; \omega, -\omega, \omega)$

$$\hat{\chi}^{(3)}_{xxxx}(-\omega; \omega, -\omega, \omega) = \frac{N}{\varepsilon_0 \hbar^3} f^{(3)} \left\{ |\bar{\mu}_{10}|^2 (\bar{\mu}_{11}-\bar{\mu}_{00})^2 \left[\frac{1}{(\hat{\bar{\omega}}_{10}-\omega)^2 \hat{\bar{\omega}}_{10}} \right. \right.$$

$$+ \frac{1}{(\hat{\bar{\omega}}_{10}^*+\omega)\hat{\bar{\omega}}_{10}(\hat{\bar{\omega}}_{10}-\omega)} + \frac{1}{(\hat{\bar{\omega}}_{10}^*+\omega)\hat{\bar{\omega}}_{10}^*(\hat{\bar{\omega}}_{10}-\omega)} + \frac{1}{(\hat{\bar{\omega}}_{10}^*+\omega)^2 \hat{\bar{\omega}}_{10}^*} \right]$$

$$-|\bar{\mu}_{01}|^4 \left[\frac{1}{(\hat{\bar{\omega}}_{10}-\omega)^3} + \frac{1}{(\hat{\bar{\omega}}_{10}^*-\omega)(\hat{\bar{\omega}}_{10}-\omega)^2} + \frac{1}{(\hat{\bar{\omega}}_{10}^*+\omega)^2(\hat{\bar{\omega}}_{10}+\omega)} \right.$$

$$+ \left. \left. \frac{1}{(\hat{\bar{\omega}}_{10}^*+\omega)^3} \right] \right\}. \tag{9.8}$$

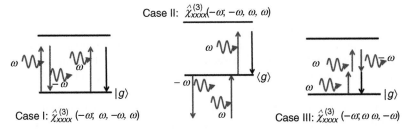

Case II: $\hat{\chi}^{(3)}_{xxxx}(-\omega; -\omega, \omega, \omega)$

Case I: $\hat{\chi}^{(3)}_{xxxx}(-\omega; \omega, -\omega, \omega)$

Case III: $\hat{\chi}^{(3)}_{xxxx}(-\omega; \omega, \omega, -\omega)$

FIGURE 9.2 The three different $\hat{\chi}^{(3)}$'s needed to evaluate nonlinear absorption and refraction.

Case II: $\hat{\chi}_{xxxx}^{(3)}(-\omega; -\omega, \omega, \omega)$

$$\hat{\chi}_{xxxx}^{(3)}(-\omega; -\omega, \omega, \omega) = \frac{N}{\varepsilon_0 \hbar^3} f^{(3)} \left\{ |\bar{\mu}_{10}|^2 (\bar{\mu}_{11} - \bar{\mu}_{00})^2 \left[\frac{1}{(\hat{\bar{\omega}}_{10}^2 - \omega^2)\hat{\bar{\omega}}_{10}} \right. \right.$$

$$+ \frac{1}{(\hat{\bar{\omega}}_{10}^* + \omega)\hat{\bar{\omega}}_{10}(\hat{\bar{\omega}}_{10} + \omega)} + \frac{1}{(\hat{\bar{\omega}}_{10}^* - \omega)\hat{\bar{\omega}}_{10}^*(\hat{\bar{\omega}}_{10} - \omega)} + \frac{1}{(\hat{\bar{\omega}}_{10}^{*2} - \omega^2)\hat{\bar{\omega}}_{10}^*} \right]$$

$$- |\bar{\mu}_{01}|^4 \left[\frac{1}{(\hat{\bar{\omega}}_{10}^2 - \omega^2)} \left(\frac{1}{(\hat{\bar{\omega}}_{10} - \omega)} + \frac{1}{(\hat{\bar{\omega}}_{10}^* + \omega)} \right) \right.$$

$$+ \frac{1}{(\hat{\bar{\omega}}_{10}^{*2} - \omega^2)} \left. \left. \left(\frac{1}{(\hat{\bar{\omega}}_{10} - \omega)} + \frac{1}{(\hat{\bar{\omega}}_{10}^* + \omega)} \right) \right] \right\}. \tag{9.9}$$

Case III: $\hat{\chi}_{xxxx}^{(3)}(-\omega; \omega, \omega, -\omega)$

$$\hat{\chi}_{xxxx}^{(3)}(-\omega; \omega, \omega, -\omega) = \frac{N}{\varepsilon_0 \hbar^3} f^{(3)} \left\{ |\bar{\mu}_{10}|^2 (\bar{\mu}_{11} - \bar{\mu}_{00})^2 \left[\frac{1}{(\hat{\bar{\omega}}_{10} - 2\omega)(\hat{\bar{\omega}}_{10} - \omega)} \right. \right.$$

$$\times \left(\frac{1}{(\hat{\bar{\omega}}_{10} - \omega)} + \frac{1}{(\hat{\bar{\omega}}_{10}^* - \omega)} \right) + \frac{1}{(\hat{\bar{\omega}}_{10}^* + 2\omega)(\hat{\bar{\omega}}_{10}^* + \omega)} \left. \left(\frac{1}{(\hat{\bar{\omega}}_{10}^* + \omega)} + \frac{1}{(\hat{\bar{\omega}}_{10}^* + \omega)} \right) \right]$$

$$- |\bar{\mu}_{01}|^4 \left[\frac{1}{(\hat{\bar{\omega}}_{10}^2 - \omega^2)} \left(\frac{1}{(\hat{\bar{\omega}}_{10} - \omega)} + \frac{1}{(\hat{\bar{\omega}}_{10}^* + \omega)} \right) \right.$$

$$+ \frac{1}{(\hat{\bar{\omega}}_{10}^{*2} - \omega^2)} \left. \left. \left(\frac{1}{(\hat{\bar{\omega}}_{10} - \omega)} + \frac{1}{(\hat{\bar{\omega}}_{10}^* + \omega)} \right) \right] \right\}. \tag{9.10}$$

The last case is the only one that gives rise to a two-photon peak, and hence we label all the terms proportional to $(\bar{\mu}_{11} - \bar{\mu}_{00})^2$ as two-photon contributions. For the two-level model, two-photon absorption requires a molecule with permanent dipole moments. Only one-photon transitions are allowed for the second set of terms in Eq. 8.53.

The typical frequency dependence of different contributions to the total third-order susceptibility is shown in Fig. 9.3. The one-photon contributions ($\propto |\bar{\mu}_{10}|^4$) to both the real and imaginary components (denoted by subscript $\Re$ and $\Im$, respectively) of $\hat{\chi}_{xxxx}^{(3)}$ are negative at all frequencies.

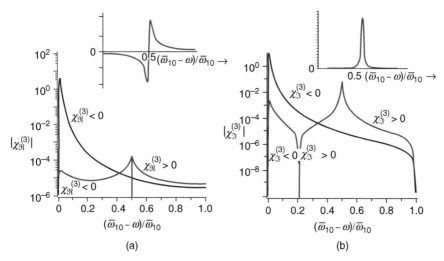

FIGURE 9.3 Generic dependence on the normalized frequency of the real and imaginary components (in arbitrary units) of the one- and two-photon terms of the third-order susceptibility in the two-level model. The blue curves are for the total one-photon terms ($\propto |\bar{\mu}_{10}|^4$), and the red curves are for the total two-photon terms ($\propto |\bar{\mu}_{10}|^2 (\bar{\mu}_{11} - \bar{\mu}_{00})^2$). The regions of positive and negative susceptibilities are identified. The upper curves show the dispersion of two-photon resonance terms on a linear scale.

The situation is more complicated for the two-photon contributions ($\propto |\bar{\mu}_{10}|^2 (\bar{\mu}_{11} - \bar{\mu}_{00})^2$). Between $\omega = \bar{\omega}_{10}$ and the dispersion resonance at $\omega = \bar{\omega}_{10}/2$, $\widehat{\chi}_{\Re}^{(3)}$ is negative and after the resonance it is positive up to the nonresonant limit $\omega = 0$. The imaginary component starts out positive at $\omega = \bar{\omega}_{10}$, changes sign before it reaches the two-photon peak at $\omega = \bar{\omega}_{20}/2$, and remains positive up to the nonresonant limit where it falls to zero. Whether the nonresonant value is positive or negative for the sum of the two contributions depends on which process, i.e., one- or two-photon transition, dominates.

The detailed formulas are complicated. However, they can be simplified to yield analytical formulas accurate up to $s = 4$ in the expansion of the third-order susceptibility in increasing values of $\bar{\tau}_{10}^{-s}$, which can be used at any frequency ω. The general results are summarized in Appendix 9.1. There are resonances for input frequencies $\omega \cong \bar{\omega}_{10}$ and $2\omega \cong \bar{\omega}_{10}$ corresponding to one-photon and two-photon transitions. They can lead to enhanced nonlinear absorption and/or enhanced nonlinear-induced changes in the refractive index. However, the strength of these resonances depends on the transition dipole moments.

It is more instructive here to examine approximate formulas that are valid in each of the four frequency regimes defined below.

Near and on resonance:

One-photon resonance: $|\omega - \bar{\omega}_{10}|\bar{\tau}_{10} \leq 5$
Two-photon resonance: $|2\omega - \bar{\omega}_{10}|\bar{\tau}_{10} \leq 5$

Off resonance: $|\omega - \bar{\omega}_{10}|\bar{\tau}_{10} > 5$ and $|2\omega - \bar{\omega}_{10}|\bar{\tau}_{10} > 5$

Nonresonant: $\omega \to 0$

9.1.2.2.1 Near and On Resonance.

For near and on one-photon resonance, i.e., $\omega \approx \bar{\omega}_{10}$, the leading terms are

$$
\hat{\chi}^{(3)}_{xxxx}(-\omega; \omega, \omega, -\omega) + \hat{x}^{(3)}_{xxxx}(-\omega; \omega, -, \omega) + \hat{\chi}^{(3)}_{xxxx}(-\omega; -\omega, \omega, \omega)
$$

$$
= -\frac{2N}{\varepsilon_0 \hbar^3} f^{(3)} \left[\frac{(\bar{\omega}_{10} - \omega)}{[(\bar{\omega}_{10} - \omega)^2 + \bar{\tau}_{10}^{-2}]^2} \left\{ |\mu_{10}|^2 (\bar{\mu}_{11} - \bar{\mu}_{00})^2 \frac{\bar{\tau}_{10}^{-2}}{\bar{\omega}_{10}^2} + |\mu_{10}|^4 \frac{(\bar{\omega}_{10} - \omega)^2 - \bar{\tau}_{10}^{-2}}{[(\bar{\omega}_{10} - \omega)^2 + \bar{\tau}_{10}^{-2}]} \right\} \right.
$$

$$
\left. - i\bar{\tau}_{10}^{-1} \frac{(\bar{\omega}_{10} - \omega)^2}{[(\bar{\omega}_{10} - \omega)^2 + \bar{\tau}_{10}^{-2}]^2} \left\{ \frac{29}{6} |\bar{\mu}_{10}|^2 (\bar{\mu}_{11} - \bar{\mu}_{00})^2 \frac{1}{\bar{\omega}_{10}^2} - |\bar{\mu}_{10}|^2 \frac{2}{[(\bar{\omega}_{10} - \omega)^2 + \bar{\tau}_{10}^{-2}]} \right\} \right]
$$

$$
\tag{9.11}
$$

Since the leading term for the $(\bar{\mu}_{11} - \bar{\mu}_{00})^2$ contribution is proportional to $\bar{\tau}_{10}^{-2}$ because of the cancellation of terms, the total response is clearly dominated by the triply resonant terms in $\hat{\chi}^{(3)}_{xxxx}(-\omega; \omega, -\omega, \omega) \propto \bar{\mu}_{10}^4$ associated with Case I unless the permanent dipole moment differences are unphysically enormous. Therefore, searching for materials with large permanent dipole moments is not expected to produce large near- and on-resonant third-order nonlinearities in this model.

However, near two-photon resonance, i.e., $2\omega \approx \omega_{10}$, only the $|\bar{\mu}_{10}|^2 (\bar{\mu}_{11} - \bar{\mu}_{00})^2$ terms in $\hat{\chi}^{(3)}_{xxxx}(-\omega; \omega, \omega, -\omega)$ are enhanced and they dominate the nonlinear response; i.e.,

$$
\hat{\chi}^{(3)}_{xxxx}(-\omega; \omega, \omega, -\omega) \cong \frac{8Nf^{(3)}}{\varepsilon_0 \hbar^3} |\bar{\mu}_{10}|^2 (\bar{\mu}_{11} - \bar{\mu}_{00})^2 \frac{(\bar{\omega}_{10} - 2\omega) + i\bar{\tau}_{10}^{-1}}{\bar{\omega}_{10}^2 [(\bar{\omega}_{10} - 2\omega)^2 + \bar{\tau}_{10}^{-2}]}. \tag{9.12}
$$

9.1.2.2.2 Off Resonance.

In this region, the damping term in the resonance denominators can be ignored, greatly simplifying the analysis. The result is

$$
\hat{\chi}^{(3)}_{xxxx}(-\omega; \omega, -\omega, \omega) + \hat{\chi}^{(3)}_{xxxx}(-\omega; \omega, \omega, -\omega) + \hat{\chi}^{(3)}_{xxxx}(-\omega; -\omega, \omega, \omega) = \frac{N}{\varepsilon_0 \hbar^3} f^{(3)}
$$

$$
\times \left\{ |\bar{\mu}_{10}|^2 (\bar{\mu}_{11} - \bar{\mu}_{00})^2 \bar{\omega}_{10} \left\{ \frac{12}{(\bar{\omega}_{10}^2 - 4\omega^2)(\bar{\omega}_{10}^2 - \omega^2)} + 4i\omega\bar{\tau}_{10}^{-1} \frac{15\bar{\omega}_{10}^4 - 39\bar{\omega}_{10}^2 \omega^2 + 24\omega^4}{(\bar{\omega}_{10}^2 - 4\omega^2)^2 (\bar{\omega}_{10}^2 - \omega^2)^3} \right\} \right.
$$

$$
\left. - |\bar{\mu}_{10}|^4 4\bar{\omega}_{10} (3\bar{\omega}_{10}^2 + \omega^2) \frac{(\bar{\omega}_{10}^2 - \omega^2) + 4i\bar{\tau}_{10}^{-1}\omega}{(\bar{\omega}_{10}^2 - \omega^2)^4} \right\}. \tag{9.13}
$$

These formulas are essentially valid for the "tails" of the response on both the low and high frequency sides of one- and two-photon resonances and the region between

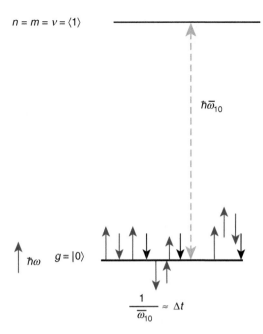

FIGURE 9.4   The nonresonant case for the interaction of small energy photons with the two-level system.

them. It is important to note two points: (1) the relative contribution of the term proportional to $|\bar{\mu}_{10}|^2(\bar{\mu}_{11}-\bar{\mu}_{00})^2$ becomes comparable to that of the $|\bar{\mu}_{10}|^4$ term off resonance; and (2) the imaginary component of the third-order susceptibility is proportional to the product of the frequency and the inverse of the excited-state lifetime.

9.1.2.2.3 Nonresonant. The relative frequencies and energy levels in the limit $\omega \to 0$ are shown in Fig. 9.4. Mathematically, the denominators in *all* the individual terms simplify to $\bar{\omega}_{10}^{-3}$. All those terms in the summation proportional to $|\bar{\mu}_{10}|^4$ contribute equally to each other, and all terms in the summation proportional to $|\bar{\mu}_{10}|^2(\bar{\mu}_{11}-\bar{\mu}_{00})^2$ also contribute equally to each other. As a result, the *relative* contribution due to the permanent dipole moments is orders of magnitude larger than that in the near- and on-resonance case. Here,

$$\tilde{\chi}_{xxxx}^{(3)}(-\omega;\,\omega,-\omega,\omega) + \tilde{\chi}_{xxxx}^{(3)}(-\omega;\,\omega,\omega,-\omega) + \tilde{\chi}_{xxxx}^{(3)}(-\omega;\,-\omega,\omega,\omega)$$

$$= 12\frac{N}{\varepsilon_0\hbar^3}f^{(3)}\frac{|\bar{\mu}_{10}|^2}{\bar{\omega}_{10}^3}\left\{(\bar{\mu}_{11}-\bar{\mu}_{00})^2 - |\bar{\mu}_{10}|^2\right\}.$$

(9.14)

Equation 9.14 indicates that the two contributions interfere, which can result in a reduced nonlinearity for molecules. Note that as $\omega \to 0$, $\Im mag(\hat{\chi}_{xxxx}^{(3)}) \propto \omega \to 0$.

9.1.3 First-Order Effect on $\hat{\chi}^{(3)}$ of Population Changes in Two-Level Systems

For very intense light inputs, the population of some of the excited states may become significant and a complete analysis in that case will involve rate equations describing the evolution of the populations in time. This will be dealt with briefly in Section 9.3. In the extreme case, the populations of the two states can approach equality with the result that the oscillator strength, which is proportional to the difference in the populations between the states, approaches zero. To this point, this case is not treated in detail and the assumption has implicitly been made that the excited-state populations are very small and that the probability of exciting an electron to a higher lying state is independent of the excited-state populations. However, this process, even in the weak excitation limit, also contributes to $\chi^{(3)}$, especially near resonances. In this section the lowest order effects of saturation will be quantified.

As shown in Fig. 9.5, when a photon of frequency $\omega \approx \bar{\omega}_{10}$ is incident on a two-level system with all its electrons initially in the ground state, it can be absorbed, thus raising one electron (per absorbed photon) to the excited state. This process is called (stimulated) absorption. Due to the finite lifetime of the electron in the excited state, it decays back to the ground state by both stimulated ($\propto I$) emission and spontaneous emission. Absorption leads to a change in the population (number density) of the ground state from N to N_0 and in the excited state to $N_1 > 0$, with the total population (number density) $N = N_0 + N_1$.

Consider input intensities well below saturation: $\Delta N = N_0 - N_1 \rightarrow N_1 = N_0 - \Delta N$. The equations describing changes in the populations due to stimulated emission and absorption as well as spontaneous emission (finite excited-state lifetime) (see Fig. 9.5) are

$$\frac{dN_1}{dt} = \bar{B}[f^{(1)}]^2 I(\omega)(N_0 - N_1) - \frac{N_1}{\bar{\tau}_{10}}, \qquad \frac{dN_0}{dt} = -\bar{B}[f^{(1)}]^2 I(\omega)(N_0 - N_1) + \frac{N_1}{\bar{\tau}_{10}}.$$
(9.15)

The rate equation for the population difference ΔN is obtained by subtracting the equations in 9.15, which gives

$$\frac{d\Delta N}{dt} = -2\Delta N \bar{B}[f^{(1)}]^2 I(\omega) + 2\frac{N_0 - \Delta N}{\bar{\tau}_{10}} \xrightarrow{\text{steady-state} \frac{d\Delta N}{dt} = 0} \Delta N \cong \frac{N}{1 + \bar{B}[f^{(1)}]^2 \bar{\tau}_{10} I(\omega)}.$$
(9.16)

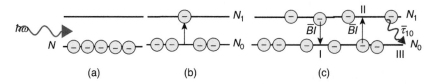

(a) (b) (c)

FIGURE 9.5 (a) Two-level system with all electrons initially in the ground state. (b) A single incoming photon is absorbed and an electron is raised to the excited state. (c) Situation after many photons have been absorbed. Process I refers to stimulated emission, Process II refers to stimulated absorption, and Process III to spontaneous emission.

Defining

$$I_{sat}(\omega) = [\bar{B}\{f^{(1)}\}^2 \bar{\tau}_{10}]^{-1} \quad \rightarrow \quad \Delta N \cong \frac{N}{1 + I(\omega)/I_{sat}(\omega)}. \quad (9.17)$$

The linear susceptibility given by Eq. 8.35 when there is a population difference ΔN between the ground and excited state for the two level system is for x-polarized light given by

$$\hat{\chi}_{xx}^{(1)}(-\omega; \omega) = \frac{N}{\hbar\varepsilon_0} f^{(1)}(\omega) \frac{N}{[1 + I(\omega)/I_{sat}(\omega)]} |\bar{\mu}_{10}|^2$$

$$\times 2\bar{\omega}_{10} \left\{ \frac{\bar{\omega}_{10}^2 - \omega^2 + \bar{\tau}_{10}^{-2} + 2i\bar{\tau}_{10}^{-1}\omega}{[(\bar{\omega}_{10} - \omega)^2 + \bar{\tau}_{10}^{-2}][(\bar{\omega}_{10} + \omega)^2 + \bar{\tau}_{10}^{-2}]} \right\} \quad (9.18a)$$

For small intensities, i.e. $I(\omega) << I_{sat}(\omega)$, Eq. 9.18a becomes

$$\hat{\chi}_{xx}^{(1)}(-\omega; \omega) = \frac{N}{\hbar\varepsilon_0} f^{(1)}(\omega) N[1 - I(\omega)/I_{sat}(\omega)] |\bar{\mu}_{10}|^2$$

$$\times 2\bar{\omega}_{10} \left\{ \frac{\bar{\omega}_{10}^2 - \omega^2 + \bar{\tau}_{10}^{-2} + 2i\bar{\tau}_{10}^{-1}\omega}{[(\bar{\omega}_{10} - \omega)^2 + \bar{\tau}_{10}^{-2}][(\bar{\omega}_{10} + \omega)^2 + \bar{\tau}_{10}^{-2}]} \right\} \quad (9.18b)$$

In general the polarization can be expanded as

$$P_x^{(1)}(\omega) + P_x^{(3)}(\omega) = \varepsilon_0 \left[\hat{\chi}_{xx}^{(1)}(-\omega; \omega)\mathcal{E}_x(\omega) + \frac{1}{4}\hat{\chi}_{xxxx,eff}^{(3)}(-\omega; \omega, -\omega, \omega) \right.$$

$$\left. \times \mathcal{E}_x(\omega)|\mathcal{E}_x(\omega)|^2 \right]. \quad (9.19)$$

where the $\hat{\chi}_{eff}^{(3)}(-\omega; \omega, -\omega, \omega)$ is the first order saturation contributions. Therefore making the substitution $I(\omega) = 0.5nc\varepsilon_0\mathcal{E}(\omega)\mathcal{E}^*(\omega)$,

$$\hat{\chi}_{xxxx,eff}^{(3)}(-\omega; \omega, -\omega, \omega) = -4\frac{ncN}{\hbar} f^{(1)}(\omega) \frac{|\bar{\mu}_{10}|^2}{I_{sat}(\omega)}$$

$$\times \bar{\omega}_{10} \left\{ \frac{\bar{\omega}_{10}^2 - \omega^2 + \bar{\tau}_{10}^{-2} + 2i\bar{\tau}_{10}^{-1}\omega}{[(\bar{\omega}_{10} - \omega)^2 + \bar{\tau}_{10}^{-2}][(\bar{\omega}_{10} + \omega)^2 + \bar{\tau}_{10}^{-2}]} \right\} \quad (9.20)$$

Furthermore, since the transfer of electrons from the ground to the excited state occurs via absorption, where the (Maxwell) field absorption coefficient is $\alpha_1(\omega) = k_{vac}(\omega)\Im mag\{\hat{\chi}_{ii}^{(1)}(-\omega; \omega)\}/2n(\omega)$, Eq. 9.15 can be re-written as

$$\frac{d}{dt}N_1 \cong \bar{B}N[f^{(1)}]^2I - \frac{N_1}{\bar{\tau}_{10}} \cong \frac{2\alpha_1 I}{\hbar\omega} - \frac{N_1}{\bar{\tau}_{10}} \rightarrow \bar{B} = \frac{1}{nN[f^{(1)}]^2\hbar c}\Im\text{mag}\left\{\hat{\chi}_{ii}^{(1)}(-\omega;\omega)\right\}$$

$$(9.21)$$

which gives on substitution of the imaginary part of the linear susceptibility

$$\frac{1}{I_{sat}(\omega)} = \frac{\{f^{(1)}\}^3}{n(\omega)c\hbar}\frac{|\vec{\mu}_{10}|^2}{\hbar\varepsilon_0}\left\{\frac{4\bar{\omega}_{10}\omega}{(\bar{\omega}_{10}^2-\omega^2)^2 + 2\bar{\tau}_{10}^{-2}(\bar{\omega}_{10}^2 + \omega^2)}\right\} \qquad (9.22a)$$

and

$$\hat{\chi}_{xxxx,eff}^{(3)}(-\omega;\omega,-\omega,\omega)$$

$$= -16\frac{N[f^{(1)}(\omega)]^4|\bar{\mu}_{10}|^4\bar{\omega}_{10}^2\omega}{\varepsilon_0\hbar^3}\left\{\frac{\bar{\omega}_{10}^2-\bar{\omega}^2 + \bar{\tau}_{10}^{-2} + 2i\bar{\tau}_{10}^{-1}\omega}{[(\bar{\omega}_{10}^2-\omega^2)^2 + 2\bar{\tau}_{10}^{-2}(\bar{\omega}_{10}^2 + \omega^2)]^2}\right\}. \qquad (9.22b)$$

The contribution off resonance decreases with decreasing ω and is zero in the nonresonant limit because the linear absorption approaches zero in this limit.

The relative contribution due to this population effect needs to be added to Eqs 9.11 and 9.13. However, since the saturation effect deals with real populations, its relative contribution decreases faster with increasing $|\bar{\omega}_{10}-\omega|$ than do the electronic non-linear terms. The contribution due to the onset of saturation is largest near one-photon resonance, as shown in Fig. 9.6 for the real (Fig. 9.6a) and imaginary (Fig. 9.6b) components of the total $\hat{\chi}_{xxxx}^{(3)}(-\omega;\omega,-\omega,\omega)$. Both figures also show that the saturation contribution decreases faster with detuning from the resonance than does the electronic contribution. In the nonresonant limit the saturation term, which are proportional to the frequency ω, approaches zero.

9.1.4 Summary for the Two-Level Model

Important lessons can be learned from the two-level model. Although this model is the simplest possible, it has been instrumental in custom designing asymmetric (nonzero permanent dipole moments) organic molecules with enhanced second-order nonlinearities, as discussed in Chapter 7. Many features predicted by the model are similar to those found in more complex cases. The Kerr nonlinearity is now defined as $\chi^{(3)}(\text{Kerr}) = \chi^{(3)}(\text{electronic}) + \chi^{(3)}(\text{saturation})$.

1. Although there are resonant enhancements in the denominator of the expressions for $\chi^{(3)}$ near one- and two-photon (and multiphoton in the general case) resonances, this does not necessarily mean that enhancement actually occurs because the numerators may vanish or be very small. For example, cancellation

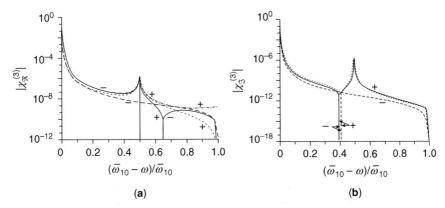

FIGURE 9.6 The relative contributions to the total nonlinearity (black line) of the Kerr electronic nonlinearity (dotted line) of the saturation contribution (dashed line), all for the case $(\bar{\mu}_{11}-\bar{\mu}_{00})^2 = |\bar{\mu}_{10}|^2$. (a) The real part, which also shows the total nonlinearity for $(\bar{\mu}_{11}-\bar{\mu}_{00})^2 = 1.2|\bar{\mu}_{10}|^2$ as a dash-dot line. (b) The imaginary part of the third order nonlinearity in arbitrary units. The $\pm$ signs identify whether the nonlinearity is positive or negative. The vertical lines indicate where the nonlinearity changes sign.

effects make the contributions of the term $|\bar{\mu}_{10}|^2(\bar{\mu}_{11}-\bar{\mu}_{00})^2$ to $\chi^{(3)}$ negligible relative to the $|\bar{\mu}_{10}|^4$ term, which does have a $[(\omega_{10}\pm\omega)^2+\bar{\tau}_{10}^{-2}]^{-3}$ resonance near one-photon resonance.

2. An intensity-dependent change in the refractive index ($\propto \Re\text{eal}[\hat{\chi}^{(3)}]$) is always accompanied by a nonlinear change in the absorption ($\propto \Im\text{mag}[\hat{\chi}^{(3)}]$).

3. The magnitude and dispersion of the nonlinear resonant change in the index and absorption near and on one-photon resonance is negligible compared with the magnitude and dispersion in the linear refractive index and absorption that occurs at exactly the same frequency and over the same spectral range. However, since two-photon absorption occurs in a frequency region far from the dominant one-photon absorption spectrum, it is resonantly enhanced and can be measured and used.

4. The contribution to the nonlinearity due to the population of the excited states is important near one-photon resonance, is less significant off resonance, and is zero in the nonresonant limit.

5. Off resonance and especially in the nonresonant limit, the sign of $\Re\text{eal}(\hat{\chi}^{(3)})$ depends on the relative magnitudes of the contributions due to $|\bar{\mu}_{10}|^2(\bar{\mu}_{11}-\bar{\mu}_{00})^2$ and $|\bar{\mu}_{10}|^4$. The relative contributions depend strongly on frequency. In contrast to the "on-resonance" case where the one-photon terms dominate, the contributions of different terms become comparable in the nonresonant regime.

6. $\Re\text{eal}(\hat{\chi}^{(3)})$ approaches a constant value with decreasing frequency, whereas $\Im\text{mag}(\hat{\chi}^{(3)})$ decreases linearly to zero with ω at $\omega=0$.

7. In the two-level model, two-photon absorption occurs only for asymmetric molecules with nonzero dipole moments. However, two-photon absorption does occur for symmetric molecules (with zero dipole moments) when three levels are used as will be discussed next.

9.2 SYMMETRIC MOLECULES

It was shown in the preceding sections that the two-level model does not predict two-photon absorption in symmetric molecules. And yet two-photon absorption does occur in practice, indicating that the two-level model is inadequate for linear molecules and symmetric molecules in general. Furthermore, since symmetric molecules have no permanent dipole moments, the lowest order nonlinearity is third order.

9.2.1 General SOS Model

The general formula given by Eq. 8.53 specific to symmetric molecules (no permanent dipole moments in the ground and excited states) is

$$\hat{\chi}_{ijkl}^{(3)}(-[\omega_p + \omega_q + \omega_r]; \omega_p, \omega_q, \omega_r)$$

$$= \frac{N}{\varepsilon_0 \hbar^3} f^{(3)} \sum_{v,n,m}' \left\{ \frac{\bar{\mu}_{gv,i}\bar{\mu}_{vn,l}\bar{\mu}_{nm,k}\bar{\mu}_{mg,j}}{(\hat{\bar{\omega}}_{vg}-\omega_p-\omega_q-\omega_r)(\hat{\bar{\omega}}_{ng}-\omega_q-\omega_p)(\hat{\bar{\omega}}_{mg}-\omega_p)} \right.$$

$$+ \frac{\bar{\mu}_{gv,j}\bar{\mu}_{vn,k}\bar{\mu}_{nm,i}\bar{\mu}_{mg,l}}{(\hat{\bar{\omega}}_{vg}^*+\omega_p)(\hat{\bar{\omega}}_{ng}^*+\omega_q+\omega_p)(\hat{\bar{\omega}}_{mg}-\omega_r)} + \frac{\bar{\mu}_{gv,l}\bar{\mu}_{vn,i}\bar{\mu}_{nm,k}\bar{\mu}_{mg,j}}{(\hat{\bar{\omega}}_{vg}^*+\omega_r)(\hat{\bar{\omega}}_{ng}-\omega_q-\omega_p)(\hat{\bar{\omega}}_{mg}-\omega_p)}$$

$$\left. + \frac{\bar{\mu}_{gv,j}\bar{\mu}_{vn,k}\bar{\mu}_{nm,l}\bar{\mu}_{mg,i}}{(\hat{\bar{\omega}}_{vg}^*+\omega_p)(\hat{\bar{\omega}}_{ng}^*+\omega_q+\omega_p)(\hat{\bar{\omega}}_{mg}^*+\omega_p+\omega_q+\omega_r)} \right\}$$

$$- \frac{N}{\varepsilon_0 \hbar^3} f^{(3)} \sum_{n,m}' \left\{ \frac{\bar{\mu}_{gn,i}\bar{\mu}_{ng,l}\bar{\mu}_{gm,k}\bar{\mu}_{mg,j}}{(\hat{\bar{\omega}}_{ng}-\omega_p-\omega_q-\omega_r)(\hat{\bar{\omega}}_{ng}-\omega_r)(\hat{\bar{\omega}}_{mg}-\omega_p)} \right.$$

$$+ \frac{\bar{\mu}_{gn,i}\bar{\mu}_{ng,l}\bar{\mu}_{gm,k}\bar{\mu}_{mg,j}}{(\hat{\bar{\omega}}_{mg}^*+\omega_q)(\hat{\bar{\omega}}_{ng}-\omega_r)(\hat{\bar{\omega}}_{mg}-\omega_p)} + \frac{\bar{\mu}_{gn,l}\bar{\mu}_{ng,i}\bar{\mu}_{gm,j}\bar{\mu}_{mg,k}}{(\hat{\bar{\omega}}_{ng}^*+\omega_r)(\hat{\bar{\omega}}_{mg}^*+\omega_p)(\hat{\bar{\omega}}_{mg}-\omega_q)}$$

$$\left. + \frac{\bar{\mu}_{gn,l}\bar{\mu}_{ng,i}\bar{\mu}_{gm,j}\bar{\mu}_{mg,k}}{(\hat{\bar{\omega}}_{ng}^*+\omega_r)(\hat{\bar{\omega}}_{mg}^*+\omega_p)(\hat{\bar{\omega}}_{ng}^*+\omega_p+\omega_q+\omega_r)} \right\}. \tag{9.23}$$

Near and on resonance, these equations are too complex for the general case to obtain simple, useful formulas and numerical methods are needed using Eq. 9.23 directly. However, for copolarized inputs and output, analytical formulas can be obtained for the real part of the nonlinear susceptibility in the limit that the state lifetimes can be neglected relative to the frequency shifts from the resonance frequency, which would be valid for the off-resonance and nonresonant regimes. Thus, in Eq. 9.23, assume that $\hat{\bar{\omega}}_{\ell g} = \hat{\bar{\omega}}_{\ell g}^* = \bar{\omega}_{\ell g}$ for $\ell = m, n, v$. For example, off resonance,

$$\Re\mathrm{eal}\left\{\hat{\chi}^{(3)}_{xxxx}(-\omega;\ \omega, -\omega, \omega) + \hat{\chi}^{(3)}_{xxxx}(-\omega;\ \omega, \omega, -\omega) + \hat{\chi}^{(3)}_{xxxx}(-\omega;\ -\omega, \omega, \omega)\right\}$$

$$= \frac{N}{\varepsilon_0 \hbar^3} f^{(3)} \left[4 {\sum_{m}}' {\sum_{v}}' {\sum_{n}}' \frac{\bar{\mu}_{gv}\bar{\mu}_{vn}\bar{\mu}_{nm}\bar{\mu}_{mg}}{\bar{\omega}_{ng}(\bar{\omega}_{ng}^2 - 4\omega^2)(\bar{\omega}_{vg}^2 - \omega^2)(\bar{\omega}_{mg}^2 - \omega^2)} \right.$$

$$\left\{ 3\bar{\omega}_{ng}^2 \bar{\omega}_{vg}\bar{\omega}_{mg} + \bar{\omega}_{ng}^2(\bar{\omega}_{vg} - \bar{\omega}_{mg})\omega + [\bar{\omega}_{ng}^2 + 2\bar{\omega}_{ng}(\bar{\omega}_{vg} + \bar{\omega}_{mg}) - 8\bar{\omega}_{vg}\bar{\omega}_{mg}]\omega^2 \right.$$

$$\left. - 4(\bar{\omega}_{vg} - \bar{\omega}_{mg})\omega^3 \right\} - 2 {\sum_{n}}' {\sum_{m}}' \frac{|\bar{\mu}_{ng}|^2 |\bar{\mu}_{mg}|^2}{(\bar{\omega}_{ng}^2 - \omega^2)^2(\bar{\omega}_{mg}^2 - \omega^2)^2} \left\{ (\bar{\omega}_{mg} + \bar{\omega}_{ng}) \right.$$

$$\left. \left. \left\{ 3\bar{\omega}_{mg}^2\bar{\omega}_{ng}^2 + (2\bar{\omega}_{ng}^2 + 2\bar{\omega}_{mg}^2 - 7\bar{\omega}_{ng}\bar{\omega}_{mg})\omega^2 - \omega^4 \right\} + \omega^2(\bar{\omega}_{mg}^3 + \bar{\omega}_{ng}^3) \right\} \right].$$

$$(9.24)$$

Depending on the spectral locations and transition dipole moments linking different states, there is the potential that the real component of the third-order nonlinearity (and hence the nonlinear index coefficient) can change sign with frequency a number of times.

The pathways corresponding to the first and second summations in Eqs 9.23 and 9.24 are shown in Fig. 9.7. The energy levels for symmetric molecules have either odd ("ungerade," B_u) or even ("gerade," A_g) symmetry. Because the transition electric dipole moments are proportional to $\int_{-\infty}^{\infty} \bar{u}_\ell^*(\vec{r})\vec{r}\bar{u}_{\ell'}(\vec{r})\, d\vec{r}$ and $\vec{r}$

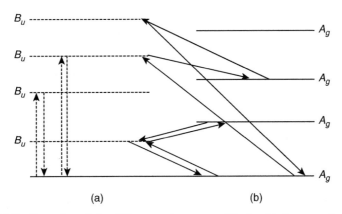

FIGURE 9.7 Examples of the different pathways possible for (a) the second summation (dashed lines) and (b) the first summation (solid lines) in Eqs 9.22a and 9.23.

is inherently antisymmetric (odd symmetry), one-photon electric dipole transitions between states with the same symmetry are not allowed. Thus, in the first summation, the symmetric excited states ℓ can be reached only by intermediate coupling to an odd symmetry state ℓ' via two electric dipole transition moments, namely, $\bar{\mu}_{\ell' g}$ and $\bar{\mu}_{\ell \ell'}$. Therefore, excited states with even symmetry can be accessed only by the simultaneous absorption of two photons and are frequently called "two-photon" states. The contribution from the second summation involves only one-photon transitions from the even symmetry ground state to odd symmetry excited states, sometimes called "one-photon" states.

For the nonresonant case, Eq. 9.24 simplifies to

$$\Re\text{eal}\{\hat{\chi}^{(3)}_{xxxx}(-\omega;\,\omega,-\omega,\omega) + \hat{\chi}^{(3)}_{xxxx}(-\omega;\,\omega,\omega,-\omega) + \hat{\chi}^{(3)}_{xxxx}(-\omega;\,-\omega,\omega,\omega)\}$$

$$= \frac{N}{\varepsilon_0 \hbar^3} f^{(3)} \left[12 \sum_{v}' \sum_{n}' \sum_{m}' \frac{\bar{\mu}_{gv}\bar{\mu}_{vn}\bar{\mu}_{nm}\bar{\mu}_{mg}}{\bar{\omega}_{ng}\bar{\omega}_{vg}\bar{\omega}_{mg}} - 6\sum_{n}' \sum_{m}' \frac{|\bar{\mu}_{ng}|^2 |\bar{\mu}_{mg}|^2}{\bar{\omega}_{ng}^2 \bar{\omega}_{mg}^2} (\bar{\omega}_{mg} + \bar{\omega}_{ng}) \right].$$

$$(9.25)$$

Note that the contribution of the one-photon transitions is always negative and that of the two-photon transitions is positive. There is interference between the two terms. Equation 9.25 shows that a positive real total nonresonant susceptibility (and therefore positive intensity-dependent refractive index coefficient) implies that the two-photon coupling to the even symmetry excited states contributes more to the third-order susceptibility than does the one-photon coupling to the odd symmetry (one-photon excited states). This distinction is important because the even symmetry states and their transition moments do not contribute to the linear susceptibility and must be evaluated by nonlinear spectroscopy. A negative nonlinear nonresonant refractive index coefficient indicates that the one-photon transitions dominate the nonlinearity.

9.2.2 Three-Level Model

The minimum number of states required to describe the nonlinear optics in symmetric molecules is 3. In such a model, the first state is the even symmetry ground state $1A_g$, the second is the odd symmetry excited state $1B_u$, and the third is the lowest lying even symmetry excited state mA_g with strong coupling to $1B_u$ via a large transition dipole moment $\bar{\mu}_{mA_g \leftarrow 1B_u}$. In some cases, mA_g may represent a clustered grouping of even symmetry excited states (3,4). The three levels are shown in Fig. 9.8. Since symmetric molecules are by definition centrosymmetric, they exhibit no second-order effects and it is for the odd-order nonlinear susceptibilities that a three-level model is required as the absolute minimum to describe these susceptibilities. Again due to the even symmetries of the wave functions for the two even symmetry states mA_g and $1A_g$, spontaneous decay to the ground state is not allowed and the state mA_g can decay only to $1B_u$ with a subsequent decay to the ground state via $\bar{\tau}_{10}$. The effective decay time $\bar{\tau}_{eff}$ via this coupling is given by $\bar{\tau}_{eff}^{-1} = \bar{\tau}_{21}^{-1} + \bar{\tau}_{210}^{-1}$.

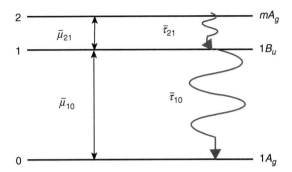

FIGURE 9.8 The three energy levels, the electric dipole matrix elements, and the excited-state lifetimes for the three-level model.

The typical frequency dependence of the different contributions to the total third-order susceptibility is shown in Fig. 9.9. Near and on one-photon resonance, $\widehat{\chi}^{(3)}_{xxxx}$ has negative real and imaginary components (denoted by subscripts $\Re$ and $\Im$, respectively). One-photon contributions are always negative. A dispersion type of resonance exists at $\omega = \bar{\omega}_{20}/2$, which leads to a positive $\widehat{\chi}^{(3)}_{\Re}$ for the two-photon contribution for frequencies below the two-photon resonance. The net result can be either a positive or a negative $\widehat{\chi}^{(3)}_{\Re}$ as the nonresonant limit approaches. Whether the nonresonant value is positive or negative depends on which process, i.e., whether the one- or two-photon transition, dominates. (The parameter range that leads to positive values will be discussed later in this section.) The two-photon contribution to $\widehat{\chi}^{(3)}_{\Re}$ is positive throughout the entire frequency range, which increases with a decrease in the frequency up to $\omega = \bar{\omega}_{20}/2$, where it peaks and then decreases as the zero frequency limit approaches.

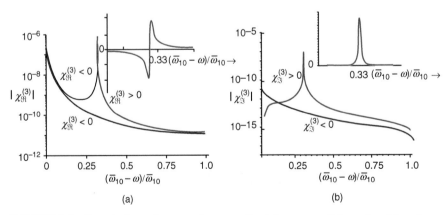

FIGURE 9.9 Generic dependence on the normalized frequency of the real and imaginary components (in arbitrary units) of the one- and two-photon terms of the third-order susceptibility in the three-level model. The blue curves are for the total of the one-photon terms, and the black (negative values) and red (positive values) curves are for the total two-photon terms. The upper curves show the dispersion of two-photon resonance terms on a linear scale.

This case, although complicated, has been reduced to analytical formulas for the general case (see Appendix 9.2). Again, it is more instructive to examine the terms in three limits. The leading terms near and on one- and two-photon resonances, respectively, are given by

$$
\hat{\chi}^{(3)}_{xxxx}(-\omega; -\omega, \omega, \omega) + \hat{\chi}^{(3)}_{xxxx}(-\omega; \omega, -\omega, \omega) + \hat{\chi}^{(3)}_{xxxx}(-\omega; \omega, \omega, -\omega) = \frac{N}{\varepsilon_0 \hbar^3} f^{(3)} |\bar{\mu}_{10}|^2
$$

$$
\times \left\{ |\bar{\mu}_{21}|^2 \left[\frac{2(\bar{\omega}_{10}^2 - \omega^2)(\bar{\omega}_{20} - \bar{\omega}_{10})(\bar{\omega}_{20} - 2\bar{\omega}_{10})[(\bar{\omega}_{10}^2 - \omega^2) + 2i\bar{\omega}_{10}\bar{\tau}_{10}^{-1}] + i\bar{\tau}_{eff}^{-1}\bar{\omega}_{20}(\bar{\omega}_{10}^2 - \omega^2)^2}{2\bar{\omega}_{10}^2 \bar{\omega}_{20}(\bar{\omega}_{20} - 2\bar{\omega}_{10})^2[(\bar{\omega}_{10} - \omega)^2 + \bar{\tau}_{10}^{-2}]^2} \right] \right.
$$

$$
\left. - |\bar{\mu}_{10}|^2 \frac{2(\bar{\omega}_{10} - \omega)^3 + 4i\bar{\tau}_{10}^{-1}(\bar{\omega}_{10} - \omega)^2}{[(\bar{\omega}_{10} - \omega)^2 + \bar{\tau}_{10}^{-2}]^3} \right\}
\tag{9.26}
$$

and

$$
\hat{\chi}^{(3)}_{xxxx}(-\omega; \omega, \omega, -\omega) = \frac{4N}{\varepsilon_0 \hbar^3} f^{(3)} |\bar{\mu}_{01}|^2 |\bar{\mu}_{12}|^2 \left\{ \frac{(\bar{\omega}_{20}^2 - 4\omega^2)}{[(\bar{\omega}_{20} - 2\omega)^2 + \bar{\tau}_{eff}^{-2}](2\bar{\omega}_{10} - \bar{\omega}_{20})^2 \bar{\omega}_{20}} \right.
$$

$$
\left. + 2i \frac{\bar{\tau}_{eff}^{-1}\bar{\omega}_{20}(4\bar{\omega}_{10}^2 - \bar{\omega}_{20}^2)(\bar{\omega}_{20} + 2\omega_{10})^2 + \bar{\tau}_{10}^{-1}(\bar{\omega}_{20}^2 - 4\omega^2)(\bar{\omega}_{20} + 2\bar{\omega}_{10})^3}{[(\bar{\omega}_{20} - 2\omega)^2 + \bar{\tau}_{eff}^{-2}](4\bar{\omega}_{10}^2 - \bar{\omega}_{20}^2)^3 \bar{\omega}_{20}} \right\},
\tag{9.27}
$$

respectively.

For the off-resonance case,

$$
\hat{\chi}^{(3)}(-\omega; \omega, \omega, -\omega) + \hat{\chi}^{(3)}(-\omega; \omega, -\omega, \omega) + \hat{\chi}^{(3)}(-\omega; -\omega, \omega, \omega)
$$

$$
= \frac{N}{\varepsilon_0 \hbar^3} f^{(3)} \left[|\bar{\mu}_{10}|^2 |\bar{\mu}_{21}|^2 \left\{ 4 \frac{3\bar{\omega}_{10}^2 \bar{\omega}_{20}^2 + \omega^2(\bar{\omega}_{20}^2 + 4\bar{\omega}_{20}\bar{\omega}_{10} - 8\bar{\omega}_{10}^2)}{\bar{\omega}_{20}(\bar{\omega}_{10}^2 - \omega^2)^2(\bar{\omega}_{20}^2 - 4\omega^2)} \right. \right.
$$

$$
+ 4i\omega \left[\bar{\tau}_{10}^{-1} \frac{\bar{\omega}_{20}\bar{\omega}_{10}^2(7\bar{\omega}_{20} + 2\bar{\omega}_{10}) + \omega^2(\bar{\omega}_{20} + 8\bar{\omega}_{10})(\bar{\omega}_{20} - 2\bar{\omega}_{10})}{\bar{\omega}_{20}(\bar{\omega}_{10}^2 - \omega^2)^3(\bar{\omega}_{20}^2 - 4\omega^2)} \right.
$$

$$
\left. \left. + 2\bar{\tau}_{eff}^{-1} \frac{(\bar{\omega}_{20} + 2\bar{\omega}_{10})(\bar{\omega}_{20}\bar{\omega}_{10} + 2\omega^2)}{(\bar{\omega}_{10}^2 - \omega^2)^2(\bar{\omega}_{20}^2 - 4\omega^2)^2} \right] \right\} - |\bar{\mu}_{10}|^4 4\bar{\omega}_{10}(3\bar{\omega}_{10}^2 + \omega^2)
$$

$$
\times \frac{(\bar{\omega}_{10}^2 - \omega^2) + 4i\omega\bar{\tau}_{10}^{-1}}{(\bar{\omega}_{10}^2 - \omega^2)^4} \right].
\tag{9.28}
$$

Finally, for the nonresonant case,

$$\hat{\chi}^{(3)}_{xxxx}(-\omega;\,\omega,-\omega,\omega)+\hat{\chi}^{(3)}_{xxxx}(-\omega;\,\omega,\omega,-\omega)+\hat{\chi}^{(3)}_{xxxx}(-\omega;\,-\omega,\omega,\omega)$$

$$= 12N\frac{f^{(3)}|\bar{\mu}_{10}|^2}{\varepsilon_0\hbar^3\bar{\omega}_{10}^2}\left(\frac{|\bar{\mu}_{21}|^2}{\bar{\omega}_{20}}-\frac{|\bar{\mu}_{10}|^2}{\bar{\omega}_{10}}\right). \tag{9.29}$$

The sign of the nonlinearity is determined by the ratio $|\bar{\mu}_{21}|^2\bar{\omega}_{10}/|\bar{\mu}_{10}|^2\bar{\omega}_{20}$. When it is greater than unity, the net nonlinearity is positive, and vice versa. This conclusion has been verified in a number of cases (5,6).

9.3 DENSITY MATRIX FORMALISM

Except Section 9.1.3 that dealt with the first-order contribution for the onset to saturation in the population of the excited state in a two-level model, it has been explicitly assumed that the electrons are initially in the ground state before the interaction with the optical field and that a negligible number of them are raised to an excited state. Since $k_BT = 0.026$ eV at room temperature and 500 nm corresponds to 2.48 eV, thermal excitation into excited states as given by the Boltzmann factor $\exp[-(E_{\mathrm{exc}}-E_g)/k_BT]$ is negligible and this initial condition is an excellent approximation. However, this corresponds to only one possible initial state of the system, s, and there exist cases where other excitation mechanisms can populate some of the excited states. For example in gases, collisions between atoms or molecules or with electrons under discharge conditions can occur, which result in the population of excited states. In this section, the more general case of an arbitrary initial condition in terms of the population of the excited states is treated. This formalism is called the "density matrix" approach.

The probability that the system is in an initial state s is given by the wave functions

$$\hat{\Psi}^{(0)}_s(\vec{r},t) = \sum_m \hat{a}^{(0)}_{m,s}\bar{u}_m(\vec{r})^{-i\hat{\omega}_m t} \quad \text{with} \quad \hat{a}^{(0)}_{m,s}(t) \geq 0. \tag{9.30}$$

The density matrix is defined as

$$\rho_{nm}(t) = \sum_s pr(s,t)\hat{a}^*_{m,s}(t)\hat{a}_{n,s}(t) \tag{9.31}$$

in which $pr(s,t)$ is the probability that the system is in state s due to some initial condition defined by $\hat{a}^{(0)}_{m,s}(t)$. The physical interpretation of the diagonal terms

$$\rho_{nn}(t) = \sum_s pr(s,t)|\hat{a}_{n,s}(t)|^2 \tag{9.32}$$

is straightforward: It gives the probability that the system is in state n. The off-diagonal elements $\rho_{nm}(t)$ are not related to excited-state populations but rather to coherence between the quantum mechanical wave functions of states n and m.

These off-diagonal elements are nonzero only if the wave function of the system of molecules can be described by a *coherent* superposition of the wave functions of the two states, i.e., $\Psi_m(t) + e^{i\theta}\Psi_n(t)$, where θ is the relative phase angle (most frequently chosen as 0 or π), determined by the technique used to generate the coherent superposition. For example, in a gas the two states of an atom can be electromagnetically coupled by the transition dipole moment until a collision occurs, which decouples the two states. This is called the *coherence time* or *dephasing time*. In condensed matter, the coherence is lost due to many effects and interactions and the coherence time can be as short as femtoseconds.

Normally the focus is on the condition $\rho_{nm}(t) = \delta_{nm}$: i.e., time scales are longer than the dephasing time. The number density of the electrons in state n is $N_n(t) = N\rho_{nn}(t)$, where N is the total number density. The general time evolution of the density matrix is described solely by the interactions with the electromagnetic fields and is obtained from Eq. 9.31 as

$$\dot{\rho}_{nm}(t) = \sum_s pr(s) \left[\hat{\bar{a}}_{m,s}^*(t) \frac{\partial}{\partial t} \hat{\bar{a}}_{n,s}(t) + \hat{\bar{a}}_{n,s}(t) \frac{\partial}{\partial t} \hat{\bar{a}}_{m,s}^*(t) \right]. \tag{9.33}$$

Substituting from Eqs 8.14 and 8.17b for a perturbation potential (with the electric dipole interaction being of interest here) gives

$$\dot{\hat{\bar{a}}}_{n,s}(t) = \frac{1}{i\hbar} \sum_m \hat{\bar{a}}_{m,s}(t) \bar{V}_{nm}(t)\, e^{i(\hat{\omega}_n - \hat{\omega}_m)t}$$

$$\text{with}\quad \bar{V}_{nm}(t) = \int_{-\infty}^{\infty} \bar{u}_n^*(\vec{r}) \bar{V}(\vec{r},t) \bar{u}_m(\vec{r})\, d\vec{r} \tag{9.34}$$

and so

$$\dot{\rho}_{nm}(t) = \frac{1}{i\hbar} \sum_s pr(s) \left[\sum_m \rho_{nm}(t) \bar{V}_{nm}(t) e^{i(\hat{\omega}_n - \hat{\omega}_m)t} - \rho_{mn}(t) \sum_m \bar{V}_{mn}(t) e^{-i(\hat{\omega}_n - \hat{\omega}_m)t} \right]. \tag{9.35}$$

As discussed in Section 8.1, Eq. 9.35 cannot be solved exactly and the same philosophy of tracking solutions in terms of the number of interactions with the electromagnetic field needs to be adopted. As shown in Boyd's textbook, this leads to equivalent definitions for nonlinear susceptibilities to those discussed in this chapter.

APPENDIX 9.1: TWO-LEVEL MODEL FOR ASYMMETRIC MOLECULES—EXACT SOLUTION

The two-level system, despite the apparent complexity that will be encountered here, is the simplest system that can be analyzed for $\hat{\chi}^{(3)}$.

A.9.1.1 Summary of General Formulas

For the two-photon terms proportional to $|\bar{\mu}_{10}|^2(\bar{\mu}_{11}-\bar{\mu}_{00})^2$,

$$\hat{\chi}^{(3)}_{xxxx}(-\omega;\,\omega,\omega,-\omega)+\hat{\chi}^{(3)}_{xxxx}(-\omega;\,\omega,-\omega,\omega)+\hat{\chi}^{(3)}_{xxxx}(-\omega;\,-\omega,\omega,\omega)=\frac{N}{\varepsilon_0\hbar^3}f^{(3)}\left\{|\bar{\mu}_{10}|^2(\bar{\mu}_{11}-\bar{\mu}_{00})^2\right.$$

$$\times\frac{\begin{pmatrix}4\bar{\omega}_{10}(\bar{\omega}_{10}^2-\omega^2)^2(\bar{\omega}_{10}^2-4\omega^2)(\bar{\omega}_{10}^2+5\omega^2)\\[4pt]-4\bar{\omega}_{10}(\bar{\omega}_{10}^2-\omega^2)[-2\bar{\omega}_{10}^4+33\bar{\omega}_{10}^2\omega^2+5\omega^4]\bar{\tau}_{10}^{-2}-12\bar{\omega}_{10}\omega^2(9\bar{\omega}_{10}^2-11\omega^2)\bar{\tau}_{10}^{-4}\\[4pt]+4i\bar{\omega}_{10}\omega\bar{\tau}_{10}^{-1}[(\bar{\omega}_{10}^2-\omega^2)(11\bar{\omega}_{10}^4-7\bar{\omega}_{10}^2\omega^2-40\omega^4)+\bar{\tau}_{10}^{-2}(\bar{\omega}_{10}^2-\omega^2)(17\bar{\omega}_{10}^2-39\omega^2)\\[4pt]+\bar{\tau}_{10}^{-4}(\bar{\omega}_{10}^2-6\omega^2)]\end{pmatrix}}{[(\bar{\omega}_{10}-2\omega)^2+\bar{\tau}_{10}^{-2}][(\bar{\omega}_{10}+2\omega)^2+\bar{\tau}_{10}^{-2}][(\bar{\omega}_{10}-\omega)^2+\bar{\tau}_{10}^{-2}]^2[(\bar{\omega}_{10}+\omega)^2+\bar{\tau}_{10}^{-2}]^2}$$

$$\left.+4\bar{\omega}_{10}\frac{\begin{pmatrix}2\bar{\omega}_{10}^2(\bar{\omega}_{10}^2-\omega^2)^2+2(\bar{\omega}_{10}^2-\omega^2)^2\bar{\tau}_{10}^{-2}-2\omega\bar{\tau}_{10}^{-4}\\[4pt]+i\omega\bar{\tau}_{10}^{-1}[4\bar{\omega}_{10}^2(\bar{\omega}_{10}^2-\omega^2)-4\omega^2\bar{\tau}_{10}^{-2}-4\bar{\tau}_{10}^{-4}]\end{pmatrix}}{[\bar{\omega}_{10}^2+\bar{\tau}_{10}^{-2}][(\bar{\omega}_{10}+\omega)^2+\bar{\tau}_{10}^{-2}]^2[(\bar{\omega}_{10}-\omega)^2+\bar{\tau}_{10}^{-2}]^2}\right\}. \tag{A.9.1.1}$$

For the one-photon terms proportional to $|\bar{\mu}_{10}|^4$,

$$\hat{\chi}^{(3)}_{xxxx}(-\omega;\,\omega,-\omega,\omega)+\hat{\chi}^{(3)}_{xxxx}(-\omega;\,\omega,\omega,-\omega)+\hat{\chi}^{(3)}_{xxxx}(-\omega;\,-\omega,\omega,\omega)=-\frac{N}{\varepsilon_0\hbar^3}f^{(3)}|\bar{\mu}_{10}|^4$$

$$\times\left\{4\bar{\omega}_{10}\frac{\begin{pmatrix}(\bar{\omega}_{10}^2-\omega^2)[(\bar{\omega}_{10}^2-\omega^2)^2(3\bar{\omega}_{10}^2+\omega^2)+2(3\bar{\omega}_{10}^4-3\omega^4-8\omega^2\bar{\omega}_{10}^2)\bar{\tau}_{10}^{-2}-12\omega^2\bar{\tau}_{10}^{-4}]\\[4pt]+4i\omega\bar{\tau}_{10}^{-1}[(\bar{\omega}_{10}^2-\omega^2)^2(3\bar{\omega}_{10}^2+\omega^2)+4\bar{\omega}_{10}^2(\bar{\omega}_{10}^2-\omega^2)\bar{\tau}_{10}^{-2}-(\bar{\omega}_{10}^2+3\omega^2)\bar{\tau}_{10}^{-4}]\end{pmatrix}}{[(\bar{\omega}_{10}+\omega)^2+\bar{\tau}_{10}^{-2}]^3[(\bar{\omega}_{10}-\omega)^2+\bar{\tau}_{10}^{-2}]^3}\right\}. \tag{A.9.1.2}$$

A.9.1.2 Numerical Calculations Near Resonances

The purpose here is to identify which terms are the most important near and on resonances.

A.9.1.2.1 One-Photon Resonance.

$$\frac{\Re eal\{\hat{\chi}^{(3)}_{xxxx}(|\bar{\mu}_{11}-\bar{\mu}_{00}|^2|\bar{\mu}_{10}|^2=1)\}}{\Re eal\{\hat{\chi}^{(3)}_{xxxx}(|\bar{\mu}_{10}|^4=1)\}}\approx 10^{-6}$$

$$\longrightarrow\quad \Re eal\left\{\hat{\chi}^{(3)}_{xxxx}(|\bar{\mu}_{11}-\bar{\mu}_{00}|^2|\bar{\mu}_{10}|^2)\right\}\text{ not plotted.}$$

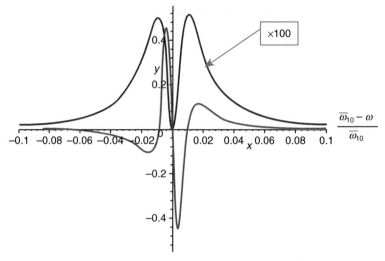

FIGURE A.9.1.1   The real part of the different contributions: $\hat{\chi}^{(3)}_{xxxx}(-\omega;\omega,-\omega,\omega)$ (red curve); $\hat{\chi}^{(3)}_{xxxx}(-\omega;\omega,\omega,-\omega) + \hat{\chi}^{(3)}_{xxxx}(-\omega;-\omega,\omega,\omega)$ multiplied by 100 (blue curve); and the sum of all the contributions, exact calculation (red curve). The triply resonant curve due to $\hat{\chi}^{(3)}_{xxxx}(-\omega;\omega,-\omega,\omega)$ clearly dominates the response.

The following normalizations were used:

$$\Re eal\{\hat{\chi}^{(3)}_{xxxx}(-\omega)\} \propto |\bar{\mu}_{10}|^4\bar{\tau}_{10}^{-3}, \qquad \Im mag\{\hat{\chi}^{(3)}_{xxxx}(-\omega)\} \propto |\bar{\mu}_{10}|^4\bar{\tau}_{10}^{-3}.$$

The real and imaginary parts of $\hat{\chi}^{(3)}_{xxxx}(-\omega)$ are shown in Figs A.9.1.1 and A.9.1.2, respectively.

A.9.1.2.2   Two-Photon Resonance.   The following normalizations were used (see Fig. A.9.1.3):

$$\hat{\chi}^{(3)}_{\Re,xxxx}(-\omega;\omega,\omega,-\omega) \propto [|\bar{\mu}_{10}|^2(\bar{\mu}_{11}-\bar{\mu}_{00})^2 = 1]\bar{\tau}_{10}^{-3},$$

$$\hat{\chi}^{(3)}_{\Im,xxxx}(-\omega;\omega,\omega,-\omega) \propto [|\bar{\mu}_{10}|^2(\bar{\mu}_{11}-\bar{\mu}_{00})^2 = 1]\bar{\tau}_{10}^{-3}.$$

APPENDIX 9.2: THREE-LEVEL MODEL FOR SYMMETRIC MOLECULES—EXACT SOLUTION

Here the general equations for an expansion of $\hat{\chi}^{(3)}_{xxxx}(-\omega;\omega,-\omega,\omega)$ $+ \hat{\chi}^{(3)}_{xxxx}(-\omega;\omega,\omega,-\omega) + \hat{\chi}^{(3)}_{xxxx}(-\omega;-\omega,\omega,\omega)$ in terms of increasing lifetime products $\bar{\tau}_{10}^{s1}\bar{\tau}_{21}^{s2}$ valid for $4 \geq s1 + s2$ are given.

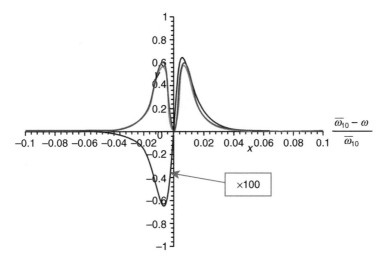

FIGURE A.9.1.2 The imaginary part of the different contributions: $\hat{\chi}^{(3)}_{xxxx}(-\omega;\omega,-\omega,\omega)$ (red curve); $\hat{\chi}^{(3)}_{xxxx}(-\omega;\omega,\omega,-\omega) + \hat{\chi}^{(3)}_{xxxx}(-\omega;-\omega,\omega,\omega)$ multiplied by 100 (blue curve); the sum of all the contributions, exact calculation (red curve). The triply resonant curve due to $\hat{\chi}^{(3)}_{xxxx}(-\omega;\omega,-\omega,\omega)$ clearly dominates the response.

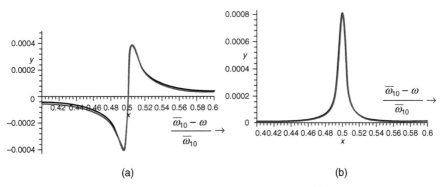

(a)   (b)

FIGURE A.9.1.3 The (a) real and (b) imaginary components of $\hat{\chi}^{(3)}_{xxxx}(-\omega;\omega,\omega,-\omega)$ versus normalized frequency. The red curve is the exact calculation, and the black curve is the approximate formula Eq. 9.12.

For the two-photon terms proportional to $|\bar{\mu}_{10}|^2|\bar{\mu}_{21}|^2$,

$$\chi^{(3)}(-\omega;\omega,\omega,-\omega) = \frac{N}{\varepsilon_0\hbar^3}f^{(3)}|\bar{\mu}_{01}|^2|\bar{\mu}_{12}|^2$$

$$\frac{\begin{pmatrix} 4(\bar{\omega}_{20}^2-4\omega^2)(\bar{\omega}_{10}^2-\omega^2)^2[\bar{\omega}_{20}(\bar{\omega}_{10}^2+\omega^2)+4\omega^2\bar{\omega}_{10}]+8\bar{\omega}_{20}(\bar{\omega}_{20}^2-4\omega^2)(\bar{\omega}_{10}^2-\omega^2)^2\bar{\tau}_{10}^{-1} \\ +4(\bar{\omega}_{20}^2-4\omega^2)[2\bar{\omega}_{20}(\bar{\omega}_{10}^2+\omega^2)-8\bar{\omega}_{10}\omega^2]\bar{\tau}_{10}^{-4} \\ -4(\bar{\omega}_{10}^2-\omega^2)[\bar{\omega}_{10}(\bar{\omega}_{20}^2+4\omega^2)(\bar{\omega}_{10}^2+3\omega^2)+4\bar{\omega}_{20}\omega^2(3\bar{\omega}_{10}^2+\omega^2)]\bar{\tau}_{eff}^{-1}\bar{\tau}_{10}^{-1} \\ -8(\bar{\omega}_{10}^2-\omega^2)[\bar{\omega}_{10}(\bar{\omega}_{20}^2+4\omega^2)+4\bar{\omega}_{20}\omega^2]\bar{\tau}_{10}^{-3}\bar{\tau}_{eff}^{-1} \\ +4(\bar{\omega}_{10}^2-\omega^2)^2[\bar{\omega}_{20}(\bar{\omega}_{10}^2+\omega^2)-4\bar{\omega}_{10}\omega^2]\bar{\tau}_{eff}^{-2} \\ -4\bar{\omega}_{10}(\bar{\omega}_{10}^2-\omega^2)(\bar{\omega}_{10}^2+3\omega^2)\bar{\tau}_{10}^{-1}\bar{\tau}_{eff}^{-3}+8\bar{\omega}_{20}(\bar{\omega}_{10}^2-\omega^2)^2\bar{\tau}_{10}^{-2}\bar{\tau}_{eff}^{-2} \end{pmatrix}}{[(\bar{\omega}_{20}-2\omega)^2+\bar{\tau}_{eff}^{-2}][(\bar{\omega}_{10}-\omega)^2+\bar{\tau}_{10}^{-2}][(\bar{\omega}_{20}+2\omega)^2+\bar{\tau}_{eff}^{-2}][(\bar{\omega}_{10}+\omega)^2+\bar{\tau}_{10}^{-2}]^2}$$

$$+\frac{\begin{pmatrix} 8i\omega\bar{\tau}_{eff}^{-1}\{(\bar{\omega}_{10}^2-\omega^2)^2[(\bar{\omega}_{20}+2\bar{\omega}_{10})(\bar{\omega}_{20}\bar{\omega}_{10}+2\omega^2)+4\bar{\omega}_{20}\bar{\tau}_{10}^{-2}+\bar{\omega}_{10}\bar{\tau}_{eff}^{-2}] \\ +(2\bar{\omega}_{10}-\bar{\omega}_{20})(\bar{\omega}_{10}\bar{\omega}_{20}-2\omega^2)\bar{\tau}_{10}^{-4}\}+4i\omega\bar{\tau}_{10}^{-1}(\bar{\omega}_{10}^2-\omega^2) \\ \times[(\bar{\omega}_{20}^2-4\omega^2)(3\bar{\omega}_{20}\bar{\omega}_{10}^2+\bar{\omega}_{20}\omega^2+2\bar{\omega}_{10}^3+6\bar{\omega}_{10}\omega^2) \\ +(3\bar{\omega}_{20}\bar{\omega}_{10}^2+\bar{\omega}_{20}\omega^2-2\bar{\omega}_{10}^3-6\bar{\omega}_{10}\omega^2)\bar{\tau}_{eff}^{-2}] \\ +4i\omega\bar{\tau}_{10}^{-3}\{(\bar{\omega}_{20}^2-4\omega^2)[2(\bar{\omega}_{10}^2-\omega^2)(\bar{\omega}_{20}+2\bar{\omega}_{10})+(2\bar{\omega}_{10}-\bar{\omega}_{20})\bar{\tau}_{10}^{-2}] \\ +2(\bar{\omega}_{10}^2-\omega^2)(\bar{\omega}_{20}-2\bar{\omega}_{10})\bar{\tau}_{eff}^{-2}\} \end{pmatrix}}{[(\bar{\omega}_{20}-2\omega)^2+\bar{\tau}_{eff}^{-2}][(\bar{\omega}_{10}-\omega)^2+\bar{\tau}_{10}^{-2}]^2[(\bar{\omega}_{20}+2\omega)^2+\bar{\tau}_{eff}^{-2}][(\bar{\omega}_{10}+\omega)^2+\bar{\tau}_{10}^{-2}]^2} \qquad (A.9.2.1)$$

$$\hat{\chi}^{(3)}(-\omega;\omega,-\omega,\omega)+\hat{\chi}^{(3)}(-\omega;-\omega,\omega,\omega) = \frac{N}{\varepsilon_0\hbar^3}f^{(3)}|\bar{\mu}_{01}|^2|\bar{\mu}_{12}|^2$$

$$\left\{8\bar{\omega}_{10}\frac{\{\bar{\omega}_{10}\bar{\omega}_{20}[(\bar{\omega}_{10}^2-\omega^2)+\bar{\tau}_{10}^{-2}]^2-\bar{\tau}_{10}^{-1}\bar{\tau}_{eff}^{-1}[(\bar{\omega}_{10}^2-\omega^2)(\bar{\omega}_{10}^2+\omega^2)+2\bar{\omega}_{10}^2\bar{\tau}_{10}^{-2}+\bar{\tau}_{10}^{-4}]\}}{[(\bar{\omega}_{10}-\omega)^2+\bar{\tau}_{10}^{-2}]^2[(\bar{\omega}_{10}+\omega)^2+\bar{\tau}_{10}^{-2}]^2[\bar{\omega}_{20}^2+\bar{\tau}_{eff}^{-2}]}\right.$$

$$\left.-16i\omega\bar{\omega}_{10}\bar{\tau}_{10}^{-1}\frac{\{\bar{\tau}_{10}^{-1}\bar{\tau}_{eff}^{-1}[2\bar{\omega}_{10}^2-\bar{\omega}_{20}\bar{\omega}_{10}+\bar{\tau}_{10}^{-2}]-\bar{\omega}_{20}\bar{\omega}_{10}(\bar{\omega}_{10}^2-\omega^2)\}}{[(\bar{\omega}_{10}-\omega)^2+\bar{\tau}_{10}^{-2}]^2[(\bar{\omega}_{10}+\omega)^2+\bar{\tau}_{10}^{-2}]^2[\bar{\omega}_{20}^2+\bar{\tau}_{eff}^{-2}]}\right\}. \qquad (A.9.2.2)$$

For the one-photon terms proportional to $|\bar{\mu}_{10}|^4$,

$$\hat{\chi}_{xxxx}^{(3)}(-\omega;-\omega,\omega,\omega)+\hat{\chi}_{xxxx}^{(3)}(-\omega;\omega,-\omega,\omega)+\hat{\chi}_{xxxx}^{(3)}(-\omega;\omega,\omega,-\omega) = -\frac{N}{\varepsilon_0\hbar^3}f^{(3)}|\bar{\mu}_{01}|^4$$

$$4\bar{\omega}_{10}\frac{\begin{pmatrix}(\bar{\omega}_{10}^2-\omega^2)[(\bar{\omega}_{10}^2-\omega^2)^2(3\bar{\omega}_{10}^2+\omega^2)+2\bar{\tau}_{10}^{-2}(3\bar{\omega}_{10}^4-3\omega^4-8\omega^2\bar{\omega}_{10}^2)-12\omega^2\bar{\tau}_{10}^{-4}] \\ +4i\omega\bar{\tau}_{10}^{-1}\{(\bar{\omega}_{10}^2-\omega^2)^2(3\bar{\omega}_{10}^2+\omega^2)+4\bar{\omega}_{10}^2(\bar{\omega}_{10}^2-\omega^2)\bar{\tau}_{10}^{-2}-(\bar{\omega}_{10}^2+3\omega^2)\bar{\tau}_{10}^{-4}\}\end{pmatrix}}{[(\bar{\omega}_{10}+\omega)^2+\bar{\tau}_{10}^{-2}]^3[(\bar{\omega}_{10}-\omega)^2+\bar{\tau}_{10}^{-2}]^3}.$$

$$(A.9.2.3)$$

PROBLEMS

1. Assume that you have a three-level system, i.e., two excited states (labeled 1 and 2) and one ground state (labeled 0), for a *centrosymmetric* molecule whose *excited*

states are also centrosymmetric. In such a case, the spatial part of the wave function of the ground state is usually symmetric, that of excited state 1 is antisymmetric, and that of excited state 2 is symmetric.

(a) Show that $\bar{\mu}_{01} \neq 0$, $\bar{\mu}_{01} \neq 0$, $\bar{\mu}_{12} \neq 0$, and $\bar{\mu}_{02} = 0$.

(b) Assume a single beam of frequency ω. Find those terms in $\Re eal\{\hat{\chi}^{(3)}_{xxxx}(-\omega; \omega, -\omega, \omega) + \hat{\chi}^{(3)}_{xxxx}(-\omega; -\omega, \omega, \omega) + \hat{\chi}^{(3)}_{xxxx}(-\omega; \omega, \omega, -\omega)\}$ in which all three levels participate. Show that when loss is neglected ("off resonance")

$$\Re eal\{\hat{\chi}^{(3)}_{xxxx}(-\omega; \omega, -\omega, \omega) + \hat{\chi}^{(3)}_{xxxx}(-\omega; -\omega, \omega, \omega) + \hat{\chi}^{(3)}_{xxxx}(-\omega; \omega, \omega, -\omega)\}$$

$$= 4 \frac{N}{\varepsilon_0 \hbar^3} |\bar{\mu}_{01}|^2 |\bar{\mu}_{12}|^2 \left\{ \frac{\bar{\omega}_{20}(\bar{\omega}_{10}^2 + \omega^2) + 4\bar{\omega}_{10}\omega^2}{(\bar{\omega}_{20}^2 - 4\omega^2)(\bar{\omega}_{10}^2 - \omega^2)^2} + \frac{2\bar{\omega}_{10}^2}{\bar{\omega}_{20}(\bar{\omega}_{10}^2 - \omega^2)^2} \right\}.$$

(c) Find the conditions on the transition dipole moments and $\bar{\omega}_{20}$ and $\bar{\omega}_{10}$, leading to a negative and a positive "nonresonant" n_2, which includes all the terms.

2. Derive the nonlinear susceptibility $\Re eal\{\hat{\chi}^{(2)}_{xxx}(0; \omega, -\omega) + \hat{\chi}^{(2)}_{xxx}(0; -\omega, \omega)\}$ for a three-level system (ground state labeled 0 and excited states labeled 1 and 2) that has a permanent dipole moment $\bar{\mu}_{11} \neq 0$ in excited state 1 and no dipole moments in the ground state and excited state 2. $\bar{\mu}_{01} \neq 0$ is the transition dipole moment between the ground state and excited state 1, and $\bar{\mu}_{12} \neq 0$ is the transition dipole moment between excited states 1 and 2. Assume a Type 1 phase-match geometry. What can you deduce about the symmetries of the three states?

3. Consider a two-level system with a permanent dipole moment $\bar{\mu}_{11} \neq 0$ in the first excited state, no dipole moment in the ground state, and a transition dipole moment $\bar{\mu}_{01} \neq 0$ and set $\bar{\tau}_{10}^{-1} = 0$. For a single-beam input polarized along a crystal axis, calculate $\hat{\chi}^{(3)}(-3\omega; \omega, \omega, \omega)$.

4. Show that the third-order susceptibility $\hat{\chi}^{(3)}_{xxxx}(-3\omega; \omega, \omega, \omega)$ for copolarized inputs and outputs and a two-level model with no permanent dipole moments is given by

$$\hat{\chi}^{(3)}_{xxxx}(-3\omega; \omega, \omega, \omega)$$

$$= -\frac{4N}{\varepsilon_0 \hbar^3} |\bar{\mu}_{10}|^4 \frac{\bar{\omega}_{10}^2 - \omega^2 - 2\bar{\tau}_{10}^{-2} - 2i\bar{\tau}_{10}^{-1}\omega}{\left(|\bar{\omega}_{10}|^2 - \omega^2 + \bar{\tau}_{10}^{-2} + 2i\bar{\tau}_{10}^{-1}\omega\right)\left(|\bar{\omega}_{10}|^2 - 9\omega^2 + \bar{\tau}_{10}^{-2} + 6i\bar{\tau}_{10}^{-1}\omega\right)}.$$

5. Consider a two-level system with a permanent dipole moment $\bar{\mu}_{11} \neq 0$ in the first excited state, no dipole moment in the ground state, and a transition dipole moment $\bar{\mu}_{01} \neq 0$. For a single-beam input polarized along a crystal axis, calculate the nonlinear susceptibility $\Re eal\{\hat{\chi}^{(3)}(-2\omega; \omega, \omega, 0) + \hat{\chi}^{(3)}(-2\omega; \omega, 0, \omega) + \hat{\chi}^{(3)}(-2\omega; 0, \omega, \omega)\}$ for the DC field-induced second harmonic generation.

6. (a) Find a general expression for the off-resonance $\Re eal\{\hat{\chi}^{(2)}_{ijk}(-\omega; \omega, 0) + \hat{\chi}^{(2)}_{ijk}(-\omega; 0, \omega)\}$.

(b) Show that for all fields polarized along the x-axis, the frequency dependence reduces to

$$f^{(2)} \frac{6\bar{\omega}_{ng}^2 \bar{\omega}_{mg}^2 - 2\omega^2 [\bar{\omega}_{ng}^2 + \bar{\omega}_{mg}^2 - \bar{\omega}_{mg}\bar{\omega}_{ng}]}{\left(\bar{\omega}_{ng}^2 - \omega^2\right)\left(\bar{\omega}_{mg}^2 - \omega^2\right)\bar{\omega}_{ng}\bar{\omega}_{mg}}$$

with $\quad f^{(2)} = \left(\dfrac{\varepsilon_r(\omega) + 2}{3}\right)^2 \left(\dfrac{\varepsilon_r(0) + 2}{3}\right)$.

(c) Show that for a two-level system this further reduces to

$$\Re\text{eal}\left\{\hat{\chi}_{xxx}^{(2)}(-\omega;\, 0, \omega) + \hat{\chi}_{xxx}^{(2)}(-\omega;\, \omega, 0)\right\}$$

$$= \frac{N}{\varepsilon_0 \hbar^2} f^{(2)} \left\{\bar{\mu}_{10,x}\left(\bar{\mu}_{11,x} - \bar{\mu}_{00,x}\right)\bar{\mu}_{01,x}\right\} \frac{2\left(3\bar{\omega}_{10}^2 - \omega^2\right)}{\left(\bar{\omega}_{10}^2 - \omega^2\right)^2}.$$

(d) Given that the electro-optics effect leads to an induced polarization of the form

$$P_i^{\text{NL}}(\omega) = -\frac{1}{2\varepsilon_0}\varepsilon_{ii}\varepsilon_{jj}r_{ijk}(-\omega;\, \omega, 0)E_j(\omega)E_k(0),$$

where $\hat{r}_{ijk}(-\omega;\, \omega, 0)$ is the electro-optic coefficient, find an expression for the frequency dispersion of $r_{ijk}(-\omega;\, \omega, 0)$. Reduce the general expression for the case of all x-polarized fields and then finally to the two-level limit.

7. Consider a three-level model in which all three states have permanent dipole moments.

(a) Assuming the nonlinear interaction puts a negligible population into the first excited state, find all the additional terms for n_2 that are introduced by these permanent dipole moments. [Hint: Do not try to simplify the resulting expression.]

Optional challenging part:

(b) What terms would need to be added if there were a significant population in the first excited state, e.g., N_1?

REFERENCES

1. D. Beljonne, J. L. Bredas, M. Cha, W. E. Torruellas, G. I. Stegeman, W. H. G. Horsthuis, and G. R. Mohlmann, "Two-photon absorption and third harmonic generation of di-alkyl-amino-nitro-stilbene (DANS): a joint experimental and theoretical study," J. Chem. Phys., **103**, 7834–7843 (1995).

2. Reviewed in S. Barlow and S. R. Marder, "Nonlinear optical properties of organic materials," in Functional Organic Materials: Syntheses, Strategies and Applications, edited by T. J. J. Muller and U. H. F. Bunz (John Wiley & Sons, Hoboken, NJ, 2007), Chap. 11.

3. Z. Soos, P. McWilliams, and G. Hayden, "Coulomb correlations and two-photon spectra of conjugated polymers," Chem. Phys. Lett., **171**, 14–18 (1990).

4. P. C. McWilliams, G. W. Hayden, and Z. G. Soos, "Theory of even-parity states and two-photon spectra of conjugated polymers," Phys. Rev. B, **43**, 9777–9791 (1991).

5. K. S. Mathis, M. G. Kuzyk, C. W. Dirk, A. Tan, S. Martinez, and G. Gampos, "Mechanisms of the nonlinear optical properties of squaraine dyes in poly(methyl methacrylate) polymer," J. Opt. Soc. Am. B, **15**, 871–883 (1998).

6. G. I. Stegeman and H. Hu, "Refractive nonlinearity of linear symmetric molecules and polymers revisited," **1**, 148–150 (2009); G. I. Stegeman, "Nonlinear optics of conjugated polymers and linear molecules," Nonlinear Opt., Quant. Opt.: Concepts in Mod. Opt., in press.

SUGGESTED FURTHER READING

R. W. Boyd, Nonlinear Optics, 3rd Edition (Academic Press, Burlington, MA, 2008).

D. N. Christodoulides, I. C. Khoo, G. J. Salamo, G. I. Stegeman, and E. W. Van Stryland, "Nonlinear refraction and absorption: mechanisms and magnitudes," Adv. Opt., **2**, 60–200 (2010).

THIRD-ORDER PHENOMENA

CHAPTER 10

Kerr Nonlinear Absorption and Refraction

The Kerr effect discussed in Chapters 8 and 9 results in nonlinear changes in the refractive index Δn^{NL} and the absorption $\Delta \alpha^{\mathrm{NL}}$ proportional to the intensity. These phenomena are "universal" in the sense that they occur in all media. They occur in every experiment, irrespective of that experiment's original goals. For example, although the purpose of an experiment may be efficient second harmonic generation, one must still consider the impact of nonlinear refraction (NLR) and nonlinear absorption (NLA) on the results. Furthermore, NLR and NLA have many applications; e.g., soliton generation via NLR and two-photon absorption (NLA) is a powerful tool in high resolution microscopy, in optical limiting, and, most recently, in local medicine delivery for light-activated drugs.

The third-order susceptibility contains both real and imaginary components; i.e., $\hat{\chi}^{(3)} = \Re\mathrm{eal}(\hat{\chi}^{(3)}) + i\,\Im\mathrm{mag}(\hat{\chi}^{(3)})$. The real part produces changes in the refractive index, and the imaginary part produces changes in absorption, both proportional to the intensities of input beams or the products of input beams. Only certain susceptibility terms will produce resonances (peaks) in the imaginary component in spectral ranges away from the usual linear absorption peaks, and these are identified as two-photon resonances. However, there are other nonzero terms in the imaginary component that also produce NLA proportional to the intensity without two-photon resonances, and they can be considered as intensity-dependent corrections to linear absorption.

Please note that in this chapter, all relations involving nonlinear susceptibilities based on the Kerr effect are not universally valid for other mechanisms that can also lead to intensity-dependent refraction and absorption.

Aside: NLA and NLR come from $\hat{\chi}^{(2m+1)}(-\omega_a;\ \omega_b\ \ldots)$, $m \geq 1$, i.e., odd order terms, with the most common case occurring for $\omega_a = \omega_b$. Some examples of susceptibilities responsible for more general NLA and NLR are as follows:

$$\hat{\chi}^{(3)}_{xxxx}(-\omega_a;\omega_b,-\omega_b,\omega_a) \rightarrow \Delta n^{\mathrm{NL}}(\omega_a) \;\; \text{and} \;\; \Delta\alpha^{\mathrm{NL}}(\omega_a) \propto I(\omega_b),$$

$$\hat{\chi}^{(5)}_{xxxxxx}(-\omega_a;\omega_a,-\omega_b,\omega_b,-\omega_b,\omega_b) \rightarrow \Delta n^{\mathrm{NL}}(\omega_a) \;\; \text{and} \;\; \Delta\alpha^{\mathrm{NL}}(\omega_a) \propto I^2(\omega_b),$$

Nonlinear Optics: Phenomena, Materials, and Devices, George I. Stegeman and Robert A. Stegeman.
© 2012 John Wiley & Sons, Inc. Published 2012 by John Wiley & Sons, Inc.

$$\hat{\chi}^{(5)}_{xxxxxx}(-\omega_b; \omega_b, -\omega_a, \omega_a, -\omega_c, \omega_c) \rightarrow \Delta n^{\mathrm{NL}}(\omega_b) \quad \text{and} \quad \Delta \alpha^{\mathrm{NL}}(\omega_b) \propto I(\omega_a)I(\omega_c),$$

$$\hat{\chi}^{(7)}_{xxxxxxxx}(-\omega_a; \omega_a, -\omega_b, \omega_b, -\omega_c, \omega_c, -\omega_c, \omega_c) \rightarrow \Delta n^{\mathrm{NL}}(\omega_a) \quad \text{and}$$

$$\Delta \alpha^{\mathrm{NL}}(\omega_a) \propto I(\omega_b)I^2(\omega_c), \text{ etc.}$$

10.1 NONLINEAR ABSORPTION

10.1.1 Single-Beam Input

For a single-beam input, e.g., x-polarized along a symmetry axis, recall from Chapter 8 that the nonlinear polarization is given by

$$\mathcal{P}^{(3)}_x(z, \omega) = \frac{3}{4} \varepsilon_0 \hat{\chi}^{(3)}_{xxxx}(-\omega; \omega, -\omega, \omega)|\mathcal{E}_x(z, \omega)|^2 \, \mathcal{E}_x(z, \omega), \qquad (10.1a)$$

$$\hat{\chi}^{(3)}_{xxxx}(-\omega; \omega, -\omega, \omega) = \frac{1}{3} \left[\hat{\chi}^{(3)}_{xxxx}(-\omega; \omega, \omega, -\omega) + \hat{\chi}^{(3)}_{xxxx}(-\omega; \omega, -\omega, \omega) \right.$$

$$\left. + \hat{\chi}^{(3)}_{xxxx}(-\omega; -\omega, \omega, \omega) \right], \qquad (10.1b)$$

with the previously defined notation that the "peaked hat" above, e.g., $\hat{A}$, identifies a complex quantity and the "rounded hat" above, e.g., $\hat{A}$, identifies the averaged susceptibility given by Eq. 10.1b. Inserting polarizations into the slowly varying envelope approximation, Eq. 2.25 gives

$$\frac{d\mathcal{E}_x(z, \omega)}{dz} = i \frac{3\omega}{8nc} \hat{\chi}^{(3)}_{xxxx}(-\omega; \omega, -\omega, \omega)|\mathcal{E}_x(z, \omega)|^2 \mathcal{E}_x(z, \omega). \qquad (10.2)$$

Writing $\mathcal{E}_x(z, \omega) = \rho(z) e^{i\phi^{\mathrm{NL}}(z)}$ and separating the resulting equations into the real and imaginary parts in Eq. 10.2 gives

$$\frac{d\phi^{\mathrm{NL}}(z)}{dz} = \frac{3\omega}{8nc} \Re\mathrm{eal}\left\{ \hat{\chi}^{(3)}_{xxxx}(-\omega; \omega, -\omega, \omega) \right\} \rho^2(z) \qquad \rightarrow \qquad \text{NLR.} \qquad (10.3a)$$

$$\frac{d\rho(z)}{dz} = -\frac{3\omega}{8nc} \Im\mathrm{mag}\left\{ \hat{\chi}^{(3)}_{xxxx}(-\omega; \omega, -\omega, \omega) \right\} \rho^3(z) \qquad \rightarrow \qquad \text{NLA}$$

$$\qquad (10.3b)$$

$$\rightarrow \quad 2\rho(z)\frac{d\rho(z)}{dz} = \frac{d\rho^2(z)}{dz} = -\frac{3\omega}{4nc} \Im\mathrm{mag}\left\{ \hat{\chi}^{(3)}_{xxxx}(-\omega; \omega, -\omega, \omega) \right\} \rho^4(z).$$

Equation 10.3a indicates that there is an additional phase shift imparted onto the beam due to the third-order susceptibility, which depends on the intensity of the beam

$[\rho^2(z) = |\mathcal{E}_x(z, \omega)|^2 = 2[n\varepsilon_0 c]^{-1}I(z)]$, to be discussed in Section 10.2. Equation 10.3b identifies two-photon absorption proportional to the intensity that is associated with the imaginary component of the third-order susceptibility. Converting to the intensity $I(z)$ and defining $\alpha_{2\|}(-\omega; \omega)$ by

$$\frac{dI(z)}{dz} = -\alpha_{2\|}(-\omega; \omega)I^2(z) \;\rightarrow\; \alpha_{2\|}(-\omega; \omega) = \frac{3\omega}{2n^2\varepsilon_0 c^2}\Im\mathrm{mag}\left\{\hat{\chi}^{(3)}_{xxxx}(-\omega; \omega, -\omega, \omega)\right\}.$$

(10.4)

The parameter $\alpha_{2\|}(-\omega; \omega)$ is called the two-photon absorption coefficient. In the literature, $\alpha_2 \equiv \alpha_{2\|}(-\omega; \omega)$ is frequently identified with NLA due to two-photon resonances, although it contains all the terms proportional to $\rho^2(z)$, which occur via third-order susceptibilities. The subscript $\|$ signifies that the polarizations of the input and output frequencies are equal. (If the polarizations are orthogonal, the two-photon absorption coefficient is written as $\alpha_{2\perp}(-\omega; \omega)$). In the early published literature, $\alpha_{2\|}(-\omega; \omega)$ was frequently called β. (Also to avoid confusion later with the multiple α_2's used in the literature, $\alpha_{2\|,2\perp}(-\omega_a; \omega_b)$ refers to two-photon absorption at frequency ω_a induced by the presence of a second beam at frequency ω_b for the copolarized and orthogonally polarized cases, respectively.)

The two-photon resonance effects that give rise to NLA peaks come from the $\Im\mathrm{mag}\{\chi^{(3)}_{xxxx}(-\omega; \omega, \omega, -\omega)\}$ susceptibility. Substituting the arguments $\omega, \omega, -\omega$ into Eq. 8.53 gives

$$\hat{\chi}^{(3)}_{ijkl}(-\omega; \omega, \omega, -\omega) = \frac{N}{\varepsilon_0\hbar^3}f^{(3)}\sum_{v,n,m}{}'$$

$$\times\left\{\frac{\bar{\mu}_{gv,i}(\bar{\mu}_{vn,l} - \bar{\mu}_{gg,l})(\bar{\mu}_{nm,k} - \bar{\mu}_{gg,k})\bar{\mu}_{mg,j}}{(\hat{\bar{\omega}}_{vg} - \omega)(\hat{\bar{\omega}}_{ng} - 2\omega)(\hat{\bar{\omega}}_{mg} - \omega)} + \frac{\bar{\mu}_{gv,j}(\bar{\mu}_{vn,k} - \bar{\mu}_{gg,k})(\bar{\mu}_{nm,i} - \bar{\mu}_{gg,i})\bar{\mu}_{mg,l}}{(\hat{\bar{\omega}}^*_{vg} + \omega)(\hat{\bar{\omega}}^*_{ng} + 2\omega)(\hat{\bar{\omega}}_{mg} + \omega)}\right.$$

$$\left. + \frac{\bar{\mu}_{gv,l}(\bar{\mu}_{vn,i} - \bar{\mu}_{gg,i})(\bar{\mu}_{nm,k} - \bar{\mu}_{gg,k})\bar{\mu}_{mg,j}}{(\hat{\bar{\omega}}^*_{vg} - \omega)(\hat{\bar{\omega}}_{ng} - 2\omega)(\hat{\bar{\omega}}_{mg} - \omega)} + \frac{\bar{\mu}_{gv,j}(\bar{\mu}_{vn,k} - \bar{\mu}_{gg,k})(\bar{\mu}_{nm,l} - \bar{\mu}_{gg,l})\bar{\mu}_{mg,i}}{(\hat{\bar{\omega}}^*_{vg} + \omega)(\hat{\bar{\omega}}^*_{ng} + 2\omega)(\hat{\bar{\omega}}^*_{mg} + \omega)}\right\}.$$

(10.5)

This susceptibility contains imaginary components that have two-photon resonances, i.e., have peaks whenever $2\omega \cong \bar{\omega}_{ng}$ for some excited states n, which are two-photon-transition-allowed excited states. From Chapter 8, it is clear that all these states are either mixed one- and two-photon states in molecules with permanent dipole moments or states with even symmetry in symmetric molecules. The energy levels and the transition diagram responsible for two-photon resonance effects are illustrated in Fig. 10.1. The other two third-order susceptibilities—$\hat{\chi}^{(3)}_{xxxx}(-\omega; -\omega, \omega, \omega)$ and $\hat{\chi}^{(3)}_{xxxx}(-\omega; \omega, -\omega, \omega)$—contain nonlinear resonances only at the fundamental frequencies, i.e., at $\omega \cong \bar{\omega}_{ng}$, and they give intensity-dependent changes in absorption over essentially all frequencies.

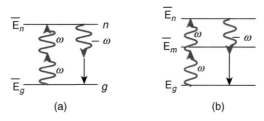

FIGURE 10.1   The resonant two-photon transitions that involve the simultaneous absorption of two photons between energy levels $\bar{E}_n$ and $\bar{E}_g$ with the photon energy $2\hbar\omega \cong \bar{E}_n - \bar{E}_g$ are shown. (a) Asymmetric molecules with permanent dipole moments. (b) Symmetric molecules that require an odd symmetry excited state $\bar{E}_m$ in the vicinity of $\bar{E}_n$ and $\bar{E}_g$.

The net amount of the incident beam transmitted after a distance L with two-photon absorption can be evaluated by integrating Eq. 10.4. This gives

$$I(L) = \frac{I(0)}{1 + \alpha_{2\parallel}(-\omega;\omega)I(0)L} \rightarrow T = \frac{I(L)}{I(0)} = \frac{1}{1 + \alpha_{2\parallel}(-\omega;\omega)I(0)L}, \quad (10.6)$$

which, when the inverse of the transmission, T^{-1}, is plotted as shown in Fig. 10.2, has a slope given by $\alpha_{2\parallel}(-\omega;\omega)$.

In many cases, the linear absorption [$\alpha_1(\omega)$] must also be taken into account.

$$\frac{dI(z)}{dz} = -\alpha_1(\omega)I(z) - \alpha_{2\parallel}(-\omega;\omega)I^2(z)$$

$$\rightarrow I(L) = \frac{I(0)\,e^{-\alpha_1(\omega)L}}{1 + (\{\alpha_{2\parallel}(-\omega;\omega)I(0)[1 - e^{-\alpha_1(\omega)L}]\}/\alpha_1(\omega))}. \quad (10.7)$$

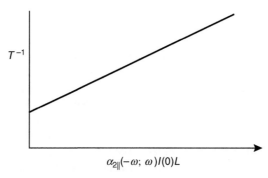

FIGURE 10.2   Plot of the inverse of the transmission, T^{-1}, versus the normalized parameter $\alpha_{2\parallel}(-\omega;\omega)I(0)L$. The slope is given by $\alpha_{2\parallel}(-\omega;\omega)$.

Defining

$$L_{\text{eff}} = \frac{1 - e^{-\alpha_1(\omega)L}}{\alpha_1(\omega)} \quad \rightarrow \quad I(L) = \frac{I(0)\, e^{-\alpha_1(\omega)L}}{1 + \alpha_{2\|}(-\omega;\omega)I(0)L_{\text{eff}}}. \tag{10.8}$$

10.1.1.1   *Two-Level Model.*

Consider the two-level model with the excited state exhibiting mixed, i.e., odd and even symmetry (nonzero dipole moments). In Chapter 9, it was shown for the terms responsible for a two-photon absorption peak that for the near- and on-resonance case,

$$\Im\mathrm{mag}\left\{\hat{\chi}^{(3)}_{xxxx}(-\omega;\omega,\omega,-\omega)\right\}$$

$$= \frac{N}{\varepsilon_0 \hbar^3} f^{(3)} |\bar{\mu}_{10}|^2 (\bar{\mu}_{11} - \bar{\mu}_{00})^2 \frac{8\bar{\tau}_{10}^{-1}}{\bar{\omega}_{10}^2([\bar{\omega}_{10} - 2\omega]^2 + \bar{\tau}_{10}^{-2})}$$

$$\rightarrow \quad \alpha_{2\|}(-\omega;\omega) = \frac{1}{n^2 \varepsilon_0 c^2} \frac{N}{\varepsilon_0 \hbar^3} f^{(3)} |\bar{\mu}_{10}|^2 (\bar{\mu}_{11} - \bar{\mu}_{00})^2 \frac{4\bar{\tau}_{10}^{-1}}{\bar{\omega}_{10}([\bar{\omega}_{10} - 2\omega]^2 + \bar{\tau}_{10}^{-2})}. \tag{10.9}$$

In the off-resonance case, all third-order susceptibilities contribute, and so

$$\alpha_{2\|}(-\omega;\omega) = \frac{2\omega^2 \bar{\omega}_{10} \bar{\tau}_{10}^{-1}}{n^2 \varepsilon_0 c^2} \frac{N}{\varepsilon_0 \hbar^3} f^{(3)} |\bar{\mu}_{10}|^2$$

$$\times \left\{ (\bar{\mu}_{11} - \bar{\mu}_{00})^2 \frac{15\bar{\omega}_{10}^4 - 39\bar{\omega}_{10}^2\omega^2 + 24\omega^4}{(\bar{\omega}_{10}^2 - 4\omega^2)^2(\bar{\omega}_{10}^2 - \omega^2)^3} - |\bar{\mu}_{10}|^2 \frac{4(3\bar{\omega}_{10}^2 + \omega^2)}{(\bar{\omega}_{10}^2 - \omega^2)^4} \right\}. \tag{10.10}$$

As $\omega \to 0$, $\alpha_{2\|}(-\omega;\omega) \propto \omega^2$ and $\alpha_{2\|}(-\omega;\omega) \to 0$. This decrease in the two-photon absorption coefficient quadratic with decreasing frequency (i.e., $\alpha_{2\|}(-\omega;\omega) \propto \omega^2$) always occurs, independent of the model for the molecular system.

10.1.1.2   *Three-Level Model.*

As noted previously in Section 9.2, a minimum of three states—one ground and two excited states—are needed for two-photon transitions in *symmetric* molecules (no permanent dipole moments) due to the symmetry properties of wave functions. The process is summarized in Fig. 10.3.

In this case, near and on one-photon resonance $\alpha_{2\|}(-\omega;\omega)$ is given by

$$\alpha_{2\|}(-\omega;\omega) = \bar{\omega}_{10} \frac{N|\bar{\mu}_{10}|^2}{n^2 \varepsilon_0^2 c^2 \hbar^3} f^{(3)}_{(\omega)} \left\{ -|\bar{\mu}_{10}|^2 \frac{2\bar{\tau}_{10}^{-1}(\bar{\omega}_{10} - \omega)^2}{[(\bar{\omega}_{10} - \omega)^2 + \bar{\tau}_{10}^{-2}]^3} + |\bar{\mu}_{21}|^2 (\omega_{10} - \omega) \right.$$

$$\left. \times \frac{2\bar{\tau}_{10}^{-1}(\bar{\omega}_{20} - \bar{\omega}_{10})(\bar{\omega}_{20} - 2\omega_{10}) + 2\bar{\tau}_{21}^{-1}\bar{\omega}_{20}(\omega_{10} - \omega)}{\omega_{20}(\bar{\omega}_{20} - 2\bar{\omega}_{10})^2[(\omega_{10} - \omega)^2 + \bar{\tau}_{10}^{-2}]^2} \right\}. \tag{10.11a}$$

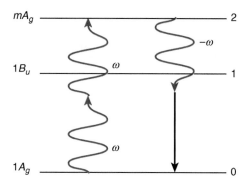

FIGURE 10.3 The electronic states needed for two-photon absorption resonances in symmetric molecules that require the participation of three distinct states.

And near and on two-photon resonance is given by

$$\alpha_{2\parallel}(-\omega;\omega) = \frac{4N\bar{\omega}_{20}}{n^2\varepsilon_0^2 c^2 \hbar} f^{(3)}(\omega)|\bar{\mu}_{01}|^2|\bar{\mu}_{12}|^2$$

$$\times\left\{\frac{[(\bar{\omega}_{20}^2 - 4\omega^2)(\bar{\omega}_{20} + 2\bar{\omega}_{10})^3\bar{\tau}_{10}^{-1} + \bar{\omega}_{20}(4\bar{\omega}_{10}^2 - \bar{\omega}_{20}^2)(4\bar{\omega}_{10}^2 + \bar{\omega}_{20}^2)\bar{\tau}_{eff}^{-1}]}{[(\bar{\omega}_{20} - 2\omega)^2 + \bar{\tau}_{eff}^{-1}](4\bar{\omega}_{10}^2 - \bar{\omega}_{20}^2)^3}\right\}.$$

$$(10.11b)$$

Note that the two-photon absorption resonance is given by the term proportional to $\bar{\tau}_{eff}^{-1}$, but the line shape is distorted from a simple Lorentzian by the dispersive term proportional to $\bar{\tau}_{10}^{-1}(\bar{\omega}_{20}^2 - 4\omega^2)$. Examples of the line shape for different lifetimes are shown in Fig. 10.4b. For $\bar{\tau}_{eff}^{-1} \geq \bar{\tau}_{10}^{-1}$, the effect of the dispersion term is small, provided $\bar{\omega}_{20}\bar{\tau}_{eff} \gg 1$. If these conditions are not satisfied, the overall two-photon resonance line shape must be analyzed to obtain $|\bar{\mu}_{21}|^2$ and $\bar{\tau}_{eff}^{-1}$ once $|\bar{\mu}_{01}|^2$ and $\bar{\tau}_{10}^{-1}$ have been obtained from the linear absorption spectrum.

Off resonance,

$$\alpha_{2\parallel}(-\omega;\omega) = \frac{2\omega^2 N}{n^2\varepsilon_0^2 c^2\hbar^3} f^{(3)}|\bar{\mu}_{10}|^2\left\{|\bar{\mu}_{21}|^2\right.$$

$$\times\left\{\bar{\tau}_{10}^{-1}\frac{\bar{\omega}_{20}\bar{\omega}_{10}^2(7\bar{\omega}_{20} + 2\bar{\omega}_{10}) + \omega^2(\omega_{20} + 8\bar{\omega}_{10})(\omega_{20} - 2\bar{\omega}_{10})}{\omega_{20}(\bar{\omega}_{10}^2 - \omega^2)^3(\bar{\omega}_{20}^2 - 4\omega^2)}\right.$$

$$\left.+ 2\bar{\tau}_{eff}^{-1}\frac{(\omega_{20} + 2\omega_{10})(\bar{\omega}_{20}\omega_{10} + 2\omega^2)}{(\bar{\omega}_{10}^2 - \omega^2)^2(\bar{\omega}_{20}^2 - 4\omega^2)^2}\right\} - |\bar{\mu}_{10}|^4 4\bar{\omega}_{10}\bar{\tau}_{10}^{-1}\frac{(3\bar{\omega}_{10}^2 + \omega^2)}{(\bar{\omega}_{10}^2 - \omega^2)^4}\right\}.$$

$$(10.12)$$

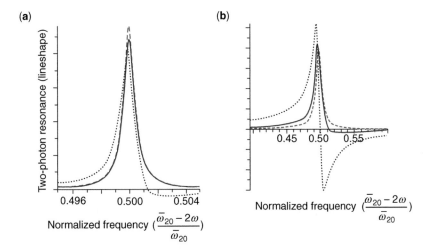

FIGURE 10.4 Examples of the two-photon resonance line shape with $\bar{\omega}_{20} = 1.333\bar{\omega}_{10}$ and various ratios of excited lifetimes calculated for Eq. 9.11b. Here, $\bar{\tau}_{eff}^{-1} = 100\bar{\tau}_{10}^{-1}$ (black dotted line), $\bar{\tau}_{eff}^{-1} = \bar{\tau}_{10}^{-1}$ (blue solid line), and $\bar{\tau}_{eff}^{-1} = 0.1\bar{\tau}_{10}^{-1}$ (red dashed line). (a) $\bar{\omega}_{20}\bar{\tau}_{eff} = 10^3$. (b) $\bar{\omega}_{20}\bar{\tau}_{21} = 100$. Note that the vertical scale is different for every case and that other nonresonant contributions to two-photon absorption are not included.

10.1.2 Two-Beam Input (Nondegenerate Case, Two Input Frequencies: ω_a and ω_b)

Consider the case when two eigenmodes of different frequencies $\omega_a > \omega_b$ are incident simultaneously on the material, both polarized along the x-axis. In addition to two-photon absorption individually at each frequency as discussed in the previous section, there will be "cross-beam" absorption at one frequency due to the intensity of the beam at the other frequency. In this case, from Eq. 8.78 the pertinent nonlinear polarization can be written as

$$P_x^{(3)}(\omega_a) = \frac{1}{4}\varepsilon_0 \left\{ \hat{\chi}_{xxxx}^{(3)}(-\omega_a;\omega_b,\omega_a,-\omega_b) + \hat{\chi}_{xxxx}^{(3)}(-\omega_a;\omega_b,-\omega_b,\omega_a) \right.$$
$$+ \hat{\chi}_{xxxx}^{(3)}(-\omega_a;\omega_a,\omega_b,-\omega_b) + \hat{\chi}_{xxxx}^{(3)}(-\omega_a;\omega_a,-\omega_b,\omega_b)$$
$$\left. + \hat{\chi}_{xxxx}^{(3)}(-\omega_a;-\omega_b,\omega_b,\omega_a) + \hat{\chi}_{xxxx}^{(3)}(-\omega_a;-\omega_b,\omega_a,\omega_b) \right\} |\mathcal{E}_x(\omega_b)|^2 \mathcal{E}_x(\omega_a),$$

$$P_x^{(3)}(\omega_b) = \frac{1}{4}\varepsilon_0 \left\{ \hat{\chi}_{xxxx}^{(3)}(-\omega_b;\omega_b,\omega_a,-\omega_a) + \hat{\chi}_{xxxx}^{(3)}(-\omega_b;\omega_b,-\omega_a,\omega_a) \right.$$
$$+ \hat{\chi}_{xxxx}^{(3)}(-\omega_b;\omega_a,\omega_b,-\omega_a) + \hat{\chi}_{xxxx}^{(3)}(-\omega_b;\omega_a,-\omega_a,\omega_b)$$
$$\left. + \hat{\chi}_{xxxx}^{(3)}(-\omega_b;-\omega_a,\omega_a,\omega_b) + \hat{\chi}_{xxxx}^{(3)}(-\omega_b;-\omega_a,\omega_b,\omega_a) \right\} |\mathcal{E}_x(\omega_a)|^2 \mathcal{E}_x(\omega_b).$$

$$(10.13)$$

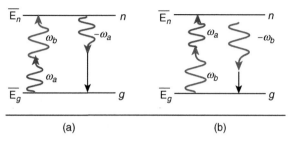

FIGURE 10.5   The two processes that lead to two-photon resonances at $\bar{\omega}_{ng} \cong \omega_a + \omega_b$ for (a) $\mathcal{P}_x^{(3)}(\omega_b)$ and (b) $\mathcal{P}_x^{(3)}(\omega_a)$, respectively.

Substituting Eq. 9.6 for each of the susceptibilities reveals that the additional two-photon resonances that occur at $\omega_{ng} = \omega_a + \omega_b$ are shown in Fig. 10.5 for the nonlinear polarization induced at the frequencies ω_a and ω_b.

Writing $\mathcal{E}_x(z, \omega_a) = \rho_a(z)\, e^{i\phi_a^{NL}(z)}$ and $\mathcal{E}(z, \omega_b) = \rho_b(z)\, e^{i\phi_b^{NL}(z)}$, substituting into the slowly varying envelope approximation, isolating the amplitude terms, and finally multiplying the resulting equations by the appropriate amplitude $\rho_{a,b}(z)$ gives

$$2\rho_a(z)\frac{d\rho_a(z)}{dz} = -\frac{6\omega_a}{8n_a c}\Im\text{mag}\left\{\hat{\chi}_{xxxx}^{(3)}(-\omega_a; \omega_a, -\omega_b, \omega_b)\right\}\rho_b^2(z)\rho_a(z)2\rho_a(z),$$

$$2\rho_b(z)\frac{d\rho_b(z)}{dz} = -\frac{6\omega_b}{8n_b c}\Im\text{mag}\left\{\hat{\chi}_{xxxx}^{(3)}(-\omega_b; \omega_b, -\omega_a, \omega_a)\right\}\rho_a^2(z)\rho_b(z)2\rho_b(z),$$

$$(10.14)$$

with $\hat{\chi}_{xxxx}^{(3)}(-\omega_a; \omega_a, -\omega_b, \omega_b)$ and $\hat{\chi}_{xxxx}^{(3)}(-\omega_b; \omega_b, -\omega_a, \omega_a)$ defined by

$$\hat{\chi}_{xxxx}^{(3)}(-\omega_a; \omega_a, -\omega_b, \omega_b) = \frac{1}{6}\Big[\hat{\chi}_{xxxx}^{(3)}(-\omega_a; \omega_a, \omega_b, -\omega_b) + \hat{\chi}_{xxxx}^{(3)}(-\omega_a; \omega_a, -\omega_b, \omega_b)$$

$$+\hat{\chi}_{xxxx}^{(3)}(-\omega_a; \omega_b, \omega_a, -\omega_b) + \hat{\chi}_{xxxx}^{(3)}(-\omega_a; \omega_b, -\omega_b, \omega_a)$$

$$+\hat{\chi}_{xxxx}^{(3)}(-\omega_a; -\omega_b, \omega_b, \omega_a) + \hat{\chi}_{xxxx}^{(3)}(-\omega_a; -\omega_b, \omega_a, \omega_b)\Big],$$

$$\hat{\chi}_{xxxx}^{(3)}(-\omega_b; \omega_b, -\omega_a, \omega_a) = \frac{1}{6}\Big[\hat{\chi}_{xxxx}^{(3)}(-\omega_b; \omega_b, \omega_a, -\omega_a) + \hat{\chi}_{xxxx}^{(3)}(-\omega_b; \omega_b, -\omega_a, \omega_a)$$

$$+\hat{\chi}_{xxxx}^{(3)}(-\omega_b; \omega_a, \omega_b, -\omega_a) + \hat{\chi}_{xxxx}^{(3)}(-\omega_b; \omega_a, -\omega_a, \omega_b)$$

$$+\hat{\chi}_{xxxx}^{(3)}(-\omega_b; -\omega_a, \omega_a, \omega_b) + \hat{\chi}_{xxxx}^{(3)}(-\omega_b; -\omega_a, \omega_b, \omega_a)\Big].$$

$$(10.15)$$

Now, let us convert Eq. 10.14 to intensities:

$$\frac{dI_a(z)}{dz} = -3\frac{\omega_a}{n_a n_b \varepsilon_0 c^2}\Im\mathrm{mag}\left\{\hat{\bar{\chi}}^{(3)}_{xxxx}(-\omega_a;\omega_a,-\omega_b,\omega_b)\right\}I_a(z)I_b(z)$$

$$= -\alpha_{2\|}(-\omega_a;\omega_b)I_a(z)I_b(z)$$

$$\Rightarrow \alpha_{2\|}(-\omega_a;\omega_b) = 3\frac{\omega_a}{n_a n_b \varepsilon_0 c^2}\Im\mathrm{mag}\left\{\hat{\bar{\chi}}^{(3)}_{xxxx}(-\omega_a;\omega_a,-\omega_b,\omega_b)\right\}.$$

Also,

$$\frac{dI_b(z)}{dz} = -\alpha_{2\|}(-\omega_b;\omega_a)I_a(z)I_b(z)$$

$$\Rightarrow \alpha_{2\|}(-\omega_b;\omega_a) = 3\frac{\omega_b}{n_a n_b \varepsilon_0 c^2}\Im\mathrm{mag}\left\{\hat{\bar{\chi}}^{(3)}_{xxxx}(-\omega_b;\omega_b,-\omega_a,\omega_a)\right\}.$$

$$(10.16)$$

The relevant terms for two-photon resonances at $\bar{\omega}_{ng} \cong \omega_a + \omega_b$ are

$$\Im\mathrm{mag}\left\{\hat{\bar{\chi}}^{(3)}_{xxxx}(-\omega_a;\omega_a,-\omega_b,\omega_b)\right\} = \frac{1}{6}\Im\mathrm{mag}\left\{\hat{\chi}^{(3)}_{xxxx}(-\omega_a;\omega_a,\omega_b,-\omega_b)\right.$$
$$\left.+\hat{\chi}^{(3)}_{xxxx}(-\omega_a;\omega_b,\omega_a,-\omega_b)\right\},$$

$$\Im\mathrm{mag}\left\{\hat{\bar{\chi}}^{(3)}_{xxxx}(-\omega_b;\omega_b,-\omega_a,\omega_a)\right\} = \frac{1}{6}\Im\mathrm{mag}\left\{\hat{\chi}^{(3)}_{xxxx}(-\omega_b;\omega_b,\omega_a,-\omega_a)\right.$$
$$\left.+\hat{\chi}^{(3)}_{xxxx}(-\omega_b;\omega_a,\omega_b,-\omega_a)\right\}. \quad (10.17)$$

The two-photon resonances in absorption are given by

$$\Rightarrow \frac{dI_a(z)}{dz}\xrightarrow{\omega_a+\omega_b\cong\bar{\omega}_{ng}} -\frac{\omega_a}{n_a n_b \varepsilon_0 c^2}$$
$$\Im\mathrm{mag}\left[\hat{\chi}^{(3)}_{xxxx}(-\omega_a;\omega_a,\omega_b,-\omega_b) + \hat{\chi}^{(3)}_{xxxx}(-\omega_a;\omega_b,\omega_a,-\omega_b)\right]I_a(z)I_b(z), \quad (10.18a)$$

$$\Rightarrow \frac{dI_b(z)}{dz}\xrightarrow{\omega_a+\omega_b\cong\bar{\omega}_{ng}} -\frac{\omega_b}{n_a n_b \varepsilon_0 c^2}$$
$$\Im\mathrm{mag}\left[\hat{\chi}^{(3)}_{xxxx}(-\omega_b;\omega_b,\omega_a,-\omega_a) + \hat{\chi}^{(3)}_{xxxx}(-\omega_b;\omega_a,\omega_b,-\omega_a)\right]I_a(z)I_b(z). \quad (10.18b)$$

Combining the total (not just the resonance terms) "self" and "cross" two-photon absorption terms gives

$$\frac{dI_a(z)}{dz} = -\alpha_{2\|}(-\omega_a;\omega_a)I_a^2(z) - \alpha_{2\|}(-\omega_a;\omega_b)I_a(z)I_b(z),$$

$$(10.19)$$

$$\frac{dI_b(z)}{dz} = -\alpha_{2\|}(-\omega_b;\omega_b)I_b^2(z) - \alpha_{2\|}(-\omega_b;\omega_a)I_b(z)I_a(z).$$

The interpretation is that in this case two-photon absorption occurs, with two photons being taken from the beams $I_a(z)$ and $I_b(z)$ via $\alpha_{2\|}(-\omega_a;\omega_a)$ and $\alpha_{2\|}(-\omega_b;\omega_b)$, respectively, and in addition one photon taken from each beam via $\alpha_{2\|}(-\omega_a;\omega_b)$ and $\alpha_{2\|}(-\omega_b;\omega_a)$, respectively.

If one beam, e.g., $I_b(z)$, is a strong beam and the other, $I_a(z)$, is a weak beam, i.e., a pump-probe geometry, the relevant equations simplify to

$$\frac{dI_b(z)}{dz} = -\alpha_{2\|}(-\omega_b;\omega_b)I_b^2(z), \tag{10.20}$$

$$\frac{dI_a(z)}{dz} = -\alpha_{2\|}(-\omega_a;\omega_b)I_a(z)I_b(z). \tag{10.21}$$

The solution for $I_b(z)$ from Eq. 10.20 is inserted into Eq. 10.21, giving

$$\frac{dI_a(z)}{dz} = -\alpha_{2\|}(-\omega_a;\omega_b)I_a(z)\frac{I_b(0)}{1+\alpha_{2\|}(-\omega_b;\omega_b)I_b(0)z}. \tag{10.22}$$

Integrating this pump-probe case gives

$$I_a(L) = \frac{I_a(0)}{[1+\alpha_{2\|}(-\omega_b;\omega_b)I_b(0)L]^{\exp}}, \quad \exp = -\frac{\alpha_{2\|}(-\omega_a;\omega_b)}{\alpha_{2\|}(-\omega_b;\omega_b)},$$

$$\frac{\alpha_{2\|}(-\omega_a;\omega_b)}{\alpha_{2\|}(-\omega_b;\omega_b)} = \frac{2\omega_a n_b \, \Im\mathrm{mag}[\hat{\chi}^{(3)}_{xxxx}(-\omega_a;\omega_a,\omega_b,-\omega_b)]}{\omega_b n_a \, \Im\mathrm{mag}[\hat{\chi}^{(3)}_{\Im,xxxx}(-\omega_b;\omega_b,\omega_b,-\omega_b)]}. \tag{10.23}$$

Note that these formulas are also valid when $\omega_a = \omega_b$ in the case where the two beams constitute two *separate* eigenmodes, e.g., traveling in different directions.

10.1.3 Two Orthogonally Polarized Beam Input: Equal or Unequal Frequencies

Since orthogonally polarized waves automatically constitute different eigenmodes, there are again six different combinations of susceptibilities involving the two polarizations. Each beam can individually experience two-photon absorption as discussed previously, and the analysis here will be limited to the "cross-beam" cases. Since beam a is x-polarized and beam b is y-polarized,

$$\mathcal{P}^{(3)}_x(\omega_b) = \frac{6}{4}\varepsilon_0 \hat{\chi}^{(3)}_{xxyy}(-\omega_b;\omega_b,-\omega_a,\omega_a)|\mathcal{E}_y(\omega_a)|^2 \mathcal{E}_x(\omega_b),$$

$$\hat{\chi}^{(3)}_{xxyy}(-\omega_b;\omega_b,-\omega_a,\omega_a) = \frac{1}{6}\Big[\hat{\chi}^{(3)}_{xxyy}(-\omega_b;\omega_b,\omega_a,-\omega_a) + \hat{\chi}^{(3)}_{xxyy}(-\omega_b;\omega_b,-\omega_a,\omega_a)$$

$$+\hat{\chi}^{(3)}_{xyxy}(-\omega_b;\omega_a,\omega_b,-\omega_a) + \hat{\chi}^{(3)}_{xyyx}(-\omega_b;\omega_a,-\omega_a,\omega_b)$$

$$+\hat{\chi}^{(3)}_{xyyx}(-\omega_b;-\omega_a,\omega_a,\omega_b) + \hat{\chi}^{(3)}_{xyxy}(-\omega_b;-\omega_a,\omega_b,\omega_a)\Big], \tag{10.24a}$$

$$\mathcal{P}_y^{(3)}(\omega_a) = \frac{6}{4}\varepsilon_0 \hat{\chi}_{yyxx}^{(3)}(-\omega_a; \omega_a, -\omega_b, \omega_b)|\mathcal{E}_x(\omega_b)|^2 \mathcal{E}_y(\omega_a),$$

$$\hat{\chi}_{yyxx}^{(3)}(-\omega_a; \omega_a, -\omega_b, \omega_b) = \frac{1}{6}\Big[\hat{\chi}_{yyxx}^{(3)}(-\omega_a; \omega_a, \omega_b, -\omega_b) + \hat{\chi}_{yyxx}^{(3)}(-\omega_a; \omega_a, -\omega_b, \omega_b)$$

$$+ \hat{\chi}_{yxyx}^{(3)}(-\omega_a; \omega_b, \omega_a, -\omega_b) + \hat{\chi}_{yxxy}^{(3)}(-\omega_a; \omega_b, -\omega_b, \omega_a)$$

$$+ \hat{\chi}_{yxxy}^{(3)}(-\omega_a; -\omega_b, \omega_b, \omega_a) + \hat{\chi}_{yxyx}^{(3)}(-\omega_a; -\omega_b, \omega_a, \omega_b)\Big].$$

$$(10.24b)$$

Again substituting the nonlinear polarizations into the slowly varying envelope approximation and converting the resulting equations to intensity gives

$$\frac{dI_a(z)}{dz} = -3\frac{\omega_a}{n_a n_b \varepsilon_0 c^2}\Im\text{mag}\Big\{\hat{\chi}_{xxyy}^{(3)}(-\omega_a; \omega_a, -\omega_b, \omega_b)\Big\}I_a(z)I_b(z)$$

$$= -\alpha_{2\perp}(-\omega_a; \omega_b)I_a(z)I_b(z),$$

$$\alpha_{2\perp}(-\omega_a; \omega_b) = 3\frac{\omega_a}{n_a n_b \varepsilon_0 c^2}\Im\text{mag}\Big\{\hat{\chi}_{xxyy}^{(3)}(-\omega_a; \omega_a, -\omega_b, \omega_b)\Big\}. \qquad (10.25a)$$

Also,

$$\frac{dI_b(z)}{dz} = -\alpha_{2\perp}(-\omega_b; \omega_a)I_a(z)I_b(z),$$

$$\alpha_{2\perp}(-\omega_b; \omega_a) = 3\frac{\omega_b}{n_a n_b \varepsilon_0 c^2}\Im\text{mag}\Big\{\hat{\chi}_{yyxx}^{(3)}(-\omega_b; \omega_b, -\omega_a, \omega_a)\Big\}. \qquad (10.25b)$$

Combining the "self" and "cross" two-photon absorption terms gives

$$\frac{dI_a(z)}{dz} = -\alpha_{2\|}(-\omega_a; \omega_a)I_a^2(z) - \alpha_{2\perp}(-\omega_a; \omega_b)I_a(z)I_b(z),$$

$$\frac{dI_b(z)}{dz} = -\alpha_{2\|}(-\omega_b; \omega_b)I_b^2(z) - \alpha_{2\perp}(-\omega_b; \omega_a)I_b(z)I_a(z).$$

$$(10.26)$$

Dispersion with frequency of the pertinent susceptibilities does not allow simple relations to be given between the different coefficients.

10.2 NONLINEAR REFRACTION

10.2.1 Single-Beam Input

It was shown in Eq. 10.3a that there is a nonlinear change in the phase of the field due to the real part of $\hat{\chi}^{(3)}$. Integrating this equation and assuming that two-photon absorption is the major loss mechanism gives

$$\phi^{\mathrm{NL}}(z) - \phi^{\mathrm{NL}}(0) = \frac{3\omega}{8nc}\,\Re\mathrm{eal}\left\{\hat{\chi}^{(3)}_{xxxx}(-\omega;\omega,-\omega,\omega)\right\}\int_0^z \rho^2(z')\,dz'$$

$$\rightarrow \quad \phi^{\mathrm{NL}}(z) - \phi^{\mathrm{NL}}(0) = \frac{3\omega}{8nc\alpha_2(-\omega;\omega)}\,\Re\mathrm{eal}\left\{\hat{\chi}^{(3)}_{xxxx}(-\omega;\omega,-\omega,\omega)\right\}$$

$$\times \ln\left[1 + \alpha_{2\|}(-\omega;\omega)\rho^2(0)z\right]. \tag{10.27}$$

NLR is typically of interest in regions where two-photon absorption is small, and neglecting linear absorption and setting $\phi^{\mathrm{NL}}(0) = 0$ gives

$$1 \gg \alpha_{2\|}(-\omega;\omega)\rho^2(0)z \rightarrow \phi^{\mathrm{NL}}(z) = \frac{3\omega}{8nc}\,\Re\mathrm{eal}\left\{\hat{\chi}^{(3)}_{xxxx}(-\omega;\omega,-\omega,\omega)\right\}\rho^2(0)z. \tag{10.28}$$

Defining $\Delta k^{\mathrm{NL}} = d\phi^{\mathrm{NL}}(z)/dz = \Delta n^{\mathrm{NL}}k_{\mathrm{vac}}$, recognizing that $|\mathcal{E}_x(0,\omega)|^2 = \rho^2(0)|$, and adopting the same notation as for the two-photon absorption coefficient gives

$$\Delta n^{\mathrm{NL}} = n_{2\|,\mathrm{E}}(-\omega;\omega)|\mathrm{E}(0,\omega)|^2, \quad n_{2\|,\mathrm{E}}(-\omega;\omega) = \frac{3}{8n}\,\Re\mathrm{eal}\left\{\hat{\chi}^{(3)}_{xxxx}(-\omega;\omega,-\omega,\omega)\right\}, \tag{10.29}$$

where $n_{2\|,\mathrm{E}}(-\omega;\omega)$ is the field-dependent nonlinear refractive index coefficient. Note that another frequently used definition for $n_{2\|,\mathrm{E}}(-\omega;\omega)$ is $\Delta n^{\mathrm{NL}} = n_{2\|,\mathrm{E}}(-\omega;\omega)$ $\overline{|E(\vec{r},t)|^2}$, where $\overline{|E(\vec{r},t)|^2}$ is the time average and so $\Delta n^{\mathrm{NL}} = \frac{1}{2}n_{2\|,\mathrm{E}}(-\omega;\omega)|\mathrm{E}(0,\omega)|^2$. Here, Eq. 10.29 will be used for $n_{2\|,\mathrm{E}}(-\omega;\omega)$. Equation 10.28 can be expressed in terms of the intensity as

$$\Delta n^{\mathrm{NL}} = n_{2\|}(-\omega;\omega)I(0), \quad n_{2\|}(-\omega;\omega) = \frac{3}{4n^2\varepsilon_0 c}\,\Re\mathrm{eal}\left\{\hat{\chi}^{(3)}_{xxxx}(-\omega;\omega,-\omega,\omega)\right\}. \tag{10.30}$$

There is also an alternate notation for the intensity-dependent refractive index coefficient $n_{2\|}(-\omega;\omega)$ in the literature, namely, $\Delta n^{\mathrm{NL}} = \gamma I(0)$, in which γ is given by Eq. 10.30.

In the Kleinman (nonresonant) limit, for the most general three-level case

$$
\Re\mathrm{eal}\left\{\hat{\chi}^{(3)}_{ijk\ell}(-0;0,-0,0)\right\}
$$

$$
= \frac{Nf^{(3)}}{\varepsilon_0\hbar^3}\left[{\sum_{v,n,m}}'\frac{1}{\bar{\omega}_{vg}\bar{\omega}_{mg}\bar{\omega}_{ng}}\{\bar{\mu}_{gv,i}(\bar{\mu}_{vn,l}-\bar{\mu}_{gg,l})(\bar{\mu}_{nm,k}-\bar{\mu}_{gg,k})\bar{\mu}_{mg,j}\right.
$$

$$
+\bar{\mu}_{gv,l}(\bar{\mu}_{vn,i}-\bar{\mu}_{gg,i})(\bar{\mu}_{nm,k}-\bar{\mu}_{gg,k})\bar{\mu}_{mg,j}
$$

$$
+\bar{\mu}_{gv,j}(\bar{\mu}_{vn,k}-\bar{\mu}_{gg,k})(\bar{\mu}_{nm,i}-\bar{\mu}_{gg,i})\bar{\mu}_{mg,l}
$$

$$
+\bar{\mu}_{gv,j}(\bar{\mu}_{vn,k}-\bar{\mu}_{gg,k})(\bar{\mu}_{nm,l}-\bar{\mu}_{gg,l})\bar{\mu}_{mg,i}\}-{\sum_{n,m}}'\left(\frac{1}{\bar{\omega}_{ng}}+\frac{1}{\bar{\omega}_{mg}}\right)
$$

$$
\left.\times\left\{\frac{\bar{\mu}_{gn,i}\,\bar{\mu}_{ng,l}\,\bar{\mu}_{gm,k}\,\bar{\mu}_{mg,j}}{\bar{\omega}_{mg}\bar{\omega}_{ng}}+\frac{\bar{\mu}_{gn,l}\,\bar{\mu}_{ng,i}\,\bar{\mu}_{gm,j}\bar{\mu}_{mg,k}}{\bar{\omega}_{ng}\bar{\omega}_{mg}}\right\}\right], \tag{10.31}
$$

which gives, e.g., for an x-polarized input

$$
\Re\mathrm{eal}\{\hat{\chi}^{(3)}_{xxxx}(-0;0,-0,0)\}=\frac{Nf^{(3)}}{\varepsilon_0\hbar^3}\left[{\sum_{v,n,m}}'\frac{4\bar{\mu}_{gv,x}(\bar{\mu}_{vn,x}-\bar{\mu}_{gg,x})(\bar{\mu}_{nm,x}-\bar{\mu}_{gg,x})\bar{\mu}_{mg,x}}{\bar{\omega}_{vg}\bar{\omega}_{mg}\bar{\omega}_{ng}}\right.
$$

$$
\left.-{\sum_{n,m}}'\left(\frac{1}{\bar{\omega}_{ng}}+\frac{1}{\bar{\omega}_{mg}}\right)\frac{2\bar{\mu}_{gn,x}\,\bar{\mu}_{ng,x}\,\bar{\mu}_{gm,x}\,\bar{\mu}_{mg,x}}{\bar{\omega}_{mg}\bar{\omega}_{ng}}\right].
$$

$$\tag{10.32}$$

Therefore, the nonresonant value of $n_{2\|}(-\omega;\omega)$ is a constant determined solely by the transition dipole moments and the excited-state energies $\hbar\bar{\omega}_{ng}$. Furthermore,

$$
n_{2\|}(-0;\,0)=\frac{3Nf^{(3)}}{n^2\varepsilon_0^2\hbar^3 c}\left[{\sum_{v,n,m}}'\frac{\bar{\mu}_{gv,x}(\bar{\mu}_{vn,x}-\bar{\mu}_{gg,x})(\bar{\mu}_{nm,x}-\bar{\mu}_{gg,x})\bar{\mu}_{mg,x}}{\bar{\omega}_{vg}\bar{\omega}_{mg}\bar{\omega}_{ng}}\right.
$$

$$
\left.-{\sum_{n,m}}'\left(\frac{1}{\bar{\omega}_{ng}}+\frac{1}{\bar{\omega}_{mg}}\right)\frac{\bar{\mu}_{gn,x}\,\bar{\mu}_{ng,x}\,\bar{\mu}_{gm,x}\,\bar{\mu}_{mg,x}}{2\bar{\omega}_{mg}\bar{\omega}_{ng}}\right]. \tag{10.33}
$$

For the two-level model, the pertinent third-order susceptibilities were given in Chapter 9 near one-photon resonance, i.e., $\omega\cong\bar{\omega}_{10}$, and inserting this result into Eq. 10.30 gives

$$
n_{2\|}(-\omega;\omega)\cong-\frac{1}{2n_x^2\varepsilon_0 c}\frac{N}{\varepsilon_0\hbar^3}f^{(3)}|\bar{\mu}_{10}|^4\frac{(\bar{\omega}_{10}-\omega)^3}{[(\bar{\omega}_{10}-\omega)^2+\tau_{10}^{-2}]^3}. \tag{10.34}
$$

Note the minus sign in Eq. 10.31, which indicates that the intensity-dependent index coefficient has the opposite index dispersion to that of the linear refractive index for the dispersion around $\bar{\omega}_{10}$. However, near two-photon resonance, i.e., $2\omega \cong \bar{\omega}_{10}$, the two-photon dispersion in $n_{2\parallel}(-\omega;\omega)$ is

$$n_{2\parallel}(-\omega;\omega) = \frac{2N}{n_x^2 \varepsilon_0^2 c\hbar^3} f^{(3)} |\bar{\mu}_{10}|^2 (\bar{\mu}_{11} - \bar{\mu}_{00})^2 \frac{(\bar{\omega}_{10} - 2\omega)}{\bar{\omega}_{10}^2 [(\bar{\omega}_{10} - 2\omega)^2 + \tau_{10}^{-2}]}, \quad (10.35)$$

which has the same sign as the dispersion in the linear refractive index. Off resonance,

$$n_{2\parallel}(-\omega;\omega) = \frac{Nf^{(3)}\bar{\omega}_{10}}{n_x^2 \varepsilon_0^2 c\hbar^3} \left\{ \frac{3|\bar{\mu}_{10}|^2 (\bar{\mu}_{11} - \bar{\mu}_{00})^2}{(\bar{\omega}_{10}^2 - 4\omega^2)(\bar{\omega}_{10}^2 - \omega^2)} - \frac{|\bar{\mu}_{10}|^4 (3\bar{\omega}_{10}^2 + \omega^2)}{(\bar{\omega}_{10}^2 - \omega^2)^3} \right\}, \quad (10.36)$$

which shows the *destructive interference* between the one- and two-photon contributions. In the Kleinman limit,

$$n_{2\parallel}(-0;0) = 3 \frac{Nf^{(3)}}{n_x^2 \varepsilon_0^2 c\hbar^3 \bar{\omega}_{10}^3} \left\{ |\bar{\mu}_{10}|^2 (\bar{\mu}_{11} - \bar{\mu}_{00})^2 - |\bar{\mu}_{10}|^4 \right\}. \quad (10.37)$$

The sign of the nonlinearity is determined by the ratio $(\bar{\mu}_{11} - \bar{\mu}_{00})^2/|\bar{\mu}_{10}|^2$. If it is greater than unity, the two-photon contribution dominates, and vice versa.

The three-level model is the simplest possible approximation for symmetric molecules. The nonlinear index coefficient near and on one-photon resonance is given by

$$n_{2\parallel}(-\omega;\omega) = \frac{Nf^{(3)}}{2n^2 c\varepsilon_0^2 \hbar^3} |\bar{\mu}_{10}|^2 \left[|\bar{\mu}_{21}|^2 \frac{2(\bar{\omega}_{10} - \omega)^2 (\bar{\omega}_{20} - \bar{\omega}_{10})}{\bar{\omega}_{20}[(\bar{\omega}_{10} - \omega)^2 + \tau_{10}^{-2}]^2 (\bar{\omega}_{20} - 2\bar{\omega}_{10})} \right.$$

$$\left. - |\bar{\mu}_{10}|^2 \frac{(\bar{\omega}_{10} - \omega)^3}{[(\bar{\omega}_{10} - \omega)^2 + \tau_{10}^{-2}]^3} \right]. \quad (10.38)$$

For the off-resonance case,

$$n_{2\parallel}(-\omega;\omega) = \frac{Nf^{(3)}}{n^2 c\varepsilon_0^2 \hbar^3} \left[|\bar{\mu}_{10}|^2 |\bar{\mu}_{21}|^2 \frac{\bar{\omega}_{20}^2 (3\bar{\omega}_{10}^2 + \omega^2) + 4\omega^2 (\bar{\omega}_{20}\bar{\omega}_{10} - 2\bar{\omega}_{10}^2)}{\bar{\omega}_{20}(\bar{\omega}_{20}^2 - 4\omega^2)(\bar{\omega}_{10}^2 - \omega^2)^2} \right.$$

$$\left. - |\bar{\mu}_{10}|^4 \bar{\omega}_{10} \frac{(3\bar{\omega}_{10}^2 + \omega^2)}{(\bar{\omega}_{10}^2 - \omega^2)^3} \right], \quad (10.39)$$

and for the nonresonant nonlinearity,

$$n_{2\parallel}(-\omega;\omega) = 3N \frac{f^{(3)}|\bar{\mu}_{10}|^2}{n^2 c \varepsilon_0^2 \hbar^3 \bar{\omega}_{10}^2} \left[\frac{|\bar{\mu}_{21}|^2}{\bar{\omega}_{20}} - \frac{|\bar{\mu}_{10}|^2}{\bar{\omega}_{10}} \right]. \tag{10.40}$$

Just as in the two-level model, the sign of the nonresonant nonlinearity in the three-level model is also determined by an interference between the one- and two-photon transitions. For $\frac{|\bar{\mu}_{21}|^2}{\bar{\omega}_{20}} > \frac{|\bar{\mu}_{10}|^2}{\bar{\omega}_{10}}$, the two-photon transitions dominate the one-photon transitions, and vice versa.

10.2.2 Two-Beam Input

10.2.2.1 Copolarized Beams ($\omega_a \neq \omega_b$) and ($\omega_a = \omega_b$, not Codirectional, i.e., Different Eigenmodes).
There is now a nonlinear phase shift due to self-refraction and cross refraction:

$$\frac{d\phi_a^{NL}(z)}{dz} = \frac{3\omega_a}{4n_a\varepsilon_0 c^2} \left[2\,\Re\text{eal}\left\{ \hat{\chi}_{xxxx}^{(3)}(-\omega_a;\omega_a,-\omega_b,\omega_b) \right\} \frac{I_b(z)}{n_b} \right.$$

$$\left. + \Re\text{eal}\left\{ \hat{\chi}_{xxxx}^{(3)}(-\omega_a;\omega_a,-\omega_a,\omega_a) \right\} \frac{I_a(z)}{n_a} \right],$$

$$\frac{d\phi_b^{NL}(z)}{dz} = \frac{3\omega_b}{4n_b\varepsilon_0 c^2} \left[2\,\Re\text{eal}\left\{ \hat{\chi}_{xxxx}^{(3)}(-\omega_b;\omega_b,-\omega_a,\omega_a) \right\} \frac{I_a(z)}{n_a} \right.$$

$$\left. + \Re\text{eal}\left\{ \hat{\chi}_{xxxx}^{(3)}(-\omega_b;\omega_b,-\omega_b,\omega_b) \right\} \frac{I_b(z)}{n_b} \right]. \tag{10.41}$$

Defining

$$\frac{d\phi_{a,b}^{NL}(z)}{dz} = k_{vac}(\omega_{a,b})\Delta n_{a,b}^{NL},$$

$$\Delta n_{a,b}^{NL} = n_{2\parallel}(-\omega_{a,b};\omega_{b,a})I_{b,a}(z) + n_{2\parallel}(-\omega_{b,a};\omega_{a,b})I_{a,b}(z) \tag{10.42}$$

gives

$$n_{2\parallel}(-\omega_a;\omega_a) = \frac{3}{4n_a^2\varepsilon_0 c}\Re\text{eal}\left\{ \hat{\chi}_{xxxx}^{(3)}(-\omega_a;\omega_a,-\omega_a,\omega_a) \right\},$$

$$n_{2\parallel}(-\omega_a;\omega_b) = \frac{6}{4n_a n_b\varepsilon_0 c}\Re\text{eal}\left\{ \hat{\chi}_{xxxx}^{(3)}(-\omega_a;\omega_a,-\omega_b,\omega_b) \right\},$$

$$n_{2\parallel}(-\omega_b;\omega_a) = \frac{6}{4n_a n_b\varepsilon_0 c}\Re\text{eal}\left\{ \hat{\chi}_{xxxx}^{(3)}(-\omega_b;\omega_b,-\omega_a,\omega_a) \right\}, \tag{10.43}$$

$$n_{2\parallel}(-\omega_b;\omega_b) = \frac{3}{4n_b^2\varepsilon_0 c}\Re\text{eal}\left\{ \hat{\chi}_{xxxx}^{(3)}(-\omega_b;\omega_b,-\omega_b,\omega_b) \right\}.$$

Note that in the Kleinman (nonresonant) limit, all the individual susceptibilities are the same, and so the ratios of the nonlinear index coefficients are determined by the *number* of nonlinear susceptibilities that contribute to each effect. For example, three susceptibilities (Eq. 10.1) contribute to the nonlinearity for a single-beam input, whereas six susceptibilities (Eq. 10.15) contribute to a two-beam input. For a single polarization, this leads to

$$n_{2\|}(-\omega_a;\omega_a) = n_{2\|}(-\omega_b;\omega_b) = \frac{1}{2}n_{2\|}(-\omega_a;\omega_b) = \frac{1}{2}n_{2\|}(-\omega_b;\omega_a), \quad (10.44)$$

a considerable simplification.

10.2.2.2 Orthogonally Polarized Beams.

Example 1

$$\omega_a \neq \omega_b$$

The appropriate $\hat{\chi}_{xxyy}^{(3)}(-\omega_a;\omega_a,\omega_b,-\omega_b)$ is written as

$$\hat{\chi}_{xxyy}^{(3)}(-\omega_a;\omega_a,\omega_b,-\omega_b) = \frac{1}{6}\left\{\hat{\chi}_{xxyy}^{(3)}(-\omega_a;\omega_a,\omega_b,-\omega_b) + \hat{\chi}_{xxyy}^{(3)}(-\omega_a;\omega_a,-\omega_b,\omega_b)\right.$$

$$+\hat{\chi}_{xyxy}^{(3)}(-\omega_a;\omega_b,\omega_a,-\omega_b) + \hat{\chi}_{xyxy}^{(3)}(-\omega_a;-\omega_b,\omega_a,\omega_b)$$

$$\left.+\hat{\chi}_{xyyx}^{(3)}(-\omega_a;-\omega_b,\omega_b,\omega_a) + \hat{\chi}_{xyyx}^{(3)}(-\omega_a;\omega_b,-\omega_b,\omega_a)\right\},$$

$$(10.45)$$

which gives

$$n_{2\perp}(-\omega_a;\omega_b) = \frac{6}{4n_a n_b \varepsilon_0 c}\Re\text{eal}\left\{\hat{\chi}_{xxyy}^{(3)}(-\omega_a;\omega_a,-\omega_b,\omega_b)\right\}. \quad (10.46a)$$

Similarly,

$$n_{2\perp}(-\omega_b;\omega_a) = \frac{6}{4n_a n_b \varepsilon_0 c}\Re\text{eal}\left\{\hat{\chi}_{xxyy}^{(3)}(-\omega_b;\omega_b,-\omega_a,\omega_a)\right\}. \quad (10.46b)$$

Example 2

$\omega_b = \omega_a$ (encountered in fibers, which frequently have a small birefringence)

$$n_{2\perp}(-\omega_a;\omega_a) = \frac{6}{4n_a^2 \varepsilon_0 c}\Re\text{eal}\left\{\hat{\chi}_{xxyy}^{(3)}(-\omega_a;\omega_a,-\omega_a,\omega_a)\right\}$$

$$= \frac{2}{4n_a^2 \varepsilon_0 c}\Re\text{eal}\left\{\hat{\chi}_{xxxx}^{(3)}(-\omega_a;\omega_a,-\omega_a,\omega_a)\right\} = \frac{2}{3}n_{2\|}(-\omega_a;\omega_a)$$

$$\rightarrow \Delta n_x = n_{2\|}(-\omega_a;\omega_a)\left\{I_x(0) + \frac{2}{3}I_y(0)\right\}. \quad (10.47)$$

Note, however, that as shown in Chapter 8, there is usually a four-wave mixing term as well in the nonlinear polarization, which leads to energy exchange between the polarizations if they are coherent with respect to each other.

10.3 USEFUL NLR FORMULAS AND EXAMPLES (ISOTROPIC MEDIA)

The evaluation of nonlinear refractive index changes will be illustrated by a number of examples for isotropic media. From Eq. 8.61 for a three-eigenmode input,

$$
\mathbf{P}_i^{(3)}(\omega_p + \omega_q + \omega_r)
$$

$$
= \frac{6}{4}\varepsilon_0 \Big\{ \hat{\chi}_{1122}^{(3)}(-[\omega_p + \omega_q + \omega_r], \omega_p, \omega_q, \omega_r)\delta_{ij}\delta_{k\ell}E_j(\omega_p)E_k(\omega_q)E_\ell(\omega_r)
$$

$$
+ \hat{\chi}_{1212}^{(3)}(-[\omega_p + \omega_q + \omega_r], \omega_p, \omega_q, \omega_r)\delta_{ik}\delta_{j\ell}E_j(\omega_p)E_k(\omega_q)E_\ell(\omega_r)
$$

$$
+ \hat{\chi}_{1221}^{(3)}(-[\omega_p + \omega_q + \omega_r], \omega_p, \omega_q, \omega_r)\delta_{i\ell}\delta_{jk}E_j(\omega_p)E_k(\omega_q)E_\ell(\omega_r) \Big\},
$$

$$(10.48)$$

where there are six terms due to permutation of the frequencies for each susceptibility; i.e.,

$$
\hat{\chi}_{1122}^{(3)}(-[\omega_p + \omega_q + \omega_r], \omega_p, \omega_q, \omega_r)
$$

$$
= \frac{1}{6}\Big\{ \hat{\chi}_{1122}^{(3)}(-[\omega_p + \omega_q + \omega_r], \omega_p, \omega_q, \omega_r) + \hat{\chi}_{1122}^{(3)}(-[\omega_p + \omega_q + \omega_r]; \omega_p, \omega_r, \omega_q)
$$

$$
+ \hat{\chi}_{1122}^{(3)}(-[\omega_p + \omega_q + \omega_r]; \omega_q, \omega_r, \omega_p) + \hat{\chi}_{1122}^{(3)}(-[\omega_p + \omega_q + \omega_r]; \omega_q, \omega_p, \omega_r)
$$

$$
+ \hat{\chi}_{1122}^{(3)}(-[\omega_p + \omega_q + \omega_r]; \omega_r, \omega_q, \omega_p) + \hat{\chi}_{1122}^{(3)}(-[\omega_p + \omega_q + \omega_r], \omega_r, \omega_p, \omega_q) \Big\}, \text{etc.}
$$

$$(10.49)$$

10.3.1 Two-Frequency Input (Three Eigenmodes with Frequencies: $\omega_a, \omega_b,$ and $-\omega_b$)

Example 1

Copolarized (*x*-axis) case

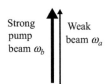

Strong pump beam ω_b Weak beam ω_a

This case is straightforward since $\delta_{ij}\delta_{k\ell} = \delta_{xx}\delta_{xx}$, $\delta_{ik}\delta_{j\ell} = \delta_{xx}\delta_{xx}$, and $\delta_{i\ell}\delta_{jk} = \delta_{xx}\delta_{xx}$.

$$
\begin{aligned}
\mathrm{P}_i^{(3)}(\omega_a) &= \frac{6}{4}\varepsilon_0\left\{\hat{\chi}_{1122}^{(3)}(-\omega_a;\omega_a,-\omega_b,\omega_b) + \hat{\chi}_{1212}^{(3)}(-\omega_a;\omega_a,-\omega_b,\omega_b)\right. \\
&\quad \left. + \hat{\chi}_{1221}^{(3)}(-\omega_a;\omega_a,-\omega_b,\omega_b)\right\}\mathrm{E}_x(\omega_a)\mathrm{E}_x(\omega_b)\mathrm{E}_x^*(\omega_b) \\
&= \frac{6}{4}\varepsilon_0\hat{\chi}_{1111}^{(3)}(-\omega_a;\omega_a,-\omega_b,\omega_b)\mathrm{E}_x(\omega_a)|\mathrm{E}_x(\omega_b)|^2.
\end{aligned}
\tag{10.50}
$$

Example 2

Birefringence induced in a weak beam

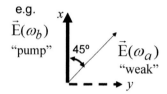

The calculation of $\mathrm{P}_x^{(3)}(\omega_a)$ is straightforward since $ijk\ell$ are all equal to x, which is identical to the previous case. However, for $\mathrm{P}_y^{(3)}(\omega_a)$, note that for $\hat{\chi}_{1122}^{(3)}$, $ijk\ell = yyxx$ and so the field product is $\mathrm{E}_y(\omega_a)\mathrm{E}_x(\omega_b)\mathrm{E}_x(-\omega_b)$, for $\hat{\chi}_{1212}^{(3)}$ the field product is $\mathrm{E}_x(\omega_a)\mathrm{E}_y(\omega_b)\mathrm{E}_x(-\omega_b)$, and for $\hat{\chi}_{1221}^{(3)}$ the field product is $\mathrm{E}_x(\omega_a)\mathrm{E}_x(\omega_b)\mathrm{E}_y(-\omega_b)$. Since $\mathrm{E}_y(\pm\omega_b) = 0$, only the first term survives.

$$
\mathrm{P}_x^{(3)}(\omega_a) = \frac{6}{4}\varepsilon_0\left\{\hat{\chi}_{1122}^{(3)} + \hat{\chi}_{1212}^{(3)} + \hat{\chi}_{1221}^{(3)}\right\}\mathrm{E}_x(\omega_a)|\mathrm{E}(\omega_b)|^2,
\tag{10.51a}
$$

$$
\begin{aligned}
\mathrm{P}_y^{(3)}(\omega_a) &= \frac{6}{4}\varepsilon_0\left\{\hat{\chi}_{1122}^{(3)}\mathrm{E}_y(\omega_a)|\mathrm{E}(\omega_b)|^2 + \hat{\chi}_{1212}^{(3)}\mathrm{E}_x(\omega_a)\mathrm{E}_y(\omega_b)\mathrm{E}_x(-\omega_b)\right. \\
&\quad \left. + \hat{\chi}_{1221}^{(3)}\mathrm{E}_x(\omega_a)\mathrm{E}_x(\omega_b)\mathrm{E}_y(-\omega_b)\right\}
\end{aligned}
\tag{10.51b}
$$

$$
\Rightarrow \mathrm{P}_y^{(3)}(\omega_a) = \frac{6}{4}\varepsilon_0\left\{\hat{\chi}_{1122}^{(3)}\mathrm{E}_y(\omega_a)|\mathrm{E}(\omega_b)|^2\right\}.
\tag{10.52}
$$

The total polarization in the medium is

$$
\begin{aligned}
\mathrm{P}_x(\omega_a) &= [\mathrm{P}_x^{(1)}(\omega_a) + \mathrm{P}_x^{(3)}(\omega_a)] \\
&= \varepsilon_0\left[(n^2-1) + \frac{6}{4}\varepsilon_0\left\{\hat{\chi}_{1122}^{(3)} + \hat{\chi}_{1212}^{(3)} + \hat{\chi}_{1221}^{(3)}\right\}|\mathrm{E}(\omega_b)|^2\right]\mathrm{E}_x(\omega_a)] \\
&= \varepsilon_0[(n+\Delta n_x)^2 - 1]\mathrm{E}_x(\omega_a) = \varepsilon_0[(n^2+2\Delta n_x n + \Delta n_x^2 - 1]\mathrm{E}_x(\omega_a)].
\end{aligned}
\tag{10.53}
$$

Assuming that Δn_x^2 is negligible relative to $2\Delta n_x n$

$$\Rightarrow \quad \Delta n_x = \frac{6}{8n}\left\{\widehat{\chi}^{(3)}_{1122} + \widehat{\chi}^{(3)}_{1212} + \widehat{\chi}^{(3)}_{1221}\right\}|E(\omega_b)|^2. \tag{10.54a}$$

Similarly,

$$\Delta n_y = \frac{6}{8n}\widehat{\chi}^{(3)}_{1122}|E(\omega_b)|^2. \tag{10.54b}$$

Therefore the induced birefringence induced at frequency ω_a by the beam of frequency ω_b is

$$\Delta n_x - \Delta n_y = \frac{3}{4n}\left\{\widehat{\chi}^{(3)}_{1212} + \widehat{\chi}^{(3)}_{1221}\right\}|E(\omega_b)|^2. \tag{10.55}$$

That is, a strong pump beam produces a nonlinear birefringence in a medium that normally does not have one at low intensities. This is another example of a concept valid in linear optics; i.e., an isotropic material exhibits no birefringence, which is broken in nonlinear optics. The reason, of course, is that the strong pump beam changes the symmetry of the isotropic medium to uniaxial. In the nonresonant limit, $\widehat{\chi}^{(3)}_{1122} = \widehat{\chi}^{(3)}_{1212} = \widehat{\chi}^{(3)}_{1221}$ and so

$$\Delta n_x - \Delta n_y = \frac{1}{2n}\widehat{\chi}^{(3)}_{1111}|E(\omega_b)|^2. \tag{10.56}$$

Example 3

What is the index change induced by a strong circularly polarized beam in a weak linearly polarized beam?

$$\vec{E}(\omega_a) = \hat{e}_x E_x(\omega_a), \qquad \vec{E}(\omega_b) = \frac{1}{\sqrt{2}}E(\omega_b)(\hat{e}_x + i\hat{e}_y).$$

It is left to the reader to fill in the details as an exercise.

$$P_x^{(3)}(\omega_a) = \frac{6}{4}\varepsilon_0\left[\widehat{\chi}^{(3)}_{1122}E(\omega_a)\left\{\frac{1}{2}|E(\omega_b)|^2 + \frac{1}{2}|E(\omega_b)|^2\right\}\right.$$

$$\left.+\widehat{\chi}^{(3)}_{1212}E(\omega_a)\frac{1}{2}|E(\omega_b)|^2 + \widehat{\chi}^{(3)}_{1221}E(\omega_a)\frac{1}{2}|E(\omega_b)|^2\right],$$

$$P_x^{(3)}(\omega_a) = \frac{6}{4}\varepsilon_0\left\{\hat{\chi}_{1122}^{(3)} + \frac{1}{2}(\hat{\chi}_{1212}^{(3)} + \hat{\chi}_{1221}^{(3)})\right\}E_x(\omega_a)|E(\omega_b)|^2$$

$$\Rightarrow \Delta n_x = \frac{3}{4n}\left\{\hat{\chi}_{1122}^{(3)} + \frac{1}{2}(\hat{\chi}_{1212}^{(3)} + \hat{\chi}_{1221}^{(3)})\right\}|E(\omega_b)|^2. \tag{10.57}$$

For two weak orthogonally polarized beams,

$$\Rightarrow \quad \Delta n_y = \frac{3}{4n}\left\{\hat{\chi}_{1122}^{(3)} + \frac{1}{2}(\hat{\chi}_{1212}^{(3)} + \hat{\chi}_{1221}^{(3)})\right\}|E(\omega_b)|^2 \rightarrow \Delta n_x - \Delta n_y = 0. \tag{10.58}$$

10.3.2 Single-Frequency Beam Input (Two Eigenmodes)

$$\hat{\chi}_{1122}^{(3)}(-\omega;\omega,\omega,-\omega) = \frac{1}{3}\Big[\hat{\chi}_{1122}^{(3)}(-\omega;\omega,\omega,-\omega) + \hat{\chi}_{1122}^{(3)}(-\omega;\omega,-\omega,\omega)$$

$$+\hat{\chi}_{1122}^{(3)}(-\omega;-\omega,\omega,\omega)\Big],$$

$$\hat{\chi}_{1212}^{(3)}(-\omega;\omega,\omega,-\omega) = \frac{1}{3}\Big[\hat{\chi}_{1212}^{(3)}(-\omega;\omega,\omega,-\omega) + \hat{\chi}_{1212}^{(3)}(-\omega;\omega,-\omega,\omega)$$

$$+\hat{\chi}_{1212}^{(3)}(-\omega;-\omega,\omega,\omega)\Big],$$

$$\hat{\chi}_{1221}^{(3)}(-\omega;\omega,\omega,-\omega) = \frac{1}{3}\Big[\hat{\chi}_{1221}^{(3)}(-\omega;\omega,\omega,-\omega) + \hat{\chi}_{1221}^{(3)}(-\omega;\omega,-\omega,\omega)$$

$$+\hat{\chi}_{1221}^{(3)}(-\omega;-\omega,\omega,\omega)\Big]. \tag{10.59}$$

Therefore

$$P_i^{(3)}(\omega) = \frac{3}{4}\varepsilon_0\Big[\hat{\chi}_{1122}^{(3)}(-\omega;\omega,\omega,-\omega)E_i(\omega)E_j(\omega)E_j(-\omega)$$

$$+\hat{\chi}_{1212}^{(3)}(-\omega;\omega,\omega,-\omega)E_j(\omega)E_i(\omega)E_j(-\omega)$$

$$+\hat{\chi}_{1221}^{(3)}(-\omega;\omega,\omega,-\omega)E_j(\omega)E_j(\omega)E_i(-\omega)\Big]. \tag{10.60}$$

Equation 10.60 can be written as

$$P_i^{(3)}(\omega) = \varepsilon_0\Big[AE_i(\omega)|E(\omega)|^2 + \frac{1}{2}BE_i(-\omega)E_j(\omega)E_j(\omega)\Big],$$

$$A = \frac{3}{4}\varepsilon_0\Big[\hat{\chi}_{1122}^{(3)}(-\omega;\omega,-\omega,\omega) + \hat{\chi}_{1212}^{(3)}(-\omega;\omega,-\omega,\omega)\Big],$$

$$B = \frac{3}{2}\varepsilon_0\hat{\chi}_{1221}^{(3)}(-\omega;\omega,-\omega,\omega), \tag{10.61}$$

where A and B are the two independent parameters that describe the nonlinearity for this process in an isotropic material. Note that in the nonresonant limit, $A = B$. The ratio B/A turns out to be useful for describing other nonresonant nonlinearities such as molecular reorientation ($B/A = 6$) and electrostriction ($B/A = 0$), discussed in Chapter 12.

Example 4

Self-phase modulation (in the nonresonant limit)

$$
\begin{aligned}
P_x^{(3)} &= \frac{3}{4}\varepsilon_0\left[\hat{\chi}_{1122}^{(3)} + \hat{\chi}_{1212}^{(3)}\right]E_x|E_x|^2 + \frac{3}{4}\varepsilon_0\hat{\chi}_{1221}^{(3)}E_x|E_x|^2 \\
&= \frac{3}{4}\varepsilon_0\left\{\hat{\chi}_{1122}^{(3)} + \hat{\chi}_{1212}^{(3)} + \hat{\chi}_{1221}^{(3)}\right\}E_x|E_x|^2 \\
&= \frac{3}{4}\varepsilon_0\hat{\chi}_{1111}^{(3)}E_x|E_x|^2,
\end{aligned}
\tag{10.62}
$$

which agrees with the previous results.

Example 5

What is the change in index experienced by circularly polarized light?

Let us first discuss some useful relations and definitions for circularly polarized light.

$$
\vec{E} = E_+\vec{\sigma}_+ + E_-\vec{\sigma}_-, \qquad \vec{\sigma}_\pm = \frac{\hat{e}_x \pm i\hat{e}_y}{\sqrt{2}}, \qquad \vec{\sigma}_\pm^* = \sigma_\mp,
$$

$$
\vec{\sigma}_\pm \cdot \vec{\sigma}_\pm = 0, \qquad \vec{\sigma}_\pm \cdot \vec{\sigma}_\mp = \frac{1}{2}(1+i)(1-i) = 1,
$$

$$
\vec{E}^* \cdot \vec{E} = (E_+^*\vec{\sigma}_+^* + E_-^*\vec{\sigma}_-^*) \cdot (E_+\vec{\sigma}_+ + E_-\vec{\sigma}_-) = E_+^*E_+ + E_-^*E_- = |E_+|^2 + |E_-|^2,
$$

$$
\vec{E} \cdot \vec{E} = (E_+\vec{\sigma}_+ + E_-\vec{\sigma}_-) \cdot (E_+\vec{\sigma}_+ + E_-\vec{\sigma}_-) = E_+E_- + E_+E_- = 2E_+E_-
\tag{10.63}
$$

$$
\to \vec{P}^{(3)} = A(|E_+|^2 + |E_-|^2)\vec{E} + B(E_+E_-)\vec{E}^*
$$

$$
= A(|E_+|^2 + |E_-|^2)(E_+\vec{\sigma}_+ + E_-\vec{\sigma}_-) + BE_+E_-(E_+^*\vec{\sigma}_+^* + E_-^*\vec{\sigma}_-^*),
$$

$$
\vec{P}^{(3)} = P_+\vec{\sigma}_+ + P_-\vec{\sigma}_-.
\tag{10.64}
$$

Equation 10.64 gives

$$
P_+ = A(|E_+|^2 + |E_-|^2)E_+ + B(E_+E_-)E_-^* = \{A|E_+|^2 + (A+B)|E_-|^2\}E_+,
$$

$$
P_- = \{A|E_-|^2 + (A+B)|E_+|^2\}E_-.
\tag{10.65}
$$

Therefore, the refractive index changes for the two circularly polarized waves are

$$\Delta n_+ = \frac{1}{2n}\left\{A|E_+|^2 + (A+B)|E_-|^2\right\}, \qquad \Delta n_- = \frac{1}{2n}\left\{A|E_-|^2 + (A+B)|E_+|^2\right\}.$$

$$(10.66)$$

The phase velocities of the two circularly polarized waves are changed by different amounts only if $|E_+|^2 \neq |E_-|^2$; i.e., they carry different powers.

PROBLEMS

1. Consider light traveling in a birefringent medium, e.g., a fiber with a birefringence of 10^{-5}. Polarization instability occurs when an intense beam is launched along one of the axis and so the birefringence disappears. What intensity is required for fused silica fibers?

2. Starting from the general expression

$$P_i^{(3)} = \varepsilon_0\left\{\frac{3}{2}\hat{\chi}_{1122}^{(3)}E_i(E_jE_j^*) + \frac{3}{4}\hat{\chi}_{1221}^{(3)}E_i^*(E_jE_j)\right\},$$

find the refractive index change induced in an isotropic medium by a *strong* elliptically polarized beam at frequency ω_a of the form

$$\vec{E} = \frac{1}{\sqrt{5}}(2\hat{e}_x + i\hat{e}_y)E_o,$$

where $\hat{e}_x$ and $\hat{e}_y$ are unit vectors.

3. Consider a strong pump beam with frequency ω_a polarized along the x-axis with a right circularly polarized probe beam of intensity I_+ traveling at a small relative angle through the pump beam in an isotropic medium.

   (a) Find the two-photon absorption coefficient in the equation

   $$\frac{dI_+(z)}{dz} = -\alpha_2 I_+ I_{\text{pump}}.$$

   (b) Show that a left circularly polarized beam at the frequency ω_b is nonlinearly generated. What does this mean for the total field at ω_b?

4. A beam of frequency ω_a (strong pump beam) is incident into an isotropic Kerr medium. It is *elliptically polarized* with polarization $(\hat{e}_x + 2i\hat{e}_y)/\sqrt{5}$.

   (a) Show that the nonlinear polarization induced in a weak beam at frequency ω_b ($\omega_b \neq \omega_a$) by the strong input beam is a complex number when the weak beam is linearly polarized at 45° between the x- and y-axes. What does this mean for the beam at ω_b? [Hint: Do *not* assume the Kleinman limit.]

(b) Show that in the Kleinman limit the net birefringence induced between the x- and y-components of the weak beam is always zero.

(c) What is the index change experienced by a weak y-polarized probe beam of frequency $\omega_b \neq \omega_a$.

(d) What is the index change experienced by a weak x-polarized probe beam of frequency $\omega_b \neq \omega_a$.

(e) Why is this nonlinear polarization along x different in (a) and (c)?

5. A beam of frequency ω_a (strong pump beam) is incident into an isotropic medium. It is *elliptically polarized* with polarization $(\hat{e}_x + 2i\hat{e}_y)/\sqrt{5}$. What is the linear birefringence induced by the strong input beam between the x and y components of a weak beam with a frequency $\omega_b \neq \omega_a$ polarized at 60° from the x-axis between the x- and y-axes.

6. Calculate from first principles the intensity-dependent refractive index coefficient $n_{2\parallel}(-\omega; \omega)$ for a two-level model in the vicinity of the two resonances that occur, one at $\bar{\omega}_{10} \cong \omega$ and one at $2\omega \cong \bar{\omega}_{10}$ valid for $1 \geq \bar{\tau}_{10}|\bar{\omega}_{10} - \omega|$ for the first case and $1 \geq \bar{\tau}_{10}|\bar{\omega}_{10} - 2\omega|$ for the second case.

(a) What is the value of n_2 at $\omega = \bar{\omega}_{10}$?

(b) What is the value of n_2 at $2\omega = \omega_{10}$?

7. Two circularly polarized beams—one rotating counterclockwise (frequency ω_a) and the other rotating clockwise (frequency ω_b)—propagate together in a Kerr isotropic material.

(a) Show that the induced intensity-dependent birefringence $\Delta n_+ - \Delta n_- = 0$, where $\Delta n_+ \propto I(\omega_a)$ and $\Delta n_- \propto I(\omega_b)$ and "+" and "−" refer to the clockwise and counterclockwise, respectively.

(b) Show that the counterclockwise beam becomes progressively more elliptical with the propagation distance. (It is sufficient to show that a clockwise beam is induced at ω_a, which grows with the distance, and is $\pi/2$ out of phase with the counterclockwise beam.) [Hint: Do *not* assume the Kleinman limit in either case.]

8. Consider a strong right circularly polarized pump beam with frequency ω_a and intensity I_+ and a weak probe beam polarized along the x-axis with intensity I_x traveling in the same direction (z-axis) in an isotropic Kerr medium.

(a) Find the two-photon absorption coefficient in the equation

$$\frac{dI_z(z)}{dz} = -\alpha_2 I_z I_+.$$

(b) Show that a left circularly polarized beam at the frequency ω_b is nonlinearly generated. What does this mean for the total field at ω_b?

SUGGESTED FURTHER READING

N. Bloembergen, Nonlinear Optics: A Lecture Note and Reprint Volume (W. A. Benjamin, New York, 1965).

R. W. Boyd, Nonlinear Optics, 3rd Edition (Academic Press, Burlington, MA, 2008).

D. N. Christodoulides, I. C. Khoo, G. J. Salamo, G. I. Stegeman, and E. W. Van Stryland, "Nonlinear refraction and absorption: mechanisms and magnitudes," Adv. Opt., **2**, 60–200 (2010).

F. A. Hopf and G. I. Stegeman, Applied Classical Electrodynamics, Volume 2: Nonlinear Optics (John Wiley & Sons, New York, 1985).

Y. R. Shen, Principles of Nonlinear Optics (John Wiley & Sons, New York, 1984).

CHAPTER 11

Condensed Matter Third-Order Nonlinearities due to Electronic Transitions

Of all the many components necessary for a successful application of third-order nonlinear optics, by far the most critical is the choice of nonlinear material.

The previous discussions were most relevant to the ultrafast local response to optical fields of the electron clouds associated with electronic levels in atoms and molecules. Probably the most important consequence of third-order nonlinearities is that they lead to an intensity-dependent change in the refractive index and absorption of a material, but not necessarily changes that are linear in the intensity. It is important to understand the physics of the various mechanisms because their nonlinear response deviates from the well-known Kerr nonlinearity discussed in Chapters 8–10, which is local and instantaneous. No material is "ideal," not even when the dominant mechanism is the Kerr nonlinearity. Here the focus is specifically on the physics of nonlinear optical mechanisms that produce a nonlinear Δn and $\Delta \alpha$. The "turn-on" and "turn-off" response times vary dramatically with the nonlinear mechanism over many orders of magnitude. The fastest are due to electronic transitions (femtoseconds), i.e., the Kerr effect, and the slowest are associated with liquid crystals, photorefractive media, or thermal effects (microseconds or even seconds).

The local molecular environment in condensed matter can lead to an actual distribution of energy levels around the isolated molecule energy level. This can result in inhomogeneous spectral broadening unrelated to the lifetime of the excited state. Furthermore, intermolecular interactions can affect the lifetimes of excited states, again leading to spectral broadening. These effects are reflected in both linear and nonlinear dispersion and absorption spectra.

Nonlinear Optics: Phenomena, Materials, and Devices, George I. Stegeman and Robert A. Stegeman.
© 2012 John Wiley & Sons, Inc. Published 2012 by John Wiley & Sons, Inc.

The following nonlinear effects due to electronic transitions in condensed matter will be discussed in this chapter:

1. Electronic nonlinearities involving discrete states (Kerr effect): molecules, conjugated molecules and polymers, charge transfer molecules, excited-state nonlinearities
2. Semiconductor nonlinearities: Passive nonlinearities (excitation of carriers), active nonlinearities (with gain), ultrafast passive nonlinearities (Kerr effect and so on), quantum confinement
3. "Glass" nonlinearities

Much of the material on condensed matter nonlinearities in this chapter is taken from the review paper by Christodoulides et al. (1), and more detailed discussions as well as many additional references can be found there.

11.1 DEVICE-BASED NONLINEAR MATERIAL FIGURES OF MERIT

The key question is: What parameters determine whether a material is useful for third-order nonlinear optics? The key trade-off is between the nonlinear index change Δn^{NL} and the total loss $\alpha = \alpha_1 + \Delta \alpha^{\mathrm{NL}}$, including scattering. For efficient all-optical phenomena such as solitons and devices such as all-optical switches, a nonlinear phase shift $\Delta \phi^{\mathrm{NL}} = \pi \rightarrow 2\pi$, depending on the interaction of interest, is needed over some sample length L. This condition can be expressed as (2)

$$\Delta \phi^{\mathrm{NL}} = k_{\mathrm{vac}} \int_0^L \Delta n^{\mathrm{NL}}[I(z)]\, dz = \pi \rightarrow 2\pi. \tag{11.1}$$

For the Kerr nonlinearity $\Delta n^{\mathrm{NL}}[I(z)] = n_{2\|}(-\omega; \omega)I(z)$, and assuming both that linear absorption (α_1) is the limiting factor and that the sample can be long enough so that $L \gg \alpha_1^{-1}$, we obtain

$$\Delta \phi^{\mathrm{NL}} = k_{\mathrm{vac}} n_{2\|}(-\omega; \omega) \int_0^L I_0\, e^{-\alpha_1 z}\, dz \xrightarrow{L \rightarrow \infty} \Delta \phi^{\mathrm{NL}} = \frac{k_{\mathrm{vac}} n_{2\|}(-\omega; \omega) I_0}{\alpha_1}. \tag{11.2}$$

Defining a figure of merit (FOM) W that requires a phase shift of 2π as

$$W = \frac{k_{\mathrm{vac}} n_{2\|}(-\omega; \omega) I_0}{\Delta \phi^{\mathrm{NL}} \alpha_1} = \frac{n_{2\|}(-\omega; \omega) I_0}{\lambda_{\mathrm{vac}} \alpha_1}, \tag{11.3}$$

we need $W > 1$ at an operating intensity I_0. For an arbitrary nonlinearity with $\Delta n^{\mathrm{NL}}(I)$,

$$W = \frac{\Delta n(I)}{\lambda_{\text{vac}} \alpha_1} > 1 \tag{11.4}$$

is needed. In all materials there is an upper limit to $\Delta n^{\text{NL}}(I) = \Delta n^{\text{NL}}_{\text{sat}}$ due to "saturation" of the index change or material damage. This inspires another similar FOM:

$$W_{\text{sat}} = \frac{\Delta n^{\text{NL}}_{\text{sat}}}{\lambda_{\text{vac}} \alpha_1} > 1. \tag{11.5}$$

If the limiting loss is due to multiphoton absorption, specifically two-photon absorption, α_1 is replaced by $\alpha_{2||}(-\omega; \omega)I(z)$ and a new FOM, T, is introduced, which in the Kerr case has the simple form

$$T = \frac{\lambda_{\text{vac}} \alpha_{2||}(-\omega; \omega)}{n_{2||}(-\omega; \omega)} < 1. \tag{11.6}$$

Higher order Kerr nonlinear refraction and absorption can lead, in principle, to additional FOMs; i.e.,

$$\begin{aligned}
\Delta n^{\text{NL}} &= n_{2||}(-\omega; \omega)I + n_{3||}(-\omega; \omega)I^2 + \cdots, \\
\Delta \alpha^{\text{NL}} &= \alpha_{2||}(-\omega; \omega)I + \alpha_{3||}(-\omega; \omega)I^2 + \cdots.
\end{aligned} \tag{11.7}$$

It is possible to define a "global" FOM as (3)

$$\text{Global FOM} = \frac{\Delta n^{\text{NL}}(I)}{\lambda_{\text{vac}} \left[\sum_{p=1} \alpha_{p||}(-\omega; \omega)I^{p-1} \right]}, \tag{11.8}$$

which should be greater than unity at the wavelength of interest. The important point is that all-optical phenomena based on nonlinear refraction need to be operated in low net absorption spectral regions.

11.2 LOCAL VERSUS NONLOCAL NONLINEARITIES IN SPACE AND TIME

The distinctions between local and nonlocal nonlinear interactions are difficult to quantify. The nonlinear polarization induced in a material was given in Eq. 1.4. Since $\Delta n_i^{\text{NL}}(\vec{r}, t)$ is proportional to $P_i^{\text{NL}}(\vec{r}, t)$, it is useful, but certainly not rigorous, to define in analogy with Eq. 1.4 the third-order nonlinear refractive index change in terms of an effective (not normalized) susceptibility $h_{ijk\ell}^{(3)}$ as

$$\begin{aligned}
\Delta n_i^{\text{NL}}(\vec{r}, t) = \int_{-\infty}^{\infty} \int_{-\infty}^{\infty} \int_{-\infty}^{\infty} \int_{-\infty}^{t} \int_{-\infty}^{t} \int_{-\infty}^{t} h_{ijk\ell}^{(3)}(\vec{r} - \vec{r}', \vec{r} - \vec{r}'', \vec{r} - \vec{r}'''; t - t', t - t'', t - t''') \\
\times E_j(\vec{r}', t') E_k(\vec{r}'', t'') E_\ell(\vec{r}''', t''') \, d\vec{r}' \, d\vec{r}'' \, d\vec{r}''' \, dt' \, dt'' \, dt'''.
\end{aligned} \tag{11.9}$$

For nonlinearity localized in space,

$$h^{(3)}_{ijk\ell}(\vec{r} - \vec{r}', \vec{r} - \vec{r}'', \vec{r} - \vec{r}'''; t - t', t - t'', t - t''')$$
$$= g^{(3)}_{ijk\ell}(\vec{r}; t - t', t - t'', t - t''')\delta(\vec{r} - \vec{r}')\delta(\vec{r} - \vec{r}'')\delta(\vec{r} - \vec{r}'''), \qquad (11.9)$$

and for locality in both space and time,

$$h^{(3)}_{ijk\ell}(\vec{r} - \vec{r}', \vec{r} - \vec{r}'', \vec{r} - \vec{r}'''; t - t', t - t'', t - t''') = f^{(3)}_{ijk\ell}(\vec{r}; t)\delta(\vec{r} - \vec{r}')\delta(\vec{r} - \vec{r}'')\delta(\vec{r} - \vec{r}''')$$
$$\times \delta(t - t')\delta(t - t'')\delta(t - t'''). \qquad (11.11)$$

However, physics requires that the actual width of the δ functions in space and time be quantified by the physical processes involved.

11.2.1 Nonlocality in Space

The simplest way to describe a "local" nonlinear optical interaction in space involving an intensity-dependent change in the refractive index or absorption is to require that the effect be confined to the atom or molecule where the fundamental process occurs. Due to the intermolecular interactions that occur in condensed matter, this is clearly applicable only to a dilute gas at absolute zero of temperature. From an operational perspective, it is better to define a "local" interaction (*1*) as one that occurs over a volume containing many atoms or molecules, but small on the scale of the wavelength of the radiation causing it and (*2*) such that the effects of the interaction diffuse only in space over distances much smaller than an optical wavelength.

An obvious example of "nonlocality" is the generation of carriers (electrons) in semiconductors due to the absorption of photons whose energy exceeds the bandgap. In this case the electrons (in the conduction band) and the positively charged holes (in the valence band) can migrate over some distance taking the effective change in the refractive index with them (as shown in Fig. 11.1) before the electrons return to the valence band. A similar process occurs in photorefractive media.

Another example involving diffusive processes is heating via photon absorption, in which the temperature changes locally, and subsequently the temperature change

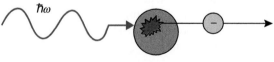

Photon absorbed

FIGURE 11.1   The absorption of a photon in a semiconductor releases an electron into the conduction band where it can propagate some distance before dropping back into the valence band.

ΔT (and hence the refractive index change via dn/dT) diffuses away in all directions, similar to the case of carrier migration discussed above. This case is discussed in Chapter 12 and is more complicated because the detailed response in the steady state depends on many factors, e.g., the shape of and the distance to the sample boundaries.

The optical excitation of propagating modes, e.g., sound waves, is yet another example of a nonlocal process in which the nonlinear interaction of light with matter at one position $\vec{r}$ is "felt" at a different position $\vec{r}'$. Probably the most interesting case is the nonlocal cascading effect involving second-order nonlinearities discussed in Chapter 12.

11.2.2 Nonlocality in Time

All the phenomena that give rise to a refractive index change in time are described by times characteristic of the phenomenon itself. For those oscillatory in time, the effects, e.g., in dilute media such as gases, are limited by the electron oscillation period in its orbit or the time needed to complete one cycle of a vibration. The two most important examples occur for atoms and molecules. For example, oscillation of an electric field induces polarization changes in an atom or molecule due to coupling between the electronic states that occurs in $10^{-14}-10^{-15}$ s (1 optical cycle) or less for the usual range of "optical" wavelengths. For molecular vibrations, the vibrational periods range over $10^{-12}-10^{-13}$ s. It would be reasonable to define the limits of locality in time for these two nonlinear contributions to the refractive index by these periods. It is noteworthy that in the last 15 years, experiments in optics have been reported that directly involve nonlocality in time, especially since the attosecond barrier has been breached (4).

Another more realistic approach to defining nonlocality in time is to ask what is the finite frequency spectrum associated with the temporal response function. For example, as noted previously, the lifetimes of excited states lead directly to a frequency spectrum in, e.g., the absorption spectrum (a response function) associated with real transitions between the ground state of the molecule or atom and the excited states. However, if there is a distribution of transition frequencies, the spectrum can be broadened by this fact alone, making it difficult to obtain the state lifetime in this way. Furthermore, such response functions are experimental entities that result in a definition that depends on instrumentation. Nevertheless, it is this definition that it is adopted here.

11.3 SURVEY OF NONLINEAR REFRACTION AND ABSORPTION MEASUREMENTS

To illustrate the range of nonlinear index and absorption coefficients in materials, a selection of third-order nonlinear refractive index and absorption material nonlinearities is given in Table 11.1 in the units that are standard in the literature. For the Kerr case those are

TABLE 11.1 Representative Kerr Dielectric Materials with Values of $n_{2\parallel}(-\omega;\omega)$ and $\alpha_{2\parallel}(-\omega;\omega)$ Ordered According to Their Bandgap Energy E_{gap} or Cutoff Wavelength

Material	$n_{2\parallel}(-\omega;\omega) \times 10^{-15}$ (cm²/W)				$\alpha_{2\parallel}(-\omega;\omega)$ (cm/GW)		
	1064 nm	532 nm	355 nm	266 nm	532 nm	355 nm	266 nm
LiF	0.081	0.061	0.061	0.13	≈0	≈0	≈0
MgF$_2$	0.057	0.057	0.066	0.15	≈0	≈0	≈0
BaF$_2$	0.14	0.21	0.27	0.31	≈0	≈0	0.06
NaCl	1.8						
SiO$_2$	0.21	0.22	0.24	0.78	≈0	≈0	0.05
Al$_2$O$_3$	0.31	0.33	0.37	0.60	≈0	≈0	0.09
BBO	0.29	0.55	0.36	0.003	≈0	0.01	0.9
KBr	0.79	1.27			≈0		
CaCO$_3$	0.29	0.29	0.37	1.2		0.018	0.8
LiNbO$_3$	0.91	8.3			0.38		
KTP	2.4	2.3			0.1		
ZnS	6.3				3.4		
Te glass	1.7	9.0			0.62		
ZnSe	29	−68			5.8		
ZnTe	120				4.2		
CdTe	−300				22		
RN glass	2.2						

Taken from Ref. 1.

The values quoted were obtained by using multiple pulse widths, nanosecond and picosecond, to isolate the ultrafast response. Blank cells indicate no measurement at that wavelength. Note that these values could have vibrational as well as Kerr contributions.

$$n_{2\parallel}(-\omega;\omega) = \frac{3}{4n^2\varepsilon_0 c}\,\Re\mathrm{eal}\left\{\widehat{\chi}^{(3)}_{xxxx}(-\omega;\omega,-\omega,\omega)\right\} \text{ in cm}^2/\text{W},$$

$$\alpha_{2\parallel}(-\omega;\omega) = \frac{3\omega}{2n^2\varepsilon_0 c^2}\,\Im\mathrm{mag}\left\{\widehat{\chi}^{(3)}_{xxxx}(-\omega;\omega,-\omega,\omega)\right\} \text{ in cm/GW}.$$

$$(11.12)$$

Note that these values depend on the effective bandgap of the material and on the measurement wavelength. In this table, two features are notable. The values of $n_{2\parallel}(-\omega;\omega)$ vary over many orders of magnitude. Furthermore, as discussed in Chapter 9, there is strong dispersion with wavelength and so measurements at one wavelength should be used only at that wavelength unless it lies in the nonresonant regime.

11.4 ELECTRONIC NONLINEARITIES INVOLVING DISCRETE STATES

The most important characteristic of this class of materials is that their response, both turn-on and turn-off times, is ultrafast (sub-femtosecond) in the off-resonance and nonresonant regimes. This spectral region is frequently discussed in terms of "virtual" levels to explain the ultrafast response. Near and on resonance, the turn-on can be ultrafast. However, since there is a transfer of population from the ground to higher lying excited states, the turn-off time is determined by the excited-state lifetimes. The non-resonant ultrafast nonlinearities range from 10^{-12} to 10^{-16} cm^2/W in condensed matter and are still smaller in dilute media such as gases in which they depend on the density of the gas molecules and/or atoms.

11.4.1 Nonlinearities in Gases

There has been a growing interest in the nonlinearities of the constituent molecules of air at atmospheric pressure when excited by high power lasers in the tens of terawatt per square centimeter range (5,6). Typically the gas density is in the range $10^{19}-10^{20}$ molecules and/or atoms per cubic centimeter. Representative numbers are listed in Table 11.2 for the most common air molecules and atoms, measured at 800 nm, which is far from any electronic resonances and can be considered as approximately nonresonant. These rank as some of the smallest nonlinearities measured to date in nonlinear optics.

11.4.2 Linear Molecules and Polymers

The case of nonlinearities arising from dipole-allowed transitions in atoms and molecules was discussed in Chapter 9, where it was shown that for asymmetric molecules the key to large third-order nonlinearities was large one-photon-allowed transition dipole moments $(|\bar{\mu}_{mg}|^2)$ and/or large changes in permanent dipole moments between the initial and final $(\bar{\mu}_{mm} - \bar{\mu}_{gg})^2$ electronic states and/or the intermediate states. However, there is a destructive interference between these two contributions, and so the best scenario is that one of these effects dominates the response. For symmetric molecules, there are no dipole moments but there is interference between contributions from the one- and two-photon-allowed transitions. The largest nonlinear response comes from materials with electron delocalization.

11.4.2.1 Conjugated Molecules and Polymers. There has been a great deal of interest in conjugated polymers for large optical nonlinearities. The physics of these molecules and polymers is very rich. Typically they consist of chains of carbon atoms with single, double, and/or triple bonds between the carbons, e.g., as shown in

TABLE 11.2 The Nonlinear Coefficient $n_{2\parallel}(-\omega;\omega)$ for Air and Its Constituents at 800 nm at Atmospheric Pressure in Units of 10^{-19} cm^2/W

Gas	N_2	O_2	Ar	Air
$n_{2\parallel}(-\omega;\omega)$	1.1 ± 0.3	2.0 ± 0.4	0.08 ± 0.3	1.3 ± 0.03

From Refs 5 and 6.

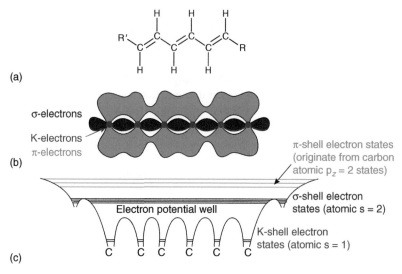

(a)

(b)

(c)

FIGURE 11.2 (a) Chemical structure of an acetylene molecule. (b) The orbitals (green) formed due to overlap of the $2p_z$ atomic orbitals of carbon atoms. (c) The electron potential well along the delocalization axis.

Fig. 11.2 for the acetylene molecule. The end groups R and R′ are usually identical and do not play a significant role in the nonlinearity. The $2p_z$ electron orbitals associated with the individual carbon atoms overlap and delocalize into new "π orbitals" that stretch along the full length of the molecule so that these electrons are shared by all the carbon bonds and can move more or less freely along the length of the molecule. As a result, the refractive index is enhanced along the chain axis and the nonlinearity is large because of the "soft" potential well associated with the π orbitals.

Since these are linear molecules, as discussed in Chapter 9, a minimum of two excited states are needed to describe their nonlinear response. These are shown in Fig. 11.3. The ground state has even symmetry (gerade), the first excited state has odd symmetry (ungerade), and the next higher excited state has even symmetry. These are labeled $1A_g$, $1B_u$, and mA_g, respectively. The electric dipole-allowed transitions along

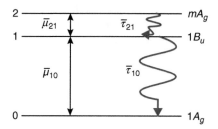

FIGURE 11.3 Three-level model for symmetric molecules. The parameters $\bar{\mu}_{10}$ and $\bar{\mu}_{21}$ are the transition dipole moments along the conjugation axis, and $\bar{\tau}_{21}$ and $\bar{\tau}_{10}$ are the decay (recombination) times between the states shown.

the chain axis are $1A_g \rightarrow 1B_u$ (transition moment $\bar{\mu}_{10}$) and $1B_u \rightarrow mA_g$ (transition moment $\bar{\mu}_{21}$) but *not* $1A_g \rightarrow mA_g$ (i.e., $\bar{\mu}_{20} = 0$) because a change in symmetry in the wave functions is required for dipole-allowed transitions. Similarly, the excited states can decay via $\bar{\tau}_{21}$ and $\bar{\tau}_{10}$ but not $\bar{\tau}_{20}$, which is not allowed; i.e., $\bar{\tau}_{20} \rightarrow \infty$ for the isolated molecule. In condensed matter, decay via $\bar{\tau}_{20}$ is weakly allowed with $\bar{\tau}_{20} \gg \bar{\tau}_{21}$.

The three-level system was discussed in Chapter 9. For applications, the most important is the nonlinear response for wavelengths longer than any of the electronic resonances. It was given for the off-resonance regime by Eqs 9.28 and for the nonresonant regime by Eq. 9.29. The key point is that there is interference between the one- and two-photon contributions to the nonlinearity for frequencies below $\bar{\omega}_{20}$. The sign of the nonresonant value of $n_{2\parallel}(-\omega; \omega)$ can be positive for $\bar{\omega}_{10}|\bar{\mu}_{21}|^2 > \bar{\omega}_{20}|\bar{\mu}_{10}|^2$ and negative for $\bar{\omega}_{20}|\bar{\mu}_{10}|^2 > \bar{\omega}_{10}|\bar{\mu}_{21}|^2$ (1,7).

Molecules such as acetylene (Fig. 11.2) with appropriate end groups do not produce good crystals but can be used in solution with the molecules distributed in random orientations (which reduces the net nonlinearity to one-fifth of that along the conjugation axis) (8). Some π-conjugated molecules can be polymerized as single crystals with good optical quality. An example of such a single crystal polymer is the polydiacetylene poly(bis(p-toluene sulfonate)) of 2,4-hexadiyne-1,6-diol whose structure and absorption spectrum is shown in Fig. 11.4 (1). The connection between parallel chains in the single crystal is via the R molecular end groups and large nonlinearities occur along the crystal b-axis. The absorption spectrum shows a number of important features, namely, that there are strong vibrational sub-bands and that the strongly coupled (to the first excited state) dominant two-photon state lies in the continuum of B_u and A_g states.

The dispersion in the nonlinear refractive index coefficients in such materials is complicated on the long wavelength side of the main linear absorption peak, i.e., the off-resonance to nonresonant regimes. The expected negative nonlinearity between

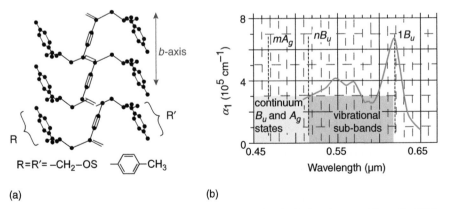

(a)   (b)

FIGURE 11.4 (a) The structure of the single crystal poly(bis(p-toluene sulfonate)) (PTS) polymer. (b) The linear absorption spectrum of PTS with the three states in the three-level model identified as well as the strong vibrational sub-bands and edge of the continuum of states (nB_u).

the one- and two-photon absorption peaks has been verified by multiple authors (1). For wavelengths longer than the two-photon absorption peak, the value of the ratio $\bar{\omega}_{10}|\bar{\mu}_{21}|^2/\bar{\omega}_{20}|\bar{\mu}_{10}|^2$ determines the sign of the nonlinearity, as discussed in Chapter 9. The spectral dependence for values of this ratio near unity is shown in Fig. 11.5 for a range of ratios for poly(bis(p-toluene sulfonate))-type parameters, *with excited state 3 in the inset playing the role of the strong two-photon state 2 in the theory*. Since $|\bar{\mu}_{31}|^2 \gg |\bar{\mu}_{21}|^2$ the coupling of the ground state to the $2A_g$ state is weak (9). The polydiacetylenes off resonance should have positive nonresonant values of $n_{2\|}(-\omega;\omega)$ from the predicted transition dipole moments whereas for squaraines (to be discussed next) the sign should be negative for short molecules (1,10). Note that the two-photon resonance "sits" on the decreasing background nonlinearity due to the one-photon resonance (proportional to $|\bar{\mu}_{10}|^4$).

The spectral dependence of both $n_{2\|}(-\omega;\omega)$ and $\alpha_{2\|}(-\omega;\omega)$ have been measured in the off-resonance regime for the two-photon absorption peak (see Fig. 11.6) (11). (Measurements over such a large off-resonance range of wavelengths typically do not exist for other molecules or crystals.) Extrapolating to the nonresonant limit,

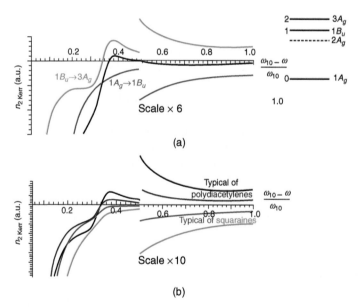

FIGURE 11.5   Calculations of the nonlinearity $n_{2\|}(-\omega;\omega)$ for frequencies below the one-photon absorption peak with $\bar{\omega}_{30} = 1.33\bar{\omega}_{10}$ and $\bar{\tau}_{10} = 10\bar{\tau}_{31}$, both typical values for poly(bis(p-toluene sulfonate)). The excited states measured experimentally are shown in the inset. (a) For $|\bar{\mu}_{10}|^2 = |\bar{\mu}_{31}|^2$ the contributions to the Kerr nonlinearity due to $|\bar{\mu}_{10}|^2$ (red line) and $|\bar{\mu}_{31}|^2$ (green line) and the sum of the two (black line) are shown. (b) The total nonlinearity is shown for different values of the ratio $|\bar{\mu}_{31}|^2/|\bar{\mu}_{10}|^2$. The purple ($|\bar{\mu}_{31}|^2 = 2|\bar{\mu}_{10}|^2$) and black ($|\bar{\mu}_{31}|^2 = 4|\bar{\mu}_{10}|^2$) curves are typical for polydiacetylenes and the brown ($2|\bar{\mu}_{31}|^2 = |\bar{\mu}_{10}|^2$) and green ($4|\bar{\mu}_{31}|^2 = |\bar{\mu}_{10}|^2$) for short squaraines (discussed in the text). Reproduced with permission from the Optical Society of America (1).

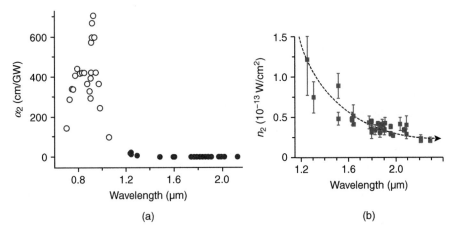

(a)   (b)

FIGURE 11.6 (a) The two-photon absorption $(\alpha_{2\parallel}(-\omega;\omega))$ spectrum of poly(bis(p-toluene sulfonate)). Note that there are strong vibrational sub-bands here as well as in the linear absorption case. (b) The off-resonance dispersion in $n_{2\parallel}(-\omega;\omega)$. Reproduced with permission from American Physical Society (11).

$n_{2\parallel}(-\omega;\omega) = (1.64 \pm 0.4) \times 10^{-14} \, \text{cm}^2/\text{W}$ (12). Comparing with Kerr values for other materials in Table 11.1, this is a large value despite the interference between the one- and two-photon terms inherent in symmetric linear systems. However, it is clear that the interference between contributions due to one- and two-photon transitions can greatly reduce the nonlinearity, which might be expected from large one-photon absorption coefficients, $\sim 10^{-6} \, \text{cm}^{-1}$ in this case.

11.4.2.2 Symmetric D-A-D Dyes. Squaraines belong to a special class of symmetric molecules containing charge transfer groups in a symmetric structure known as D-A-D, where D refers to an electron-donor group and A to an electron-acceptor group (see discussion on charge transfer in Chapter 7). Charge transfer is facilitated by delocalized π-electron linkages such as linear carbon–carbon bonds or benzene rings between the groups, as discussed in Chapter 7. These molecules derive their names from their common central acceptor group as identified in Fig. 11.7. They

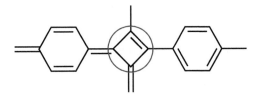

FIGURE 11.7 Chemical structure of a squaraine molecule. Identical donor or acceptor groups are added onto opposite ends so that the molecule is symmetric. The red circled "box" group is the source of the name squaraines.

are linear symmetric molecules and hence the three-level model discussed previously should provide a good description of their nonlinearities. (Although such molecules have zero dipole moments, the charge transfer can lead to large quadrupole moments.) Short molecules with limited conjugation linkage lengths exhibit electric dipole transition moments $\bar{\mu}_{10} \cong 10D$ and $\bar{\mu}_{21} \cong 4D$ (10). Equations 9.28 and 9.29 and Fig. 11.5 predict a negative $n_{2\|}(-\omega; \omega)$ off resonance, and nonresonant values and magnitudes in the range -5×10^{-12} to $-10^{-14} \, \mathrm{cm^2/W}$ have been measured for $1500 \, \mathrm{nm} > \lambda > 1064 \, \mathrm{nm}$ (1). In most cases, significant two-photon absorption was also found (1). Squaraine-based molecules have been successfully doped into polymer films and low loss fibers and demonstrated for nonlinear optics (13,14). For long, complicated conjugation lengths, the transition dipole moments are not known although large positive nonlinearities have been measured in the off-resonance regime, e.g., $n_{2\|}(-\omega; \omega) = 8 \times 10^{-13} \, \mathrm{cm^2/W}$ at $\lambda = 1.33 \, \mu\mathrm{m}$ for the long squaraine molecule (15).

Another class of promising D-A-D molecules is polymethines. For a molecule with large electron delocalization lengths, large values of $n_{2\|}(-\omega; \omega)$ have been measured at two wavelengths presumably in the off-resonance regime, namely, $-4.9 \times 10^{-12} \, \mathrm{cm^2/W}$ at $\lambda = 1.3 \, \mu\mathrm{m}$ and $-2.9 \times 10^{-12} \, \mathrm{cm^2/W}$ at $\lambda = 1.55 \, \mu\mathrm{m}$ (16). The linear absorption maximum is $\sim 1 \, \mu\mathrm{m}$.

To date, the nonlinear third-order properties of many organic molecules have been measured. Extensive tables of values can be found in Refs 17 and 18.

11.4.3 Excited-State Absorption and Reverse Saturable Absorption

Chapter 9 dealt with transitions from the ground state to excited states. At sufficiently high intensities, the population of the first excited state can subsequently act as an effective ground state for further transitions to even higher lying levels (17,19). This is a different scenario from two-photon absorption that is instantaneous and does not rely on the population in the first excited state, although it can be enhanced if the incident photon energy is near resonance with the transition from the ground state to the first excited state.

As discussed in Chapter 9, in general, more than one excited state can have significant population due to transitions from the ground state or even other excited states. This ensemble of excited states interacts with the incident light, which leads to changes in the population densities of these excited states. There is a well-developed approach to treating this problem, namely, the density matrix formalism discussed briefly in Chapter 9. For times longer than the coherence time that can be very short in condensed matter, i.e., femtoseconds, the transitions between the nth and mth states with $\bar{\omega}_{mn} > 0$ can be adequately described by the time evolution of the population densities $N_n(t)$ and $N_m(t)$ due to their coupling via the dipole transition matrix elements $|\bar{\mu}_{mn}|^2$ and the decay time from state m to n, i.e., $\bar{\tau}_{mn}$, as

$$\frac{dN_m}{dt} = \frac{\bar{\sigma}_{mn}N_n}{\hbar\omega}I - \frac{N_m}{\tau_{mn}}, \qquad (11.13)$$

with

$$\bar{\sigma}_{mn} = \frac{\alpha_{mn}}{N_n} = \frac{\omega}{2cn(\omega)\hbar\varepsilon_0} \left[\frac{\varepsilon_r + 2}{3}\right] |\vec{\mu}_{nm}|^2 \frac{\bar{\tau}_{mn}^{-1}}{(\bar{\omega}_{mn} - \omega)^2 + \bar{\tau}_{mn}^{-2}}$$ (11.14)

and the linear absorption coefficient α_{mn} is given by the simplified resonance term in Eq. 8.35 with the ground state replaced by the excited state n.

Linear absorption can promote species to excited states that serve as the lower state of a second electric dipole-allowed transition that occurs before electrons in the first excited state decay back to the ground state, as shown in Fig. 11.8 (1). Successive linear absorption processes from excited states are called *excited-state absorption* (ESA). There is a vibrational manifold associated with each excited state. Normally, any transition into states in the manifold results in the electron relaxing into the lowest vibrational level on a very fast temperature-dependent timescale so that the net result is to populate this lowest vibrational level in the excited states. Hence, reference to excited-state populations is to the ground vibrational state for timescales greater than 100 fs. Such a quasi-three-level system is a good approximation for many organic dyes.

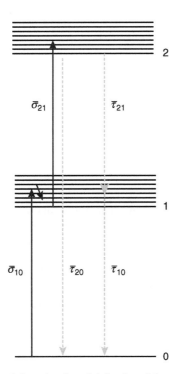

FIGURE 11.8 The quasi-three-level model for describing excited-state absorption.

The detailed evolution of the population densities for the quasi-three-level case is described by the equations (20).

$$
\frac{dN_0}{dt} = -\frac{\bar\sigma_{10}N_0}{\hbar\omega}I + \frac{N_1}{\bar\tau_{10}}, \qquad
\frac{dN_1}{dt} = \frac{\bar\sigma_{10}N_0}{\hbar\omega}I - \frac{\bar\sigma_{21}N_1}{\hbar\omega}I - \frac{N_1}{\bar\tau_{10}} + \frac{N_2}{\bar\tau_{21}},
$$

$$
\frac{dN_2}{dt} = \frac{\bar\sigma_{21}N_1}{\hbar\omega}I - \frac{N_2}{\bar\tau_{21}} - \frac{N_2}{\bar\tau_{20}}, \qquad
\frac{dI}{dz} = -\bar\sigma_{10}N_0 I - \bar\sigma_{21}N_1 I,
\tag{11.15}
$$

with $N = N_0 + N_1 + N_2$.

This is a useful model that can be used to describe ESA in many organic molecules and can also describe some free-carrier absorption phenomena in semiconductors. Although this is a complicated system of equations that must be solved simultaneously, for many molecules simplifications are possible. Typically, $\bar\tau_{20} \gg \bar\tau_{21}$ and $\bar\tau_{10}$. In the limit $N_1 \gg N_2$ and, for pulse widths (Δt), $\bar\tau_{10} \gg \Delta t$,

$$
\frac{dN_1}{dt} = \frac{\alpha_{10}I}{\hbar\omega} \quad \text{and} \quad \frac{dI}{dz} = -\bar\sigma_{10}N_0 I - \bar\sigma_{21}N_1 I
\tag{11.16}
$$

are the relevant equations. For short pulses, the population density N_1 is

$$
N_1(t) = \frac{\alpha_{10}}{\hbar\omega}\int_{-\infty}^{t} I(t')\,dt' \equiv \frac{\alpha_{10}}{\hbar\omega}F(t) \quad \Rightarrow \quad \frac{dF(t)}{dz} = -\bar\sigma_{10}N_0 F(t) - N_0\frac{\bar\sigma_{10}\bar\sigma_{21}}{2\hbar\omega}F^2(t),
\tag{11.17}
$$

in which α_{10} is the absorption coefficient for state $0 \rightarrow$ state 1 and $F(t)$ is the integrated fluence (total energy) up to time t. In this limit, the sequential absorption processes that include ESA via $\sigma_{10}\sigma_{21} \propto |\bar\mu_{21}|^2|\bar\mu_{10}|^2$ *look like* two-photon absorption. For $\bar\tau_{10} \gg \bar\tau_{21}$, a single excited-state absorber can efficiently absorb multiple times, even for pulsed inputs. If $\bar\sigma_{21} > \bar\sigma_{10}$, the absorption process is referred to as *reverse saturable absorption* (RSA) and can lead to increasing loss with increasing input (21).

Additional refractive index and absorption changes can be produced by the redistribution of population densities of two or more excited states via successive linear absorption processes from excited states. The refractive index changes occur both through the reduction of oscillator strength due to saturation as discussed in Chapter 9 and through the production of new absorbers as discussed in this section.

In cases where the input pulses deplete the population of the lower level and decay of the upper state is possible, Eq. 11.16 governing the process simplify to (21)

$$
\frac{dN_0}{dt} = -\frac{\bar\sigma_{10}N_0}{\hbar\omega}I + \frac{N_1}{\bar\tau_{10}}, \qquad
\frac{dN_1}{dt} = \frac{\bar\sigma_{10}N_0}{\hbar\omega}I - \frac{\bar\sigma_{21}N_1}{\hbar\omega}I - \frac{N_1}{\bar\tau_{10}}, \qquad
\frac{dI}{dz} = -\bar\sigma_{10}N_0 I - \bar\sigma_{21}N_1 I.
\tag{11.18}
$$

For short pulses the transmittance depends on the incident fluence. RSA occurs first and then ground-state depletion eventually kills the RSA. At that point it is possible to even saturate the upper level absorption. For sufficiently high inputs, all the populations will eventually equilibrate; i.e., saturation always "wins" at high input powers. The refraction from these absorption processes is simply related to the redistribution of the population of levels when absorbing species are created and/or removed. Depending on which side of resonance the operating frequency occurs, the refractive index can be increased or lowered. For pulsed input, the refractive index change follows the populations in time. For short pulses, compared to the population decay time the refractive index change follows the integrated energy, which has the shape of an error function.

In many situations, molecular triplet states (electronic state of molecules whose total spin angular momentum quantum number is equal to 1, i.e., two paired electron spins aligned instead of counter-aligned) are excited and the appropriate level structure to describe this scenario is a five-level system. Solutions to the equations show an increasing loss with increasing intensity that eventually turns into saturation for high inputs. More sophisticated approximations yield overall saturable absorption for $\sigma_{21} < \sigma_{10}$ and RSA (i.e., increasing loss with increasing intensity) for $\sigma_{21} > \sigma_{10}$, as seen in Fig. 11.9 (21).

Refractive changes occur as a result of the changes in the linear absorption by creating excited states and removing population from the ground state. They can be calculated from the Kramers–Kronig relations (discussed in the next section). When the creation of the excited-state absorbers dominates the absorption changes (as opposed to the loss

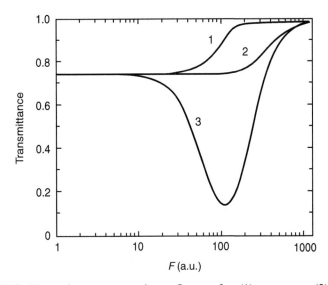

FIGURE 11.9 Transmittance versus input fluence for (1) $\sigma_{21} < \sigma_{10}$, (2) $\sigma_{21} = \sigma_{10}$, (3) $\sigma_{21} > \sigma_{10}$ (1).

TABLE 11.3 Parameters of RSA Dyes

Material/Solvent	σ_{1g} (cm^2)	σ_{21} (cm^2)	σ_{21}/σ_{1g}	σ_R (cm^2)	τ (ns)
Polymethine/ethanol (21)	0.7×10^{-17}	60×10^{-17}	81		0.3
CAP/methanol (19)	2.2×10^{-18}	2.3×10^{-17}	10	1.8×10^{-17}	7.0
SiNc/toluene (19)	2.8×10^{-18}	3.9×10^{-17}	14	4.7×10^{-18}	3.2

of ground-state absorbers), $\bar{\omega}_{10} < \omega \rightarrow \Delta n^{NL} > 0$ and $\bar{\omega}_{10} > \omega \rightarrow \Delta n^{NL} < 0$. This should be the case where RSA dominates saturable absorption. The changes are also usually fluence dependent versus intensity dependent. The refractive index changes can be written as

$$k_{vac}\Delta n^{NL} = \bar{\sigma}_R N_1, \qquad (11.19)$$

and values of $\bar{\sigma}_R$ for a few ESA dyes are given in Table 11.3.

11.5 OVERVIEW OF SEMICONDUCTOR NONLINEARITIES

In general, semiconductor nonlinearities exhibit very rich physics, which has been studied extensively because of the technological importance of semiconductors in optics. In semiconductors there are multiple mechanisms that contribute to $\Delta n^{NL}(I)$. Near and on resonance their nonlinear optics response can be very complicated because their states consist of quasi-continuous bands and not discrete states, and saturation effects can set in at low power levels. For details, see Ref. 22.

The largest nonlinear optics effects occur due to the absorption of photons of energy $\hbar\omega > E_g(E_{gap})$, which move electrons from the valence band to the conduction band. The nonlinear change in the refractive index $\Delta n^{NL} \propto N_e$ (electron density in the conduction band) is a function of the intensity for excitation times longer than the time τ_r it takes electrons in the conduction band to recombine with holes in the valence band. For pulses of width Δt, the refractive index change is proportional to the integrated fluence (intensity), i.e., $\int I(t)\, dt$. In this case, the turn-on time depends on the pulse width and the turn-off time is determined by τ_r. In addition, the electrons in the conduction band interact via Coulomb forces with positively charged "holes" in the valence band to produce hydrogen-like states just below the conduction band called *excitons*. They can also be populated by electrons from the valence band via photon absorption and hence are also responsible for nonlinearities.

Furthermore, it is possible to have gain by initially pumping electrons into the conduction band, either by light absorption or by electron injection. Such inversion of the conduction band electron population relative to some part of the valence band can lead to the amplification of incident light by stimulated emission. Semiconductor optical amplifiers with gain are currently the most versatile all-optical signal processing elements available (23).

For photon energies far enough below the bandgap, multiple (weaker) mechanisms contribute to a sub-picosecond third-order nonlinearity.

Quantum confinement of the electrons (conduction band) and holes (valence band) in two-dimensional (quantum wells), one-dimensional (quantum wires), and zero-dimensional (quantum dots) leads to new discrete bound states in the previous bulk gap of the semiconductor, which can result in either enhancing or diminishing the nonlinearity, depending on the wavelength.

11.5.1 Charge Carrier-Related Nonlinearities

11.5.1.1 Bandgap Renormalization (Band Filling). Consider the simplest possible model for a semiconductor with a single direct bandgap separating a valence and a conduction band (see Fig. 11.10a). At $T = 0°$K, the conduction band is empty and the valence band is full. In both bands, the electron states lie on the surface of "bowls" with electron energies that are functions of the gap energy E_{gap} and the electron "momenta" $\hbar k_x$, $\hbar k_y$, and $\hbar k_z$. The transition dipole moments for electrons between the two bands are more complicated than for the discrete energy level case since there is a *quasi-continuum* of initial and final states with wave functions ψ_{val} and ψ_{con}, respectively. When photons of energy $\hbar\omega$ are incident (Fig. 11.10b), a photon may be absorbed, raising an electron to the conduction band (Fig. 11.10c). A positively charged hole is left behind in the valence band. The electron's momentum is conserved when a photon is absorbed and the induced polarization is proportional to

$$\left\langle \psi_{con}, \vec{k}' |\vec{\mu}| \psi_{val}, \vec{k} \right\rangle = \vec{\mu}_{c\text{-}v} \delta_{\vec{k}\vec{k}'}, \tag{11.20}$$

where $\hbar\vec{k}$ and $\hbar\vec{k}'$ are the initial (valence band) and final (conduction band) electron momenta, respectively.

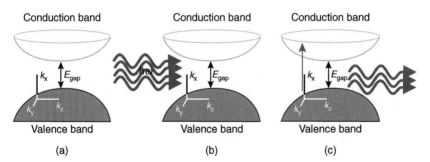

FIGURE 11.10 (a) Valence and conduction bands of a semiconductor in the $\vec{k}_e - E$ space at $T = 0$ K. (b) Photon flux incident on the semiconductor. (c) One photon is absorbed and an electron is raised from the valence band to the conduction band.

The following are the key differences between how a semiconductor versus atoms and molecules interacts with light:

1. In semiconductors, the transitions occur between a quasi-continuum of states with well-defined densities of states versus in molecules in which they occur between well-defined discrete states.

2. In a semiconductor the electrons see a periodic potential but can move more or less freely throughout the semiconductor. In atoms and molecules, the electrons in excited states continue to be bound to the molecules. (The exception in the molecular case is conjugated polymers where the electrons can move more or less freely along the conjugation chain, as discussed in Section 11.4.2.1, which explains why conjugated polymers were initially treated in semiconductor "language." However, even in conjugated polymers there is a coherence length that limits the distance.)

There is a universal relation (the Kramers–Kronig relation) due to causality between the change in the refractive index and the change in absorption. Since the nonlinear change in the refractive index is relatively more difficult to calculate than the change in absorption, typically the change in absorption is calculated (or measured) and the Kramers–Kronig relation (24),

$$\Delta n^{NL}(\omega) = \frac{c}{\pi} \, \text{P} \int_0^\infty \frac{\Delta \alpha^{NL}(\Omega)}{(\Omega^2 - \omega^2)} \, d\Omega, \tag{11.21}$$

is used to calculate the corresponding change in the refractive index. Here Ω is the frequency at which $\Delta \alpha^{NL}(\Omega)$ occurs, ω is the frequency at which $\Delta n^{NL}(\omega)$ is calculated, and P is the principal value of the integral.

Consider Fig. 11.11a that illustrates what happens to the valence and conduction bands after the absorption of a strong photon flux. The electrons can be initially raised from anywhere in the valence band to unpopulated states with the same momentum in the conduction band where collisions cause them to fall down (thermalize) on femtosecond timescales to empty states at the bottom of the conduction band. Similar thermalization occurs in the valence band, leading to the electron occupation of the two bands shown in Fig. 11.11a. This results in a change in the bandgap, which becomes larger and hence the names "bandgap renormalization" and "band filling" used to describe this effect. When $E'_{gap} \cong \hbar\omega$ for the incident light frequency, steady-state saturation in $\Delta \alpha^{NL}$ and Δn^{NL} occurs because there are no more empty states in the conduction band available. The change in the absorption spectrum is shown in Fig. 11.11b and c. This change is then used via the Kramers–Kronig relation to calculate the refractive index change shown in Fig. 11.11d.

11.5.1.2 *Exciton Bleaching.*
Excitons are bound states due to the Coulomb interaction between a hole (positively charged quasiparticle in the valence band) and an excited conduction-band electron (see Fig. 11.12a) (22). They exist near

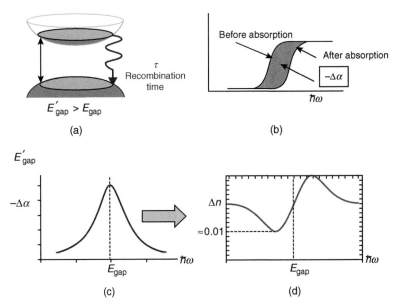

FIGURE 11.11 (a) The "renormalized" valence and conduction bands after absorption of a large number of photons. (b) The consequent change in the absorption spectrum. (c) The change $\Delta\alpha$, which is inserted into the Kramers–Kronig relation. (d) The calculated dispersion in the change in the refractive index. The peak and minimum values shown here correspond to approximately the maximum and minimum values that can be induced in bulk GaAs.

$\vec{k} = 0$ with a maximum binding energy E_0 and have hydrogen-like energy levels of the form

$$E_n = E_{\text{gap}} - E_0 \frac{1}{n^2} - \text{Coulomb correction}, \qquad E_0 = \frac{\hbar^2}{2m_r a_B^2},$$

$$a_B = \frac{\hbar^2 \varepsilon_0}{e^2 m_r} \quad (a_B = \text{exciton Bohr radius } (n = 1)), \tag{11.22}$$

$$\frac{1}{m_r} = \frac{1}{m_e} + \frac{1}{m_h} \quad (m_r = \text{electron} - \text{hole reduced mass}),$$

extending upward in energy toward the bottom of the conduction band (see Fig. 11.12b). These features in the absorption spectrum can be very narrow, of order millielectronvolts, at low temperatures. They are broadened via the thermal excitation of valence-band electrons for $k_B T > E_0$. There is also an "Urbach" tail that extends below the conduction band (in energy), also due to the thermal excitation of carriers from the valence band. It overlaps the exciton features, tending to further broaden the spectral features below the zero-temperature bandgap. Figure 11.13 shows the values of E_0 for various semiconductors. The larger the value of E_0,

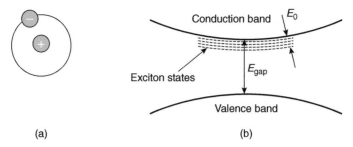

Conduction band E_0

Exciton states

E_{gap}

Valence band

(a)   (b)

FIGURE 11.12 (a) A simple model of an exciton. (b) The location of the exciton states below the conduction band.

the higher the temperature at which thermal effects tend to broaden the exciton spectral lines.

Detailed theoretical calculations including both band filling and exciton bleaching reproduced in Fig. 11.14 illustrate the changes in the absorption spectrum and refractive index change just discussed using GaAs as the specific example. The exciton line saturates very quickly with increasing N_e (caused by increasing light intensity).

At room temperature, only a weak broad peak remains of the exciton state near the bandgap edge in GaAs. Note the large refractive index changes of the order of 0.3, which are calculated for the refractive index at 10°K and the existence of gain at very high excitation densities as shown by a negative absorption coefficient. These calculations have been verified by an experiment (25)].

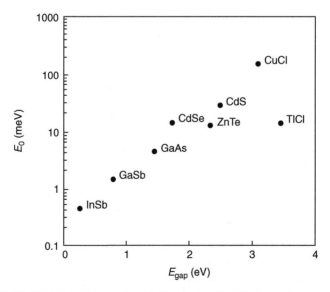

FIGURE 11.13 Variation of the exciton binding energy E_0 with the bandgap energy E_{gap} for various semiconductors. Reproduced with permission from World Scientific Press (22).

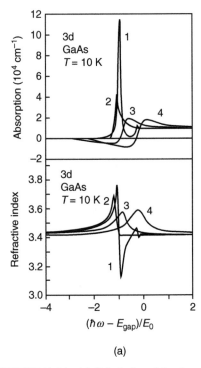

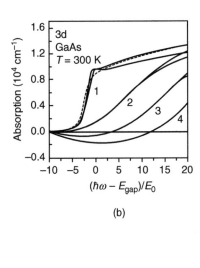

(b)

(a)

FIGURE 11.14   (a) Calculation of the change in the refractive index and absorption in GaAs at 10 K for four different electron densities in the conduction band: (*1*) $N_e = 0$; (*2*) $N_e = 5 \times 10^{15}$ cm^{-3}; (*3*) $N_e = 3 \times 10^{16}$ cm^{-3}; (*4*) $N_e = 8 \times 10^{16}$ cm^{-3}. (b) Calculation of the absorption spectrum at room temperature (300 K) for various excited electron densities: (*1*) $N_e = 1 \times 10^{16}$ cm^{-3}; (*2*) $N_e = 1 \times 10^{18}$ cm^{-3}; (*3*) $N_e = 2 \times 10^{18}$ cm^{-3}; (*4*) $N_e = 3 \times 10^{18}$ cm^{-3}. Here $m_e = 0.0665 m_0$ is the effective electron mass in the conduction band, $m_h = 0.457 m_0$ is the effective mass of the holes in the valence band, $a_B = 12.5$ nm, $E_0 = 4.2$ meV, and m_0 is the mass of a free electron. Reproduced with permission from World Scientific Press (22).

11.5.1.3   *Active Nonlinearities (with Gain).*   Consider the case in Fig. 11.15a, in which electrons are pumped into the conduction band by electrical injection (26). When photons of frequency ω are incident, they stimulate emission from an occupied electron state in the conduction band to an unoccupied state with the same $\vec{k}$ in the valence band, separated by the energy of the photon (Fig. 11.15b). This reduces the population in the conduction band so that the bandgap E'_{gap} decreases toward E_{gap}. This leads to an increase in the absorption coefficient because more conduction-band states are available (Fig. 11.15c). From the Kramers–Kronig relation, the change in the refractive index can be calculated (see Fig. 11.15d). Note that the dispersion in refractive index is opposite to the case for bandgap renormalization. And, of course, there is an increase in the signal beam, which is amplified. These are all carrier-related events.

At the transparency point, the losses are balanced by gain and so carrier generation by absorption is no longer the dominant nonlinear mechanism for the refractive index

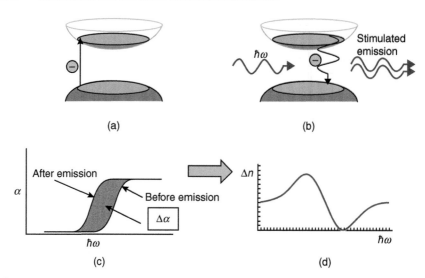

FIGURE 11.15   (a) External electron injection produces an "inversion" between the bottom of the conduction band and the top of the valence band. (b) An incident photon flux stimulates electrons to drop back to the valence band with the emission of photons at the frequency ω. (c) This causes the bandgap to shrink, and there is an increase in the local absorption coefficient, which in turn results in index dispersion (d).

change. At this point the dynamics of the weaker, ultrafast, electron-related, nonlinear emission and absorption processes can now be investigated. There is the usual instantaneous Kerr effect as well as other sub-picosecond phenomena, which now dominate the response of the electron populations. Before the external photon arrives, there is a Fermi electron distribution characterized by a steady-state temperature in the conduction band, as shown in Fig. 11.16a. When an intense 10–100-fs pulse passes through, "spectral hole burning" occurs first in the conduction band due to stimulated emission at the maximum gain wavelength determined by the maximum product of the density of occupied states in the conduction band and density of unoccupied states in the valence band (see Fig. 11.16b). Electron–electron scattering over approximately the next 100 fs relaxes this dip into a new electron distribution and new effective electron temperature. Subsequently, the electron temperature relaxes over a few picoseconds to a new equilibrium Fermi distribution at a lower electron temperature (Fig. 11.16c). Finally, continuing charge injection returns the electron distribution back to the original one before the passage of the pulse.

Because all these processes affect the density of the conduction-band electrons, they are accompanied by nonlinear index changes. Therefore, if a time-delayed weak probe beam passes through the semiconductor, it will experience changes in its phase ($\Delta \phi^{NL}$) due to these nonlinear index changes and the time evolution of different processes. Calculations of these phase changes are shown in Fig. 11.17, and the total phase change shown in this figure was verified experimentally (28). However, the key point is that all these ultrafast effects are small and the nonlinear optics of semiconductors is

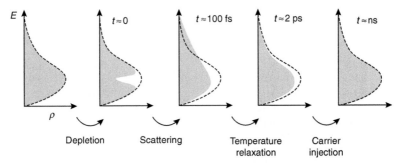

FIGURE 11.16 (a) The steady-state Fermi electron distribution in the conduction band due to electron pumping. (b) Passage of an intense femtosecond pulse at the transparency point produces a dip in the distribution due to stimulated emission and recombination of electrons into the valence band. (c) This dip is smoothed out by electron–electron scattering. (d) Temperature relaxation returns the electron distribution to a new Fermi distribution. (e) Steady-state pumping returns the electrons to the initial Fermi distribution. Reproduced with permission from the Optical Society of America (27).

dominated on longer (multi-picosecond) timescales by the band-filling and exciton effects.

11.5.1.4 Off Bandgap Charge Carrier Nonlinearities, $\hbar\omega > E_{gap}$. This spectral region essentially corresponds to the low-energy tail of the absorption band.

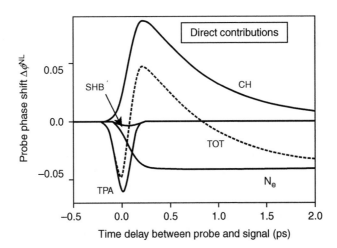

FIGURE 11.17 The calculated contributions of various sub-picosecond processes to the nonlinear optics of semiconductors as indicated by the nonlinear phase shift of a probe beam at the transparency point. TPA, two-photon absorption (Kerr effect); SHB, spectral hole burning; CH, carrier "heating" (temperature relaxation); TOT, total of all the ultrafast effects. Reproduced with permission from the Optical Society of America (29).

TABLE 11.4 Parameters $E_{gap}, \bar{\sigma}_R, \tau_r, n$ Listed for Selected Semiconductors

	InSb	GaAs	ZnSe	ZnS	CdS	CdTe
E_{gap} (eV)	0.18	1.42	2.67	3.66	2.42	1.44
$\bar{\sigma}_R$ (cm^2)	2.4×10^{-15}	3.8×10^{-16}	4.7×10^{-17}		3.8×10^{-16}	3.0×10^{-16}
τ_r	50 ns	Approximately few ns	1 ns	$\approx$1 ns	3.6 ns	
λ	10 μm	1.06 μm	1.06 μm	1.06 μm	0.53 μm	1.06 μm
n	4.0	3.43	2.7	2.4	2.6	2.7

See Ref. 1 for the references to the values quoted in this table.

At room temperature, there is a thermally induced smearing of the band boundaries by about $k_B T$, the so-called Urbach tail. For the simplest case of a two-band model (30),

$$\Delta n^{NL}(\omega) = \frac{\bar{\sigma}_r N_e}{k_{vac}} = \frac{\hbar^2 e^2}{2\varepsilon_0 n_0 m_r} \frac{1}{E_{gap}^2} \frac{1}{x^2(x^2 - 1)} N_e, \tag{11.23}$$

where $\bar{\sigma}_R$ is the absorption cross section per electron and $x = \hbar\omega/E_{gap}$. The parameters needed to evaluate Eq. 11.23 are given in Table 11.4 for a few semiconductors.

To define an equivalent $n_{2\|,eff}(-\omega; \omega)$ for this spectral region, it is convenient to write the time evolution of N_e as

$$\frac{dN_e}{dt} = \alpha_1(\omega)\frac{I(t)}{\hbar\omega} - \frac{N_e}{\tau_r}, \tag{11.24}$$

which includes the generation of electrons into the conduction band via absorption of light and the decay back to the valence band after an average time τ_r. In the steady case (for which $n_{2\|,eff}(-\omega; \omega)$ is normally defined), $dN_e/dt = 0$ and the steady-state electron density N_{ess} is

$$N_{ess} = \frac{\alpha_1(\omega)\tau_r}{\hbar\omega}I(t) \to \Delta n = \frac{N_{ess}\bar{\sigma}_R}{k_{vac}} = \frac{\bar{\tau}_r \alpha_1(\omega)\bar{\sigma}_R}{\hbar\omega k_{vac}}I \to n_{2\|,eff}(-\omega;\omega) = \frac{\bar{\tau}_r \alpha_1(\omega)\bar{\sigma}_R}{\hbar\omega k_{vac}}. \tag{11.25}$$

This relation is valid only at intensities well below saturation since it is only in that regime that $N_e \propto I$. Furthermore, once the electron is in the conduction band, there is a change in the refractive index that is experienced at all other frequencies. Due to diffusion, this $n_{2,eff}(-\omega; \omega)$ is also nonlocal.

For pulses (Δt) longer than the characteristic "dephasing" time (defined as coherence time τ_{coh} in Section 9.3) but shorter than the recombination time ($\tau_r \gg \Delta t$),

$$N_e(t) \cong \int_{-\infty}^{t} \frac{\alpha_1(\omega)}{\hbar\omega}I(t') \, dt' \xrightarrow{t \cong \text{many } (\Delta t)s} \frac{\alpha_1(\omega)}{\hbar\omega}\frac{\text{pulse energy}}{\text{area}} = \frac{\alpha_1(\omega)}{\hbar\omega}E_p.$$

Therefore, $\Delta n_{max}^{NL} = \dfrac{\alpha_1(\omega)\bar{\sigma}_R E_p}{\hbar\omega k_{vac}}$. $\tag{11.26}$

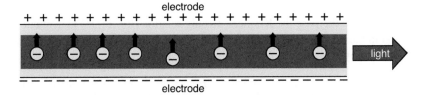

FIGURE 11.18   A method to reduce the effective carrier decay time by sweeping the conduction-band electrons out of the optical beam path.

Here ΔE_p is the pulse energy. After the passage of the pulse, Δn_{max}^{NL} decays exponentially with time τ_r. Note that the effective "decay" time τ_{eff} can be reduced to picoseconds by "sweeping out" the carriers from the optical field region with an electric field, as indicated in Fig. 11.18 (31). Note, however, that the $n_{2,eff}(-\omega; \omega)$ in Eq. 11.25 is also reduced by the same amount.

11.5.2   Semiconductor Response for Photon Energies Below the Bandgap

As the photon frequency decreases below the bandgap, the contribution to the electron population due to absorption decreases rapidly since the absorption coefficient decays toward zero. Thus other mechanisms become important. For photon energies less than the bandgap energy, a number of passive ultrafast (femtosecond turn-on and turn-off times) nonlinear mechanisms contribute to $n_{2\parallel}(-\omega; \omega)$ and $\alpha_{2\parallel}(-\omega; \omega)$. Even though first-order perturbation theory is used for their calculation, the density of states for both the valence and conduction bands is essentially assumed to be continuous. The detailed calculation is beyond the scope of this textbook (32,33). The simplest theory for the Kerr effect is based on the single valence and conduction bands with the electromagnetic field altering the energies of both the electrons and holes.

There are four ultrafast processes that contribute, namely, the Kerr effect, the Raman effect, the linear Stark effect, and the quadratic Stark effect. The three most important processes are shown schematically in Fig. 11.19.

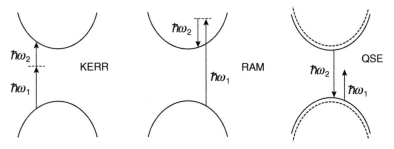

FIGURE 11.19   The most important contributions to the ultrafast nonlinearities below the bandgap for the nondegenerate case. RAM, Raman effect; QSE, quadratic Stark effect.

The theoretical approach is to calculate the nonlinear change in the absorption spectrum first and then use the Kramers–Kronig relation (Eq. 11.21) to evaluate the dispersion in the nonlinear index change (33). The nonlinear absorption for the Kerr effect is given by

$$\alpha_{2\|}(-\omega_1;\omega_2)=\alpha_{2\|}(-\omega_2;\omega_1)=K\frac{\sqrt{E_p}}{n_0(\omega_1)n_0(\omega_2)E_g^3}F_2(x_1,x_2),\quad K=\frac{2^5}{5\pi\varepsilon_0^2}\frac{1}{\sqrt{m_0c^2}}\frac{e^4}{},$$

$$x_i=\frac{\hbar\omega_i}{E_{gap}},\quad x_1-x_2>1:\ F(x_1,x_2)=\frac{(x_1+x_2-1)^{3/2}}{2^7x_1x_2^2}\left(\frac{1}{x_1}+\frac{1}{x_2}\right)^2,\qquad (11.27)$$

where E_p ("Kane energy") is given in terms of the semiconductor's properties and the constant $K\approx 3100\ \text{cm/GW eV}^{5/2}$. The nonlinear index coefficient is

$$n_{2\|}(-\omega_1;\omega_2)=n_{2\|}(-\omega_2;\omega_1)\ =\ \frac{\hbar cK}{2}\frac{\sqrt{E_p}}{n_0(\omega_1)n_0(\omega_2)E_g^4}G_2(x_1,x_2),$$

$$G_2(x_1,x_2)=\frac{2}{\pi}\int_0^\infty\frac{F_2(x';x_2)}{x'^2-x_1^2}dx',\qquad (11.28)$$

and the function $G_2(x_1,x_2)$ is given in Appendix 11.1. The detailed results for the Raman effect and the quadratic Stark effect are also given in Appendix 11.1.

The different contributions, along with the small contribution due to the linear Stark effect, are shown in Fig. 11.20. Near the bandgap, the quadratic Stark effect is the most important, whereas around half the bandgap, there is a strong peak due to the Kerr effect. Experiments have confirmed the validity of this theory. Clearly, the nonlinearity in this region of the spectrum is well understood.

Because accurate models exist for the detailed nonlinear optics of semiconductors for photon energies below the bandgap, it is possible to evaluate the material FOMs discussed in Section 11.1. The results are shown in Fig. 11.21. There are regions with $W>2$ for photon energies below the bandgap and with $T<1$ for $E_{gap}/2>\hbar\omega$ where semiconductors have favorable properties for ultrafast nonlinear optics (34,35).

11.5.3   Quantum Confined Semiconductors

When the translational degrees of freedom of electrons in both the valence and conduction bands are confined to distances of the order of the exciton Bohr radius a_B (defined in Eq. 11.22), the strength of the transition dipole moments is redistributed in frequency, the density of states $\rho_e(E)$ changes, and new bound states appear. The density of states for the different degrees of spatial confinement is shown in Fig. 11.22.

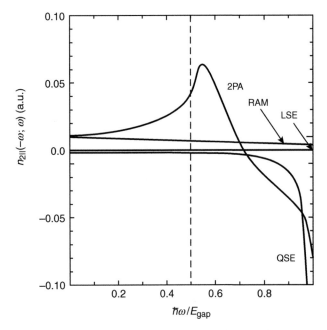

FIGURE 11.20 The relative contributions of the four different effects versus normalized frequency: two-photon absorption (TPA), Raman effect (RAM), quadratic Stark effect (QSE), and linear Stark effect (LSE), which lead to the ultrafast nonlinearity $n_{2\parallel}(-\omega; \omega)$ for photon energies below the bandgap energy. Reproduced with permission from the Institute of Electrical and Electronic Engineers (33).

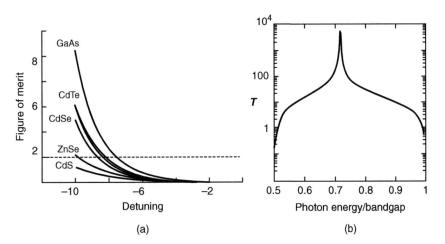

FIGURE 11.21 (a) The room temperature figure of merit W for a number of common semiconductors versus the detuning from the bandgap $(E_{\text{photon}} - E_{\text{bandgap}})/\Delta E_{\text{exciton linewidth}}$. (b) The figure of merit T for semiconductors based on Eqs 11.27 and 11.28 versus $\hbar\omega/E_{\text{gap}}$. Reproduced with permission from American Physical Society (34,35).

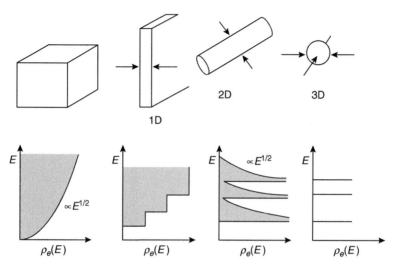

FIGURE 11.22   The density of states in the conduction band for different spatial degrees of confinement of the electrons (36).

As a result, there is the possibility that the nonlinear optical properties can be enhanced (or reduced) in some spectral regions.

11.5.3.1 Multiple Quantum Wells (MQW).
Very high quality quantum wells of GaAs–AlGaAs have been grown and hence the most complete nonlinear measurements have been made in that system. A typical structure made by molecular beam epitaxy using Ga, As, and Al beams for the deposition in high vacuum of multilayers is shown in Fig. 11.23a, and some measurements of the absorption spectrum near the bandgap are shown in Fig. 11.23b. The redistribution of the absorption spectrum relative to bulk GaAs is clear. The peaks in the quantum well sample correspond to bound states associated with thin slabs of AlGaAs surrounded by regions of GaAs, which has a smaller bandgap than that of AlGaAs (see Fig. 11.23a). The wave functions in the GaAs wells are sinusoidal-like, and they decay exponentially into the surrounding AlGaAs media.

Very careful, detailed measurements have been made of the dependence of the refractive index change per electron excited into the conduction-band states in the bandgap regions where peaks are available due to the creation of bound states, as illustrated previously in Fig. 11.23b (38). The data are shown in Fig. 11.24, and they reveal an enhancement as large as a factor of 3. Unfortunately there are no measurements in the off-resonance regime where all-optical applications would be likely.

11.5.3.2 Quantum Dots.
Quantum dot effects become important when the crystallite size $r_0 \lesssim a_B$ (exciton Bohr radius). For example, the exciton Bohr radius a_B is 3.2 nm for CdS, 5.6 nm for CdSe, 7.4 nm for CdTe, and 12.5 nm for GaAs (1).

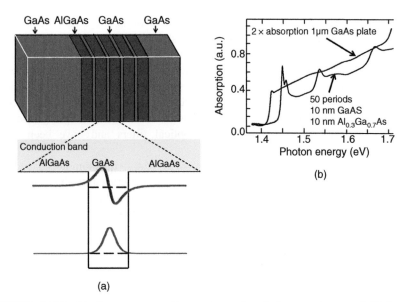

FIGURE 11.23 (a) A quantum well sample made of alternate layers of GaAs and $Al_{0.3}Ga_{0.7}As$ layers. The lower part shows the energy well with two bound states (dashed black lines) and their wave functions in the conduction band. (b) Comparison of the absorption spectrum of bulk GaAs (magnified by a factor of 2) and that of an $Al_{0.3}Ga_{0.7}As$ quantum well sample. Reproduced with permission from Taylor and Francis Publishers (37).

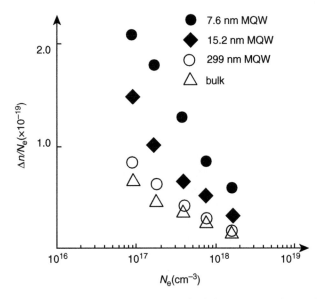

FIGURE 11.24 The refractive index change per excited electron versus the number of excited electrons for bulk GaAs and three quantum wells of different thicknesses. Reproduced with permission from the American Institute of Physics (38).

Stimulated by the enhancements found in quantum wells, there have been many theoretical predictions that GaAs quantum dots should exhibit even larger enhancements over the usual GaAs nonlinearities. Currently, there has been a great deal of progress in growth conditions and sample characterization, but to date there have been no consistent measurements in this material system. Similarly, there are predictions of enhancements in II–VI semiconductors, such as CdTe, CdS_xSe_{1-x}, and other semiconductors doped into glasses. The Cd-based semiconductor-doped glasses are commercially available as edge-blocking optical filters and have been well-characterized over the years.

Most theoretical approaches to quantum confinement in quantum dots have been based on the following two critical assumptions: (*1*) the crystallites have spherical symmetry; (2) the crystallites form a periodic lattice in 3 dimensions in the host materials. Thus the wave functions are typically written as

$$\psi(\vec{r}, \vec{k}) = \sum c_k \varphi(\vec{k}, \vec{r}) \cong \varphi(0, \vec{r}) \sum c_k \, e^{i\vec{k}\cdot\vec{r}} = \varphi(0, \vec{r}) F_c(\vec{r}). \tag{11.29}$$

Here $F_c(\vec{r})$ is the envelope function and $\varphi(0, \vec{r})$ is the Bloch wave function at $\vec{k} \cong 0$. $F_c(\vec{r})$ is given as

$$F_c(\vec{r}) = B_{\ell p} j_\ell(r/r_0) Y_\ell^m(\theta, \phi), \tag{11.30}$$

where $j_\ell(r/r_0)$ is the spherical Bessel function of order ℓ, $Y_\ell^m(\theta, \phi)$ are spherical harmonics, and r_0 is the crystallite radius. This leads to bound states with the absorption peaks located at

$$\hbar\omega_{\ell p} = E_g + \frac{\hbar^2 \beta_{\ell p}^2}{2m_h r_0^2} + \frac{\hbar^2 \beta_{\ell p}^2}{2m_e r_0^2} - b\frac{e^2}{\varepsilon_r r_0}. \tag{11.31}$$

$\beta_{\ell p}$ is the pth zero of the Bessel function and the constant $b = 1.8$ quantifies the electron–hole interaction.

The absorption spectra of three CdSe-doped glasses are shown in Fig. 11.25 for different crystallite sizes, all smaller than the exciton Bohr radius. Hence strong quantum effects should be visible (39). The small vertical lines indicate the assignment of peaks according to equations similar to Eq. 11.31, in which bound states clearly exist. However, the peaks are broad and many can be defined only by taking the second derivative of the spectrum. This shows a key fabrication problem, namely, that a distribution of crystallite sizes of only 20% is sufficient to produce such broad peaks.

There are many other spurious effects contributing to optical nonlinearities in II–VI doped glasses, e.g., photodarkening due to electron trapping in the glass host, free carrier absorption, and interface trapping states. This has led to many contradictory reports on enhancement in nonlinear optics properties, summarized by Banfi et al. (40).

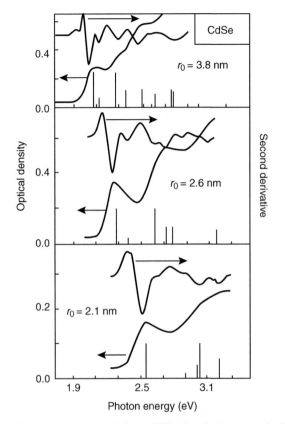

FIGURE 11.25 Absorption spectra of three CdSe-doped glasses with different average crystallite sizes. The vertical lines correspond to the assignment of various bound states. Reproduced with permission from the Optical Society of America (39).

There have been definitive nonlinear absorption measurements performed on very well-characterized II–VI samples by Banfi et al. for $a_B \cong r_0 \leq 3a_B$ (41). The results comparing the normalized two-photon absorption versus crystallite size are shown in Fig. 11.26.

Within the experimental uncertainty, no evidence exists for significant enhancement. On the contrary, for $a_B \geq r_0$ evidence exists for a decrease in the nonlinearity (which subsequently was also observed by other researchers).

This question of enhancement in quantum dots has not been resolved.

11.6 GLASS NONLINEARITIES

Glasses are the most common optical media, known primarily for their excellent transmission properties in the visible region. They are amorphous media, and their optical properties—linear and nonlinear—may vary from sample to sample. The reasons are primarily twofold. First, glass properties vary with the *details* of the

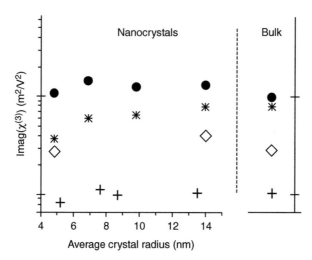

FIGURE 11.26 The normalized two-photon absorption versus crystallite size for a number of different samples at different wavelengths. The measurement wavelengths were at $1.2\,\mu m$ ($\bullet$), $1.4\,\mu m$ ($*$), and $1.58\,\mu m$ ($\lozenge$) for CdTe and at $0.79\,\mu m$ ($+$) for $CdS_{0.9}Se_{0.1}$. Reproduced with permission from the Optical Society of America (41).

preparation technique, which is usually proprietary to each commercial supplier. Different complexes can form on a local scale, especially for multicomponent glasses. Second, the optical properties depend on the purity of the starting materials and on small amounts of "impurities" added to the fabrication for stability. In addition, the glass properties may depend on the location in the melt from which the sample was taken, e.g., the center and the edges.

Fused silica is a very important glass for optics applications because of its high optical quality and low loss, e.g., when pulled into fibers. Its nonlinear optical properties have been investigated in great detail and are summarized in Fig. 11.27a (42,43). Silica glass is available commercially from many sources with different ingredients—purity and quality—prepared in proprietary and different ways so that their properties are not identical (see the table in Fig. 11.27b). The accepted value of $n_{2\parallel}(-\omega;\omega)$ at 1550 nm is $2.5 \times 10^{-16}\,cm^2/W$ (42,43).

Glasses are normally used far from their absorption regions since one of their most attractive properties is their low absorption. Their nonresonant response is ultrafast. Heavy glasses have larger (than silica) nonlinearities and can have very complex compositions. An example of the typical dispersion with the wavelength of both the two-photon absorption coefficient and $n_{2\parallel}(-\omega;\omega) \propto \chi^{(3)}_{\Re,1111}(-\omega;\omega,-\omega,\omega)$ is shown for a "heavy" glass in Fig. 11.28 (44). Note that sufficiently far from the electronic resonances in the visible region, this glass can be used for nonlinear refractive applications with negligible two-photon absorption.

Over the last 10–20 years many new glasses, mostly heavy oxides and chalcogenides, have been synthesized. The range of values along with the glass classification is documented in Fig. 11.29. However, the values for loss (α_1 in cm^{-1}) achieved to

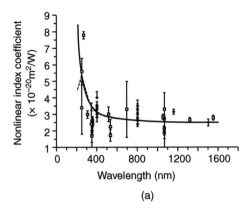

(a)   (b)

FIGURE 11.27   (a) The wavelength dispersion of the nonlinearity $n_{2\|}(-\omega;\omega)$ in fused silica. (b) The nonlinearity measured at 800 nm for glasses from different sources. Reproduced with permission from the Optical Society of America (42,43).

date in these materials have led to lower net FOMs defined by $n_{2\|}(-\omega;\omega)/\alpha_1(\omega)$ than those in fused silica. In general, the larger the $n_{2\|}(-\omega;\omega)$, the more the absorption edge due to shifts in electronic transitions to longer wavelengths and hence the larger the residual absorption in the near-infrared and $1-1.5$-μm regions relative to that in fused silica.

The nonlinear properties of a number of glasses in standard glass catalogs were investigated in the late 1980s and early 1990s. The range of values found for $n_{2\|}(-\omega;\omega)$ was $10^{-15}-10^{-14}$ cm^2/W. The inclusion of metal oxides of, e.g., Te, Ti, Th, and Nb, in

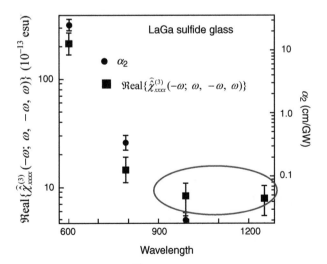

FIGURE 11.28   The dependence of $\chi^{(3)}_{\Re,1111}(-\omega;\omega,-\omega,\omega)$ and $\alpha_{2\|}(-\omega;\omega)$ on the wavelength of a typical heavy glass. The refractive nonlinearity dominates in the elliptical region. Reproduced with permission from the Optical Society of America (44).

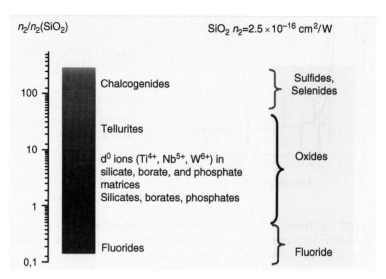

FIGURE 11.29 Summary of the trends in $n_{2\parallel}(-\omega;\omega)$ for different families of glasses. Courtesy of Dr Thierry Cardinal, ICBM Bourdeaux.

glasses produced $n_{2\parallel}(-\omega;\omega)$ in the range 10^{-15}–6×10^{-14} cm^2/W (see Ref. 1 and references therein). The chalcogenide glass family has been of special interest because of its high nonlinearities in the near- and mid-infrared regions. Values are typically in the 10^{-14}–2×10^{-13} cm^2/W range, and a number of promising FOMs have been identified (45). Many of these glasses have absorption cutoffs in the near-infrared region. A great deal of research has gone into eliminating optical damage in chalcogenides.

The BGO (Boling, Glass, and Owyoung) equation is frequently used (with some success) to predict $n_{2\parallel}(-\omega;\omega)$ in glasses with absorption edges in the ultraviolet and blue-visible regions (46). It has the form

$$n_{2\parallel}(-\omega;\omega)(\text{cm}^2/\text{W}) = \frac{0.29(n_d - 1)(n_d^2 + 2)^2}{n(\lambda)v_d\sqrt{[1.52 + (n_d - 1)(n_d^2 + 2)v_d]/6n_d}} \times 10^{-13},$$

(11.32)

where n_d is the refractive index at 0.486 µm and v_d is the Abbé number.

APPENDIX 11.1: EXPRESSIONS FOR THE KERR, RAMAN, AND QUADRATIC STARK EFFECTS

A.11.1.1 Nonlinear Absorption

$$\alpha_{2\parallel}(-\omega_1;\omega_2) = \alpha_{2\parallel}(-\omega_2;\omega_1) = K\frac{\sqrt{E_p}}{n_0(\omega_1)n_0(\omega_2)E_g^3}F_2(x_1, x_2); \quad x_i = \frac{\hbar\omega_i}{E_{\text{gap}}}$$

Nonlinear Process	$F_2(x_1, x_2)$
Kerr $x_1 + x_2 > 1$	$\frac{(x_1+x_2-1)^{3/2}}{2^7 x_1 x_2^2}\left(\frac{1}{x_1}+\frac{1}{x_2}\right)^2$
Raman $x_1 - x_2 > 1$	$\frac{(x_1-x_2-1)^{3/2}}{2^7 x_1 x_2^2}\left(\frac{1}{x_1}-\frac{1}{x_2}\right)^2$
QSE $x_1 > 1$	$-\frac{1}{2^9 x_1 x_2^2 (x_1-1)^{1/2}}\left(\frac{x_1}{x_1^2-x_2^2}-\frac{2(x_1-1)(x_1^2+x_2^2)}{(x_1^2-x_2^2)^2}+\frac{8(x_1-1)^2}{x_2^2}\right)$

A.11.1.2 Nonlinear Refraction

$$n_{2\parallel}(-\omega_1;\omega_2) = n_{2\parallel}(-\omega_2;\omega_1) = \frac{\hbar c K}{2}\frac{\sqrt{E_p}}{n_0(\omega_1)n_0(\omega_2)E_g^4}G_2(x_1,x_2);$$

Nonlinear Process	$G_2(x_1; x_2)$
Kerr	$H(x_1, x_2) + H(-x_1, x_2)$
Raman	$H(x_1, -x_2) + H(-x_1, -x_2)$
QSE $x_1 \neq x_2$	$\frac{1}{2^9 x_1^2 x_2^2}\left[\begin{array}{l}-\frac{1}{2}-\frac{4}{x_1^2}+\frac{4}{x_2^2}-\frac{x_2^2[(1-x_1)^{-1/2}-(1+x_1)^{-1/2}]}{x_1(x_1^2-x_2^2)}\\[2mm]+\frac{2x_1^2(3x_2^2-x_1^2)}{x_2^2(x_1^2-x_2^2)^2}\left[(1-x_2)^{1/2}+(1+x_2)^{1/2}\right]\\[2mm]-\frac{2x_2^2(3x_1^2-x_2^2)}{x_1^2(x_1^2-x_2^2)^2}\left[(1-x_1)^{1/2}+(1+x_1)^{1/2}\right]\end{array}\right]$
$x_1 = x_2$	$\frac{1}{2^9 x_1^4}\left[\frac{3}{4}\frac{(1-x_1)^{-1/2}-(1+x_1)^{-1/2}}{x_1}-\frac{(1-x_1)^{-3/2}+(1+x_1)^{-3/2}}{8}-\frac{1}{2}\right]$

Here

$$H(x_1, x_2) = \frac{1}{2^6 x_1^4 x_2^4}$$

$$\left[\begin{array}{l}\frac{5}{16}x_2^3 x_1^2 + \frac{9}{8}x_2^2 x_1^2 - \frac{9}{4}x_2 x_1^2 - \frac{3}{4}x_2^3 - \frac{1}{32}x_2^3 x_1^2(1-x_1)^{-3/2}\\[2mm]+\frac{1}{2}(x_2+x_1)^2\left[(1-x_2-x_1)^{3/2}-(1-x_1)^{3/2}\right]\\[2mm]-\frac{3}{16}x_2^2 x_1^2\left[(1-x_1)^{-1/2}+(1-x_2)^{-1/2}\right]+\frac{3}{2}x_2 x_1^2(1-x_2)^{1/2}+\frac{3}{2}x_2^2 x_1(1-x_1)^{1/2}\\[2mm]+\frac{3}{4}x_2(x_2^2+x_1^2)(1-x_1)^{1/2}-\frac{3}{8}x_2^3 x_1(1-x_1)^{-1/2}+\frac{1}{2}(x_2^2+x_1^2)\left[1-(1-x_2)^{3/2}\right]\end{array}\right]$$

PROBLEMS

1. Various approximate formulas have been reported in the literature for calculating the nonlinear refractive index coefficient for glasses in the visible region. Probably the most simple is

$$n_{2\|}(-\omega;\omega) = 1.64 \frac{(n_d - 1)(n_d + 1)}{n(\lambda)[\sqrt[4]{v_d}]^5} \times 10^{-13},$$

where v_d is the Abbé number defined as $v_d = (n_d - 1)/(n_F - n_c)$. Here the linear refractive indices n_d, n_F, and n_c are measured at 0.48613, 0.58756, and 0.65627 μm, respectively. Look up the relevant parameters in the literature and compare the estimated and measured values for $n_{2\|}(-\omega;\omega)$ for fused silica, SF6, and BK7 glass to determine whether this formula is useful. Compare the results with those generated via the BGO (Boling, Glass, and Owyoung) formula given by Eq. 11.32.

2. Calculate Δn at $0.8E_{gap}$ for the semiconductors GaAs, CdTe, CdS, and ZnS assuming the formula

$$\Delta n = \frac{\sigma_R N_e}{k_{vac}} = \frac{\hbar^2 e^2}{2\varepsilon_0 n_0 m_{eh}} \frac{1}{E_{gap}^2} \frac{1}{x^2(x^2 - 1)} N_e$$

at 1.06 μm for an electron excitation of 10^{18} cm^{-3}. What is the light intensity required in each case? [Hint: Find the absorption spectrum in the literature for each case.]

REFERENCES

1. D. N. Christodoulides, I. C. Khoo, G. J. Salamo, G. I. Stegeman, and E. W. Van Stryland, "Nonlinear refraction and absorption: mechanisms and magnitudes," Adv. Opt. Photon., **2**, 60–200 (2010).

2. G. I. Stegeman, "Material figures of merit and implications to all-optical switching," SPIE Proc. Nonlinear Opt. Prop. Adv. Mater., **1852**, 75–89 (1993).

3. F. Yoshino, S. Polyakov and G. I. Stegeman, "All-optical multi-photon absorption figures of merit: polydiacetylene poly(bis(*para*-toluene sulfonate)) of 2,4-hexadyine-1,6 diol," Appl. Phys. Lett., **84**, 5362–5364 (2004).

4. F. Krausz and M. Ivanov, "Attosecond physics," Rev. Mod. Phys., **81**, 163–234 (2009).

5. E. T. J. Nibbering, G. Grillon, M. A. Franco, B. S. Prade, and A. Mysyrowicz, "Determination of the inertial contribution to the nonlinear refractive index of air, N$_2$, and O$_2$ by use of unfocused high-intensity femtosecond laser pulses," J. Opt. Soc. Am. B, **14**, 650–660 (1997).

6. V. Loriot, E. Hertz, O. Faucher, and B. Lavorel, "Measurement of high order Kerr refractive index of major air components," Opt. Exp., **17**, 13429–13434 (2009).

7. G. I. Stegeman and H. Hu, "Refractive nonlinearity of linear symmetric molecules and polymers revisited," Photonics Lett. Poland, **1**, 148–150 (2009).

8. F. Kajzar and J. Messier, "Third-harmonic generation in liquids," Phys. Rev. A, **32**, 2352–2363 (1985).

9. F. Kajzar and J. Messier, "Resonance enhancement in cubic susceptibility of Langmuir–Blodgett multilayers of polydiacetylenes," Thin Solid Films, **132**, 11–19 (1985).

10. K. S. Mathis, M. G. Kuzyk, C. W. Dirk, A. Tan, S. Martinez, and G. Gampos, "Mechanisms of the nonlinear optical properties of squaraine dyes in poly(methyl methacrylate) polymer," J. Opt. Soc. Am. B, **15**, 871–883 (1998).

11. W. E. Torruellas, B. L. Lawrence, G. I. Stegeman and G. Baker, "Two-photon saturation in the bandgap of a molecular quantum wire," Opt. Lett., **21**, 1777–1779 (1996); S. Polyakov, F. Yoshino, M. Liu, and G. Stegeman, "Nonlinear refraction and multiphoton absorption in polydiacetylenes from 1200 to 2200 nm, " Phys. Rev. B, **69**, 1154211 (2004).

12. G. I. Stegeman, "Nonlinear optics of conjugated polymers and linear molecules," Nonlinear Opt. Quant. Opt.: Concepts Mod. Opt., in press.

13. M. G. Kuzyk, U. C. Paeka, and C. W. Dirk, "Guest–host polymer fibers for nonlinear optics," Appl. Phys. Lett., **59**, 902–904 (1991).

14. D. W. Garvey, Q. Li, M. G. Kuzyk, Carl W. Dirk, and S. Martinez, "Sagnac interferometric intensity-dependent refractive-index measurements of polymer optical fiber," Opt. Lett., **21**, 104–106 (1996).

15. S.-J. Chung, S. Zheng, T. Odani, L. Beverina, J. Fu, L. Padilha, A. Biesso, J. Hales, X. Zhan, K. Schmidt, A. Ye, E. Zojer, S. Barlow, D. Hagan, E. Van Stryland, Y. Yi, Z. Shuai, G. Pagani, J.-L. Bredas, J. Perry, and S. Marder "Extended squaraine dyes with large two-photon absorption cross-sections," J. Am. Chem. Soc., **128**, 14444–14445 (2006).

16. J. M. Hales, S. Zheng, S. Barlow, S. R. Marder, and J. W. Perry, "Bisdioxaborine polymethines with large third-order nonlinearities for all-optical signal processing," J. Am. Chem. Soc., **128**, 11362–11363 (2006).

17. H. S. Nalwa and S. Miyata, Nonlinear Optics of Organic Molecules and Polymers (CRC Press, Boca Raton, FL, 1997).

18. A. E. Garito and M. G. Kuzyk, "Nonlinear refractive index: organic materials," in Handbook of Laser Science and Technology, Supplement 2: Optical Materials, edited by M. J. Weber (CRC Press, Boca Raton, FL, 1995).

19. E. W. Van Stryland, M. Sheik-Bahae, A. A. Said, and D. J. Hagan, "Characterization of nonlinear optical absorption and refraction," Prog. Cryst. Growth Charact. Mater, **27**, 279–311 (1993).

20. T. Xia, D. Hagan, A. Dogariu, A. Said, and E. Van Stryland, "Optimization of optical limiting devices based on excited-state absorption," Appl. Opt., **36**, 4110–4122 (1997).

21. T. H. Wei, D. J. Hagan, M. J. Sence, E. W. Van Stryland, J. W. Perry, and D. R. Coulter, "Direct measurements of nonlinear absorption and refraction in solutions of phthalocyanines," Appl. Phys. B, **54**, 46–51 (1992).

22. H. Haug and S. W. Koch, Quantum Theory of the Optical and Electronic Properties of Semiconductors (World Scientific, Singapore, 1990).

23. R. J. Manning, A. D. Ellis, A. J. Poustie, and K. J. Blow, "Semiconductor laser amplifiers for ultrafast all-optical signal processing," J. Opt. Soc. Am. B, **14**, 3204–3216 (1997).

24. N. Bloembergen, Nonlinear Optics, 4th Edition (World Scientific Press, Singapore, 1996).

25. Y. H. Lee, A. Chavez-Pirson, S. W. Koch, H. M. Gibbs, S. H. Park, J. Morhange, A. Jeffery, N. Peyghambarian, L. Banyai, A. C. Gossard, and W. Wiegmann, "Room-temperature optical nonlinearities in GaAs," Phys. Rev. Lett., **57**, 2446–2449 (1986).

26. A. Othonos, "Probing ultrafast carrier and phonon dynamics in semiconductors," J. Appl. Phys., **83**, 1789–1830 (1998).

27. J. Mørk, M. L. Nielsen, and T. W. Berg, "The dynamics of semiconductor optical amplifier: modeling and applications," Opt. Photonics News, **14**(7), 42–48 (2003).

28. C. T. Hultgren, D. J. Dougherty, and E. P. Ippen, "Above- and below-band femtosecond nonlinearities in active AlGaAs waveguides," Appl. Phys. Lett., **61**, 2767–2769 (1992).

29. J. Mørk and A. Mecozzi, "Theory of the ultrafast optical response of active semiconductor waveguides," J. Opt. Soc. Am. B, **13**, 1803–1815 (1996).

30. M. Sheik-Bahae, "Optical nonlinearities in the transparency region of bulk semiconductors," in Nonlinear Optics in Semiconductors, edited by E. Garmire and A. Kost. (Academic Press, San Diego, 1999) Chap. 4, pp. 257–318.

31. P. LiKamWa, A. Miller, J. S. Roberts, and P. N. Robson, "130 ps recovery of all-optical switching in a GaAs multi-quantum well directional coupler," Appl. Phys. Lett., **58**, 2055–2057 (1991).

32. D. C. Hutchings, M. Sheik-Bahae, D. J. Hagan, and E. W. Van Stryland, "Kramers–Krönig relations in nonlinear optics," Opt. Quantum Electron., **24**, 1–30 (1992).

33. M. Sheik-Bahae, D. C. Hutchings, D. J. Hagan, and E. W. Van Stryland, "Dispersion of bound electronic nonlinear refraction in solids," IEEE J. Quantum Electron., **QE-27** 1296–1309 (1991).

34. E. M. Wright, S. W. Koch, J. E. Ehrlich, C. T. Seaton, and G. I. Stegeman, "Semiconductor figure of merit for nonlinear directional couplers," Appl. Phys. Lett., **52**, 2127–2129 (1988).

35. K. DeLong and G. I. Stegeman, "Dispersion of the two photon absorption parameter for all-optical switching," Appl. Phys. Lett., **57**, 2063–2064 (1990).

36. A. D. Yoffe, "Low-dimensional systems: quantum size effects and electronic properties of semiconductor microcrystallites (zero-dimensional systems) and some quasi-two-dimensional systems," Adv. Phys., **51**, 799–890 (2002).

37. S. Schmitt-Rink, D. S. Chemla, and D. A. B. Miller, "Linear and nonlinear optical properties of semiconductor quantum wells," Adv. Phys., **38**, 89–188 (1989).

38. S. H. Park, J. F. Morhange, A. D. Jefferey, R. A. Morgan, A. Chavez-Pirson, H. M. Gibbs, S. W. Koch, N. Peyghamberian, M. Derstine, A. C. Gossard, J. H. English, and W. Weigmann, "Measurements of room temperature, band-gap-resonant optical nonlinearities of GaAs/AlGaAs multiple quantum wells and bulk GaAs," Appl. Phys. Lett., **52**, 1201–1203 (1988).

39. A. I. Ekimov, F. Hache, M. C. Schanne-Klein, D. Ricard, C. Flytzanis, I. A. Kudryavtsev, T. V. Yazeva, A. V. Rodina, and A. L. Efros, "Absorption and intensity-dependent photoluminescence measurements on CdSe quantum dots: assignment of the first electronic transitions," J. Opt. Soc. Am. B, **10**, 100–107 (1993).

40. G. P. Banfi, V. Degiorgio, and D. Ricard, "Nonlinear optical properties of semiconductor nanocrystals," Adv. Phys., **47**, 447–510 (1998).

41. G. P. Banfi, V. Degiorgio, D. Fortusini, and M. Bellini, "Measurement of the two-photon absorption coefficient of semiconductor nanocrystals by using tunable femtosecond pulses," Opt. Lett., **21**, 1490–1492 (1996).

42. S. Santran, L. Canioni, L. Sarge, Th. Cardinal, and E. Fargin, "Precise and absolute measurements of the complex third-order optical susceptibility," J. Opt. Soc. Am. B, **21**, 2180–2190 (2004).

43. D. Milan, "Review and assessment of measured values of the nonlinear refractive index coefficient of fused silica," Appl. Opt., **37**, 546–550 (1998).

44. I. Kang, T. D. Krauss, F. W. Wise, B. G. Aitken, and N. F. Borrelli, "Femtosecond measurement of enhanced optical nonlinearities of sulfide glasses and heavy-metal-doped oxide glasses," J. Opt. Soc. Am. B, **12**, 2053–2059 (1995).

45. J. M. Harbold, F. Ö. Ilday, F. W. Wise, and B. G. Aitken, "Highly nonlinear Ge–As–Se and Ge–As–S–Se glasses for all-optical switching," IEEE Photonics Technol. Lett., **14**, 822–824 (2002).

46. N. L. Boling, A. J. Glass, and A. Owyoung, "Empirical relations for predicting nonlinear refractive index changes in optical solids," IEEE J. Quantum Electron., **QE-14** 601–608 (1978).

SUGGESTED FURTHER READING

G. P. Banfi, V. Degiorgio, and D. Ricard, "Nonlinear optical properties of semiconductor nanocrystals," Adv. Phys. **47**, 447–510 (1998).

S. Schmitt-Rink, D. S. Chemla, and D. A. B. Miller, "Linear and nonlinear optical properties of semiconductor quantum wells," Adv. Phys. **38**, 89–188 (1989).

A. D. Yoffe, "Low-dimensional systems: quantum size effects and electronic properties of semiconductor microcrystallites (zero-dimensional systems) and some quasi-two-dimensional systems," Adv. Phys., **51**, 799–890 (2002).

CHAPTER 12

Miscellaneous Third-Order Nonlinearities

There are many mechanisms through which light interacts with matter. In Chapter 11, the consequences to nonlinear optics of light-induced electronic transitions in atoms and molecules were examined. In this chapter the third-order nonlinearities associated with other nonlinear phenomena in condensed matter are discussed. Some are due to the coupling of light to molecular degrees of freedom such as vibrational and rotational molecular motions, are relatively fast, and occur on the timescales of hundreds of femtoseconds to tens of picoseconds. They are local in space.

Interesting physics happens in the light–matter interaction in the transition from single molecules to condensed matter, which can introduce new nonlinearities. Van der Waals, Coulomb, and other interactions can lead to cooperative behavior on the scale of optical wavelengths or larger and/or to the weak breaking of molecular symmetries. For example, in liquid crystals, strong intermolecular interactions and geometry effects lead to strong intermolecular coupling in the relative orientation of molecules and the nonlinear response can be highly nonlocal and very slow. In photorefractive materials, charge transport due to Coulomb interactions subsequent to the absorption of light also results in nonlocality. The nonlinearities in these two classes of materials can be very large and hence can be very accessible for simple experiments at low powers. Unfortunately, in some cases they have erroneously been referred to as Kerr nonlinearities.

The following nonlinear effects in condensed matter will be discussed in this chapter:

1. Molecules with anisotropic polarizability: molecular reorientation and cooperative effects in liquid crystals
2. Photorefractive effects
3. Nuclear (vibrational) contributions to n_2
4. Electrostriction
5. Thermal nonlinearities
6. Cascading of second-order nonlinearities

Nonlinear Optics: Phenomena, Materials, and Devices, George I. Stegeman and Robert A. Stegeman.
© 2012 John Wiley & Sons, Inc. Published 2012 by John Wiley & Sons, Inc.

Electrostriction, thermal effects, and cascading of second-order nonlinearities are nonlinear phenomena that can effectively change the refractive index (or the phase of light beams) nonlocally. They are characterized by the propagation of mechanical effects (sound waves), thermal effects (heat), and the coupling between optical waves at different frequencies via $\chi^{(2)}$, respectively.

12.1 MOLECULAR REORIENTATION EFFECTS IN LIQUIDS AND LIQUID CRYSTALS

12.1.1 Single Molecule Reorientation

Light incident onto an anisotropic molecule induces a dipole moment in that molecule due to its anisotropic polarizability. This dipole moment interacts with the optical field to produce a torque on the molecule. The molecules reorient in response to this light-induced torque, hindered by viscosity and randomized by thermal fluctuations in the positional, rotational, and vibrational degrees of freedom. The "turn-on" time depends on the strength of the applied field, the liquid viscosity, and the molecular shape. The "turn-off" time depends only on the liquid viscosity and molecular shape and is typically in the range of picoseconds to nanoseconds.

Consider the example of the linear CS_2 molecule illustrated in Fig. 12.1a. Its polarizability tensor is shown in Fig. 12.1b.

Since molecules in a liquid have definite axes for their polarizability tensor but are randomly oriented in the liquid, it is necessary to project the induced dipoles onto the field direction to obtain a macroscopic response when averaged over a wavelength. The Euler angles defined in Fig. 12.2 are convenient for relating the molecule's coordinate system $(\bar{x}, \bar{y}, \bar{z})$ to the laboratory frame of reference (x, y, z), in which the electric field $\vec{E}(\vec{r}, t) = \frac{1}{2}\hat{e}_z\mathcal{E}_z\,f^{(1)}(\omega)\,e^{i(kz-\omega t)} + \text{c.c.}$ lies along the z-axis.

The Euler transformations for relating vectors in the (x, y, z) system to the $(\bar{x}, \bar{y}, \bar{z})$ system and vice versa are given by

$$
\begin{bmatrix} \hat{e}_{\bar{x}} \\ \hat{e}_{\bar{y}} \\ \hat{e}_{\bar{z}} \end{bmatrix} = \begin{bmatrix} -\cos\psi\,\sin\phi - \cos\theta\,\cos\phi\,\sin\psi & \cos\phi\,\cos\psi - \cos\theta\,\sin\phi\,\sin\psi & \sin\theta\,\sin\psi \\ \sin\phi\,\sin\psi - \cos\theta\,\cos\phi\,\cos\psi & -\cos\phi\,\sin\psi - \cos\theta\,\sin\phi\,\cos\psi & \sin\theta\,\cos\psi \\ \sin\theta\,\cos\phi & \sin\theta\,\sin\phi & \cos\theta \end{bmatrix}
$$
$$
\times \begin{bmatrix} \hat{e}_x \\ \hat{e}_y \\ \hat{e}_z \end{bmatrix}, \tag{12.1a}
$$

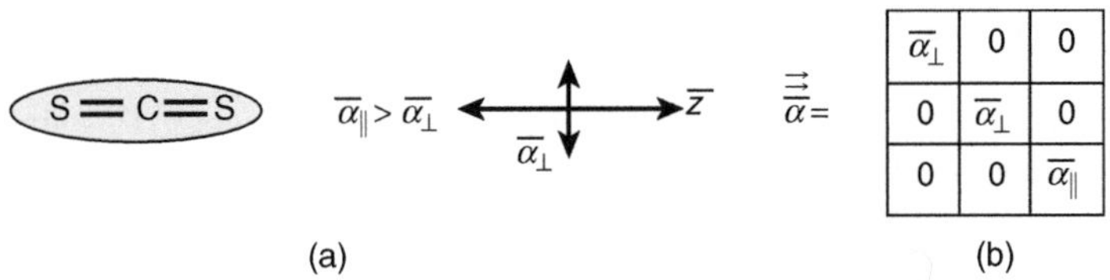

(a)   (b)

FIGURE 12.1 (a) CS_2 molecule and its polarizability. (b) Polarizability tensor for CS_2.

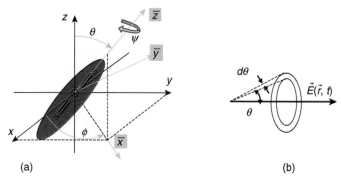

(a)   (b)

FIGURE 12.2   (a) Definition of the Euler angles and the laboratory (x, y, z) and molecular $(\bar{x}, \bar{y}, \bar{z})$ frames of reference. (b) The circular area element lying on a sphere for averaging over angles.

$$
\begin{bmatrix} \hat{e}_x \\ \hat{e}_y \\ \hat{e}_z \end{bmatrix} = \begin{bmatrix} -\sin\phi\cos\psi - \cos\theta\cos\phi\sin\psi & \sin\phi\sin\psi - \cos\theta\cos\phi\cos\psi & \sin\theta\cos\phi \\ \cos\phi\cos\psi - \cos\theta\sin\phi\sin\psi & -\cos\phi\sin\psi - \cos\theta\sin\phi\cos\psi & \sin\theta\sin\phi \\ \sin\theta\sin\psi & \sin\theta\cos\psi & \cos\theta \end{bmatrix}
$$
$$
\times \begin{bmatrix} \hat{e}_{\bar{x}} \\ \hat{e}_{\bar{y}} \\ \hat{e}_{\bar{z}} \end{bmatrix}, \tag{12.1b}
$$

respectively. Therefore the incident field experienced by the molecule has the components

$$
\mathcal{E}_{\bar{x}} = [-\sin\phi\cos\psi - \cos\theta\cos\phi\sin\psi]\mathcal{E}_x + [\cos\phi\cos\psi - \cos\theta\sin\phi\sin\psi]\mathcal{E}_y + \sin\theta\sin\psi\,\mathcal{E}_z,
$$
$$
\mathcal{E}_{\bar{y}} = [\sin\phi\sin\psi - \cos\theta\cos\phi\cos\psi]\mathcal{E}_x + [-\cos\phi\sin\psi - \cos\theta\sin\phi\cos\psi]\mathcal{E}_y + \sin\theta\cos\psi\,\mathcal{E}_z,
$$
$$
\mathcal{E}_{\bar{z}} = \sin\theta\cos\phi\,\mathcal{E}_x + \sin\theta\sin\phi\,\mathcal{E}_y + \cos\theta\,\mathcal{E}_z
$$

$$\tag{12.2}$$

for molecules with symmetry axes at angles (θ, ϕ, ψ) relative to the applied field. Therefore the induced polarization is

$$
\vec{\mathcal{P}} = \overleftrightarrow{\bar{\alpha}} \cdot \vec{\mathcal{E}}_{\mathrm{loc}} = [\hat{e}_{\bar{x}}\bar{\alpha}_\perp \sin\theta\sin\psi\,\mathcal{E}_z + \hat{e}_{\bar{y}}\bar{\alpha}_\perp \sin\theta\cos\psi\,\mathcal{E}_z + \hat{e}_{\bar{z}}\bar{\alpha}_\| \cos\theta\,\mathcal{E}_z]f^{(1)}(\omega),
$$
$$
\vec{\mathcal{P}} = \hat{e}_{\bar{x}}\bar{\mathcal{P}}_{\bar{x}} + \hat{e}_{\bar{y}}\bar{\mathcal{P}}_{\bar{y}} + \hat{e}_{\bar{z}}\bar{\mathcal{P}}_{\bar{z}}, \tag{12.3}
$$

and transforming back to the laboratory frame of reference yields for one molecule

$$
\mathcal{P}_x = [-\sin\phi\cos\psi - \cos\theta\cos\phi\sin\psi]\mathcal{P}_{\bar{x}} + [\sin\phi\sin\psi - \cos\theta\cos\phi\cos\psi]\mathcal{P}_{\bar{y}} + \sin\theta\cos\phi\,\mathcal{P}_{\bar{z}},
$$
$$
\mathcal{P}_y = [\cos\phi\cos\psi - \cos\theta\sin\phi\sin\psi]\mathcal{P}_{\bar{x}} + [-\cos\phi\sin\psi - \cos\theta\sin\phi\cos\psi]\mathcal{P}_{\bar{y}} + \sin\theta\sin\phi\,\mathcal{P}_{\bar{z}},
$$
$$
\mathcal{P}_z = \sin\theta\sin\psi\,\mathcal{P}_{\bar{x}} + \sin\theta\cos\psi\,\mathcal{P}_{\bar{y}} + \cos\theta\,\mathcal{P}_{\bar{z}}.
$$

$$\tag{12.4}$$

To find the total macroscopic polarization $\vec{\mathcal{P}}$ in the laboratory frame, it is necessary to average over all angles $\langle \theta, \phi, \psi \rangle$ and multiply by N, the number of molecules per unit volume, which gives

$$\mathcal{P}_x = N\{\bar{\alpha}_\perp \langle -\sin\theta \sin\psi \sin\phi \cos\psi - \sin\theta \cos\theta \cos\phi \sin^2\psi + \sin\theta \sin\phi \sin\psi$$
$$-\sin\theta \cos\theta \cos\phi \cos^2\psi \rangle + \bar{\alpha}_\| \langle \sin\theta \cos\theta \cos\phi \rangle \} \mathcal{E}_z [f^{(1)}(\omega)]^2$$
$$= N\langle \bar{g}_x(\theta, \phi, \psi) \rangle \mathcal{E}_z [f^{(1)}(\omega)]^2,$$

(12.5a)

$$\mathcal{P}_y = N\{\bar{\alpha}_\perp \langle \sin\theta \cos\phi \cos\psi \sin\psi - \sin\theta \cos\theta \sin\phi \sin^2\psi - \sin\theta \cos\phi \sin\psi \cos\psi$$
$$-\sin\theta \cos\theta \sin\phi \cos^2\psi \rangle + \bar{\alpha}_\| \langle \sin\theta \cos\theta \sin\phi \rangle \} \mathcal{E}_z [f^{(1)}(\omega)]^2$$
$$= N\langle \bar{g}_y(\theta, \phi, \psi) \rangle \mathcal{E}_z [f^{(1)}(\omega)]^2,$$

(12.5b)

$$\mathcal{P}_z = N\{\bar{\alpha}_\perp \langle \sin^2\theta \sin^2\psi + \sin^2\theta \cos^2\psi \rangle + \bar{\alpha}_\| \langle \cos^2\theta \rangle \} \mathcal{E}_z [f^{(1)}(\omega)]^2$$
$$= N\langle \bar{g}_z(\theta, \phi, \psi) \rangle \mathcal{E}_z [f^{(1)}(\omega)]^2.$$

(12.5c)

To facilitate the averaging process, define $d\,\mathrm{Pr}(\theta, \phi, \psi) = \{$probability of the molecular axis $(\alpha_\|)$ in a cone at angle θ to the z-axis (field direction) in a cone segment $d\phi\,d\psi$, with a cone of width $d\theta \propto \{\sin\theta\,d\theta$ (i.e., $d\theta \propto -d\cos\theta)\} \times \{$net effect of ordering by the applied field$\}$ (see Fig. 12.2b).

There are two factors that determine the molecule's ordering by the field—the potential well $\overline{V}_{\mathrm{int}}$ created at the molecule by the applied field and the thermal fluctuations (energy $k_B T$, where k_B is the Boltzmann constant)—which tend to return the molecular angular distribution back to a random state. The external field produces a potential well created by the torque of the form

$$\overline{V}_{\mathrm{int}} = -\int \vec{P}(\vec{r}, t) \cdot d\vec{E}_{\mathrm{loc}}(\vec{r}, t) = -\frac{1}{8}\left[\hat{e}_{\bar{x}} \bar{\alpha}_\perp \sin\theta \sin\psi \mathcal{E}_z + \bar{\alpha}_\perp \hat{e}_{\bar{y}} \sin\theta \cos\psi \mathcal{E}_z + \hat{e}_{\bar{z}} \bar{\alpha}_\| \cos\theta \mathcal{E}_z \right]$$
$$\times [\sin\theta \sin\psi \hat{e}_{\bar{x}} + \sin\theta \cos\psi \hat{e}_{\bar{y}} + \cos\theta \hat{e}_{\bar{z}}][f^{(1)}(\omega)]^2 \mathcal{E}_z \{e^{2i\omega t} + 2 + e^{-2i\omega t}\}.$$

(12.6)

The molecular relaxation times (e.g., picoseconds in CS_2) for returning an oriented molecule back to a random distribution are much larger than a period of the electromagnetic (EM) field's oscillation. Hence the molecular reorientation process cannot follow the field at 2ω and responds only to the time average. This gives for the time averaged potential $\overline{\overline{V}}_{\mathrm{int}}$

$$\overline{\overline{V}}_{\mathrm{int}} = -\frac{1}{4}\left[\bar{\alpha}_\| \cos^2\theta + \bar{\alpha}_\perp \sin^2\theta \cos^2\psi + \bar{\alpha}_\perp \sin^2\theta \sin^2\psi \right][f^{(1)}(\omega)]^2 \mathcal{E}_z^2$$

$$= -\frac{1}{4}\left[\bar{\alpha}_\| \cos^2\theta + \bar{\alpha}_\perp (1 - \cos^2\theta) \right][f^{(1)}(\omega)]^2 \mathcal{E}_z^2$$

(12.7)

$$= -\frac{1}{4}\left[(\bar{\alpha}_\| - \bar{\alpha}_\perp) \cos^2\theta + \bar{\alpha}_\perp \right][f^{(1)}(\omega)]^2 \mathcal{E}_z^2$$

$$\rightarrow \quad d\Pr(\theta, \phi, \psi) \propto -\exp\left(-\frac{\overline{\overline{V}}_{\text{int}}}{k_B T}\right) d\{\cos\theta\}\, d\phi\, d\psi. \tag{12.8}$$

Therefore the polarization induced in the laboratory frame is given by Eq. 12.9 for which $\bar{g}_{x,y,z}(\theta, \phi, \psi)$ was defined in Eq. 12.5a and the total probability is normalized to unity by the integral in the denominator:

$$\mathcal{P}_{x,y,z} = N\mathcal{E}_z\left[f^{(1)}(\omega)\right]^2 \frac{\int_0^{2\pi}\int_0^{2\pi}\int_1^0 \bar{g}_{x,y,z}(\theta, \phi, \psi)\exp[-\overline{V}_{\text{int}}/k_B T]\, d[\cos\theta]\, d\phi\, d\psi}{\int_0^{2\pi}\int_0^{2\pi}\int_1^0 \exp[-\overline{V}_{\text{int}}/k_B T]\, d[\cos\theta]\, d\phi\, d\psi}. \tag{12.9}$$

For the small reorientation angles typical of condensed matter, the $\exp[-\overline{V}_{\text{int}}/k_B T]$ term can be expanded as

$$\exp\left[-\frac{\overline{\overline{V}}_{\text{int}}}{k_B T}\right] \cong 1 + \frac{1}{4k_B T}\left[(\bar{\alpha}_{\parallel} - \bar{\alpha}_{\perp})\cos^2\theta + \bar{\alpha}_{\perp}\right][f^{(1)}(\omega)]^2\mathcal{E}_z^2, \tag{12.10}$$

and so all the terms in Eq. 12.10 become simple trigonometric functions. After some manipulations, this gives $\mathcal{P}_z = \mathcal{P}_z^L + \mathcal{P}_z^{\text{NL}}$, with $\mathcal{P}_z^L = N\mathcal{E}_z f^{(1)}(\omega)$ $\frac{1}{3}(\bar{\alpha}_{\parallel} + 2\bar{\alpha}_{\perp})$ and

$$\mathcal{P}_z^{\text{NL}} = N\mathcal{E}_z[f^{(1)}(\omega)]^4 \frac{(\bar{\alpha}_{\parallel} - \bar{\alpha}_{\perp})^2}{45k_B T}\mathcal{E}_z^2 = N\mathcal{E}_z[f^{(1)}(\omega)]^4 \frac{2(\bar{\alpha}_{\parallel} - \bar{\alpha}_{\perp})^2}{45k_B Tnc\varepsilon_0}I. \tag{12.11}$$

An extra local field correction has been added as explained in Section 8.2.2.1. The $n_{2\parallel,\text{or}}(-\omega; \omega)$ for this process is then given by

$$n_{2\parallel,\text{or}}(-\omega; \omega) = \frac{N}{n_0^2\varepsilon_0^2 c}\frac{(\bar{\alpha}_{\parallel} - \bar{\alpha}_{\perp})^2}{45k_B T}[f^{(1)}]^4. \tag{12.12}$$

Using a similar analysis, nonlinear index coefficients for orthogonal and circular polarizations are

$$n_{2\perp,\text{or}}(-\omega; \omega) = -\frac{N[f^{(1)}]^4}{n_0^2\varepsilon_0^2 c}\frac{(\bar{\alpha}_{\parallel} - \bar{\alpha}_{\perp})^2}{90k_B T} = \frac{1}{2}n_{2\parallel,\text{or}}(-\omega; \omega),$$

$$\tag{12.13}$$

$$n_{2,\text{cp}}(-\omega; \omega) = \frac{N[f^{(1)}]^4}{n_0^2\varepsilon_0^2 c}\frac{(\bar{\alpha}_{\parallel} - \bar{\alpha}_{\perp})^2}{180k_B T},$$

respectively. By definition, the net index change saturates when all the molecules are lined up—a situation never achieved for single molecules in condensed matter.

This situation in practice occurs only for liquid crystals, as discussed in Section 12.1.2. For an anisotropic molecule in which $\bar{\alpha}^2_{\bar{x}\bar{x}} \neq \bar{\alpha}^2_{\bar{y}\bar{y}} \neq \bar{\alpha}^2_{\bar{z}\bar{z}}$, a more complex analysis gives

$$n_{2\|,\text{or}}(-\omega; \omega) = \frac{N}{n_0^2 \varepsilon_0^2 c} \frac{[f^{(1)}]^4}{45 k_B T} \left\{ \left[\bar{\alpha}^2_{\bar{x}\bar{x}} + \bar{\alpha}^2_{\bar{y}\bar{y}} + \bar{\alpha}^2_{\bar{z}\bar{z}}\right] - \left[\bar{\alpha}_{\bar{x}\bar{x}}\bar{\alpha}_{\bar{z}\bar{z}} + \bar{\alpha}_{\bar{x}\bar{x}}\bar{\alpha}_{\bar{y}\bar{y}} + \bar{\alpha}_{\bar{y}\bar{y}}\bar{\alpha}_{\bar{z}\bar{z}}\right] \right\}.$$

$$(12.14)$$

Approximate turn-on and turn-off times can be obtained from the Debye rotational diffusion equation in terms of the "order parameter" Q (1):

$$Q = \left\langle \frac{3}{2}\cos^2\theta - \frac{1}{2} \right\rangle = \frac{\int_1^0 \left\{ \frac{3}{2}\cos^2\theta - \frac{1}{2} \right\} \exp\left\{ [f^{(1)}]^2 \frac{\mathcal{E}_z^2}{4k_B T}(\bar{\alpha}_\| - \bar{\alpha}_\perp)\cos^2\theta \right\} d[\cos\theta]}{\int_1^0 \exp\left([f^{(1)}]^2 \frac{\mathcal{E}_z^2}{4k_B T}(\bar{\alpha}_\| - \bar{\alpha}_\perp)\cos^2\theta \right) d[\cos\theta]}.$$

$$(12.15)$$

The order parameter Q quantifies the amount of the order present (1). When the molecules are randomly oriented, $Q = 0$, and when they are all aligned, $Q = 1$. The molecules undergo a diffusive random walk when they reorient, as illustrated in Fig. 12.3. Mathematically, this is described by the Debye rotational diffusion equation for Q in terms of the Debye time $\tau_D = \eta/5k_B T$ and the viscosity η as

$$\frac{\partial Q}{\partial t} = -\frac{Q}{\tau_D} + [f^{(1)}]^2 \frac{2\mathcal{E}_z^2}{3\eta}(\bar{\alpha}_\| - \bar{\alpha}_\perp) \quad \Rightarrow \quad Q = \frac{2}{3}[f^{(1)}]^2 \frac{\tau_D \mathcal{E}_z^2}{\eta}(\bar{\alpha}_\| - \bar{\alpha}_\perp)\{1 - e^{-t/\tau_D}\}.$$

$$(12.16)$$

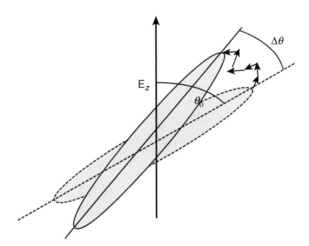

FIGURE 12.3 Molecular reorientation driven by an external field occurs via stochastic diffusion of the molecular orientation due to collisions and other interaction with neighboring molecules.

Also, when the external field is turned off, Q decays as $Q = Q_{max}\, e^{-t/\tau_D}$. Typical times range from a few picoseconds for simple molecules such as CS_2 to nanoseconds for large molecules.

12.1.2 Liquid Crystals

Liquid crystals are a novel form of matter with some liquidlike and some crystallike properties that are strongly temperature dependent (1,2). Strong intermolecular forces between molecules in the liquid state can lead to molecular "clusters," aligned along a specific direction in space ("director"). It is this alignment that leads to the name "liquid crystal." The orientational correlation exists only over a limited temperature range above the melting point. Note that in contrast to the solid state where X-ray diffraction patterns also reveal positional correlation, there is no such positional correlation between the molecules. The nonlinear optics of liquid crystals is in some ways closely related to the previous liquid molecule case.

12.1.2.1 General Optical Properties of Liquid Crystals. Many "families" of liquid crystals exist (1,2). Most of the liquid crystal molecules can be considered to have ellipsoidal shapes, as shown in Fig. 12.4, and are basically charge transfer systems of the type discussed in Chapter 7. The most commonly used and extensively studied molecule is 5CB, also shown in Fig. 12.4. Examples of R and R′ are C_nH_{2n+1} and $C_nH_{2n+1}O$, with nitro groups and cyano groups (e.g., 5CB). The most common types of liquid crystal classes are shown in Fig. 12.5.

A single molecule can take on different liquid crystal ordering as the temperature or the side groups are changed. For example, nCB is not a liquid crystal for $n \leq 4$; it is nematic for $n = 5-7$ and then smectic for larger n. Although some molecules may exhibit a permanent dipole moment, the net alignment can average the dipole moment to zero over optical wavelengths. Note that the alignment is not perfect and is described by an order parameter Q similar to the single molecule reorientation case. This "average" direction is called the "director" $\hat{n}$.

The preponderance of nonlinear optics experiments in liquid crystals has been performed with nematic liquid crystals. As the temperature is raised, a second-order phase transition to an isotropic liquid occurs, the order is lost, and the birefringence

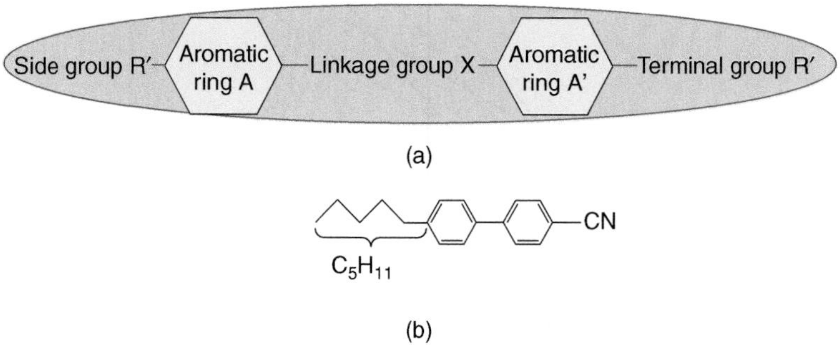

(a)

(b)

FIGURE 12.4   (a) Typical internal structure of a liquid crystal molecule. (b) Structure of 5CB.

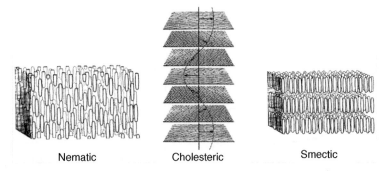

FIGURE 12.5 Internal molecular order in nematic, cholesteric, and smectic liquid crystals. Reproduced with permission from Elsevier Publishing (3).

disappears. In the transition region, Q approaches zero and the correlation distance over which orientational order exists decreases, resulting in a great deal of spurious light scattering, a region normally avoided for optical experiments. At appropriately prepared single boundaries, as well as between two plates with prepared surfaces, it is possible to anchor the orientation of the molecules at the boundary and maintain this order for some distance away from the boundary. There are two possible directions for the director $\hat{n}$ at the boundary: parallel to (planar boundary condition) or orthogonal to (homeotropic boundary condition) the boundary (see Fig. 12.6).

12.1.2.2 Liquid Crystals Nonlinear Optics: Reorientation Effects. As the temperature of a liquid crystal medium is increased near the nematic–isotropic phase transition, limited orientational order persists over subwavelength volumes with directors ($\hat{n}$) not parallel to each other. These clusters behave like large, highly polarizable molecules and can be oriented by strong optical fields as discussed before for the single molecule case. The larger the cluster size, the larger the nonlinearity and the slower the response time (see Fig. 12.7a). As the temperature is increased, the cluster size decreases until this reorientational nonlinearity reaches the single molecule value ($\sim 10^{-13}\,\mathrm{cm^2/W}$) (see Fig. 12.7b).

The combination of a planar boundary condition, an applied zero frequency (DC) field, and a copolarized (to the DC field) optical field can lead to a large optical

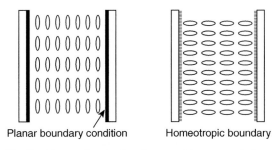

Planar boundary condition   Homeotropic boundary

FIGURE 12.6 The liquid crystal order for planar and homeotropic boundary conditions.

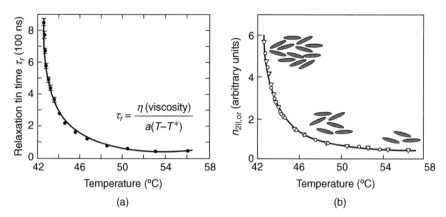

FIGURE 12.7 (a) Variation of the relaxation time with temperature for the liquid crystal MBBA (n-4'-Methoxy Benzilidene-n Butyl Anilin). (b) Variation of the nonlinearity of MBBA versus temperature. Reproduced with permission from Elsevier Publishing (3)

nonlinearity $n_{2\|,\text{or}}(-\omega;\omega)$ (4). Consider a sample with a DC field applied orthogonally to the boundaries (see Fig. 12.8a). As the field is increased, at some critical field E_F the boundary molecules become decoupled from the boundary (see Fig. 12.8b) and start to rotate toward the field direction (see Fig. 12.8c). This decoupling occurs at the "Fréedericksz transition." The DC field E_F is given by

$$E_F = \frac{\pi}{d}\sqrt{\frac{K_1}{\Delta\varepsilon}}, \tag{12.17}$$

where d is the plate separation and K_1 is the "splay" Frank elastic constant ($\approx 10^{-11}$ N).

Since the molecules have essentially uniaxial symmetry, the extraordinary refractive index along the field direction becomes a function of angle θ. If now an optical field is applied with polarization parallel to the DC field, it increases the angle θ, as shown in Fig. 12.9. The net result is that optical powers of just a few milliwatts are sufficient to obtain appreciable index changes.

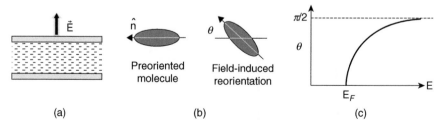

FIGURE 12.8 (a) E_{DC} field orthogonal to planar boundaries. (b) Liquid crystal molecules and their director prior to and after the Fréedericksz transition. (c) The direct current field dependence of the orientation angle. Courtesy of Prof. G. Assanto, University Rome 3.

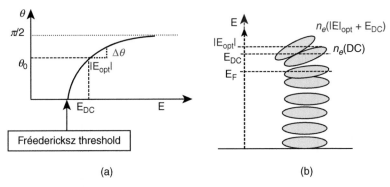

FIGURE 12.9   (a) Combination of both a DC field and an optical field used to increase the director angle by $\Delta\theta$ and hence the refractive index by Δn orthogonal to the plate boundary. (b) The orientation of the liquid crystal molecules as a function of field. Courtesy of Prof. G. Assanto, University Rome 3.

12.1.2.3   *Giant Orientational Optical Nonlinearities in Doped Nematic Liquid Crystals.*   The addition of photosensitive dye molecules as *dopants* can be used to mediate, facilitate, and enhance the reorientation process (5). The largest effects are obtained with azobenzenes that can undergo light-induced trans–cis isomerization, which is the key enabling mechanism. These molecules can exist in two states: cis and trans (see Fig. 12.10a). When illuminated with blue light, the geometry of the azobenzene molecule changes from a linear shape to a bent shape, causing a disruption in the liquid crystal order; i.e., there is a transformation from a liquid crystal to an amorphous medium with randomized orientation of the liquid crystal molecules. This change, when written in terms of an effective n_2, gives $n_{2\|,\mathrm{or}}(-\omega;\omega) \approx -10^3\,\mathrm{cm}^2/\mathrm{W}$. The azobenzene can usually be returned to the trans form by illuminating with ultraviolet light.

12.1.2.4   *Thermal Nonlinearities.*   Liquid crystals with large absorption coefficients (more than a few cm^{-1}) have very large nonlinearities since the effective $|\partial n/\partial T|$ is orders (>2) of magnitude larger than for normal liquids or solids due to the effect of

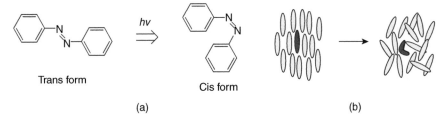

FIGURE 12.10   (a) The conformational change induced in an azobenzene molecule when it is illuminated with light in the blue region of the spectrum. (b) The change in the liquid order that occurs due to the cis–trans isomerization of the azobenzene molecule. Reproduced with permission from the Optical Society of America (6).

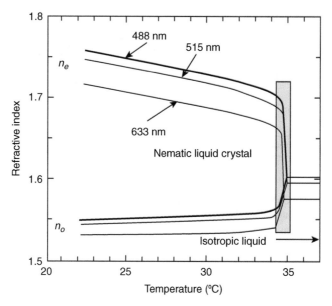

FIGURE 12.11 The variation in the refractive indices of a nematic liquid crystal at various optical wavelengths as it passes through the nematic–isotropic phase transition (3).

temperature on the order parameter Q (1). The dependence of refractive index on temperature is shown in Fig. 12.11. The large $\partial n/\partial T$ in the blue box rectangle region is a consequence of the rapid decrease with increasing temperature of the size of the aligned regions as the nematic to isotropic liquid crystal transition is approached. At the temperature at which the birefringence $n_e - n_o$ vanishes, the domains become subwavelength in size and the material strongly scatters light. These effects can be enhanced by including other strongly absorbing molecules. The resulting nonlinearities can attain values of $1\ \text{cm}^2/\text{W}$, usually with large attenuation coefficients.

12.1.2.5 Summary of Liquid Crystal Nonlinear Refractive Index Coefficients. This summary is given in Table 12.1.

12.2 PHOTOREFRACTIVE NONLINEARITIES

These *nonlocal* nonlinearities occur due to a combination of physical phenomena that indirectly lead to an index change that depends on the input intensity or integrated fluence in electro-optic active media (7). They have proven very useful for investigating nonlinear phenomena at sub-watt powers, even as low as microwatts. However, their detailed physics can be very complicated.

Photorefractive materials have electron donor and acceptor states (defects, dopants, and so on) that are located between the valence and conduction bands, as shown in Fig. 12.12.

TABLE 12.1 The Effective $n_{2||,\text{eff}}(-\omega; \omega)$ Available in Liquid Crystals for Different Physical Mechanisms

Mechanism	$n_{2||,\text{eff}}(-\omega; \omega) \ (\text{cm}^2/\text{W})$
Kerr electronic nonresonant nonlinearities	$10^{-14}-10^{-13}$
Isotropic phase, molecular reorientation	$10^{-14}-10^{-12}$
Nematic phase, thermal and order parameter change	$10^{-6}-10^{-4}$
Nematic phase, optically induced reorientation	$10^{-4}-10^{-3}$
Nematic phase, photorefractive, doped	$10^{-3}-10^{-1}$
Nematic phase, excited dopant (dye molecule) assisted	$10^{-3}-10^{-1}$
Nematic phase, azobenzene liquid crystal (BMAB) doped	$10^{-2}-10^{0}$
Nematic phase, azobenzene trans–cis isomerization	$10^{0}-2 \times 10^{3}$

From Ref. 3.
BMAB = 4.4 butylmethoxy azobenzene.

The magnitude of $n_{2||,\text{pr}}(-\omega; \omega)$ can be very large. However, nonlinearity $\times$ response time $\cong$ constant trade-off since it is the accumulated energy *absorbed* from a beam, which is the key parameter. A number of different photorefractive mechanisms are possible. The most commonly used is the steady-state response due to the diffusion and screening mechanisms, which leads to index changes in electro-optic active materials.

12.2.1 Screening Nonlinearity

12.2.1.1 No Applied DC Field. The photorefractive effect is usefully described in terms of the steps that occur to establish it.

1. Electrons are promoted from neutral electron donor states (N_D) into the conduction band by the *absorption* of light, resulting in electrons more or less free to migrate, and in ionized electron donor states, N_D^+ (see Fig. 12.13).

   The equations describing this process are the rate of generation of ionized donor states and the charge continuity condition (6):

$$\frac{\partial N_D^+}{\partial t} = \frac{\alpha_1}{hv}(I + I_d)(N_D - N_D^+) - \frac{N_e N_D^+}{\tau_R}, \qquad \frac{\partial(N_D^+ - N_e)}{\partial t} + \frac{1}{e}\nabla \cdot \vec{J} = 0.$$

$$(12.18)$$

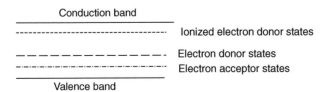

FIGURE 12.12 The band and impurity state structure of a typical photorefractive material.

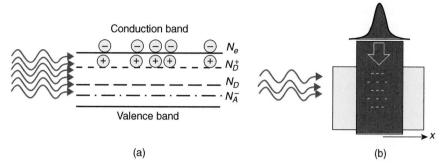

FIGURE 12.13 (a) Photons are absorbed and electrons are excited from the donor state (N_D) into the conduction band, reducing the density of neutral donor states and producing positively charged electron donor states (N_D^+). (b) Initial charge distribution relative to the incident beam.

In these equations, τ_R is the recombination time and I_d is the equivalent "dark current" that quantifies the steady-state thermal excitation of electrons into the conduction band. Initially the density of charges is highest at the center of the beam, as indicated in Fig. 12.13b.

2. Electrons travel due to Coulomb repulsion forces via diffusion to regions of lower electron density N_e in the conduction band. This creates a local space charge field E_{SC}. The relevant current density and the $\vec{\nabla} \cdot \vec{D} = \rho$ equations, respectively, describe the charge motion and evolution of the space charge field $\vec{E}_{SC}$:

$$J_x = eD\frac{\partial N_e}{\partial x} = k_B T \mu \frac{\partial N_e}{\partial x}, \qquad \vec{\nabla} \cdot (\varepsilon \vec{E}_{SC}) = e(N_D^+ - N_e - N_A). \quad (12.19)$$

Here D is the diffusion constant and μ is the mobility.

3. Subsequently, the electrons fall back into acceptor states, essentially "freezing in" the charge distribution in the various states, as shown in Fig. 12.14a. Since electrons are trapped in new sites (see Fig. 12.14b), the charge separation produces a steady-state "space-charge" field E_{SC} (see Fig. 12.14c), which in turn produces an index change since the medium is electro-optic active.

The pertinent equations for this steady state are

$$J_x = e\mu[N_e + N_{th}]E_{sc} + k_B T\mu\frac{\partial N_e}{\partial x}$$

$$\text{Because } J_x = 0 \quad \Rightarrow \quad E_{SC} = -\frac{k_B T}{e[N_e + N_{th}]}\frac{\partial N_e}{\partial x}$$

$$\text{Because } N_e \propto I \quad \Rightarrow \quad E_{SC} \cong -\frac{k_B T}{e[I + I_d]}\frac{\partial I}{\partial x} \qquad (12.20)$$

$$\Rightarrow \quad \Delta n = n_0^3 r_{\text{eff}} E_{SC}.$$

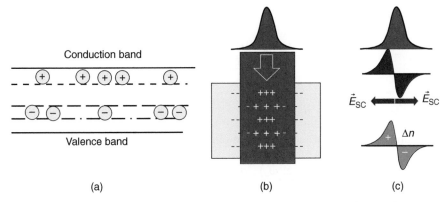

(a)   (b)   (c)

FIGURE 12.14 (a) Equilibrium distribution of charges in various states after the electrons in the conduction band are retrapped. (b) Spatial distribution of charges across the optical beam. (c) The spatial variation of the space charge field (middle) and the index change induced by the space charge field created (lower) by a combination of the space charge field and the electro-optic effect. The sign of the coefficient determines the sign of the index change in each half.

Here r_{eff} is the appropriate electro-optic coefficient for the beam geometry. The turn-on and turn-off response times are usually *very slow*. The turn-on time, ranging from microseconds to seconds, is determined by the input intensity, i.e., the integrated absorbed flux and the carrier diffusion time. The turn-off time depends on the thermal excitation rate of the carriers and their diffusion time.

The nonlocal nature of the nonlinearity leads to a novel "*self-bending*" effect of the propagating beam (8). In simple physical terms, the index change in Fig. 12.14c will cause light to bend into the region of higher index and away from the region of lower index, similar to refraction at the interface between two media with different refractive indices (see Fig. 12.15a). Mathematically, since $I_d = f(T) \neq f(I) \neq f(x)$,

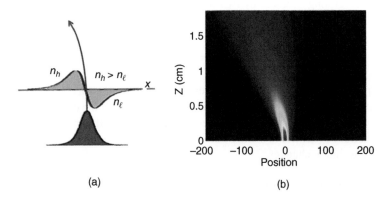

(a)   (b)

FIGURE 12.15 (a) Simple model for the self-bending process. (b) Simulation of self-bending. A $w_0 = 4\,\mu\text{m}$ beam self-deflects by $5.7\,\mu\text{m}$ along the x-axis after a distance of 10 diffraction lengths.

Eq. 12.20 can be rewritten as

$$\vec{E}_{SC} = -\frac{k_B T}{e} \frac{\nabla I}{I + I_d} = -\frac{k_B T}{e} \nabla \left(\frac{|\overline{E_{opt}}|}{|E_d|} \right)^2 \bigg/ \left[1 + \left(\frac{|\overline{E_{opt}}|}{|E_d|} \right)^2 \right],$$

$$\vec{E}_{SC} \xrightarrow{\overline{E}_{opt} \gg |E_d|} -\frac{k_B T}{e} \frac{\nabla |\overline{E_{opt}}|^2}{|E_d|^2}.$$

$$(12.21)$$

Substituting $\Delta n = n_0^3 r_{eff} E_{SC}$ into the two-dimensional wave equation gives

$$\frac{d^2 E(x, z; \omega_a)}{dx^2} + \frac{d^2 E(x, z; \omega_a)}{dz^2} - \frac{(n + \Delta n)^2}{c^2} \frac{d^2 E(x, z; \omega_a)}{dt^2} = 0,$$

$$\text{SVEA}: \quad i \frac{\partial E_{opt}}{\partial z} + \frac{1}{2k} \frac{\partial^2 E_{opt}}{\partial x^2} + \frac{k_0}{2} n_0^3 r_{eff} \frac{k_B T}{e} \left(\frac{\partial(|\overline{E_{opt}}|^2)/\partial x}{|E_{opt}|^2} \right) \overline{E}_{opt} = 0.$$

$$(12.22)$$

The solution will depend on the input beam's field distribution. Assuming a Gaussian beam $E(z = 0, x) = A \exp[-x^2/2w_0^2]$, it can be shown that the beam bends a distance x_d after propagating a distance z (8):

$$x_d = \frac{k_0^2 w_0^2 n_0^4 r_{eff} k_B T}{2e} \left[2 \frac{z}{k w_0^2} \tan^{-1} \left(\frac{z}{k w_0^2} \right) - \ln \left(1 + \left(\frac{z}{k w_0^2} \right)^2 \right) \right]. \quad (12.23)$$

A numerical simulation of Eq. 12.22 is shown in Fig. 12.41b. Both diffraction and self-bending occur simultaneously. With diffraction, the intensity decreases and therefore so does the bending effect. This phenomenon has been observed experimentally.

Spatially periodic interference between two beams in a photorefractive medium exhibits yet another property unique to photorefractive media (7). Consider the interference "grating" created by two beams crossing at small angles, as shown in Fig. 12.16a. The time average intensity pattern is given by

$$\overline{I(\vec{r}, t)} = \frac{1}{2} c\varepsilon_0 n \{ |\mathcal{E}_1|^2 + |\mathcal{E}_2|^2 + 2\mathcal{E}_1 \mathcal{E}_2 \cos[2k \cos\theta \, x] \}. \quad (12.24)$$

The key point is the $\pi/2$ phase shift induced between the intensity maxima and the index maxima see Fig. 12.16b. This is a direct consequence of the charge motion that occurs, i.e., the nonlocality. In a Kerr medium that is local, the intensity and index grating maxima occur at the same position and there is no energy transfer between beams due to index changes.

12.2.1.2 Screening Nonlinearity: Applied DC Bias Field. The shape of the induced index change shown in Fig. 12.14c is "inconvenient" for many applications. Another possibility is to use a strong bias field $\cong E_0$ to "overcome" diffusion effects (see Fig. 12.17a). In this regime the net field across the beam is much larger than the

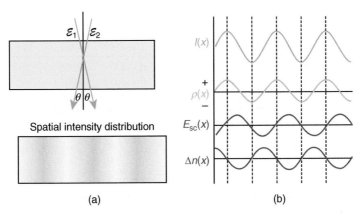

(a)   (b)

FIGURE 12.16 (a) The interference between two crossing light fields produces a spatial intensity grating. (b) From top to bottom are the intensity grating, the charge distribution induced, the space charge field E_{SC} induced by the charge distribution, and the resulting index change produced by the electro-optic effect with a negative coefficient.

small diffusion effects and hence the net space charge field varies with the optical intensity (not with its derivative as with diffusion). The space charge field opposes the applied field, reducing the net field in the region of the optical beam. Furthermore, the steady-state turn-on time can also be reduced by illuminating the whole sample uniformly, called I_∞, which contributes an extra uniform background charge density N_{e0}. In steady state, J is constant, which gives for the space charge field

$$
\begin{aligned}
E_{SC} = & E_0 \frac{(I_\infty + I_d)}{(I + I_d)} \left(1 + \frac{\varepsilon}{eN_A} \frac{\partial E_{SC}}{\partial x} \right) - \frac{k_B T}{e} \frac{(\partial I/\partial x)}{(I + I_d)} \\
& + \frac{k_B T}{e} \frac{\varepsilon}{eN_A} \left(1 + \frac{\varepsilon}{eN_A} \frac{\partial E_{SC}}{\partial x} \right)^{-1} \frac{\partial^2 E_{SC}}{\partial x^2}.
\end{aligned}
\tag{12.25}
$$

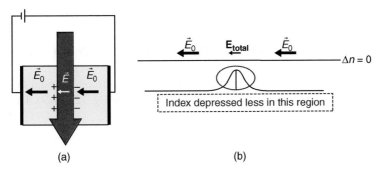

(a)   (b)

FIGURE 12.17 (a) Beam geometry and applied external bias. (b) The index change induced by the field distribution shown in (a).

TABLE 12.2 Parameters for a Number of Photorefractive Materials

Material	Dopant	λ (μm)	$n^3 r_{eff}$ (pm/V)	$^a\tau_{diel}$(sec)	Δn_{max}	E_{DC} (KV/cm)
$Sr_{0.75}Ba_{0.25}Nb_2O_6$	Ce	0.4–0.6	17390	0.1–1.0	0.005	3
$Sr_{0.6}Ba_{0.4}Nb_2O_6$	Ce	0.4–0.6	3000	0.1–1.0	0.0014	3
$BaTiO_3$	Fe	0.4–0.9	21,500	0.1–1.0	0.005	2.5
InP	Fe	0.9–1.3	52	10^{-6}–10^{-4}	$^b5\times10^{-5}$	8

From Ref. 6.
^{a} At an intensity of 1 W/cm^2.
^{b} With enhancement can reach 5×10^{-4}.

The gradient terms can be neglected for strong bias fields, yielding the very simple result for E_{SC}:

$$E_{SC} \cong E_0 \frac{I_\infty + I_d}{I + I_d} \quad \Rightarrow \quad \Delta n = -\frac{1}{2} n_0^3 r_{eff} E_{SC} = -\frac{1}{2} n_0^3 r_{eff} E_0 \frac{I_\infty + I_d}{I + I_d}. \quad (12.26)$$

The properties of a number of common photorefractive materials are listed in Table 12.2.

12.2.2 Photovoltaic Nonlinearity

This nonlinearity arises due to a strong photovoltaic current that is induced by absorption of light alone (no bias field required) in a direction determined by the crystalline symmetry. It occurs naturally in doped noncentrosymmetric crystals, such as $LiNbO_3$ (Table 12.3). Its strength is inversely proportional to the mobility of free carriers, μ. An electric field is created along a specific crystal axis where $E_p = \kappa/e\mu\tau_R$ is the maximum steady-state photovoltaic field, with κ being the photovoltaic constant, e the charge on the electron, and τ_R the carrier's recombination time. The direction of the field, i.e., whether it occurs along the positive or negative axis, depends on the material. This field drives the electrons in the conduction band from the region of the photoionized states and adds to the current density J discussed previously. A typical beam geometry and direction of index change are shown in Fig. 12.18.

12.3 NUCLEAR (VIBRATIONAL) CONTRIBUTIONS TO $n_{2||}(-\omega; \omega)$

Light couples via electric dipole effects to other (than electronic) normal modes in matter, e.g., molecular vibrations for the CO_2 molecule (see Fig. 12.19) (9). This leads

TABLE 12.3 Parameters of Photovoltaic Effect in $LiNbO_3$

Photovoltaic material	Dopant	r_{33} (pm/V)	Δn_{max}	Intensity (W/cm^2)	$\tau_{response}$
$LiNbO_3$	Fe	30	0.002	$\approx$W/cm^2	$\approx$min

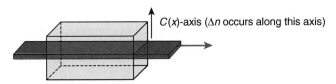

$C(x)$-axis (Δn occurs along this axis)

FIGURE 12.18 Optical beam and induced field geometry for the photovoltaic effect in $LiNbO_3$.

to significant contributions to $n_{2\parallel}(-\omega; \omega)$ (10−20% in glasses and can actually be larger than the electronic Kerr effect (see Fig. 13.1). The formulation given here is for the continuous-wave case, normally valid for pulse widths > 1 ps.

In the presence of vibrating modes with amplitudes $\bar{q}_n^\beta$, the polarizability tensor is given by

$$\bar{\alpha}_{\ell m} = \bar{\alpha}_{\ell m}^L + \sum_\beta \bar{q}_n^\beta \frac{\partial \bar{\alpha}_{\ell mn}^\beta}{\partial \bar{q}_n^\beta}\bigg|_{\bar{q}_n^\beta = 0}, \qquad (12.27)$$

where $\dfrac{\partial \bar{\alpha}_{\ell mn}^\beta}{\partial \bar{q}_n^\beta}\bigg|_{\bar{q}_n^\beta = 0}$ is the Raman hyperpolarizability tensor. The summation over β is over all the vibrational modes, which modulate the linear polarizability, i.e., Raman-active modes. The Raman hyperpolarizability has dispersion with *optical* frequency similar to that of other nonlinear coefficients but is usually considered a constant in Raman spectroscopy (9,10). When an optical field of frequency ω is applied, this gives rise to a nonlinear polarizability *in the molecule* of the form

$$\bar{p}_\ell^{NL} = \sum_\beta \bar{q}_n^\beta \frac{\partial \bar{\alpha}_{\ell\ell n}^\beta}{\partial \bar{q}_n^\beta}\bigg|_{\bar{q}_n^\beta = 0} [f^{(1)}] E_\ell. \qquad (12.28)$$

The potential energy associated with the interaction of the induced dipole with the field is given by

$$\bar{V}_{int} \propto -\int \bar{p}_\ell \, dE_\ell = -\frac{1}{2} \sum_\beta \bar{q}_n^\beta \frac{\partial \bar{\alpha}_{\ell\ell n}^\beta}{\partial \bar{q}_n^\beta}\bigg|_{\bar{q}_n^\beta = 0} [f^{(1)}]^2 E_\ell E_\ell. \qquad (12.29)$$

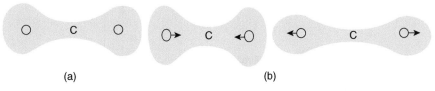

(a)   (b)

FIGURE 12.19 (a) CO_2 molecule at rest with its electron cloud distribution. (b) CO_2 with its electron cloud distorted by the vibration along the molecular axis.

From classical mechanics, there is an all-optical force that can drive the vibration of the βth mode with frequency Ω_β (approximated as a simple harmonic oscillator) in the molecule:

$$
\begin{aligned}
\bar{F}^\beta_n &= -\frac{\partial}{\partial \bar{q}^\beta_n} V_{\text{int}} = \frac{1}{2} \frac{\partial \bar{\alpha}_{\ell\ell n}}{\partial \bar{q}^\beta_n}\bigg|_{\bar{q}^\beta_n = 0} [f^{(1)}]^2 E_\ell E_\ell \\
&\Rightarrow \quad \bar{m}_\beta \left[\dot{\bar{q}}^\beta_n \ddot{\bar{q}}^\beta_n + 2\bar{\tau}_\beta^{-1} \dot{\bar{q}}^\beta_n + \bar{\Omega}^2_\beta \bar{q}^\beta_n \right] = \frac{1}{2} \frac{\partial \bar{\alpha}_{\ell\ell n}}{\partial \bar{q}^\beta_n}\bigg|_{\bar{q}^\beta_n = 0} [f^{(1)}]^2 E_\ell E_\ell,
\end{aligned}
\tag{12.30}
$$

where $\bar{\tau}_\beta$ is the decay time of the βth vibration mode.

The field product

$$
E_\ell(\vec{r}, t) E_\ell(\vec{r}, t) = \frac{1}{4} \mathcal{E}^2(\omega) \, e^{2i(\vec{k}\cdot\vec{r} - \omega t)} + \frac{1}{4} |\mathcal{E}(\omega)|^2 + \text{c.c.}
$$

can drive the vibration at frequencies 0 and 2ω. Since typically $2\omega \gg \bar{\Omega}_\beta$ (see Fig. 12.20), the $|\mathcal{E}(\omega)|^2$ term is the important driving term so that

$$
\bar{q}^\beta_n(\vec{r}, t) = \frac{1}{2} \bar{Q}^\beta_n + \text{c.c.} \quad \rightarrow \quad \bar{Q}^\beta_n(\vec{r}, t) = \frac{|\mathcal{E}(\omega)|^2}{4\bar{m}_\beta D(0)} \frac{\partial \bar{\alpha}^\beta_{\ell\ell n}}{\partial \bar{q}^\beta_n}\bigg|_{\bar{q}^\beta_n = 0} [f^{(1)}]^2.
\tag{12.31}
$$

Substituting $\bar{q}^\beta_n$ into the nonlinear polarization gives

$$
\bar{p}^{\text{NL}}_\ell(\vec{r}, t) = \sum_\beta \frac{1}{8\bar{m}_\beta \bar{\Omega}^2_\beta} \left| \frac{\partial \bar{\alpha}^\beta_{\ell\ell n}}{\partial \bar{q}^\beta_n}\bigg|_{\bar{q}^\beta_n = 0} \right|^2 [f^{(1)}]^4 \left\{ |\mathcal{E}_\ell(\omega)|^2 \mathcal{E}_\ell(\omega) \, e^{i(\vec{k}\cdot\vec{r} - \omega t)} + \text{c.c.} \right\}.
\tag{12.32}
$$

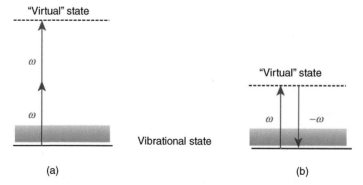

FIGURE 12.20   The all-optical interaction for the (a) 2ω and (b) 0 frequency cases.

Proceeding in the usual way of obtaining $n_{2\parallel}(-\omega; \omega)$ from the nonlinear polarization gives

$$n_{2\parallel,\text{nuc}}(-\omega; \omega) = \sum_{\beta} \frac{N}{4n^2(\omega)\bar{m}_\beta \varepsilon_0^2 c\bar{\Omega}_\beta^2} \left| \frac{\partial \bar{\alpha}_{\ell\ell n}^\beta}{\partial \bar{q}_n^\beta} \right|_{\bar{q}_n^\beta=0} \Big|^2 [f^{(1)}]^4. \tag{12.33}$$

The analysis at this point has assumed an ordered medium such as a crystal. However the vibrational contributions to the nonlinear refraction are also important for amorphous media such as glasses. For random media an angular average over all possible orientations of the molecules is needed; i.e.,

$$n_{2\parallel,\text{nuc}}(-\omega; \omega) = \sum_{\beta} \frac{N}{4n^2(\omega)\bar{m}_\beta \varepsilon_0^2 c\bar{\Omega}_\beta^2} \left\langle \left| \frac{\partial \bar{\alpha}_{\ell\ell n}^\beta}{\partial \bar{q}_n^\beta} \right|_{\bar{q}_n^\beta=0} \Big|^2 \right\rangle [f^{(1)}]^4. \tag{12.34}$$

The problem now is to experimentally find values for the $\left\langle \left| \frac{\partial \bar{\alpha}_{\ell\ell n}^\beta}{\partial \bar{q}_n^\beta} \right|_{\bar{q}_n^\beta=0} \Big|^2 \right\rangle$ term. It turns out that it is the same coefficient that gives the spontaneous Raman scattering spectrum (see Appendix 12.1). The expression for $n_{2\parallel,\text{nuc}}(-\omega; \omega)$ now becomes

$$n_{2\parallel,\text{nuc}}(-\omega; \omega) = \frac{P_{x,\text{Tot}}^{\text{RS}}}{\Delta\Omega I_x(\omega_P)} \frac{c^3(4\pi)^2}{n^2(\omega)\omega_P^4 k_B T}, \tag{12.35}$$

where $P_{x,\text{Tot}}^{\text{RS}}$ is the spontaneous Raman spectrum integrated over frequency, measured over the "solid angle" $\Delta\Omega$ subtended at the detector for the incident intensity $I_x(\omega_P)$ (Table 12.4).

A key question is: "How fast does the vibrational contribution 'kick in' for pulsed lasers?" A specific example is shown in Fig. 13.1 for a nonglass case. For a typical glass vibration, modeling has shown that the full value of $n_{2\parallel,\text{nuc}}(-\omega; \omega)$ is included in $n_{2\parallel}(-\omega; \omega)$ for pulses $>1-10$ ps long, depending on the material (12).

TABLE 12.4 Fractional Contribution Measured of $n_{2\parallel,\text{nuc}}(-\omega; \omega)$ to the Total n_2 for a Few Glasses

Glass	Wavelength (nm)	Nuclear Fraction	Method
Fused silica (SiO_2) (9)	Visible	$\approx$15–18%	Raman
87% GeS_2–13%Ga_2S_3 (11)	825	13$\pm$5%	35 fs OKE
64%PbO–14%Bi_2O_3–7%B_2O_3–15%SiO_2 (12)	825	12$\pm$5%	35 fs OKE
50%GeO_2 in GeO_2–SiO_2 (13)	800	13–18%	18 fs SRTBC
20%Nb_2O_5–80%TeO_2 (13)	800	20%	100 fs OKE

From Ref. 6.
SRTBC, spectrally resolved two beam coupling; OKE, optical Kerr effect.

12.4 ELECTROSTRICTION

Electrostriction is a *universal* mechanism that always has the same sign (>0) for $n_{2\|,\text{str}}(-\omega; \omega)$.

To better understand this effect, consider an electrical capacitor with two parallel plates. When an electric field is applied, charges are induced on the plates with opposite signs on opposite plates. Due to the presence of the positive and negative charges, there is a compressive force squeezing the medium that produces a strain field S_{xx} in the case of an x-polarized electric field. This results in a material contraction $\vec{u}$ and hence a volume change along the x-axis of the form

$$\frac{\Delta V}{V_0} = -\frac{\Delta \rho}{\rho} \quad \Rightarrow \quad \frac{\Delta \rho}{\rho} = -\vec{\nabla} \cdot \vec{u} = -\frac{\partial u_x}{\partial x_x} = -S_{xx},$$

$$\text{Generalized strain } S_{k\ell} = \frac{1}{2}\left[\frac{\partial u_k}{\partial x_\ell} + \frac{\partial u_\ell}{\partial x_k}\right].$$

(12.36)

The density change leads to an increase in the local EM field energy density, i.e., the number of atoms or molecules per unit volume increases. The time average work done in compressing the medium (ΔU) is equal to the increase in EM energy density (ΔW). Taking p_{st} as the pressure (force/area) exerted by the EM field, we obtain

$$\Delta W = p_{\text{st}}\frac{\Delta V}{V} = -p_{\text{st}}\frac{\Delta \rho}{\rho}.$$

(12.37)

Now consider an optical field of the form $\vec{E} = \hat{e}_x \mathcal{E}_0 \cos(\omega t)$

$$\Delta U = \Delta\overline{\left(\frac{1}{2}\varepsilon_0 \varepsilon_{r,x}[\mathcal{E}_0 \cos(\omega t)]^2\right)} = \frac{1}{2}\varepsilon_0\overline{[\mathcal{E}_0 \cos(\omega t)]^2}\Delta\varepsilon_{r,x} = \varepsilon_0\frac{\mathcal{E}_0^2}{4}\frac{\partial \varepsilon_{r,x}}{\partial \rho}\Delta\rho,$$

$$\Delta W = \Delta U$$

$$\Rightarrow \quad p_{\text{st}} = -\rho\frac{\varepsilon_0\mathcal{E}_0^2}{4}\frac{\partial \varepsilon_{r,x}}{\partial \rho} \quad \rightarrow \quad \Delta\rho = \frac{\partial \rho}{\partial p_{\text{st}}}p_{\text{st}} = -\rho\frac{\partial \rho}{\partial p_{\text{st}}}\frac{\mathcal{E}_0^2}{4}\varepsilon_0\frac{\partial \varepsilon_{r,x}}{\partial \rho}.$$

(12.38)

The material compressibility K^{-1} is given by

$$K^{-1} = \frac{1}{\rho}\frac{\partial \rho}{\partial p_{\text{st}}} \quad \rightarrow \quad \frac{\Delta\rho}{\rho} = -S_{xx} = -\frac{\varepsilon_0\rho}{4K}\mathcal{E}_0^2\frac{\partial \varepsilon_{r,x}}{\partial \rho},$$

(12.39)

where $K = (c_{11} + 2c_{12})/3$ is the bulk modulus for a pure compressive force and dilation along the field direction in an isotropic medium, and c_{11} and c_{12} are elastic constants in the Voigt notation. Also, there is a nonlinear polarization created due to the elasto-optic interaction given by

$$P_i^{\text{NL}}(\vec{r}, t) = -\varepsilon_0 n_i^2 n_j^2 p_{ijk\ell}S_{k\ell}E_j(\vec{r}, t),$$

(12.40)

where $p_{ijk\ell}$ are elasto-optic coefficients. This gives

$$\frac{P_x^{NL}(\vec{r},t)}{\varepsilon_0 E_x(\vec{r},t)} = \Delta\varepsilon_{r,x} = -n_x^4 p_{xxxx} S_{xx} = -n_x^4 p_{xxxx}\frac{\varepsilon_0\rho}{4K}\mathcal{E}_0^2\frac{\partial\varepsilon_{r,x}}{\partial\rho}.$$

$$\text{But } \rho\frac{\partial\varepsilon_{r,x}}{\partial\rho} \cong \frac{\Delta\varepsilon_{r,x}}{\Delta\rho/\rho} = -\frac{\Delta\varepsilon_{r,x}}{S_{xx}} = -n_x^4 p_{xxxx} \rightarrow \frac{P_x^{NL}(\vec{r},t)}{\varepsilon_0 E_x(\vec{r},t)} = n_x^8 p_{xxxx}^2\frac{\varepsilon_0}{4K}\mathcal{E}_0^2.$$

$$(12.41)$$

Since

$$\frac{P_1^{NL}(\vec{r},t)}{\varepsilon_0 E_1(\vec{r},t)} \cong 2\Delta n_x^{NL} n_x \rightarrow \Delta n_x^{NL} = \frac{\varepsilon_0 n_x^7 p_{xxxx}^2 \mathcal{E}_0^2}{8K}$$

$$\rightarrow n_{2\|,str}(-\omega;\omega) = \frac{n_x^6 p_{xxxx}^2}{4Kc}.$$

$$(12.42)$$

Note that $\varepsilon_0 n_x^4 p_{xxxx}^2 = \gamma_e$ is the electrostrictive constant. For a weak y-polarized probe beam in the presence of the strong x-polarized beam,

$$n_{2\perp,str}(-\omega;\omega) = \frac{n_y^3 n_x^3 p_{xxxx} p_{yyxx}}{4Kc}.$$

$$(12.43)$$

For anisotropic media, K^{-1} is more complicated but can still be approximated by $K = (c_{11} + c_{12})/3$ along crystal axes with appropriate substitutions for 1 and 2 in p_{11}, p_{12}, c_{11}, and c_{12}. For propagation not along crystal axes, the situation is considerably more complex and the effective $p_{ijk\ell}$ and K require detailed calculation (Table 12.5).

Turn-on and "turn-off" times are complex issues because turning on or off an optical beam involves compressive forces. They lead to the generation of a spectrum of acoustic waves (see Fig. 12.21). The acoustic decay time $\tau_s(\Omega_s) \propto \Omega_s^{-2}$ and the details of beam shape, sample boundaries, and so on, influence the acoustic spectrum

TABLE 12.5 Material Parameters Required to Calculate $n_{2\|,str}(-\omega;\omega)$ for Various Materials

Material	Polarization λ (μm)	Elasto-optic coefficient	K (10^{10} m^2/N)	n	$n_{2\|,str}(-\omega;\omega)$ (cm^2/W)
Fused silica	(0.63)	$p_{11} = 0.12$	3.69	1.46	0.4×10^{-16}
GaAs	[110] (1.15)	≈ 0.14	≈ 7.6	3.37	1.6×10^{-13}
MgO	[100] (0.59)	$p_{11} = 0.08$	≈ 15.3	1.74	1.0×10^{-17}
Al$_2$O$_3$	[001] (0.63)	$p_{33} = 0.20$	≈ 27.0	1.76	3.7×10^{-17}
Polystyrene	(0.63)	$p_{11} = 0.31$	0.54	1.59	2.4×10^{-15}
Acetone		$p_{11} = 0.35$	0.080	1.36	8.5×10^{-15}
Methanol		$p_{11} = 0.32$	0.083	1.33	5.7×10^{-15}

From Ref. 6.
Additional elasto-optic constants can be found in Ref. 14.

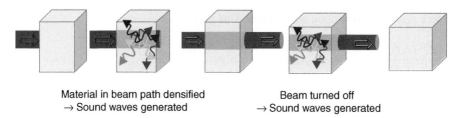

Material in beam path densified
→ Sound waves generated

Beam turned off
→ Sound waves generated

FIGURE 12.21 Light beam entering and leaving a sample. When light first enters the sample, the material along the optical path is compressed (darker blue shaded region) and sound waves are generated. When the beam departs, the compression is released, accompanied again by the generation of sound waves.

generated, which includes both compressional and shear waves with finite optical beams. In an "infinite" medium, the *shortest possible* turn-on and turn-off times are given by the acoustic transit time across the optical beam [beam diameter]/v_S, with $v_S \approx 1\,\mu\text{m/s}$ giving times in the range of microseconds to nanoseconds.

12.5 THERMO-OPTIC EFFECT

This is the dominant nonlinearity for continuous-wave optical excitation on the time scales of 100 ns and longer in most materials and was erroneously interpreted as the Kerr term in measurements in the early days of nonlinear optics. It is nonlocal in both space and time. Absorption of light leads to a temperature change δT, which in turn also leads to change in local density $\delta \rho$. Both δT and $\delta \rho$ lead to changes in the refractive index; i.e.,

$$\delta n = \left(\frac{\partial n}{\partial \rho}\right)_T \delta\rho + \left(\frac{\partial n}{\partial T}\right)_\rho \delta T. \tag{12.44}$$

Also a change in the density leads to the generation of sound waves that relieve the stresses induced by the density changes.

Evaluating effective nonlinear parameters due to heating of a material via absorption is a complex problem (15). Fortunately estimates valid to about a factor of 2 can be made by simplifying the equations. Defining Q as the absorbed power per unit volume per unit time, the temperature change and its evolution is given by the thermal diffusion equation

$$\frac{\partial\,\delta T}{\partial t} - \frac{\kappa}{\rho C_p}\nabla^2(\delta T) = \frac{Q}{\rho C_p} = \frac{\alpha_1}{\rho C_p}I, \tag{12.45}$$

where C_p is the specific heat and κ is the thermal diffusion constant. Note that $\kappa\nabla^2/\rho C_p$ has the units of inverse time defined here as τ_{th}^{-1}, which clearly depends on the beam size and shape via the ∇^2 term. Assuming the incident beam to be a Gaussian pulse of the form $I(\vec{r}, t) = I_0(z) \exp[-r^2/w_0^2 - t^2/\tau_{\text{opt}}^2]$, the maximum temperature distribution with $\tau_{\text{th}} \gg \tau_{\text{opt}}$ has the spatial distribution

$$\delta T_{\max}(\vec{r}) = \frac{\alpha_1}{\rho C_p} I(r) \int_{-\infty}^{\infty} e^{-t^2/\tau_{\text{opt}}^2} \, dt = \sqrt{\pi} \tau_{\text{opt}} \frac{\alpha_1}{\rho C_p} I(\vec{r}). \qquad (12.46)$$

This temperature distribution now relaxes via the diffusion equation 12.45 and therefore

$$\nabla^2[\delta T_{\max}(r)] = \left[\frac{1}{r}\frac{\partial}{\partial r} + \frac{\partial^2}{\partial r^2}\right]\delta T_{\max}(r) = -\frac{4}{w_0^2}\left\{1 - \frac{r^2}{w_0^2}\right\}\delta T_{\max}(r), \qquad (12.47)$$

which does not have a simple solution because of the r^2/w_0^2 term. Ignoring this term to obtain an approximate solution,

$$\frac{\partial\, \delta T_{\max}(\vec{r}, t)}{\partial t} = -\frac{4\kappa}{w_0^2 \rho C_p}\delta T_{\max}(\vec{r}, t) \quad \Rightarrow \quad \delta T_{\max}(\vec{r}, t) = \delta T_{\max}(0, t)\, e^{-t/\tau_{\text{th}}},$$

$$(12.48)$$

with $\tau_{\text{th}} = w_0^2 \rho C_p / 4\kappa$. Table 12.6 gives τ_{th} for a number of materials.

For $\tau_{\text{th}} \gg \Delta t_{\text{opt}}$ (where Δt_{opt} is the pulse width), the response of the index change is shown in Fig. 12.22 and the maximum index change $\delta n_{\max}$ is given by

$$\delta n_{\max}(\vec{r}) = \left[\frac{\partial n}{\partial T}\right]\delta T_{\max}(\vec{r}) = \sqrt{\pi}\left[\frac{\partial n}{\partial T}\right]\Delta t_{\text{opt}}\frac{\alpha_1}{\rho C_p}I(\vec{r})$$

$$\Rightarrow \quad n_{2,\text{th}}(-\omega; \omega) \cong \sqrt{\pi}\left[\frac{\partial n}{\partial T}\right]\Delta t_{\text{opt}}\frac{\alpha_1}{\rho C_p}. \qquad (12.49)$$

In terms of pulse energy, $\delta n_{\max}$ is given as follows:

$$\Delta E_{\text{pulse}} = \sqrt{\pi^3}\tau_{\text{opt}}I_{\text{peak}}w_0^2$$

$$\Rightarrow \quad \delta n_{\max} = \frac{1}{\pi w_0^2}\left[\frac{\partial n}{\partial T}\right]\frac{\alpha_1}{\rho C_p}\Delta E_{\text{pulse}}. \qquad (12.50)$$

TABLE 12.6 Material Parameters Needed to Calculate τ_{th} and dn/dT for Various Materials

Material	GaAs	Al$_2$O$_3$	NaCl	ZnO	Acetone	C$_6$H$_6$	Methanol
κ (W/(cm °C))	0.55	0.024	0.065	0.30	0.0019	0.0016	0.0020
C_p (J/(g °C))	0.33	0.75	0.85	0.83	2.2	1.7	2.4
ρ (g/cm^3)	5.32	3.98	2.2	5.5	0.79	0.90	0.80
τ_{th} (ms)	0.080	3.1	0.72	0.39	45	24	20
$dn/dT \times 10^{-4}$(°C^{-1})	1.6–2.7	0.13	0.25	0.1	−5.6	−6.2	−4.0

From Ref. 6.

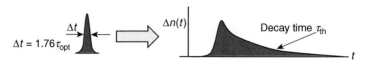

FIGURE 12.22 The time dependence of the nonlinear index change for an incident pulse much shorter than the carrier recombination time.

It is instructive to evaluate $n_{2,\text{th}}(-\omega; \omega)$ for GaAs, a common material for many optical applications: $n_{2,\text{th}}(-\omega; \omega) = 3 \times 10^{-4} \times (\alpha \text{ in cm}^{-1}) \times (\Delta t \text{ in s})$. For $\alpha_1 = 1 \text{ cm}^{-1}$ and for $\Delta t = 1 \text{ μs}$, 1 ns, and 1 ps, values of $n_{2,\text{th}}(-\omega; \omega)$ are 3×10^{-10}, 3×10^{-13}, and 3×10^{-16}, respectively, all in cm^2/W. These values should be compared to the Kerr nonlinearity at 1550 nm; namely, $n_{2,\text{el}}(-\omega; \omega) = 1.5 \times 10^{-13} \text{ cm}^2/\text{W}$. For the high repetition rates typical of mode-locked lasers, the key parameter is the energy accumulation over all the pulses within the time window τ_{th}. For example, for a mode-locked laser operating with 1-ps pulses at a repetition rate of 100 MHz, the energy accumulates from 10^3 pulses over τ_{th}, giving a cumulative value of $n_{2,\text{th}}(-\omega, \omega) = -1.2 \times 10^{-12} \text{ cm}^2/\text{W}$ (bigger than the Kerr nonlinearity).

The important point is that pulsed lasers need to be used with single pulses to ensure that the Kerr nonlinearity is truly measured and used.

12.6 $\chi^{(3)}$ VIA CASCADED $\chi^{(2)}$ NONLINEAR PROCESSES: NONLOCAL

Coherent nonlinear optical processes based on even-order nonlinear susceptibilities involve additional nonlinear phase shifts when not phase matched. For example, as discussed in Chapter 4, in second harmonic generation a nonlinear phase shift in the fundamental beam occurs when the non-phase-matched second harmonic generated by the fundamental interacts with it on propagation. That is, propagation is required, making the process *nonlocal* (16,17). The equations for second harmonic generation approximately in the nonresonant limit are

$$\omega + \omega \to 2\omega, \quad \frac{d\mathcal{E}(z, 2\omega)}{dz} = i\frac{2\omega}{n(2\omega)c}\tilde{\chi}^{(2)}_{\text{eff}}(-2\omega; \omega, \omega)\mathcal{E}^2(\omega)\, e^{i\Delta kz},$$

$$2\omega - \omega \to \omega, \quad \frac{d\mathcal{E}(z, \omega)}{dz} = i\frac{2\omega}{n(\omega)c}\tilde{\chi}^{(2)}_{\text{eff}}(-\omega; 2\omega, -\omega)\mathcal{E}(2\omega)\mathcal{E}^*(\omega)\, e^{-i\Delta kz},$$

$$\zeta_1 = \frac{2k_{\text{vac}}(\omega)}{n(\omega)}\tilde{\chi}^{(2)}_{\text{eff}}(-\omega; 2\omega, -\omega), \qquad \zeta_2 = \frac{2k_{\text{vac}}(\omega)}{n(2\omega)}\tilde{\chi}^{(2)}_{\text{eff}}(-2\omega; \omega, \omega).$$

$$(12.51)$$

A simple plane wave model for the processes implied by these equations is shown in Fig. 12.23. Energy is converted to the harmonic, which travels at a different phase velocity relative to the fundamental when $\Delta k \neq 0$. Down-conversion back to the

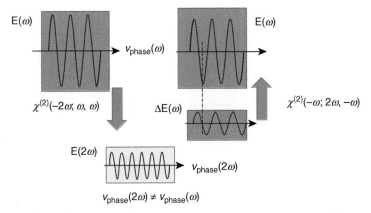

FIGURE 12.23   The sequential processes by which a nonlinear phase shift $\Delta\phi^{NL}$ is imparted onto the fundamental beam. Here $\Delta E(\omega)$ is the contribution to the fundamental down-converted field from the second harmonic.

fundamental occurs after some effective propagation distance and the down-converted fundamental is no longer in phase with the unconverted fundamental. Hence the net fundamental is shifted in phase. The larger the input fundamental, the more efficient the process and the larger the phase shift $\Delta\phi^{NL}$.

Calculations of the detailed evolution of the nonlinear phase shift with propagation distance based on Eqs 12.51 are shown in Fig. 12.24 for different values of phase mismatch ΔkL (18). The phase shift occurs essentially in steps, with maximum step size of $\pi/2$ for small ΔkL. The rate of increase of the phase shift is maximum near the minima in the fundamental transmission, which is to be expected since the

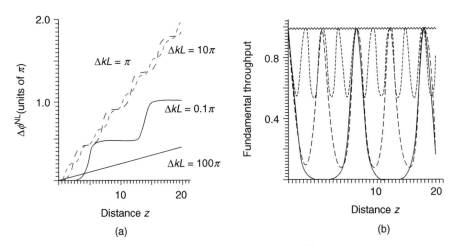

FIGURE 12.24   (a) Evolution with propagation distance of $\Delta\phi^{NL}$ for different values of ΔkL. (b) The transmission of the fundamental beam intensity with propagation distance. Solid line: $\Delta kL = 0.1\pi$. Long dashes: $\Delta kL = \pi$. Short dashed: $\Delta kL = 10\pi$. "Rippled" line: $\Delta kL = 100\pi$ Reproduced with permission from the Optical Society of America (18).

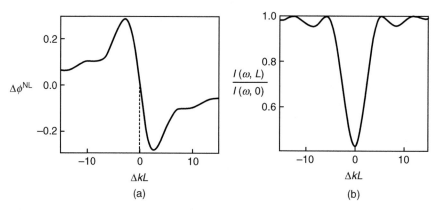

FIGURE 12.25 (a) The variation in the nonlinear phase shift with the phase mismatch for a fixed L. (b) The variation of the fundamental transmission with phase mismatch. Reproduced with permission from the Optical Society of America (20).

unconverted fundamental is smallest there and the impact of the down-converted harmonic is the largest. The process is clearly nonlocal and complicated. To be effective, this process requires an accumulation over at least one coherence length.

The dependence of the nonlinear phase shift and the fundamental transmission on phase-mismatch are shown in Fig. 12.25. Note that the features are very similar to those exhibited by nonlinear refraction and absorption, as discussed in Chapters 8 and 10 (19).

In *analogy* with the Kerr case, writing

In the low fundamental depletion limit, it is possible to obtain simple analytical formulas for the nonlinear phase shift and the fundamental transmission. Assuming that the fundamental field is essentially undepleted and inserting the weak harmonic field into the equation for the fundamental gives

$$\mathcal{E}(2\omega) \cong \frac{2\omega}{n(2\omega)c}\tilde{\chi}_{\text{eff}}^{(2)}(-2\omega;\omega,\omega)\mathcal{E}^2(\omega)\frac{e^{i\Delta kz}-1}{\Delta k}$$

$$\Rightarrow \quad \frac{d\mathcal{E}(z,\omega)}{dz} = i\frac{2\omega}{n(\omega)c}\tilde{\chi}_{\text{eff}}^{(2)}(-\omega;2\omega,-\omega)\mathcal{E}(2\omega)\mathcal{E}^*(\omega)\,e^{-i\Delta kz} \qquad (12.52)$$

$$= \frac{\zeta_1\zeta_2}{\Delta k}\{-\sin(\Delta kz) + i[1-\cos(\Delta kz)]\}|\mathcal{E}(\omega)|^2\mathcal{E}(\omega).$$

In *analogy* with the Kerr case, writing

$$\frac{d\mathcal{E}(z,\omega)}{dz} = ik_{\text{vac}}n_{2,\text{Kerr}}(-\omega;\omega)I(\omega)\mathcal{E}(z,\omega)$$

$$\Rightarrow \quad n_{2,\text{nlcas}}(-\omega;\omega:z) = \frac{4\omega[d_{\text{eff}}^{(2)}]^2}{c^2\varepsilon_0n^2(\omega)n(2\omega)\Delta k}\sin^2\left(\frac{\Delta kz}{2}\right). \qquad (12.53)$$

Note that the sign of nonlinearity depends on the sign of Δk, an externally controlled parameter; i.e., the nonlinearity can be effectively self-focusing or self-defocusing. But, as was shown previously, this is not really an n_2 process, since there is no refractive index change that can be measured with an arbitrary additional probe beam. The only beams that are affected are the fundamental and harmonic, which are strongly coupled by $\chi^{(2)}$. What is measurable is the nonlinear phase shift $\Delta\phi^{\mathrm{NL}}$. Integrating over z, the nonlinear phase shift and the fundamental transmission are given, respectively, by

$$\Delta\phi^{\mathrm{NL}} = \int_0^L k_{\mathrm{vac}} n_{2,\mathrm{nlcas}}\, dz\, I(\omega) = \frac{2\omega^2 [d_{\mathrm{eff}}^{(2)}]^2}{c^2 \varepsilon_0 n^2(\omega) n(2\omega)\Delta k} L\{1 - \mathrm{sinc}[\Delta k L]\} I(\omega),$$

$$T = 1 - \frac{I(0,\,\omega) - I(L,\,\omega)}{I(0,\omega)} \simeq 1 - \frac{2[d_{\mathrm{eff}}^{(2)}]^2 k_{\mathrm{vac}}^2(\omega)}{n(2\omega) n^2(\omega)\varepsilon_0 c} L^2 I(0,\,\omega)\,\mathrm{sinc}^2\left(\frac{\Delta k L}{2}\right).$$

$$(12.54)$$

The maximum nonlinearity occurs at $\Delta k L \cong \pi$ and has the value

$$n_{2,\mathrm{nlcas}}(-\omega;\omega) \cong \mathrm{sign}(\Delta k)\, 0.36 \frac{4[k_{\mathrm{vac}(\omega)} L][\tilde{d}_{\mathrm{eff}}^{(2)}(-2\omega;\omega,\omega)]^2}{n_\omega^2 n_{2\omega} c \varepsilon_0}. \qquad (12.55)$$

This nonlinearity can be quite large under the right conditions.

Example 1: DSTMS, $\ell_{\mathrm{coh}} = 3.6\,\mu\mathrm{m}$, maximum $n_{2,\mathrm{nlcas}}(-\omega;\omega) = 4 \times 10^{-13}$ cm^2/W

Example 2: QPM LiNbO$_3$, $L = 1$ cm, maximum $n_{2,\mathrm{nlcas}}(-\omega;\omega) = 2 \times 10^{-12}$ cm^2/W

For these examples, the nonlinearity is actually larger than the Kerr nonlinearity. It is noteworthy that all-optical switching devices and even solitons based on this nonlinearity have been reported (17).

APPENDIX 12.1: SPONTANEOUS RAMAN SCATTERING

Light interacts with vibrational modes in molecules that leads to the scattering of light at frequencies shifted from the incident light by the vibrational frequencies. The interaction is quantified by the Raman hyperpolarizability tensor $\dfrac{\partial \bar{\alpha}_{ijm}^\beta}{\partial \bar{q}_m^\beta}\Big|_{\bar{q}_m = 0}$.

This interaction also leads to a number of nonlinear phenomena such as stimulated Raman scattering and contributions to $n_{2\|,\mathrm{nuc}}(-\omega;\omega)$. The scattering of light by thermally excited vibrations, known as spontaneous Raman scattering, is a technique available in most chemistry laboratories, and with appropriate calibrations, the Raman hyperpolarizability can be measured routinely. In this appendix, expressions are derived for the Raman spectrum so that the wealth of data available on spontaneous Raman scattering can be used for nonlinear optics (21).

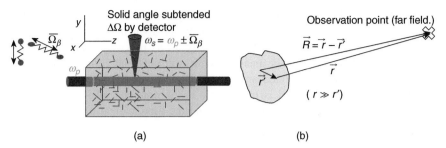

FIGURE A.12.1.1 (a) Spontaneous Raman scattering geometry. The vibrating molecules are considered to be simple harmonic oscillators. (b) General scattering geometry for a single molecule located at $\vec{r}'$ inside a sample.

The basic spontaneous scattering geometry is shown in Fig. A.12.1.1. Light scattered (frequency ω_s) at 90° from the incident light direction (frequency ω_p) contains components at frequencies $\omega_s = \omega_p \pm \bar{\Omega}_\beta$. $\bar{\Omega}_\beta$ is the vibrational frequency of the βth mode. The vibrations are driven by thermal noise; i.e., each mode has the energy $k_B T$. Furthermore, the vibrations are uncorrelated from molecule to molecule since they are driven by noise and the individual vibrations $\bar{\Omega}_\beta$ at different frequencies are uncorrelated. The light scattered into a solid angle $\Delta\Omega$ (not to be confused with the vibrational frequencies $\bar{\Omega}_\beta$) is collected. A spectrometer is then typically used to separate out different frequency components and the whole system is calibrated for the measurement of the total light scattered by a given vibration into a predetermined (by geometry, apertures, and so on) solid angle.

Let us first discuss some useful definitions involving Fourier transforms employed in this appendix.

$$E(t) = \int_{-\infty}^{\infty} E(\omega) e^{-i\omega t}\, d\omega, \qquad E(\omega) = \frac{1}{2\pi} \int_{-\infty}^{\infty} E(t') e^{i\omega t'}\, dt',$$

$$\delta(t - t') = \frac{1}{2\pi} \int_{-\infty}^{\infty} e^{-i\omega(t-t')}\, d\omega, \qquad \delta(\omega - \omega_a) = \frac{1}{2\pi} \int_{-\infty}^{\infty} e^{i(\omega-\omega_a)t'}\, dt',$$

$$\delta(t - t') = \frac{1}{2\pi} \left\{ \int_{-\infty}^{0} e^{-i\omega(t-t')}\, d\omega + \int_{0}^{\infty} e^{-i\omega(t-t')}\, d\omega \right\} = \frac{1}{\pi} \int_{0}^{\infty} e^{-i\omega(t-t')}\, d\omega,$$

$$\delta(\omega - \omega_a) = \frac{1}{2\pi} \left\{ \int_{-\infty}^{0} e^{i(\omega-\omega_a)t'}\, dt' + \int_{0}^{\infty} e^{i(\omega-\omega_a)t'}\, dt' \right\} = \frac{1}{\pi} \int_{0}^{\infty} e^{i(\omega-\omega_a)t'}\, dt'.$$

$$(A.12.1.1)$$

A.12.1.1 Single Noninteracting Molecules

The natural approach to describing light scattered into a solid angle, as is the case for spontaneous Raman scattering, is the Green's function approach; the complete details of this approach are beyond the scope of this appendix. The starting point is the vector potential $\vec{A}(\vec{r}', t')$, which is driven by an effective current source in the medium $\vec{J}(\vec{r}', t')$ via

$$-\nabla^2 \vec{A} + \frac{n^2}{c^2}\frac{\partial^2 \vec{A}}{\partial t'^2} = \mu_o \vec{J}. \tag{A.12.1.2}$$

The general Green's function solution for $\vec{A}(\vec{r}, t)$ *with polarization parallel* (along the x-axis) *to the incident field* (along the x-axis) in the far (at the detector) field is

$$A_x(\vec{r}, t) = \frac{\mu_0}{4\pi}\int d\vec{r}'\int dt'\,\frac{J_x(\vec{r}', t')}{r}f(\omega)\delta(t' + nR/c - t) = \frac{\mu_0}{4\pi}\int d\vec{r}'\,\frac{J_x(\vec{r}', t - nR/c)}{r}, \tag{A.12.1.3}$$

where the δ function allows for the propagation from the dipole source at $(\vec{r}', t')$ to the detector at $(\vec{r}, t)$. Writing $A_x = \frac{1}{2}\mathsf{A}_x + \text{c.c.}$ (in the far field) and $J_x(\vec{r}') = \frac{1}{2}\mathsf{J}_x(\vec{r}') + \text{c.c.}$, for $r \cong R$ and $\vec{e}_r$ as the unit vector along, $\vec{r}$

$$\mathsf{A}_x = \frac{\mu_0}{4\pi r}\int \mathsf{J}_x(\vec{r}')\,e^{-ink_{vac}\vec{e}_r\cdot\vec{r}'}\,d\vec{r}'. \tag{A.12.1.4}$$

Since

$$J^{RS}_{x,\text{eff}}(\vec{r}', t') = \frac{\partial P^{RS}_x(\vec{r}', t')}{\partial t'} \quad\rightarrow\quad \mathsf{J}^{RS}_x(\vec{r}') = -i\omega\sum_m \mathsf{p}^{RS}_{x,m}(\vec{r}')\delta(\vec{r}'_m), \tag{A.12.1.5}$$

with $p^{RS}_{x,m} = \frac{1}{2}\mathsf{p}^{RS}_{x,m}\delta(\vec{r}'_m)\,e^{-i\omega t} + \text{c.c.}$ For m excited molecules in the scattering volume ΔV defined by incident and collecting lenses,

$$\mathsf{A}_x = -i\omega\frac{\mu_0}{4\pi r}\sum_m \mathsf{p}^{RS}_{x,m}\,e^{-ink_{vac}\vec{e}_r\cdot\vec{r}'_m}. \tag{A.12.1.6}$$

Note that $\mathsf{p}^{RS}_{x,m}$ is the vibrating dipole and has not yet been averaged over all the molecular orientations typical of, e.g., a disordered glass. $\mathsf{p}^{RS}_{x,m}$ is the projection along the polarization of the incident field.

The propagating (scattered) electric field $\vec{E}^{RS}(\vec{r}, t)$ is then obtained from $\vec{E}^{RS}(\vec{r}, t) = -\partial\vec{A}(\vec{r}, t)/\partial t'$ as

$$\mathsf{E}^{RS}_x = \omega^2\frac{\mu_0}{4\pi r}\sum_m \mathsf{p}^{RS}_{x,m}\,e^{-\frac{in\omega}{c}\vec{e}_r\cdot\vec{r}'_m}. \tag{A.12.1.7}$$

The polarization induced at the frequency ω by the βth vibrational mode is given by

$$p^{RS,\beta}_x(t') = \sum_m \left.\frac{\partial\bar{\alpha}^\beta_{xxx}}{\partial q^\beta_x}\right|_{\bar{q}^\beta=0}\bar{q}^\beta_x(t')f^{(1)}(\omega)f^{(1)}(\omega_P)E_x(\omega_P, t'),$$

$$\bar{q}^\beta_x(\vec{r}', t') = \frac{1}{2}\bar{Q}^\beta_x(\bar{\Omega}_\beta)\delta(\vec{r}')\,e^{-t'/\bar{\tau}_\beta}\,e^{-i\bar{\Omega}_\beta t' + i\phi^\beta_m} + \text{c.c. for } t' \geq 0 \tag{A.12.1.8}$$

$$\text{and } \bar{q}^\beta_{m,x}(\vec{r}', t') = 0 \text{ for } t' < 0.$$

That is, the fluctuation was "born" at $t' = 0$. This gives

$$
E_x^{RS} = \frac{1}{4}\omega^2 \frac{\mu_0}{4\pi r} \sum_m f^{(1)}(\omega) f^{(1)}(\omega_P) \frac{\partial \bar{\alpha}_{xxx}^\beta}{\partial \bar{q}_{x,m}^\beta}\Big|_{\bar{q}_{x,m}^\beta=0} e^{-t/\bar{\tau}_\beta} \left[\bar{Q}_{x,m}^\beta(\bar{\Omega}_\beta) e^{-ink_{vac}\vec{e}_r\cdot\vec{r}'_m} \right.
$$

$$
\times e^{-i\bar{\Omega}_\beta t + i\phi_m^\beta} + \bar{Q}_{x,m}^{*\beta}(\bar{\Omega}_\beta) e^{ink_{vac}\vec{e}_r\cdot\vec{r}'_m} e^{i\bar{\Omega}_\beta t - i\phi_m^\beta} \Big] [E_x(\omega_P) e^{-i\omega_P t} + E_x^*(\omega_P) e^{+i\omega_P t}],
$$

$$
E_x^{RS}(t) = \frac{1}{4}\omega^2 \frac{\mu_0}{4\pi r} \sum_m f^{(1)}(\omega) f^{(1)}(\omega_P) \frac{\partial \bar{\alpha}_{xxx}^\beta}{\partial \bar{q}_{x,m}^\beta}\Big|_{\bar{q}_{x,m}^\beta=0} e^{-t/\bar{\tau}_\beta} \left[\bar{Q}_{x,m}^\beta(\bar{\Omega}_\beta) E_x(\omega_P) \right.
$$

$$
\times e^{-ink_{vac}\vec{e}_r\cdot\vec{r}'_m} e^{-i(\omega_P + \bar{\Omega}_\beta)t + i\phi_m^\beta} + \bar{Q}_{x,m}^{*\beta}(\bar{\Omega}_\beta) E_x(\omega_P) e^{ink_{vac}-e_r\cdot\vec{r}'_m} e^{-i(\omega_P - \bar{\Omega}_\beta)t - i\phi_m^\beta} + \text{c.c.} \Big].
$$

(A.12.1.9)

Taking the Fourier transform

$$
E_x^{RS}(\omega) = \frac{1}{2\pi}\int_{-\infty}^\infty E_x^{RS}(t)\, e^{i\omega t}\, dt = \frac{1}{\pi}\int_0^\infty E_x^{RS}(t)\, e^{i\omega t}\, dt
$$

$$
\Rightarrow \quad E_x^{RS}(\omega) = \frac{1}{4\pi}\omega^2 \frac{\mu_0}{4\pi r} \sum_m f^{(1)}(\omega) f^{(1)}(\omega_P)
$$

$$
\times \frac{\partial \bar{\alpha}_{xxx}^\beta}{\partial \bar{q}_{x,m}^\beta}\Big|_{\bar{q}_{x,m}^\beta=0} E_x(\omega_P) \left[\bar{Q}_{x,m}^\beta(\bar{\Omega}_\beta) \int_0^\infty dt\, e^{-[i(\omega_P + \bar{\Omega}_\beta - \omega) + \bar{\tau}_\beta^{-1}]t} \right.
$$

$$
\times e^{+i\phi_m^\beta - ink_{vac}\vec{e}_r\cdot\vec{r}'_m} + \int_0^\infty dt\, \bar{Q}_{x,m}^{*\beta}(\bar{\Omega}_\beta) e^{-[i(\omega_P - \bar{\Omega}_\beta - \omega) + \bar{\tau}_\beta^{-1}]t - i\phi_m^\beta + ink_{vac}\vec{e}_r\cdot\vec{r}'_m} \Big] + \text{c.c.}
$$

$+$ terms which will eventually time average to zero.

(A.12.1.10)

Evaluating the integral

$$
\int_0^\infty dt\, e^{-[i(\omega_P - \bar{\Omega}_\beta - \omega) + \bar{\tau}_\beta^{-1}]t - i\phi_m^\beta + ink_{vac}\vec{e}_r\cdot\vec{r}'_m} = \left(\frac{e^{-[i(\omega_P - \bar{\Omega}_\beta - \omega) + \bar{\tau}_\beta^{-1}]t - i\phi_m^\beta + ink_{vac}\vec{e}_r\cdot\vec{r}'_m}}{-[i(\omega_P - \bar{\Omega}_\beta - \omega) + \bar{\Gamma}_\beta]} \right)_0^\infty
$$

$$
\Rightarrow \quad E_x^{RS}(\omega) = \frac{1}{4\pi}\omega^2 \frac{\mu_0}{4\pi r} \sum_m f^{(1)}(\omega) f^{(1)}(\omega_P) \frac{\partial \bar{\alpha}_{xxx}^\beta}{\partial \bar{q}_{x,m}^\beta}\Big|_{\bar{q}_{x,m}^\beta=0} E(\omega_P)
$$

$$
\times \left[\frac{\bar{Q}_{x,m}^\beta(\bar{\Omega}_\beta) e^{i\phi_m^\beta - ink_{vac}\vec{e}_r\cdot\vec{r}'_m}}{i(\omega_P + \bar{\Omega}_\beta - \omega) + \bar{\tau}_\beta^{-1}} + \frac{\bar{Q}_{x,m}^{*\beta}(\bar{\Omega}_\beta) e^{-i\phi_m^\beta + ink_{vac}\vec{e}_r\cdot\vec{r}'_m}}{i(\omega_P - \bar{\Omega}_\beta - \omega) + \bar{\tau}_\beta^{-1}} + \text{c.c.} \right].
$$

(A.12.1.11)

Summing over all the Raman-active vibrational modes β, we obtain

$$
E_x^{RS}(\omega) = \frac{1}{4\pi}\omega^2 \frac{\mu_0}{4\pi r} E_x(\omega_P) \sum_\beta \sum_m f^{(1)}(\omega) f^{(1)}(\omega_P) \frac{\partial \bar{\alpha}_{xxx}^\beta}{\partial \bar{q}_{x,m}^\beta}\Big|_{\bar{q}_{x,m}^\beta=0}
$$

$$
\times \left[\frac{\bar{Q}_{x,m}^\beta(\bar{\Omega}_\beta) e^{i\phi_m^\beta - ink_{vac}\vec{e}_r\cdot\vec{r}'_m}}{i(\omega_P + \bar{\Omega}_\beta - \omega) + \bar{\tau}_\beta^{-1}} + \frac{\bar{Q}_{x,m}^{*\beta}(\bar{\Omega}_\beta) e^{-i\phi_m^\beta + ink_{vac}\vec{e}_r\cdot\vec{r}'_m}}{i(\omega_P - \bar{\Omega}_\beta - \omega) + \bar{\tau}_\beta^{-1}} + \text{c.c.} \right]
$$

(A.12.1.12)

Defining

$$g(\omega,\beta,m) = \frac{\partial \bar{\alpha}^{\beta}_{xxx}}{\partial \bar{q}^{\beta}_{x,m}}\bigg|_{\bar{q}^{\beta}_{x,m}=0} \left[\frac{\bar{Q}^{\beta}_{x,m}(\bar{\Omega}_{\beta})\, e^{i\phi^{\beta}_{m}-ink_{vac}\vec{e}_r \cdot \vec{r}'_m}}{i(\omega_P + \bar{\Omega}_{\beta} - \omega) + \bar{\tau}^{-1}_{\beta}} + \frac{\bar{Q}^{*\beta}_{x,m}(\bar{\Omega}_{\beta})\, e^{-i\phi^{\beta}_{m}+ink_{vac}\vec{e}_r \cdot \vec{r}'_m}}{i(\omega_P - \bar{\Omega}_{\beta} - \omega) + \bar{\tau}^{-1}_{\beta}} \right]$$

$$\rightarrow\ E^{RS}_x(\omega) = \frac{1}{4\pi}\omega^2\frac{\mu_0}{4\pi r}E_x(\omega_P)\sum_{\beta}\sum_{m}f^{(1)}(\omega)f^{(1)}(\omega_P)g(\omega,\beta,m) + \text{c.c.}$$

$$\rightarrow\ E^{RS}_x(t) = \frac{1}{4\pi}\frac{\mu_0}{4\pi r}E_x(\omega_P)\sum_{\beta}\sum_{m}f^{(1)}(\omega)f^{(1)}(\omega_P)\int_{-\infty}^{\infty} d\omega\,\omega^2 g(\omega,\beta,m)\,e^{-i\omega t} + \text{c.c.}$$

$$(A.12.1.13)$$

Normally one would take the intensity as $nc\varepsilon_0\overline{E^{RS}_x(t)E^{*RS}_x(t)}$. This requires dealing with the statistical properties of the fluctuations in the vibrational modes, uncorrelated from one another, occurring randomly in time and then decaying. To take this into account, noting that $\overline{xxx} \equiv \langle xxx \rangle_{st}$ for quantities related to noise variables, define

$$I_x(t) = nc\varepsilon_0\overline{E^{RS}_x(t)E^{RS}_x(t')} \equiv nc\varepsilon_0 \lim_{T\to\infty}\frac{1}{T}\int_0^{\infty} E^{RS}_x(t')E^{RS}_x(t'+\tau)\,dt',$$

$$nc\varepsilon_0 E^{RS}_x(t')E^{RS}_x(t'+\tau) = \left[\frac{1}{4\pi}\frac{\mu_0}{4\pi r}\right]^2 nc\varepsilon_0$$

$$\times \left\{ E_x(\omega_P)\sum_{\beta}\sum_{m}[f^{(1)}(\omega)f^{(1)}(\omega_P)]^2\int_{-\infty}^{\infty} d\omega'\,\omega'^2 g(\omega',\beta,m)\,e^{-i\omega't'} + \text{c.c.} \right\}$$

$$\times \left\{ E_x(\omega_P)\sum_{\beta'}\sum_{m'}f^{(1)}(\omega)f^{(1)}(\omega_P)\int_{-\infty}^{\infty} d\omega''\,\omega''^2 g(\omega'',\beta'',m')\,e^{-i\omega''(t'+\tau)} + \text{c.c.} \right\}.$$

$$(A.1.12.14)$$

The rapidly oscillating terms average to zero when the time average is taken, leaving

$$nc\varepsilon_0 E^{RS}_x(t)E^{RS}_x(t+\tau) = \left[\frac{1}{4\pi}\frac{\mu_0}{4\pi r}\right]^2 nc\varepsilon_0 E_x(\omega_P)E^{*}_x(\omega_P)\sum_{\beta}\sum_{m}\sum_{\beta'}\sum_{m'}$$

$$\times [f^{(1)}(\omega)f^{(1)}(\omega_P)]^2$$

$$\times \int_{-\infty}^{\infty} d\omega'\,\omega'^2 g(\omega',\beta,m)\,e^{-i\omega't'}\int_{-\infty}^{\infty} d\omega''\,\omega''^2 g^{*}(\omega'',\beta',m')\,e^{i\omega''(t'+\tau)} + \text{c.c.}$$

$$= \frac{1}{8\pi^2}\left[\frac{\mu_0}{4\pi r}\right]^2 nc\varepsilon_0 I_{P,x}\sum_{\beta}\sum_{m}\sum_{\beta'}\sum_{m'}[f^{(1)}(\omega)f^{(1)}(\omega_P)]^2$$

$$\times \int_{-\infty}^{\infty} d\omega'\,\omega'^2 g(\omega',\beta,m)\,e^{-i\omega't'}\int_{-\infty}^{\infty} d\omega''\,\omega''^2 g^{*}(\omega'',\beta',m')\,e^{i\omega''(t'+\tau)} + \text{c.c.}$$

$$(A.12.1.15)$$

Defining

$$G_x(\tau) = nc\varepsilon_0 \overline{E_x^{RS}(t)E_x^{RS}(t)} \equiv \lim_{T\to\infty} nc\varepsilon_0 \frac{1}{T}\int_0^\infty E_x^{RS}(t)E_x^{RS}(t+\tau)\,dt \quad \text{(A.12.1.16)}$$

$$\to G_x(\tau) = \frac{1}{8\pi^2}\left[\frac{\mu_0}{4\pi r}\right]^2 I_P \sum_\beta \sum_m \sum_{\beta'} \sum_{m'} [f^{(1)}(\omega)f^{(1)}(\omega_P)]^2$$

$$\times \lim_{T\to\infty}\frac{1}{T}\int_0^T \left\{ \int_{-\infty}^\infty d\omega'\,\omega'^2 g(\omega',\beta,m)e^{-i\omega't}\int_{-\infty}^\infty \omega''^2 g^*(\omega'',\beta',m')e^{i\omega''(t+\tau)}\right\}dt + \text{c.c.}$$

$$\text{(A.12.1.17)}$$

For T many cycles long, $\langle e^{i(\omega''-\omega')t}\rangle = \lim_{T\to\infty}\frac{1}{T}\int_0^\infty e^{i(\omega''-\omega')t}\,dt = \frac{\pi}{T}\delta(\omega''-\omega')$, and since the phases of different modes and the time of excitation of different molecules are uncorrelated,

$$\sum_\beta \sum_m \sum_{\beta'} \sum_{m'} = \sum_\beta \delta_{\beta\beta'} \sum_m \delta_{mm'},$$

$$\int_0^\infty d\omega''\,\omega''^2 g^*(\omega'',\beta',m')\delta(\omega''-\omega')e^{i\omega''\tau} = \int_{-\infty}^\infty d\omega''\,\omega''^2 g^*(\omega'',\beta',m')\delta(\omega''-\omega')e^{i\omega''\tau}$$
$$= \omega'^2 g^*(\omega',\beta,m)e^{i\omega'\tau}$$

$$\text{(A.12.1.18)}$$

$$\to G_x(\tau) = \frac{1}{8\pi}\left[\frac{\mu_0}{4\pi r}\right]^2 I_P \sum_\beta \int_{-\infty}^\infty d\omega'\,\omega'^4 \left\{ \lim_{T\to\infty}\frac{1}{T}\sum_m [f^{(1)}(\omega)f^{(1)}(\omega_P)]^2 \right.$$

$$\left. \left\langle |g(\omega',\beta,m)|^2 \right\rangle \right\} e^{i\omega'\tau} + \text{c.c.}$$

$$\text{(A.12.1.19)}$$

Using the definition of $g(\omega,\beta,m)$ from Eq. A.12.1.13, we obtain

$$\langle g(\omega',\beta,m)g^*(\omega',\beta,m)\rangle = \left.\left|\frac{\partial \bar{\alpha}_{xxx}^\beta}{\partial \bar{q}_{x,m}^\beta}\right|_{\bar{q}_{x,m}^\beta=0}\right|^2 \left\langle \bar{Q}_{x,m}^\beta(\bar{\Omega}_\beta)\bar{Q}_{x,m}^{*\beta}(\bar{\Omega}_\beta)\right\rangle$$

$$\times \left[\frac{1}{(\omega_P+\bar{\Omega}_\beta-\omega)^2+\tau_\beta^{-2}} + \frac{1}{(\omega_P-\bar{\Omega}_\beta-\omega)^2+\tau_\beta^{-2}}\right]. \qquad \text{(A.12.1.20)}$$

The fluctuations are uncorrelated between molecules; however, since the time evolution is the same,

$$\sum_m \frac{1}{T}\left\langle \bar{Q}_{x,m}^\beta(\bar{\Omega}_\beta)\bar{Q}_{x,m}^{*\beta}(\bar{\Omega}_\beta)\right\rangle = \left\langle \bar{Q}_x^\beta(\bar{\Omega}_\beta)\bar{Q}_x^{*\beta}(\bar{\Omega}_\beta)\right\rangle\frac{1}{T}\sum_m, \qquad \text{(A.12.1.21)}$$

where $\frac{1}{T}\sum_m$ is just the number of fluctuations occurring in time window T divided by T, which gives the rate at which fluctuations appear. However, the rate at which fluctuations appear is a constant and in the steady state it is equal to the inverse of the decay time of an individual fluctuation, i.e., $\frac{1}{T}\sum_m = N\bar{\tau}_\beta^{-1}$. Furthermore, since $k_BT \gg \hbar\bar{\Omega}_\beta$, the energy per mode (k_BT) is $2\times$ kinetic energy $= \left\langle \bar{Q}_x^\beta(\bar{\Omega}_\beta)\bar{Q}_x^{*\beta}(\bar{\Omega}_\beta)\right\rangle m_\beta\bar{\Omega}_\beta^2$ and therefore

$$\left\langle \bar{Q}_x^\beta(\bar{\Omega}_\beta)\bar{Q}_x^{*\beta}(\bar{\Omega}_\beta)\right\rangle = \frac{k_BT}{m_\beta\bar{\Omega}_\beta^2}. \tag{A.12.1.22}$$

This now gives

$$G_x(\tau) = \frac{1}{4}\frac{1}{\varepsilon_0^2 c^4(4\pi)^2 r^2}I_{P,x}\sum_\beta N[f^{(1)}(\omega)f^{(1)}(\omega_P)]^2$$

$$\left|\frac{\partial\bar{\alpha}_{xxx}^\beta}{\partial\bar{q}_x^\beta}\Big|_{\bar{q}_x^\beta=0}\right|^2 \frac{k_BT}{m_\beta\bar{\Omega}_\beta^2}\int_{-\infty}^{\infty}(e^{i\omega'\tau}+e^{-i\omega'\tau})\,d\omega'\omega'^4 \tag{A.12.1.23}$$

$$\times\left[\frac{1/\bar{\tau}_\beta\pi}{(\omega_P+\bar{\Omega}_\beta-\omega')^2+\bar{\tau}_\beta^{-2}}+\frac{1/\bar{\tau}_\beta\pi}{(\omega_P-\bar{\Omega}_\beta-\omega')^2+\bar{\tau}_\beta^{-2}}\right].$$

Since $I_x(\omega) = \frac{1}{4\pi}\int_{-\infty}^{\infty}d\tau\,G_x(\tau)\,e^{i\omega\tau}$ and $\int_{-\infty}^{\infty}(e^{i(\omega'+\omega)\tau}+e^{-i(\omega'-\omega)\tau})\,d\tau = 2\pi[\delta(\omega'+\omega)+\delta(\omega-\omega')]$, we obtain

$$I_x(\omega) = \frac{1}{8}\frac{1}{\varepsilon_0^2 c^4(4\pi)^2 r^2}NI_{P,x}\sum_\beta[f^{(1)}(\omega)f^{(1)}(\omega_P)]^2\left|\frac{\partial\bar{\alpha}_{xxx}^\beta}{\partial\bar{q}_x^\beta}\Big|_{\bar{q}_x^\beta=0}\right|^2\frac{k_BT}{m_\beta\bar{\Omega}_\beta^2}\omega^4$$

$$\times\left[\frac{1/\bar{\tau}_\beta\pi}{(\omega_P+\bar{\Omega}_\beta-\omega)^2+\bar{\tau}_\beta^{-2}}+\frac{1/\bar{\tau}_\beta\pi}{(\omega_P-\bar{\Omega}_\beta-\omega)^2+\bar{\tau}_\beta^{-2}}\right.$$

$$\left.+\frac{1/\bar{\tau}_\beta\pi}{(\omega_P+\bar{\Omega}_\beta+\omega)^2+\bar{\tau}_\beta^{-2}}+\frac{1/\bar{\tau}_\beta\pi}{(\omega_P-\bar{\Omega}_\beta+\omega)^2+\bar{\tau}_\beta^{-2}}\right].$$

$$\tag{A.12.1.24}$$

Since the last two terms show no peaks and hence are very small, they are neglected. This formula is useful for ordered media such as crystals.

For disordered media such as glasses, it is necessary to average over all orientations; i.e.,

$$I_x(\omega) = \frac{N}{8\varepsilon_0^2 c^4(4\pi)^2 r^2}I_{P,x}\sum_\beta[f^{(1)}(\omega)f^{(1)}(\omega_P)]^2\left\langle\left|\frac{\partial\bar{\alpha}_{xxx}^\beta}{\partial\bar{q}_{\bar{x}}^\beta}\Big|_{\bar{q}_{\bar{x}}^\beta=0}\right|^2\right\rangle_{\text{angles}}\frac{k_BT}{m_\beta\bar{\Omega}_\beta^2}\omega^4$$

$$\times\left[\frac{1/\bar{\tau}_\beta\pi}{(\omega_P+\bar{\Omega}_\beta-\omega)^2+\bar{\tau}_\beta^{-2}}+\frac{1/\bar{\tau}_\beta\pi}{(\omega_P-\bar{\Omega}_\beta-\omega)^2+\bar{\tau}_\beta^{-2}}\right].$$

$$\tag{A.12.1.25}$$

For converting Eqs A.12.1.24 and A.12.1.25 to scattered power, the area subtended by the detector is $A = r^2 \Delta\Omega$, which gives $P_x(\omega) = I_x(\omega) r^2 \Delta\Omega$. Noting that ω refers to the scattered frequency ω_S, we obtain

$$
\frac{P_x(\omega_S)}{\Delta\Omega I_{P,x}} = \frac{N}{8\varepsilon_0^2 c^4 (4\pi)^2} \sum_\beta [f^{(1)}(\omega_S) f^{(1)}(\omega_P)]^2 \left\langle \left| \frac{\partial\bar{\alpha}_{xxx}^\beta}{\partial q_{x,m}^\beta} \right|_{\bar{q}_m^\beta=0} \right|^2 \right\rangle_{\text{angles}} \frac{k_B T}{m_\beta \bar\Omega_\beta^2} \omega_S^4
$$

$$
\times \left[\frac{1/\bar\tau_\beta \pi}{(\omega_P + \bar\Omega_\beta - \omega_S)^2 + \bar\tau_\beta^{-2}} + \frac{1/\bar\tau_\beta \pi}{(\omega_P - \bar\Omega_\beta - \omega_S)^2 + \bar\tau_\beta^{-2}} \right].
$$

$$
(A.12.1.26)
$$

The first of the resonant terms leads to the anti-Stokes side of the spectrum and the second to the Stokes side. The total scattered power per unit solid angle per unit intensity is $\displaystyle\int_{-\infty}^{\infty} \frac{P_x(\omega_S)}{\Delta\Omega I_{P,x}} d\omega_S$. Noting that $|\omega_P - \omega_S| \bar\tau_\beta \gg 1$, we obtain

$$
\frac{1/\bar\tau_\beta \pi}{[(\omega_P - \omega_S) \pm \bar\Omega_\beta]^2 + \bar\tau_\beta^{-2}} = \delta(\omega_P - \omega_S \pm \bar\Omega_\beta) \qquad (A.12.1.27)
$$

$$
\rightarrow \quad \frac{P_{x,\text{Tot}}^{\text{RS}}}{\Delta\Omega I_x(\omega_P)} = \frac{N}{8\varepsilon_0^2 c^4 (4\pi)^2} \sum_\beta [f^{(1)}(\omega_S) f^{(1)}(\omega_P)]^2 \left\langle \left| \frac{\partial\bar{\alpha}_{xxx}^\beta}{\partial q_{x,m}^\beta} \right|_{\bar{q}_m^\beta=0} \right|^2 \right\rangle_{\text{angles}}
$$

$$
\times \frac{k_B T}{m_\beta \bar\Omega_\beta^2} [\{\omega_P + \bar\Omega_\beta\}^4 + \{\omega_P - \bar\Omega_\beta\}^4].
$$

$$
(A.12.1.28)
$$

Since $\{\omega_P + \bar\Omega_\beta\}^4 + \{\omega_P - \bar\Omega_\beta\}^4 \cong 2\omega_P^4 \cong 2\omega_S^4$, we obtain

$$
\frac{P_{x,\text{Tot}}^{\text{RS}}}{\Delta\Omega I_x(\omega_P)} \cong \frac{N\omega_P^4}{4\varepsilon_0^2 c^4 (4\pi)^2} [f^{(1)}(\omega_P)]^4 \sum_\beta \left\langle \left| \frac{\partial\bar{\alpha}_{xxx}^\beta}{\partial q_{x,m}^\beta} \right|_{\bar{q}_m^\beta=0} \right|^2 \right\rangle_{\text{angles}} \frac{k_B T}{m_\beta \bar\Omega_\beta^2}.
$$

$$
(A.12.1.29)
$$

Alternatively,

$$
N[f^{(1)}(\omega_P)]^4 \sum_\beta \left\langle \left| \frac{\partial\bar{\alpha}_{xxx}^\beta}{\partial q_{x,m}^\beta} \right|_{\bar{q}_m^\beta=0} \right|^2 \right\rangle_{\text{angles}} \cong \frac{P_{x,\text{Tot}}^{\text{RS}}}{\Delta\Omega I_x(\omega_P)} \frac{4\varepsilon_0^2 c^4 (4\pi)^2 m_\beta \bar\Omega_\beta^2}{\omega_P^4 k_B T}.
$$

$$
(A.12.1.30)
$$

PROBLEMS

1. Show from first principles that the effective intensity-dependent nonlinear index change $\Delta n_\perp$ induced in a direction orthogonal to the applied optical field in a cigar-shaped molecule such as CS_2 is given by

$$n_{2\perp,\mathrm{or}}(-\omega;\omega) = -\frac{N}{n_0^2 \varepsilon_0^2 c}\frac{(\bar{\alpha}_\parallel - \bar{\alpha}_\perp)^2}{90 k_B T}f^{(3)}, \quad \text{with } f^{(3)} = [f^{(1)}]^4 = \left(\frac{\varepsilon_r(\omega)+2}{3}\right)^4.$$

2. Show for the electrostrictive effect that the nonlinear refractive index coefficient for an orthogonally polarized probe beam in isotropic media is given by

$$n_{2\perp,\mathrm{eff}} = \frac{\varepsilon_0 n_y^3 n_x^3 p_{11} p_{21}}{4Kc}.$$

3. Evaluate the thermal contribution to $n_{2,\mathrm{eff}}$ for fused silica at 1.55 μm, GaAs at 0.82 and 1.55 μm, and ZnSe at 0.82 μm excited with

   (a) single ns, ps, and 10-fs pulses; and

   (b) 76-MHz trains of 1-ps pulses.

   The relevant material constants can be found in the literature.

4. Two-wave mixing in photorefractive media can lead to energy exchange between two beams. Consider two equal frequency plane waves crossing at angles $\pm\theta$ from a crystal axis. The combination of polarization and propagation gives $r_{\mathrm{eff}} \neq 0$.

   (a) Show that one beam grows and the other depletes with the propagation distance. What happens for $\theta = 0$? Interpret this result.

   (b) Assuming $I_1 \gg I_2$, solve for $I_1(z)$ and find the gain coefficient. Evaluate the gain for $BaTiO_3$, given $n_0 = 2.5$, $r_{\mathrm{eff}} = 8.2 \times 10^{-10}$ m/V, $\lambda_{\mathrm{vac}} = 0.5$ μm, and $E_D = 2n_0 k_{\mathrm{vac}} \sin\theta\, k_B T/e = 5 \times 10^2$ V/cm.

   (c) Show that energy exchange does not occur for a Kerr nonlinearity. What occurs for two beam mixing?

5. The nonlocal nature of the nonlinear index change induced by beams of finite width propagating in photorefractive media leads to many fascinating properties, including self-deflection of the beam. A beam nonlinearly induces a refractive index distribution the index change "spills out" past the beam and the index change across the beam is approximated by two regions separated by the line $\partial n/\partial x = $ constant. Show that after a propagation distance L along the z-axis (into the page), the deflection angle around the z-axis corresponds to $\theta(L) = n_0^{-1}(\partial n/\partial x)L$ for small deflection angles. Note that a unique property of photorefractive media is that the index distribution "follows" the beam as it deflects. [Hint: Try a judicious use of Snell's law.]

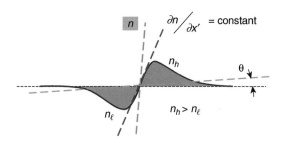

6. The nondegenerate nuclear (vibrational) contribution to $n_{2,\text{nuc}}(-\omega_b; \omega_a)$ cannot be related to the Raman spectrum as elegantly as in the degenerate case, where $\omega_b = \omega_a$. However, an approximate formula can be obtained with a number of approximations, some better than others. Initially assume an arbitrary distribution $f(\bar{\Omega}_\beta - \bar{\Omega}_{\beta 0})$ for the optical phonon frequencies associated with the βth vibrational mode and that the linewidth $\bar{\Gamma}_v^\beta = 2[\bar{\tau}_v^\beta]^{-1}$, including the effect of the randomness of a glass, is much smaller than both $\bar{\Omega}_{\beta 0}$ and the width of the distribution $f(\bar{\Omega}_\beta - \bar{\Omega}_{\beta 0})$. Assume that for a single isolated molecule,

$$n_{2||,\text{nuc}}(-\omega_b; \omega_a) = \sum_\beta \frac{1}{4n_b n_a \bar{m}_\beta \varepsilon_0^2 c} \left[\frac{\partial \bar{\alpha}_{iin}^\beta}{\partial \bar{q}_n^\beta} \bigg|_{\bar{q}_n^\beta = 0} \right]^2$$

$$\frac{\bar{\Omega}_\beta^2 - (\omega_b - \omega_a)^2}{[\bar{\Omega}_\beta^2 - (\omega_b - \omega_a)^2]^2 + ([\omega_a - \omega_b]\bar{\Gamma}_\beta)^2}.$$

(a) Show that in the glassy state

$$n_{2||,\text{nuc}}(-\omega_b; \omega_a) = \sum_\beta \frac{P_x(\omega_b)}{\Delta\Omega I(\omega_a)} \frac{c^3 (4\pi)^2 (\omega_b - \omega_a)^2}{n_b n_a \omega_b^4 k_B T} \frac{1}{f(\omega_b - \omega_a - \bar{\Omega}_{\beta 0})}$$

$$\times \int_0^\infty f(\bar{\Omega}_\beta - \bar{\Omega}_{\beta 0}) \left\{ \frac{\bar{\Omega}_\beta^2 - (\omega_b - \omega_a)^2}{[\bar{\Omega}_\beta^2 - (\omega_b - \omega_a)^2]^2 + ([\omega_a - \omega_b]\bar{\Gamma}_\beta)^2} \right\} d\bar{\Omega}_\beta.$$

(b) Plot the dispersion term with the sample parameters phonon field period/phonon field lifetime $= 0.1$.

(c) Assume that the dispersion term can be written approximately as the difference between two δ functions; i.e.,

$$\frac{\bar{\Omega}_\beta^2 - (\omega_b - \omega_a)^2}{[\bar{\Omega}_\beta^2 - (\omega_b - \omega_a)^2]^2 + ([\omega_b - \omega_a]\bar{\Gamma}_\beta)^2}$$

$$\cong C \left\{ \delta \left(\bar{\Omega}_\beta - (\omega_a - \omega_b) - \frac{\bar{\Gamma}_\beta}{2} \right) - \delta \left(\bar{\Omega}_\beta - (\omega_b - \omega_a) + \frac{\bar{\Gamma}_\beta}{2} \right) \right\}.$$

(Note that it is not a very good assumption because of the asymmetry of the distribution around the frequency $\bar{\Omega}_\beta = \omega_b - \omega_a$ and the fact that the constant C will be different for the two cases $\omega_b - \omega_a = 0 \to \omega_b - \omega_a = \bar{\Omega}_\beta$ and $\omega_b - \omega_a = \bar{\Omega}_\beta \to \infty$.) Show that for $\omega_b - \omega_a = 0 \to \omega_b - \omega_a = \bar{\Omega}_\beta$,

$$C = 2(\omega_b - \omega_a)\ln\left(1 + \left[\frac{2(\omega_b - \omega_a)}{\bar{\Gamma}_x^\beta}\right]^2\right)$$

(d) Assuming that this normalization is also valid for $\omega_b - \omega_a = \bar{\Omega}_\beta \to \infty$, show that

$$\int_0^\infty f(\bar{\Omega}_\beta - \bar{\Omega}_{\beta 0}) \frac{\bar{\Omega}_\beta^2 - (\omega_b - \omega_a)^2}{[\bar{\Omega}_\beta^2 - (\omega_b - \omega_a)^2]^2 + ([\omega_a - \omega_b]\bar{\Gamma}_\beta)^2} d\bar{\Omega}_\beta$$

$$= 4(\omega_b - \omega_a)\bar{\Gamma}_\beta\left[\ln\left(1 + \left[\frac{2(\omega_b - \omega_a)}{\bar{\Gamma}_\beta}\right]^2\right)\right]^{-1} f'((\omega_b - \omega_a) - \bar{\Omega}_{\beta 0}),$$

where $f'((\omega_b - \omega_a) - \bar{\Omega}_{\beta 0})$ is the derivative of the distribution.

(e) Assuming the Gaussian distribution $f(\bar{\Omega}_\beta - \bar{\Omega}_{\beta 0}) \propto \exp\{-[\omega_b - \omega_a - \bar{\Omega}_{\beta 0}]^2 / \Delta\bar{\Omega}_{1/2}^2\}$, show that

$$n_{2||\text{nuc}}(-\omega_b; \omega_a) = \frac{16c^3(4\pi)^2(\omega_a - \omega_b)^3}{n_b n_a k_B T \omega_b^4} \sum_\beta \frac{\bar{\Gamma}_\nu^\beta}{\Delta\bar{\Omega}_{1/2}^2} \ln\left[4\left\{\frac{\omega_b - \omega_a}{\bar{\Gamma}_\nu^\beta}\right\}^2 + 1\right][\omega_b - \omega_a = \bar{\Omega}_{\beta 0}]$$

$$\times \frac{P_x(\omega_b)}{\Delta\Omega I_x(\omega_a)}.$$

Note that despite all the approximations, this result has the correct general features, i.e., it passes through 0 at $\bar{\Omega}_{\beta 0} = \omega_b - \omega_a$; there is a strong dependence on the Raman spectrum modulated by the dispersion; and so on. However, this can be used (as an approximate result) only within a few $\Delta\bar{\Omega}_{1/2}$ of $\bar{\Omega}_{\beta 0} = \omega_b - \omega_a$.

7. Consider a liquid composed of molecules with the linear polarizability $\bar{\alpha}_1 \neq \bar{\alpha}_2 \neq \bar{\alpha}_3$.

  (a) Show from first principles that the effective intensity-dependent nonlinear index coefficient $n_{2||,\text{or}}(-\omega; \omega)$ induced in a direction parallel to an applied optical field is

$$n_{2||,\text{or}}(-\omega; \omega) = \frac{N}{n_0^2 \varepsilon_0^2 c} \frac{[f^{(1)}]^4}{45 k_B T}\{[\bar{\alpha}_1^2 + \bar{\alpha}_2^2 + \bar{\alpha}_3^2] - [\bar{\alpha}_1\bar{\alpha}_3 + \bar{\alpha}_1\bar{\alpha}_2 + \bar{\alpha}_2\bar{\alpha}_3]\},$$

  where $f^{(1)} = [\varepsilon^r(\omega) + 2]/3$.

(b) Find $n_{2\|,\text{or}}(-\omega; \omega)$ for the cases

    (i) $\bar{\alpha}_1 = \bar{\alpha}_2 \neq \bar{\alpha}_3$ (linear molecules, e.g., CO_2);

    (ii) $\bar{\alpha}_1 = 0, \bar{\alpha}_2 = \bar{\alpha}_3$ (disk molecules, isotropic in the plane of the disk, e.g., benzene); and

    (iii) $\bar{\alpha}_1 = \bar{\alpha}_2 = \bar{\alpha}_3 = 0$ (spherically symmetric molecules).

8. Show from first principles that the effective intensity-dependent nonlinear index change induced for a circularly polarized applied optical field for a liquid of cigar-shaped molecules such as CS_2 is given by

$$n_{2+,\text{or}}(-\omega; \omega) = +\frac{N}{n_0^2 \varepsilon_0^2 c} \frac{(\bar{\alpha}_\| - \bar{\alpha}_\perp)^2}{180 k_B T} [f^{(1)}]^4.$$

9. Show from first principles that the effective intensity-dependent nonlinear index change induced for a left circularly polarized applied optical field by a right circularly polarized strong beam for a liquid of a cigar-shaped molecule such as CS_2 is given by

$$n_{2-,\text{or}}(-\omega_-; \omega_r) = +\frac{N}{n_0^2 \varepsilon_0^2 c} \frac{(\bar{\alpha}_\| - \bar{\alpha}_\perp)^2}{180 k_B T} [f^{(1)}]^4.$$

REFERENCES

1. P. G. de Gennes, The Physics of Liquid Crystals (Clarendon Press, Oxford, 1974).

2. I. C. Khoo, Liquid Crystals 2nd Edition (Wiley-Interscience, Hopoken N.J., 2007).

3. I. C. Khoo, "Nonlinear optics of liquid crystalline materials," Phys. Rep. **471**, 221–267 (2009).

4. M. Peccianti, A. De Rossi, G. Assanto, A. De Luca, C. Umeton, and I. C. Khoo, "Electrically assisted self-confinement and waveguiding in planar nematic liquid crystal cells," Appl. Phys. Lett., **77**, 7–9 (2000).

5. I. C. Khoo, P. H. Chen, M. Y. Shih, A. Shishido, S. Slussarenko, and M. V. Wood, "Supra optical nonlinearities (SON) of methyl red- and azobenzene liquid crystal-doped nematic liquid crystals," Mol. Cryst. Liq. Cryst. Sci. Technol. A Mol. Cryst. Liq. Cryst., **358**, 1–13 (2001).

6. D. N. Christodoulides, I. C. Khoo, G. J. Salamo, G. I. Stegeman, and E. W. Van Stryland, "Nonlinear refraction and absorption: mechanisms and magnitudes," Adv. Opt. Photon., **2**, 60–200 (2010).

7. P. Yeh, Introduction to Photorefractive Nonlinear Optics (John Wiley & Sons, New York, 1993).

8. D. N. Christodoulides and M. I. Carvalho, "Compression, self-bending, and collapse of Gaussian beams in photorefractive crystals," Opt. Lett., **19**, 1714–1716 (1994)

9. R. Hellwarth, J. Cherlow, and T.-T. Yang, "Origin and frequency dependence of nonlinear optical susceptibilities of glasses," Phys. Rev. B, **1**, 964–967 (1975).

10. C. Rivero, R. Stegeman, M. Couzi, Th. Cardinal, K. Richardson, and G. Stegeman, "Resolved discrepancies between visible spontaneous Raman cross-section and direct NIR Raman gain measurements in TeO_2-based glasses," Opt. Express, **13**, 4759–4769 (2005).

11. I. Kang, S. Smolorz, T. Krauss, F. Wise, B. G. Aitken, and N. F. Borrelli, "Time-domain observation of nuclear contributions to the optical nonlinearities of glasses," Phys. Rev. B, **54**, R12641–R12644 (1996).

12. S. Smolorz, F. Wise, and N. F. Borrelli, "Measurement of the nonlinear optical response of optical fiber materials by use of spectrally resolved two beam coupling," Opt. Lett., **24**, 1103–1105 (1999).

13. S. Montant, A. Le Calvez, E. Freysz, and A. Ducasse, "Time-domain separation of nuclear and electronic contributions to the third-order nonlinearity in glasses," J. Opt. Soc. Am. B, **15**, 2802–2807 (1998)

14. D. A. Pinnow, "Elastooptical materials," in Handbook of Lasers with Selected Data on Optical Technology, edited by R. J. Pressley (CRC Press, Cleveland, 1971), Chap. 17.

15. D. I. Kovsh, D. J. Hagan, and E.W. Van Stryland, "Numerical modeling of thermal refraction in liquids in the transient regime," Opt. Express, **4**, 315–327 (1999).

16. L. A. Ostrovskii, "Self-action of light in crystals," JETP Lett., **5**, 272–275 (1967).

17. G. I. Stegeman, D. J. Hagan, and L. Torner, "$\chi^{(2)}$ cascading phenomena and their applications to all-optical signal processing, mode locking, pulse compression and solitons," Opt. Quantum Electron., **28**, 1691–1740 (1996).

18. G. I. Stegeman, M. Sheik-Bahae, E. W. Van Stryland, and G. Assanto, "Large nonlinear phase shifts in second-order nonlinear-optical processes," Opt. Lett., **18**, 13–15 (1993).

19. E. W. Van Stryland, "Third-order and cascaded nonlinearities," in Laser Sources and Applications, Scottish Graduate Series, edited by A. Miller and D. M. Finlayson (Institute of Physics Publishing Bristol U.K., 1997), pp. 15–62.

20. R. J. DeSalvo, D. J. Hagan, M. Sheik-Bahae, G. Stegeman, H. Vanherzeele, and E. W. Van Stryland, "Self-focusing and defocusing by cascaded second order effects in KTP," Opt. Lett., **17**, 28–30 (1991).

21. F. A. Hopf and G. I. Stegeman, Applied Classical Electrodynamics, Volume 1: Linear Optics (John Wiley & Sons, New York, 1985).

SUGGESTED FURTHER READING

B. A. Auld, Acoustic Fields and Waves in Solids (John Wiley & Sons, New York, 1973).

P. G. de Gennes, The Physics of Liquid Crystals (Clarendon Press, Oxford, 1974).

P. Günter and J.-P. Huignard, eds, Photorefractive Materials and Their Applications, I: Basic Effects (Springer-Verlag, Heidelberg, 1988).

P. Günter and J.-P. Huignard, eds, Photorefractive Materials and Their Applications, II: Survey of Applications (Springer-Verlag, Heidelberg, 1988).

I. C. Khoo, "Nonlinear optics of liquid crystalline materials," Phys. Rep., **471**, 221–267 (2009).

G. I. Stegeman, D. J. Hagan, and L. Torner, "$\chi^{(2)}$ cascading phenomena and their applications to all-optical signal processing, mode locking, pulse compression and solitons," Opt. Quantum Electron., **28**, 1691–1740 (1996).

P. Yeh, Introduction to Photorefractive Nonlinear Optics (John Wiley & Sons, New York, 1993).

CHAPTER 13

Techniques for Measuring Third-Order Nonlinearities

It was shown in Chapters 11 and 12 that there are many different contributions to an intensity and/or integrated fluence-dependent refractive index and absorption. They have different properties, and thus great care must be taken to separate the individual contributions. The key experiments that can be used are the time response of an effect and its frequency dispersion. A number of experimental techniques have been developed to evaluate the macroscopic $\chi^{(3)}_{ijk\ell}$ or the molecular $\overline{\gamma}^{(3)}_{ijk\ell}$ nonlinearities. Typically the ultimate goal is to evaluate or at least estimate $n_{2||,\mathrm{eff}}(-\omega;\ \omega)$. As high power lasers with progressively shorter pulses became available, the measured nonlinearities became smaller. In retrospect, this is exactly what can be expected given the number of phenomena that lead to an intensity- or flux-dependent change in the refractive index on different time scales, as discussed in Chapters 11 and 12.

Characterization of the nonlinearity at the wavelength and laser pulse width of interest is very important to understanding the nature of the nonlinearity, its response time, and so on, and hence how it can be used. To illustrate this point, consider the molecule CS_2 in liquid form, which is a classic case in which the nonlinearity has been measured via a number of techniques dating back to the early days of nonlinear optics. The following processes should all contribute to $n_{2||,\mathrm{eff}}(-\omega;\ \omega)$, namely, the Kerr effect, the nuclear (vibrational) nonlinearity, the reorientational nonlinearity, the electrostrictive effect, and finally thermal effects. Until recently, the first three listed have been difficult to resolve from one another. The first successful attempt by L. Canioni, with 100-fs pulses used a pump-probe nonlinear Mach Zehnder interferometer at 800 nm and obtained the following contributions to $n_{2||,\mathrm{eff}}(-\omega;\ \omega)$ (1):

Kerr 19%, response time shorter than pulse width (<100 fs)

Vibrational 64%, decay time $\sim$170 fs

Rotational 17%, decay time 880 fs

More recent Z-scan experiments by Hui et al. using 30-fs pulses have been able to resolve the electronic Kerr nonlinearity to be $2.8 \times 10^{-15}\,\mathrm{cm}^2/\mathrm{W}$ and to observe the

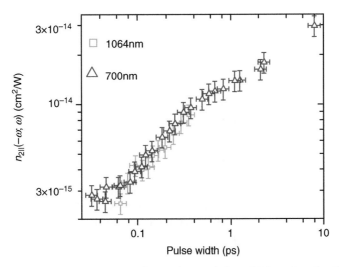

FIGURE 13.1 Z-scan measurement of $n_{2\|,\text{eff}}(-\omega;\ \omega)$ from 30 fs to 9 ps for liquid CS_2. Courtesy of Prof. E. VanStryland and Dr. H. Hu, University of Central Florida (2).

detailed rise of the nuclear and then the reorientational contributions with pulse width, as shown in Fig. 13.1 (2).

Ganeev et al. also reported results with both single pulses and pulse trains with wavelengths varying from 1054 to 532 nm with pulse widths from 100 fs to 0.1 μs (3). Their values are in excellent agreement with those of Hui et al., where their data overlapped down to 100 fs. At their longest pulse width (>300 ps), the reorientational effect has definitely reached its asymptotic value of 3×10^{-14}, indicating that the data in Fig. 13.1 at 9 ps represent the plateau for the reorientational mechanism. From the agreement between the results taken up to 10 ps at various wavelengths in Fig. 13.1, it appears that the nonresonant limit has effectively been reached at 700 nm. The electrostrictive effect based on calculations from Chapter 12 is expected to be at least about an order of magnitude smaller. Thermal nonlinearities due to linear loss either were not observed because the linear loss is too small in the near infrared or were separated from the ultrafast response by using single-pulse measurements (3). Ganeev et al. also observed significant multiphoton losses, thermal effects with mode-locked pulses due to the accumulation of the thermally induced index change as discussed in Chapter 12, and various other thermal effects, such as thermal lensing, that directly impact Z-scan measurements.

These results show that the nonlinearity $n_{2\|,\text{eff}}(-\omega;\ \omega)$ in a material can strongly depend on the pulse width, especially in the 50 fs–20 ps range. A value measured at one pulse width cannot be assumed to be the same at a different pulse width even when the same frequency is being used. Furthermore, there is dispersion with frequency so that values measured at one frequency cannot, in general, be assumed to be the same at a different frequency, as discussed in Chapter 9. Only when the frequency approaches the nonresonant regime does the nonlinearity not undergo significant frequency dispersion.

Although there have been many characterization techniques reported, only the most common generic ones will be discussed here. The following measurement techniques are discussed:

1. Z-scan (the most frequently used technique)
2. Third harmonic generation
3. Optical Kerr effect measurement of $n_{2\|}(-\omega; \omega)$
4. Nonlinear interferometry
5. Degenerate four-wave mixing (details of analysis are discussed in Chapter 15)

It is noteworthy that most measurement techniques that yield $n_{2\|}(-\omega; \omega)$ and $\alpha_{2\|}(-\omega; \omega)$ are usually accurate to $\pm5\%$ (at best). Only the two most frequently used techniques—Z-scan and third harmonic generation—will be discussed in detail.

13.1 Z-SCAN

Z-scan was developed at CREOL, University of Central Florida, to measure directly $n_{2\|}(-\omega; \omega)$ and $\alpha_{2\|}(-\omega; \omega)$ (4). It is the most powerful and yet simple technique to implement that has been developed to date. Its principal virtues are threefold: (*1*) It requires no calibration; (*2*) it does not require sophisticated instrumentation and is relatively easy to implement; and (*3*) when used properly it can yield directly both $n_{2\|}(-\omega; \omega)$ and $\alpha_{2\|}(-\omega; \omega)$. Z-scan is sensitive to *any* nonlinear absorption and *any* nonlinear refraction. In the Kerr case, its primary limitations are that it is difficult to unravel the order of the multiphoton absorption being measured from the data, e.g., two or three photons, if both are present. Furthermore, spurious effects, such as thermal lensing, can affect the results. Z-scan pump-probe experiments with the probe orthogonally polarized to the pump also lead to useful information, e.g., to distinguish the Kerr effect from the reorientational effect. The probe nonlinearity is two-thirds of the pump nonlinearity for the Kerr case and is one half for the reorientational case. However, the detailed discussion here is limited to the single-beam case. The simplest case to analyze is for a "thin" sample in which there is no change in beam profile on transit through the sample, just a small change in phase-front curvature. "Thick" samples usually require some additional analysis of data to take into account changes in beam shape inside the sample.

The experimental geometry is shown in Fig. 13.2. z is the position of the sample relative to the position of the focal point of the lens and the coordinate inside the sample is z', which varies as $z' = 0 \rightarrow L$. Furthermore, T(0) and T(L) are the Fresnel coefficients for the intensity at the entrance and exit sample boundaries, respectively. The sample is moved through the focal point of the lens, and the transmitted light is detected under two different conditions: "open" and "closed" aperture. In the open aperture geometry, the total transmitted beam is measured and the two-photon absorption coefficient can be obtained. For closed

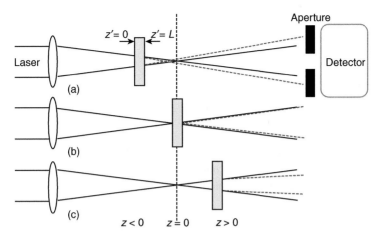

FIGURE 13.2 Z-scan measurement layout and the trajectory of the light in the absence of the sample (solid black) and after passing through a $n_{2\parallel}(-\omega; \omega) > 0$ medium at different sample positions z. The dashed red line indicates the trajectory due to nonlinear focusing. (a) Sample located before the focal point of the lens. (b) Sample centered on the focal point of the lens. (c) Sample beyond the focal point of the lens.

aperture, only a few tens of percent of light is transmitted and measured through an aperture centered on axis. The nonlinear index coefficient is also deduced by scanning the sample over z.

Typically an incident beam, Gaussian in space and time, is used in the analysis:

$$E_{inc}(r_\perp, z, t) = \frac{1}{2} \mathcal{E}_{inc}(r_\perp = 0, z, t = 0) \frac{w_0}{w(z)} e^{i(kz-\omega t) - \frac{r_\perp^2}{w^2(z)} - i\frac{kr_\perp^2}{2R(z)} - \frac{t^2}{\tau_0^2}} + c.c.$$

$$\rightarrow I_{inc}(r_\perp, z, t) = \frac{1}{2} n(\omega) c \varepsilon_0 |\mathcal{E}_{inc}(0, z, 0)|^2 \frac{w_0^2}{w^2(z)} e^{-\frac{2r_\perp^2}{w^2(z)} - \frac{2t^2}{\tau_0^2}}$$

$$= I_{inc}(0, z, 0) \frac{w_0^2}{w^2(z)} e^{-\frac{2r_\perp^2}{w^2(z)} - \frac{2t^2}{\tau_0^2}}. \tag{13.1}$$

The detailed notation for a Gaussian beam is given in Eqs 5.1 and 5.2.

13.1.1 Nonlinear Absorption: Open Aperture (Thin Sample)

The Kerr case will be used as the example. The aperture blocking light from the detector is removed or opened until all the transmitted light falls on the detector (4). Since $\Delta\alpha(z) = \alpha_{2\parallel}(-\omega; \omega)I(z)$, two-photon absorption is a maximum when the intensity is a maximum, which occurs when the sample center is located at the focal point of the lens ($z=0$) (Fig. 13.2b). Assuming $z_0 > 2L$ ("thin sample"; where

$z_0(\omega) = \pi w_0^2(\omega) n(\omega)/\lambda_{\text{vac}}(\omega))$ and nonlinear absorption, the intensity at z', from Eq. 10.8 (adapted for the pulsed case), is

$$I(r_\perp, z, t, z') = \frac{w_0^2}{w^2(z)} \frac{T(0) I_{\text{inc}}(0, z, 0, z') e^{-2\frac{r_\perp^2}{w^2(z)} - \frac{2t^2}{\tau_0^2}}}{1 + \alpha_{2\|}(-\omega; \omega) T(0) \left(1 + \frac{z^2}{z_0^2}\right)^{-2} I_{\text{inc}}(0, z, 0, z') e^{-2\frac{r_\perp^2}{w^2(z)} - \frac{2t^2}{\tau_0^2}} z'},$$

(13.2)

where $I_{\text{inc}}(0, 0, 0, z')$ is the maximum on-axis intensity at $z = 0$. The power P^{NL} at the sample exit surface outside the sample is given by

$$P^{\text{NL}}(z, t, L) = T(L) \int_0^\infty I(r_\perp, z, t, L) 2\pi r_\perp \, dr_\perp. \tag{13.3a}$$

When the sample is located far outside the focal point of the lens ($|z| \gg z_0$; Fig. 13.2a and c), there is effectively zero nonlinear loss and the transmitted power is

$$P^{\text{LIN}}(\infty, t, L) = T(L) \int_0^\infty I(r_\perp, \pm\infty, t, L) 2\pi r_\perp \, dr_\perp. \tag{13.3b}$$

A typical open aperture continuous-wave Z-scan is shown in Fig. 13.3. The relative transmission, normally given in terms of the pulse energies (parameter usually measured in pulsed experiments) $\Delta E_{\text{tr}}^{\text{NL}}(z)$ and $\Delta E_{\text{tr}}^{\text{LIN}}(z)$, is given by

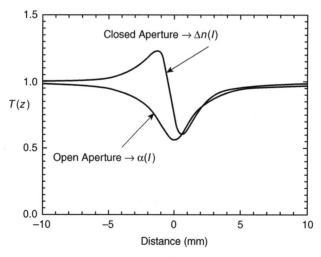

FIGURE 13.3 The characteristic shape of the transmission versus z curves for open and closed apertures for $n_{2\|}(-\omega; \omega) > 0$.

$$T(z) = \frac{\int_{-\infty}^{\infty} P^{NL}(z,t,L)\, dt}{\int_{-\infty}^{\infty} P^{LIN}(z \gg L, t, L)\, dt} = \frac{\Delta E_{tr}^{NL}(z)}{\Delta E_{tr}^{LIN}(|z| \gg L)}, \tag{13.4}$$

and $T(|\infty|) = 1$ gives automatic normalization. After some analysis, this measurement gives $\alpha_{2\parallel}(-\omega; \omega)$.

The situation is considerably simplified if the net nonlinear absorption is small, i.e., $1 \gg \alpha_{2\parallel}(-\omega; \omega)T(0)I_{inc}(r_\perp, 0, 0, z')z'$ for $z' = 0 \to L$, and so

$$\left[1 + \alpha_{2\parallel}(-\omega; \omega)T(0)I(0,z,0,L)e^{-2\frac{r^2}{w_0^2} - \frac{2t^2}{\tau_0^2}} z' \right]^{-1} \cong 1 - \alpha_{2\parallel}(-\omega; \omega)T(0)$$

$$\times I(0,z,0,L)e^{-2\frac{r^2}{w_0^2} - \frac{2t^2}{\tau_0^2}} z' \tag{13.5}$$

$$\to I(r_\perp, z, t, z') = T(0)I(0,z,0,z')\left[e^{-2\frac{r^2}{w_0^2} - \frac{2t^2}{\tau_0^2}} - \alpha_{2\parallel}(-\omega; \omega)T(0) \right.$$

$$\left. \times I(0,z,0,0)\, e^{-4\frac{r^2}{w_0^2} - \frac{4t^2}{\tau_0^2}} z' \right]. \tag{13.6}$$

Noting that the pulse energy is given by

$$\Delta E(z) = I(0,z,0,0)\int_{-\infty}^{\infty} e^{-\frac{2t^2}{\tau_0^2}}\, dt \int_0^{\infty} e^{-2\frac{r^2}{w^2(z)}} 2\pi r_\perp\, dr_\perp = \frac{\pi^{3/2} w^2(z)\tau_0}{2\sqrt{2}} I(0,z,0,0),$$

$$\tag{13.7}$$

$$T(z) = \frac{\Delta E_{tr}^{NL}(z > L/2)}{\Delta E_{tr}^{LIN}(z > L/2)} = 1 - \frac{1}{2\sqrt{2}}\alpha_{2\parallel}(-\omega; \omega)T(0)I_{inc}(0,z,0,L)L$$

$$= 1 - \alpha_{2\parallel}(-\omega; \omega)T(0)\Delta E_{inc}\frac{L}{\pi^{3/2} w_0^2 \tau_0}. \tag{13.8}$$

Equation 13.8 gives a direct measurement of the two-photon absorption coefficient under the stated assumptions. Note that these include the assumption of beams Gaussian in space and time.

13.1.2 Nonlinear Refraction: Closed Aperture

In this case, a small aperture, centered on the incident beam, is used to limit light falling on the detector (4). How much light is collected by that aperture depends on whether the sample is to the right or to the left of the focal point (see Fig. 13.2). Consider the case $n_{2\parallel,\mathrm{eff}}(-\omega; \omega) > 0$. When the sample is located before the focal point, light undergoes self-focusing before the focal point of the lens, which results in more light missing the aperture than for the sample at the focal point. On the other hand, if the sample is past the focal point, more light is incident on the open aperture than for the sample at the focal point. Therefore as z is scanned, the signal transmitted through the aperture, is shown in Fig. 13.3. The order of the maxima and minima is reversed if $n_{2\|,\text{eff}}(-\omega; \omega) < 0$.

Thus the sign of the nonlinearity is obtained. Furthermore, after some analysis, the magnitude of the peak–valley separation in the transmission gives the magnitude of $n_{2\|,\text{eff}}(-\omega; \omega)$.

To evaluate the magnitude of $n_{2\|,\text{eff}}(-\omega; \omega)$, it is necessary to analyze the phase of the field after the sample (neglecting linear loss). The accumulated nonlinear phase shift is calculated by integrating over the sample (from $z' = 0$ to $z' = L$) the equation

$$\frac{d\phi^{\text{NL}}(r_\perp, z, t, z')}{dz'} = n_{2\|}(-\omega; \omega)k_{\text{vac}}I(r_\perp, z, t, z'). \tag{13.9}$$

Evaluation of this integral can be complicated because the intensity distribution inside the sample can become non-Gaussian on transit through "thick" samples ($L > z_0/2$) or when strong two-photon absorption is present, or both. Including two-photon absorption in the *thin sample limit*,

$$I(r_\perp, z, t, z') = \frac{T(0)I(0, z, 0, z')\, e^{-2\frac{r_\perp^2}{w^2(z)} - \frac{2t^2}{\tau_0^2}}}{1 + \alpha_{2\|}(-\omega; \omega)T(0)I(0, z, 0, z')\, e^{-2\frac{r_\perp^2}{w^2(z)} - \frac{2t^2}{\tau_0^2}} z'}$$

$$\rightarrow \Delta\phi^{\text{NL}}(r_\perp, z, t, L) = \frac{n_{2\|}(-\omega; \omega)k_{\text{vac}}}{\alpha_{2\|}(-\omega; \omega)}$$

$$\ln\left[1 + \alpha_{2\|}(-\omega; \omega)T(0)I(0, z, 0, L)\, e^{-2\frac{r_\perp^2}{w^2(z)} - \frac{2t^2}{\tau_0^2}} L\right]. \tag{13.10}$$

In the limit that the effects of two-photon absorption are small,

$$\Delta\phi^{\text{NL}}(r_\perp, z, t, L) \xrightarrow{\ 1 \gg \alpha_{2\|}(-\omega;\, \omega)T(0)I(0,z,0,L)L\ } k_{\text{vac}}Ln_{2\|}(-\omega; \omega)T(0)$$

$$\times I(0, z, 0, L)e^{-2\frac{r_\perp^2}{w_0^2(z)} - \frac{2t^2}{\tau_0^2}} = k_{\text{vac}}Ln_{2\|}(-\omega; \omega)T(0)I(r_\perp, z, t, L). \tag{13.11}$$

That is, with the assumption of no loss, the shape of the nonlinearity introduced a change in the phase front shape that mirrors the Gaussian beam intensity distribution.

The nonlinearity of the sample alters some aspects of the beam. Hence the next part of the analysis is to propagate the non-Gaussian beam field exiting the sample to the detector:

$$\vec{E}(r_\perp, z, t, L) \cong \frac{1}{2}\sqrt{\frac{2T(L)I(0, z, t, L)}{n(\omega)c\varepsilon_0}}\frac{w_0}{w(z)}e^{-\frac{r_\perp^2}{w^2(z)}-\frac{t^2}{\tau_0^2}+i\left[kz-\omega t-\frac{kr_\perp^2}{2R(z)}+\Delta\phi^{\mathrm{NL}}(r_\perp, 0, t, L)\right]} + \mathrm{c.c.}$$

(13.12)

For example, Huyghen's principle can be used numerically. Alternately, the following approach that uses the cylindrical symmetry of the beam can be employed:

1. Decompose the beam field at the output sample facet into Gaussian beams.
2. Propagate each Gaussian separately to the detector.
3. Sum the Gaussian fields at the detector, giving the total field $E_a(r_\perp, z_a, t, L)$.
4. Integrate the intensity over the aperture radius a.

The energy passing through the aperture is written as

$$\Delta E(z_a) = \frac{1}{2\sqrt{2}}cn\varepsilon_0\sqrt{\pi}\tau_0\int_0^a |E_a(r_\perp, z_a, 0, L)|^2\,2\pi r_\perp\,dr_\perp.$$

(13.13)

The relative transmission given in terms of the pulse energies $\Delta E_{\mathrm{tr}}^{\mathrm{NL}}(z_a)$ and $\Delta E_{\mathrm{tr}}^{\mathrm{LIN}}(z_a)$ is

$$T(z) = \frac{\Delta E_{\mathrm{tr}}^{\mathrm{NL}}(z)}{\Delta E_{\mathrm{tr}}^{\mathrm{LIN}}(z)}.$$

(13.14)

As the sample is scanned along z, the shape of the transmission curve obtained is as shown in Fig. 13.2b. The difference in the "peak" (maximum) to "valley" (minimum) transmission is defined as

$$\Delta T_{\mathrm{p\text{-}v}}(z, t) = \frac{\Delta E_{\mathrm{tr}}^{\mathrm{NL}}(z)|_{\mathrm{peak}} - \Delta E_{\mathrm{tr}}^{\mathrm{NL}}(z)|_{\mathrm{valley}}}{\Delta E_{\mathrm{tr}}^{\mathrm{LIN}}(\infty)}.$$

(13.15)

For small peak (p) to valley (v) differences, it has been shown that $\Delta T_{\mathrm{p\text{-}v}} \cong 0.406|\Delta\phi_{\mathrm{max}}^{\mathrm{NL}}|$, which gives the nonlinear phase shift valid to 5% for $\pi \geq |\Delta\phi_{\mathrm{max}}^{\mathrm{NL}}|$ [4].

13.1.3 Nonlinear Absorption and Refraction Together

The open aperture results are not affected by the nonlinear phase change in the example of ZnSe at 532 nm (see Fig. 13.4a) [4]. However, the closed aperture results

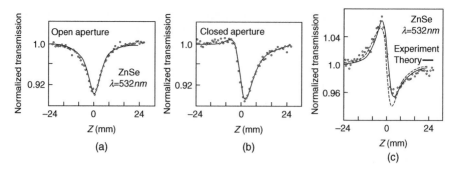

FIGURE 13.4   (a) Open aperture result. (b) Closed aperture result. (c) Closed aperture result analyzed by using the approximate procedure. Reproduced with permission from the Institute of Electronic and Electrical Engineering (4).

can be strongly affected by two-photon absorption (see Fig. 13.4b). Strictly, the problem must be solved numerically, starting with Eq. 13.9 with two-photon absorption included. A simpler approximate approach is to divide, point by point (along z), the closed aperture result by the open aperture result and then analyze that result for $n_{2\|}(-\omega; \omega)$. Figure 13.4c indicates that this approximate procedure works surprisingly well.

13.1.4   Z-Scan Solution Measurements

Z-scan is frequently used to estimate an average molecular third-order susceptibility $\langle \tilde{\gamma}_{\text{mol}}^{(3)}(-\omega; \omega, -\omega, \omega)\rangle$ of randomly oriented molecules dissolved in an appropriate solvent (see Problem 13.1). The solvent should be "nonpolar"; i.e., the solvent molecules should not have permanent dipole moments since the dipole moments can lead to a redistribution of charges in the target molecule of interest and hence potentially a change in the nonlinearity. Since the molecules are randomly oriented, only an average over all orientations is measured. Furthermore, $\langle \tilde{\gamma}_{\text{sol}}^{(3)}(-\omega; \omega, -\omega, \omega)\rangle$ should preferably be less than $\langle \tilde{\gamma}_{\text{mol}}^{(3)}(-\omega; \omega, -\omega, \omega)\rangle$ since it is the sum $N_{\text{mol}}\langle \tilde{\gamma}_{\text{mol}}^{(3)}(-\omega; \omega, -\omega, \omega)\rangle + N_{\text{sol}}\langle \tilde{\gamma}_{\text{sol}}^{(3)}(-\omega; \omega, -\omega, \omega)\rangle$ that is measured experimentally, where N_{mol} and N_{sol} are the number densities of the two species. The nonlinearity is measured simply both in the "neat" (pure) solvent and in the solution, and the values are subtracted with appropriate weighting, provided nonlinearities are in the nonresonant regime for both the solvent and the solute. Note that frequently the maximum useful N_{mol} is limited by the solubility in the solvent of the molecules. Since the molecules are randomly oriented, it is possible to deduce values for the molecular nonlinearities from solution measurements only if the target molecules exhibit symmetries in which certain values of the tensor components of $\tilde{\gamma}_{ijk\ell}^{(3)}(-\omega; \omega, -\omega, \omega)$ dominate the nonlinear response. A number of examples are given below for dilute solutions of target molecules.

Linear molecules in which $\tilde{\bar{\gamma}}^{(3)}_{xxxx}(-\omega;\ \omega,-\omega,\omega)$ dominates $\tilde{\bar{\gamma}}^{(3)}_{ijk\ell}(-\omega;\ \omega,-\omega,\omega)$;

e.g., CS_2: $\tilde{\chi}^{(3)}_{xxxx,mol}(-\omega;\ \omega,-\omega,\omega) = N_{mol} \times \frac{1}{5} \times \langle\tilde{\bar{\gamma}}^{(3)}_{mol}(-\omega;\ \omega,-\omega,\omega)\rangle$
$[f^{(1)}_{sol}]^4$.

Disklike molecules with $\tilde{\bar{\gamma}}^{(3)}_{xxxx}(-\omega;\ \omega,-\omega,\omega) = \tilde{\bar{\gamma}}^{(3)}_{yyyy}(-\omega;\ \omega,-\omega,\omega)$; e.g.,

C_6H_6: $\tilde{\chi}^{(3)}_{xxxx}(-\omega;\ \omega,-\omega,\omega) = N_{mol} \times \frac{8}{15} \times \langle\tilde{\bar{\gamma}}^{(3)}_{xxxx}(-\omega;\ \omega,-\omega,\omega)\rangle[f^{(1)}_{sol}]^4$.

The numerical factors 1/5 and 8/15 are the result of the orientation averaging, and their derivation is given as problems at the end of the chapter. Note that the relations involve the fourth power of the local field enhancement factor with all its approximations. A 5% uncertainty there leads to a 21% uncertainty in the measured molecular parameters.

13.2 THIRD HARMONIC GENERATION

This technique is routinely used to measure $\hat{\chi}^{(3)}_{ijkl}(-3\omega;\ \omega,\omega,\omega)$ of liquids and thin solid films (5). In liquids and films with randomly oriented molecules, the same averaging over random orientation considerations apply as for the Z-scan measurements discussed in Section 13.1.4. To obtain reasonable accuracy, the refractive index needs to be known at both the frequency of the input beam (ω) and the third harmonic signal (3ω). The concept and implementation is similar to that of the Maker Fringe measurements discussed for $\chi^{(2)}$ in Chapter 7. In the nonresonant regime of both $\hat{\chi}^{(3)}_{ijkl}(-3\omega;\ \omega,\omega,\omega)$ and $\hat{\chi}^{(3)}_{ijkl}(-\omega;\ \omega,-\omega,\omega)$, $3\hat{\chi}^{(3)}_{ijkl}(-3\omega;\ \omega,\omega,\omega) = \hat{\chi}^{(3)}_{ijkl}(-\omega;\ \omega,-\omega,\omega)$. However, due to the dramatically different frequency dispersion of the two susceptibilities, $3\hat{\chi}^{(3)}_{ijkl}(-3\omega;\ \omega,\omega,\omega) \neq \hat{\chi}^{(3)}_{ijkl}(-\omega;\ \omega,-\omega,\omega)$, and $\hat{\chi}^{(3)}_{ijkl}(-3\omega;\ \omega,\omega,\omega)$ cannot, in general, be used to estimate $\hat{\chi}^{(3)}_{ijkl}(-\omega;\ \omega,-\omega,\omega)$. This is not to imply that third harmonic generation is not a useful tool. It is routinely used to evaluate the effect of modifications in chemical structure on the nonlinearity in material families.

A typical liquid sample geometry is shown in Fig. 13.5. The wedge angle is normally quite small, typically less than 1°, so that well-defined interference fringes can be formed without the necessity of highly coherent sources. The separation

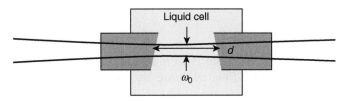

FIGURE 13.5 A typical liquid cell configuration for third harmonic generation.

between the wedges is $\sim 100-200\,\mu m$, which is usually less than the Rayleigh range (z_0) of the fundamental. The wedge is translated relative to the beam, and the interference pattern is recorded. Because of the dispersion in refractive index between the fundamental and its third harmonic, the conversion to third harmonic is usually very small so that the fundamental is essentially unattenuated.

The analysis proceeds as follows. Assume a fundamental Gaussian beam of the form

$$E_x(\omega, z, r_\perp, t) = \frac{1}{2}\hat{e}_x \mathcal{E}(\omega, z, 0, t)\frac{1}{\sqrt{1+z^2/z_1^2}}\, e^{-\frac{r_\perp^2}{w_1^2(z)} - i\frac{k(\omega)}{2R_1(z)} - i\psi_1(z) + ik(\omega)z - i\omega t} + \text{c.c.},$$

$$z_1 = \frac{\pi w_0^2}{\lambda(\omega)}, \quad \psi_1 = \tan^{-1}\left(\frac{z}{z_1}\right), \quad w_1^2(z) = w_1^2\left(1+\frac{z^2}{z_1^2}\right), \quad R_1(z) = z\left(1+\frac{z^2}{z_1^2}\right),$$

$$(13.16)$$

where subscript 1 refers to the fundamental input beam. Similarly, subscript 3 will identify the third harmonic. The sample is centered at $z=0$. The third harmonic polarization induced in the sample is

$$\mathcal{P}_x^{(3)}(3\omega, z, r_\perp, t)\, e^{ik(3\omega)z} = \frac{1}{4}\varepsilon_0 \hat{\chi}_{xxxx}^{(3)}(-3\omega; \omega, \omega, \omega)$$

$$\times \frac{\mathcal{E}_x^3(\omega, z, 0, t)}{(1+z^2/z_0^2)^{3/2}}\, e^{-\frac{3r_\perp^2}{w^2(z)} - 3i\frac{k(\omega)r_\perp^2}{2R(z)} - 3i\psi(z) + 3ik(\omega)z}. \qquad (13.17)$$

This polarization will generate a 3ω field that satisfies the wave equation; i.e.,

$$E_x(3\omega, z, r_\perp, t) = \frac{1}{2}\hat{e}_x \mathcal{E}(3\omega, z, 0, t)\frac{1}{\sqrt{1+z^2/z_3^2}}\, e^{-\frac{r_\perp^2}{w_3^2(z)} - i\frac{k(3\omega)}{2R_3(z)} - i\psi_3(z) + ik(3\omega)z - 3i\omega t} + \text{c.c.},$$

$$w_3^2(z) = \frac{w_1^2(z)}{3}, \quad z_3 = \frac{\pi w_3^2}{\lambda(3\omega)} = \frac{\pi n(3\omega)w_1^2/3}{\lambda_{vac}(\omega)/3} \approx \frac{\pi n(\omega)w_1^2}{\lambda_{vac}(\omega)} = z_1 = z_0,$$

$$\psi_3 = \tan^{-1}\left(\frac{z}{z_1}\right) = \tan^{-1}\left(\frac{z}{z_1}\right) = \psi, \quad R_3(z) = z\left(1+\frac{z_3^2}{z^2}\right) = \left(1+\frac{z_1^2}{z^2}\right) = R(z).$$

$$(13.18)$$

Although the third harmonic beam is narrower than the fundamental beam, the "Rayleigh range" and phase-front curvature are approximately the same. Since the term

$$\frac{1}{\sqrt{1 + z^2/z_0^2}} e^{-\frac{r_\perp^2}{w_3^2(z)} - i\frac{k(3\omega)r_\perp^2}{2R_3(z)} - i\psi_3(z)} \tag{13.19}$$

varies slowly over the distance z_0, its derivative with z is small compared to $d\mathcal{E}(3\omega, z, 0, t)/dz$; the slowly varying envelope approximation (Eq. 2.25) with $\Delta k = 3k(\omega) - k(3\omega)$ gives

$$\frac{d\mathcal{E}(3\omega, z, r_\perp, t)}{dz} = i\frac{3\omega}{8n(3\omega)c}\hat{\chi}^{(3)}_{xxxx}(-3\omega; \omega, \omega, \omega)$$

$$\times \frac{\mathcal{E}^3(\omega, 0, r_\perp, t)}{1 + z^2/z_0^2} e^{-i\frac{\Delta k r_\perp^2}{R(z)} - 2i\psi(z) + i\Delta k z}$$

$$\rightarrow \quad \mathcal{E}\left(3\omega, \frac{d}{2}, r_\perp, t\right) = i\frac{3\omega\hat{\chi}^{(3)}_{xxxx}(-3\omega; \omega, \omega, \omega)\mathcal{E}^3(\omega, 0, r_\perp, t)}{8n(3\omega)c}$$

$$\int_{-d/2}^{d/2} \frac{e^{-i\frac{\Delta k r_\perp^2}{R(z)} - 2i\psi(z) + i\Delta k z}}{1 + z^2/z_0^2} dz. \tag{13.20}$$

Here $\mathcal{E}(3\omega, d/2, 0, t)$ and $\mathcal{E}(\omega, 0, 0, t)$ are the on-axis fields; i.e., $r_\perp = 0$. Assuming no depletion of the fundamental field, the *peak* on-axis intensity is

$$I\left(3\omega, \frac{d}{2}, 0, 0\right) = \frac{1}{2}n(3\omega)c\varepsilon_0 \left|\mathcal{E}\left(3\omega, \frac{d}{2}, 0, 0\right)\right|^2,$$

$$I(\omega, 0, 0, 0) = \frac{1}{2}n(\omega)c\varepsilon_0 |\mathcal{E}(\omega, 0, 0, t)|^2 \rightarrow |\mathcal{E}(\omega, 0, 0, t)|^6 = \frac{2^3}{n^3(\omega)c^3\varepsilon_0^3}I^3(\omega, 0, 0, t). \tag{13.21}$$

This finally gives

$$I\left(3\omega, \frac{d}{2}, 0, t\right) = \frac{(3\omega)^2 |\hat{\chi}^{(3)}_{xxxx}(-3\omega; \omega, \omega, \omega)|^2}{16n^3(\omega)n(3\omega)c^4\varepsilon_0^2}I^3(\omega, 0, 0, t)\left|\int_{-d/2}^{d/2} \frac{e^{-i\frac{\Delta k r_\perp^2}{2R(z)} - 2i\psi(z) + i\Delta k z}}{(1 + z^2/z_0^2)}dz\right|^2. \tag{13.22}$$

Usually it is the total energy in the third harmonic relative to the fundamental cubed that is measured experimentally. To first calculate the total power in the fundamental and the harmonic, it is necessary to integrate over the spatial profiles of the beams. For the third harmonic,

$$I\left(3\omega,\frac{d}{2},r_\perp,t\right) = I\left(3\omega,\frac{d}{2},0,t\right) e^{-6\frac{r_\perp^2}{w_1^2(d/2)}}$$

$$\rightarrow P\left(3\omega,\frac{d}{2},t\right) = I\left(3\omega,\frac{d}{2},0,t\right) \int_0^\infty 2\pi r_\perp\, e^{-6\frac{r_\perp^2}{w_1^2(d/2)}}\, dr_\perp = \frac{\pi w^2(d/2)}{6} I\left(3\omega,\frac{d}{2},0,t\right).$$

$$(13.23a)$$

Furthermore, for the fundamental,

$$I(\omega,0,r_\perp,t) = I(\omega,0,0,t)\, e^{-2\frac{r_\perp^2}{w_0^2}} \rightarrow P(\omega,0,t) = I(\omega,0,0,t) \int_0^\infty 2\pi r_\perp\, e^{-2\frac{r_\perp^2}{w_0^2}}\, dr_\perp$$

$$= \frac{\pi w_0^2}{2} I(\omega,0,0,t). \qquad (13.23b)$$

This gives for the general case

$$P\left(3\omega,\frac{d}{2},t\right) = \frac{3\omega^2 w^2(d/2)|\hat{\chi}_{xxxx}^{(3)}(-3\omega;\omega,\omega,\omega)|^2}{4\pi^2 w_0^6 n^3(\omega)n(3\omega)c^4\varepsilon_0^2} P^3(\omega,0,t) \left| \int_{-d/2}^{d/2} \frac{e^{-i\frac{\Delta k r_\perp^2}{2R(z)}-2i\psi(z)+i\Delta kz}}{1+z^2/z_0^2} dz \right|^2.$$

$$(13.24)$$

For $z_0 \gg d/2$ (the usual case), $\psi(z)=0$, $R(z)\rightarrow\infty$, and $z/z_0\rightarrow0$,

$$P\left(3\omega,\frac{d}{2},t\right) = \frac{3\omega^2 |\hat{\chi}_{xxxx}^{(3)}(-3\omega;\omega,\omega,\omega)|^2}{\pi^2 w_0^4 n^3(\omega)n(3\omega)c^4\varepsilon_0^2 \Delta k^2} \sin^2\left(\frac{\Delta kd}{2}\right) P^3(\omega,0,t). \qquad (13.25)$$

To measure the pulse energy, a specific form of the time dependence of the pulse is needed. Here Gaussian input pulses were assumed, and integrating the fundamental and third harmonic powers over time gives

$$\Delta E(\omega) = 2P\left(\omega,\frac{d}{2},0\right) \int_0^\infty e^{-\frac{2t^2}{\tau_1^2}} dt \rightarrow P^3\left(\omega,\frac{d}{2},0\right) = \frac{1}{\sqrt[3]{2\pi\tau_1^3}}\Delta E^3(\omega),$$

$$\Delta E(3\omega) = 2P\left(3\omega,\frac{d}{2},0\right) \int_0^\infty e^{-\frac{2t^2}{\tau_3^2}} dt \xrightarrow{\tau_3=\tau_1/\sqrt{3}} P\left(3\omega,\frac{d}{2},0\right) = \frac{\sqrt{3}}{\tau_1\sqrt{2\pi}}\Delta E(3\omega),$$

$$(13.26)$$

where $\Delta E(3\omega)$ is the harmonic pulse energy and $\Delta E(\omega)$ is the fundamental pulse energy, such that

$$\Delta E(3\omega) = \frac{\sqrt{3}\omega^2 |\hat{\chi}_{xxxx}^{(3)}(-3\omega;\omega,\omega,\omega)|^2}{2\pi^3\tau_1^2 w_0^4 n^3(\omega)n(3\omega)c^4\varepsilon_0^2 \Delta k^2} \sin^2\left(\frac{\Delta kd}{2}\right)\Delta E^3(\omega). \qquad (13.27)$$

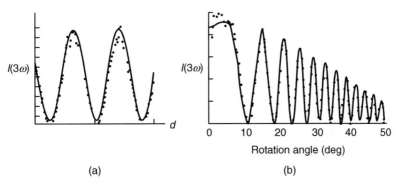

$I(3\omega)$

$I(3\omega)$

0   10   20   30   40   50

Rotation angle (deg)

(a)

(b)

FIGURE 13.6 (a) The interference fringes obtained by varying d in a third harmonic experiment. (b) The interference fringes obtained by rotating the sample relative to the incident beam. Reproduced with permission from American Physical Society (5).

The parameter d is varied by scanning the beam up and down, which results in fringes as expected from the sine function squared and shown in Fig. 13.6a for a liquid sample. For a solid sample, the incidence angle is varied and Maker Fringes are observed, as discussed in Chapter 7. A typical result for a silica plate is shown in Fig. 13.6b. The oscillation period gives the index dispersion and the maxima yield $|\hat{\chi}_{xxxx}^{(3)}(-3\omega; \omega, \omega, \omega)|^2$. It is very important to realize that it is the modulus of $\hat{\chi}_{xxxx}^{(3)}(-3\omega; \omega, \omega, \omega)$ that is measured. To obtain information about $\Re eal\{\hat{\chi}_{xxxx}^{(3)}(-3\omega; \omega, \omega, \omega)\}$, it is necessary to avoid resonances at ω and 3ω, which explains why many experiments on organic materials are performed with incident wavelengths of $1.9\,\mu m$ typically obtained by Raman shifting a Nd:YAG laser in a hydrogen gas cell.

13.3 OPTICAL KERR EFFECT MEASUREMENTS

The term "optical Kerr effect" covers a broad range of techniques for measuring $n_{2\|,\text{eff}}(-\omega; \omega)$, which depend either on intensity-induced interference effects or on optically induced birefringence. Although initially meant for Kerr media only, this technique has also been used to measure vibrational and reorientational nonlinearities. Typically the optical Kerr effect is used for optically isotropic media, such as glasses and liquids. The same comments on solution measurements as made for Z-scan previously also apply here.

A generic experimental setup is shown in Fig. 13.7. The polarizers P_1 and P_2 are arranged to be orthogonal to block the output signal, with no probe-beam present. Pump (ω_p) and probe (signal, ω_s) wavelengths may be identical or different. The signal polarization is oriented at 45° in the x–y plane so that the pump beam induces a different change in the refractive index (and hence phase shift on transit through the sample) for the polarizations parallel and perpendicular to the pump beam. The index change is then probed by the signal beam at different delay times to obtain information about the magnitude of $n_{2\|,\text{eff}}(-\omega; \omega)$ and the characteristic relaxation times associated with the nonlinear index change.

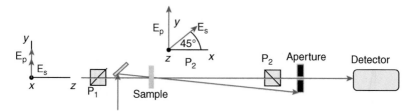

FIGURE 13.7   Typical layout for a simplified optical Kerr effect experiment.

For the simplest case of an optically isotropic medium, the input signal is written as

$$\vec{E}_{s,\text{inc}} = \frac{\mathcal{E}_s}{2\sqrt{2}} \left[\hat{e}_x + \hat{e}_y\right] e^{i[nk_{\text{vac}}z - \omega_s t]} + \text{c.c.}, \tag{13.28}$$

where the signal refractive index is $n_{s,x} = n + n_{2\perp,\text{eff}}(-\omega_s; \omega_p)I_p$ and $n_{s,y} = n + n_{2\parallel,\text{eff}}(-\omega_s; \omega_p)I_p$ and I_p is the pump intensity. After the cross-polarizer, defining $\varphi = \{n_{2\parallel,\text{eff}}(-\omega_s; \omega_p) - n_{2\perp,\text{eff}}(-\omega_s; \omega_p)\}k_{\text{vac}}I_pL$, the output is given by

$$E_{s,\text{out}} = \frac{\mathcal{E}_s}{2\sqrt{2}} e^{i[nk_{\text{vac}}z - \omega_s t]} \left\{\left[\hat{e}_x e^{in_{2\perp}(-\omega_s; \omega_p)k_{\text{vac}}I_pL} + \hat{e}_y e^{in_{2\parallel}(-\omega_s; \omega_p)k_{\text{vac}}I_pL}\right] \cdot \left[\hat{e}_x - \hat{e}_y\right]\right\} + \text{c.c.}$$

$$\rightarrow \frac{I_{s,\text{trans}}}{I_{s,\text{inc}}} = \frac{1}{2}\left|e^{in_{2\perp}(-\omega_s; \omega_p)k_{\text{vac}}I_pL} - e^{in_{2\parallel}(-\omega_s; \omega_p)k_{\text{vac}}I_pL}\right|^2 = 1 - \cos\varphi \xrightarrow{1 \gg \varphi} \frac{1}{2}\varphi^2.$$

$$\tag{13.29}$$

For Kerr media, $\varphi = \frac{2}{3}n_{2\parallel}(-\omega_s; \omega_p)k_{\text{vac}}I_pL$. For the molecular orientational nonlinearity, $\varphi = \frac{1}{2}n_{2\parallel,\text{or}}(-\omega_s; \omega_p)k_{\text{vac}}I_pL$. By introducing an adjustable delay line in one of the input beams, the decay with time of the index change induced by the pump beam can be measured.

13.4   NONLINEAR OPTICAL INTERFEROMETRY

Although very simple conceptually, these techniques require precise alignment of optical elements and careful design against vibration as well as thermal and mechanical creep problems in mounts and so on. The basic concept is to split a beam into two parts. One beam is either sent through a reference sample or air. The second beam travels through the sample to be investigated and acquires a nonlinear phase shift linear with the input intensity (for the Kerr case) and a spatial distribution across the beam that mirrors the input pulse shape. The two beams are colinearly recombined to produce fringes with circular symmetry. An aperture samples the central part of the interfering beams, and the shift of the pattern with increasing input intensity is monitored. Figure 13.8 shows the basic configuration. The period Λ of the intensity fringes produced by an intensity increase ΔI corresponds to a nonlinear phase change of 2π, which gives the nonlinearity $n_{2\parallel}(-\omega; \omega) = 2\pi/k_{\text{vac}}(\omega)L\Delta I$,

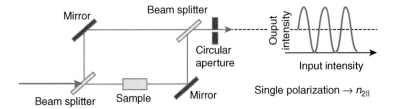

FIGURE 13.8   Basic layout for measuring interferometrically nonlinear phase shifts.

where L is the sample length for continuous-wave beams. For pulsed beams, $n_{2\parallel}(-\omega; \omega)$ can still be evaluated by taking into account the pulse parameters as just discussed in the two preceding examples.

There have been a number of implementations of this basic idea. Very sophisticated versions are capable of measuring very small fringe shifts and hence small nonlinearities. Two examples are by Rochford et al. (6) and Santran et al. (7). The unique feature in the first reference is that a tunable number of pulses from a mode-locked laser could be used to separate out thermal from ultrafast effects. In the second case, the system was stabilized so that very small phase shifts could be measured.

13.5 DEGENERATE FOUR-WAVE MIXING

This case is treated in detail in Chapter 15 on wave mixing and will only be briefly discussed here. There are three input beams, $E^{(p1)}$, $E^{(p2)}$, and $E^{(s)}$, and one output beam, $E^{(c)}$, as shown in Fig. 13.9. Beam alignment is critical, i.e., pump beams are collinear, and the beams should be optimally overlapped. The relative angles between the pump beams and the signal beam should be small. The ratio $R = I_c(0)/I_s(0)$ is measured.

Depending on the polarizations, either $n_{2\parallel}(-\omega_p; \omega_p)$ or $n_{2\perp}(-\omega_p; \omega_p)$ can be measured. Note, however, that this technique does not yield $n_{2\parallel}(-\omega_p; \omega_p)$ unless $\alpha_{2\parallel}(-\omega_p; \omega_p)$ can be neglected or independently measured (see Chapter 15). It is primarily because of this limitation that degenerate four-wave mixing is no longer frequently used for nonlinear characterization.

PROBLEMS

1. A liquid is an optically isotropic medium made up of randomly oriented molecules with N molecules per unit volume. Calculate the macroscopic third-order

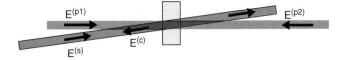

FIGURE 13.9   The beam geometry for degenerate four-wave mixing.

nonlinear coefficients $\tilde{\chi}^{(3)}_{1111}(-\omega;\,\omega,-\omega,\omega)$, $\tilde{\chi}^{(3)}_{2112}(-\omega;\,\omega,-\omega,\omega)$, and $\tilde{\chi}^{(3)}_{1221}(-\omega;\,\omega,-\omega,\omega)$ in the Kleinman limit for "rodlike" molecules with molecular electronic nonlinearity $\bar{\gamma}^{(3)}_{1111}\neq 0$ and all other $\bar{\gamma}^{(3)}_{ijk\ell}=0$.

2. A liquid is an optically isotropic medium made up of randomly oriented molecules with N molecules per unit volume.

   (a) For planar "disklike" molecules with molecular nonlinearities isotropic in the plane of the disk, i.e., $\bar{\gamma}^{(3)}_{1111}=\bar{\gamma}^{(3)}_{2222}\neq 0$, show that $\bar{\gamma}^{(3)}_{\overline{xxxx}}=\bar{\gamma}^{(3)}_{\overline{xyyx}}+\bar{\gamma}^{(3)}_{\overline{xyxy}}+\bar{\gamma}^{(3)}_{\overline{xxyy}}$.

   (b) Assuming $\bar{\gamma}^{(3)}_{1212}=\bar{\gamma}^{(3)}_{1221}=\bar{\gamma}^{(3)}_{1122}=\bar{\gamma}^{(3)}_{2211}=\bar{\gamma}^{(3)}_{2112}=\bar{\gamma}^{(3)}_{2121}$, calculate in the Kleinman limit $\tilde{\chi}^{(3)}_{1111}(-\omega;\,\omega,-\omega,\omega)$ and $\tilde{\chi}^{(3)}_{1212}(-\omega;\,\omega,-\omega,\omega)$ in the laboratory frame of reference.

   (c) Show that for a strong x-polarized pump beam and a coherent weak orthogonally polarized (y) beam, the ratio $\frac{<n_{2\perp}(-\omega;\,\omega)>}{<n_{2\parallel}(-\omega;\,\omega)>}=2/3$ for a solution of disk molecules and the ratio $<\!D4WM\!>/<\!n_{2\parallel}(-\omega;\,\omega)\!>\,=1/3$ as it should be for an isotropic medium where D4WM (Degenerate 4 wave mixing) is discussed in section 8.2.3.3.2.1.2.

REFERENCES

1. L. Canioni, Liquids and Glasses Nonlinearities Properties Analyzed by Femtosecond Interferometry, Ph.D. Thesis (University of Bordeaux, Bordeaux, France, 1994).

2. H. Hui, S. Webster, D. Hagan, and E. Van Stryland, to be published.

3. R. A. Ganeev, A. I. Ryasnyanskiĭ, and H. Kuroda, "Nonlinear optical characteristics of carbon disulfide," Opt. Spectrosc., **100**(1), 116–128 (2006).

4. M. Sheik-Bahae, A. A. Said, T.-H. Wei, D. J. Hagan, and E. W. Van Stryland, "Sensitive measurement of optical nonlinearities using a single beam," IEEE J. Quantum Electron., **26**, 760–769 (1990).

5. F. Kajzar and J. Messier, "Third-harmonic generation in liquids," Phys. Rev. A, **32**, 2352–2363 (1985).

6. K. B. Rochford, G. I. Stegeman, R. Zanoni, W. Krug, E. Miao, and M. W. Beranek, "Pulse-modulated interferometer for measuring intensity-induced phase shifts," IEEE J. Quantum Electron., **28**, 2044–2050 (1992).

7. S. Santran, L. Canioni, L. Sarge, Th. Cardinal, and E. Fargin, "Precise and absolute measurements of the complex third-order optical susceptibility," J. Opt. Soc. Am. B, **21**, 2180–2190 (2004).

SUGGESTED FURTHER READING

M. G. Kuzyk and C. W. Dirk, editors, Characterization Techniques and Tabulations for Organic Nonlinear Optical (Marcel Dekker, New York, 1998).

P. N. Prasad and D. J. Williams, Introduction to Nonlinear Optical Effects in Molecules and Polymers (John Wiley & Sons, New York, 1991).

CHAPTER 14

Ramifications and Applications of Nonlinear Refraction

Third-order nonlinear optics offers a wide range of interesting phenomena that are very different from what is expected from linear optics. The most important are due to changes in the properties of beams in space, time, and polarization due to nonlinear refraction. If $n_{2\|,\text{eff}}(-\omega; \omega) > 0$, beams can collapse in space or time, or both space and time. In the space domain, this leads to a broadening of the wave-vector spectrum; in time, to spectral broadening. In the worst case, this can lead to material damage when the local intensity exceeds the damage threshold.

This beam collapse can lead to either beam instabilities or stable wave packets. Over certain ranges of intensity and input beam width, unstable, high local intensity filaments can form due to the amplification of noise associated with real beams. Such instabilities are inherent to plane-wave nonlinear optics. For narrower beams, very stable wave packets whose shape is either invariant on propagation or periodically recurring can form due to the same physics as the instabilities. These are called solitons; solitons are the modes of nonlinear optics and exhibit unique properties.

Since nonlinear refraction leads to additional nonlinear phase shifts as discussed in Chapter 10, this property can be used in interference-based phenomena and devices. Signals can be manipulated depending on their intensity, i.e., all-optical switching. The classic cases are optical bistability, in which the output transmission of light from a cavity can be either high or low for the same input intensity, and integrated optics devices such as nonlinear directional couplers that can be used for optical logic or intensity controlled routing of signals.

The following topics with specific examples are discussed in this chapter:

1. Self-focusing and defocusing of beams in space
2. Self-phase modulation and spectral broadening in time
3. Plane-wave instabilities
4. Solitons
5. Optical bistability
6. All-optical switching

Nonlinear Optics: Phenomena, Materials, and Devices, George I. Stegeman and Robert A. Stegeman.
© 2012 John Wiley & Sons, Inc. Published 2012 by John Wiley & Sons, Inc.

14.1 SELF-FOCUSING AND DEFOCUSING OF BEAMS

This phenomenon is associated with beams of finite cross section. As shown in Chapter 10, an intensity-dependent change in the refractive index can be created by a beam. This change affects the propagation of the beam that created the index change, as well as other beams that travel through the same region in space. Hence the beam introduces an effective lens in the medium. This explanation works well for phenomena that produce actual refractive index changes, called Kerr-like changes. However for cascading of second-order effects that does not lead to an actual change in the refractive index, the appropriate parameter is the nonlinear phase shift, as discussed in Chapter 12 in the section on cascading. This nonlinear phase shift arises from the strong coupling between beams undergoing second-order nonlinear interactions but does not affect any other beams.

14.1.1 Self-Focusing and Defocusing of Beams: Kerr Nonlinearities

Consider the simplest case of Kerr nonlinearities that are local in space, are characterized by an instantaneous response, and for which the nonlinear index change is linear in intensity. The induced index change is described by the very simple formula $\Delta n^{\mathrm{NL}} = n_{2\|}(-\omega; \omega)I$ (Chapter 10). As a result, the phase velocity of a beam responds instantaneously to the local intensity as

$$V_p = \frac{c}{n_0 + n_{2\|}(-\omega; \omega)I}. \tag{14.1}$$

The consequence is that the phase front changes shape on propagation, as shown in Fig. 14.1. For the most interesting case of $n_{2\|}(-\omega; \omega) > 0$, the larger the net refractive index n, the slower the phase velocity and so the high intensity region of a beam travels slower than the wings. This leads to self-focusing, which works against diffraction and, if strong enough, can lead to beam collapse. The induced index distribution $\Delta n^{\mathrm{NL}}(x, y, z)$ mirrors the intensity distribution $I(x, y, z)$. For self-defocusing with $n_{2\|}(-\omega; \omega) < 0$, the beam defocuses, i.e., spreads, and augments normal diffraction. Although the Kerr effect has only a simple relation between the index change and the intensity, it is noteworthy that these effects occur for *all* third-order nonlinearities $n_{2\|,\mathrm{eff}}(-\omega; \omega)$.

(a)   (b)

FIGURE 14.1 The effects of an intensity-dependent phase velocity for (a) $n_{2\|,\mathrm{eff}}(-\omega; \omega) > 0$, which leads to a focusing behavior for the phase front (dashed line), and (b) $n_{2\|,\mathrm{eff}}(-\omega; \omega) < 0$, which leads to a defocusing behavior of the phase front.

14.1.1.1 *External Self-Action: Thin Sample.*

In the "thin" sample approximation for diffraction in two dimensions in space, the shape of the beam is not significantly changed inside the sample and only the beam's phase front $\phi(r_\perp, z)$ is augmented by $\Delta\phi^{\mathrm{NL}}(r_\perp, z)$ on transmission. Assuming a lossless Gaussian beam and no diffraction inside the sample, i.e., $z_0 \gg L$, gives

$$I(r_\perp, z) = I(z)\, e^{-2r_\perp^2/w_0^2} \xrightarrow{z=L} \Delta\phi^{\mathrm{NL}}(r_\perp, L) = k_{\mathrm{vac}}(\omega) L n_{2\parallel}(-\omega;\omega) I(0,z)\, e^{-2r_\perp^2/w_0^2}$$

$$\rightarrow\ \mathcal{E}(r_\perp, L) = \mathcal{E}(0, L)\, e^{-r_\perp^2/w_0^2}\, e^{i k_{\mathrm{vac}}(\omega) n_{2\parallel}(-\omega;\omega) L I(0,L) e^{-2r_\perp^2/w_0^2}}.$$

$$(14.2)$$

Clearly this field will not continue to propagate as a simple Gaussian beam outside the sample. In principle, the propagation needs to be analyzed numerically. However, a simple approximate theory has been used that does embody the main features of the problem. For small phase distortion, the exponential in the nonlinear phase can be expanded as $1 - r_\perp^2/Cw_0^2$, where numerical simulations have shown that $C \sim 3.77$ is a good approximation valid over a range of small phase shifts and so

$$\Delta\phi^{\mathrm{NL}}(r_\perp, L) \approx k_{\mathrm{vac}}(\omega) L n_{2\parallel}(-\omega;\omega) I(0,L)\left[1 - \frac{r_\perp^2}{Cw_0^2}\right] \qquad (14.3)$$

is a useful approximation. Comparing with the formula for the transmitted phase change introduced by a simple lens of focal length f, we obtain

$$T(r_\perp) \propto \exp\left[-i\frac{k_{\mathrm{vac}} r_\perp^2}{2f}\right] = \exp\left[-i\Delta\phi^{\mathrm{NL}}(r_\perp, L)\right]\ \rightarrow\ f \cong \frac{Cw_0^2}{2n_{2\parallel}(-\omega;\omega) I(0,L) L}$$

$$(14.4)$$

for the focal length of the effective lens produced by self-focusing, subject to the approximations that the beam's Rayleigh range $z_0 \gg L, |f| \gg L$ and the stated truncation of the exponential.

14.1.1.2 *External Self-Action: Thick Sample.*

In this case the beam profile can change significantly inside the nonlinear medium due to the nonlinear interaction and needs to be taken into account, which greatly complicates the analysis ("thick" sample approximation). The beam diameter gets smaller due to self-focusing and consequently the peak intensity increases; therefore, the self-focusing becomes progressively stronger as the beam collapses on propagation in the absence of diffraction (see Fig. 14.2). This can, in principle, lead to catastrophic self-focusing (in the absence of diffraction, which also becomes stronger as the beam collapses) and material damage.

The first key question is pertaining to the value of the critical power P_c (or intensity I_c) at which self-focusing dominates and catastrophic self-focusing can occur (see Fig. 14.2). To illustrate this case, consider the focal region of a Gaussian beam.

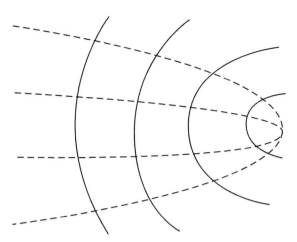

FIGURE 14.2   Illustration of catastrophic self-focusing collapse in the absence of diffraction. The evolution of the phase front is represented by solid lines and of the beam waist by dashed lines.

Assume diffraction occurs over the characteristic distance $z_0 = k_{vac}(\omega)n(\omega)w_0^2$. When self-focusing occurs over this distance and produces a nonlinear phase shift on the axis of approximately π, the critical intensity I_c required for the collapse on the axis is defined by

$$k_{vac}(\omega)n_{2||}(-\omega;\,\omega)I_c 2z_0 = \pi. \tag{14.5}$$

For two-dimensional (2D) cross-sectional beams, writing $P_c \cong \pi w_0^2 I_c$ gives

$$P_c \cong \frac{\lambda_{vac}^2(\omega)}{8n(\omega)n_{2||}(-\omega;\,\omega)}. \tag{14.6}$$

Note that this result does not depend on the beam width, but just on the power and beam shape. Therefore even very broad beams can eventually self-focus in a self-focusing medium if their total power exceeds P_c. Fortunately, much of nonlinear optics can be done with powers and distances that do not lead to catastrophic damage. More accurate numerical solutions for 2D Gaussian beams have been given by Marburger (1) and Fidich and Gaeta (2) as

$$P_c \cong \frac{3.8}{\pi} \frac{\lambda_{vac}^2}{8nn_{2||,eff}(-\omega;\,\omega)}, \qquad z_f = \frac{0.734k_{vac}(\omega)nw_0^2}{\sqrt{\left\{[P/P_c]^{1/2}-0.852\right\}^2-0.0219}}. \tag{14.7}$$

In Eq. 14.7, z_f is the distance to the singularity in the *paraxial* approximation (which of course is not valid for beam widths of the order of a few wavelengths) and in the absence of other effects discussed later in this section. For $P > P_c$, self-focusing overcomes diffraction and catastrophic self-focusing occurs. For $P_c > P$, diffraction dominates and the beam spreads. At exactly $P = P_c$, the beam propagates without any change in the shape, which is unstable against noise fluctuations in power or phase, which ultimately will lead to either beam spreading or catastrophic self-focusing. Some typical numbers for the critical power are as follows: CS_2—$n_{2\|,\text{eff}}(-\omega; \omega) \cong 4.5 \times 10^{-14}$ cm²/W and $n \sim 1.6$ at $\lambda_{\text{vac}} = 1\,\mu\text{m}$ so that $P_c \cong 30$ KW; fused silica—$n_{2\|}(-\omega; \omega) \cong 2.5 \times 10^{-16}$ cm²/W and $n \sim 1.5$ at $\lambda_{\text{vac}} \cong 1\,\mu\text{m}$ so that $P_c \cong 3.5$ MW.

In practice, catastrophic self-focusing is arrested by other physical nonlinear effects such as ionization, stimulated scattering, multiphoton absorption, and plasma formation in air. Furthermore, the scalar and paraxial models break down when $w_0 \to \lambda$ and the $\partial^2/\partial z^2$ term ignored in the paraxial approximation must be included (3). The term $\vec{\nabla} \cdot \vec{D} = \vec{\nabla} \cdot [n + n_{2\|,\text{E}}(-\omega; \omega)|\vec{E}|^2]\vec{E} = 0$ also mixes all three field components, and so for a polarized input, radiation rings of both transverse polarizations are emitted. Periodically foci occur with distance, as shown in Fig. 14.3; the higher the input intensity, the more frequent the focal spots. A detailed discussion of this complex problem can be found in a review by Couairon and Mysyrowicz (4).

The self-focusing phenomenon clearly changes the wave-vector spectrum associated with the beam. For the Gaussian beam example, for $z \gg z_0$ the divergence angle θ_{div} of a beam due to the spread of wave vectors $\Delta k_\perp$ describing the beam is given by $\Delta k_\perp/k \propto \theta_{\text{div}} = w(z)/z_0$. Therefore, the narrower the beam is, the broader the wave-vector spectrum becomes.

14.1.2 Other Nonlinearities

The preceding discussion was based on Kerr nonlinearities. The same self-focusing phenomenon occurs under appropriate conditions with all the nonlinearities discussed in Chapters 11 and 12. In fact, it has been observed experimentally in many cases.

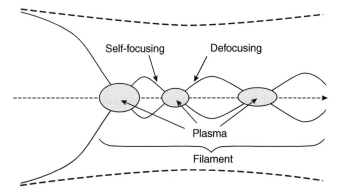

FIGURE 14.3 Simulation of the complex filaments that form in the region of catastrophic collapse in air. In addition to the positive $n_{2\|}(-\omega; \omega)$ of air, electrons are multiphoton ionized from their parent air molecules to form an electron plasma that has a negative refractive index that tends to counteract self-focusing. Reproduced with permission from Elsevier Publishing (4).

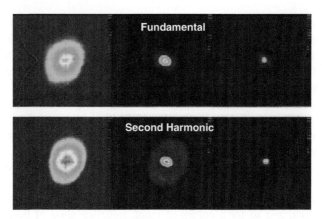

FIGURE 14.4 The output beam profiles of the fundamental and the harmonic due to the cascading nonlinearity, as the input power into a second harmonic crystal is increased from left to right.

However, the governing equations can be more complex and need to be tailored to the individual nonlinear mechanisms. Nonlocality involving diffusion of the index change in space as found in liquid crystals, photorefractive media, thermal effects, charge carrier nonlinearities in semiconductors, and so on, tends to counteract self-focusing and raise the power required for self-focusing. Self-focusing also occurs for the "cascading" nonlinearity due to second-order nonlinearities discussed in Chapter 12 for which there is no index change but for which the nonlinear phase shift is the appropriate parameter. For example, the nonlinear phase shift increases with increasing intensity in second-order nonlinear crystals near phase match for harmonic generation. Figure 14.4 shows the simultaneous self-focusing of both a fundamental and its harmonic beam due to the cascading nonlinearity.

The conclusion is that self-focusing is a universal phenomenon associated with nonlinearities that produce positive nonlinear changes in the phase on propagation.

14.2 SELF-PHASE MODULATION AND SPECTRAL BROADENING IN TIME

The consequences of self-focusing and defocusing nonlinearities in the time domain are analogous to those just described in the space domain. Just as in the space domain self-focusing led to energy localization in space, in the time domain localization occurs in time; i.e., the pulse width collapses. Again, consider the simplest case of a Kerr nonlinearity. For a temporal pulse, Eq. 14.1 now describes the phase velocity across the pulse's temporal envelope and Fig. 14.1 illustrates the changes along the temporal envelope on propagation. Pulse broadening occurs for $n_{2||}(-\omega; \omega) < 0$ and pulse narrowing occurs for $n_{2||}(-\omega; \omega) > 0$. In the absence of nonlinearity, pulse broadening always occurs due to group velocity dispersion (GVD), which is the equivalent of diffraction in the space domain. Therefore to actually obtain pulse narrowing in time, the nonlinear effect must be stronger than that due to GVD.

Again, the nonlinear phase shift is the cause of self-focusing—in this case in time. Assuming a lossless Kerr medium, the pulse's field can be written as

$$E(z,t) = E(0,T)\, e^{ikz - i[\omega_a T - \Delta\phi^{NL}(T,z)]}, \quad \Delta\phi^{NL}(T,z) = n_{2||}(-\omega;\omega)I(0,T)k_{vac}(\omega)z,$$
$$(14.8)$$

where T is the time in the pulse's frame of reference that is moving at the group velocity v_g and ω_a is the pulse's carrier (center) frequency. Because the nonlinear phase shift $\Delta\phi^{NL}(T,z)$ varies across the pulse profile, there is a corresponding nonlinear frequency change $\delta\omega^{NL}(T)$ given by

$$\delta\omega^{NL}(T,z) = -\frac{\partial\Delta\phi^{NL}(T,z)}{\partial T} = \frac{2T}{T_0^2}n_{2||}(-\omega;\omega)I(0,0)\,e^{-T^2/T_0^2}k_{vac}(\omega)z. \quad (14.9)$$

Note that the sign of $\delta\omega^{NL}$ changes from the leading ($T < 0$) to the trailing ($T > 0$) edge of the pulse (see Fig. 14.5). The minima and maxima in the frequency change occur when $\partial\,\delta\omega(T)/\partial T = 0$, which gives

$$\frac{T}{T_0} = \pm\frac{1}{\sqrt{2}}, \quad \delta\omega|_{max} = -\delta\omega|_{min} \;\rightarrow\; \delta\omega|_{max} = \frac{\sqrt{2}}{T_0\sqrt{e}}k_{vac}(\omega_a)n_{2||}(-\omega;\omega)I(0,0)z.$$
$$(14.10)$$

This spectral broadening can be quite large. For example, $\Delta\phi^{NL}_{max} = 10\pi$ gives $\Delta\omega \cong 26\Delta\omega(0)$. In fibers, $\Delta\phi^{NL}_{max}$ of many tens of π is quite feasible.

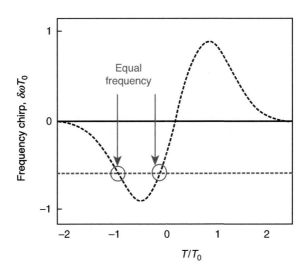

FIGURE 14.5 The normalized nonlinear frequency chirp induced across the pulse. The horizontal dashed lines identify positions of equal frequency in the pulse.

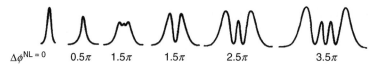

$\Delta\phi^{NL}=0$ 0.5π 1.5π 1.5π 2.5π 3.5π

FIGURE 14.6 The evolution of the output spectrum with increasing input power and the corresponding nonlinear phase shift $\Delta\phi^{NL}(L)$. Reproduced with permission from American Physical Society (5).

Such large spectral broadening in fibers is easily measurable with a spectrum analyzer. The frequency spectrum is given by $S(z, \omega, \omega_a) \propto |E(z, \omega-\omega_a)|^2$, and so

$$E(z, \omega-\omega_a) = \frac{1}{2\pi}\int_{-\infty}^{\infty} E(z, T)\, e^{i[\omega-\delta\omega(T,z)-\omega_a]T}\, dT,$$

$$S(z, \omega-\omega_a) = \frac{1}{4\pi^2}\left|\int_{-\infty}^{\infty} E(0, T)\, e^{i[\omega-\delta\omega(T,z)-\omega_a]T}\, dT\right|^2. \tag{14.11}$$

Therefore there are points on the pulse envelope for which interference can occur at the same frequency with a phase difference as shown in Fig. 14.5. As a result, the spectrum is periodic in $\omega-\omega_a$ (see Fig. 14.6).

It was shown in Chapter 13 that the time response due to a third-order nonlinearity depends on the physical mechanism responsible. For self-phase modulation to occur, the response time of the nonlinearity should be much shorter than the pulse width. For the CS_2 case shown previously in Fig. 13.1, for example, only the Kerr nonlinearity would be operative for pulses below 50 fs. Around 1 ps, both the vibrational and the Kerr nonlinearity would be appropriate.

14.3 INSTABILITIES

14.3.1 Instabilities in Space

Beam instabilities are well known to occur in all nonlinear wave phenomena, and optics provides a well-defined platform to investigate such effects. Except in this chapter and Chapter 5, most of the nonlinear optics discussed to this point has been focused on plane-wave phenomena and solutions. Plane waves are the normal modes of linear optics, but as will be shown here they are subject to instabilities under the appropriate conditions on intensity in nonlinear optics. The results derived here are valid for plane waves or very wide (relative to a wavelength) beams.

It will now be shown that a plane wave in a self-focusing $n_{2||}(-\omega; \omega) > 0$ Kerr medium is actually unstable at sufficiently high powers. For simplicity, a 1D plane wave is assumed. Using Eq. 10.3a, which is valid for plane waves (no diffraction), and replacing $\frac{3\omega}{8nc}\hat{\chi}^{(3)}_{xxxx}(-\omega; \omega, -\omega, \omega)$ by $n_{2||,E}(-\omega; \omega)$ from Eq. 10.29 gives

$$\frac{d\mathcal{E}_x(z, \omega)}{dz} = ik_{vac}(\omega)n_{2||,E}(-\omega; \omega)|\mathcal{E}_x(z, \omega)|^2\mathcal{E}_x(z, \omega). \tag{14.12}$$

This equation has the solution for the lossless case $|\mathcal{E}_x(z,\omega)|^2 = |\mathcal{E}_x(0,\omega)|^2 =$ constant:

$$\mathcal{E}_x(z,\omega) = \mathcal{E}_x(0,\omega)\, e^{ik_{vac}n_{2||,E}(-\omega;\omega)|\mathcal{E}(\omega)|^2 z}, \qquad (14.13)$$

and the term $k_{vac}n_{2||,E}(-\omega;\omega)|\mathcal{E}(0,\omega)|^2 z$ has been discussed previously in section 10.2 as an example of the nonlinear phase shift, which is characteristic of nonlinear optical phenomena. However, there is always "noise" on any real beam that can be Fourier analyzed to give a noise magnitude $\delta(\kappa)$ at the spatial frequency κ. Therefore a real beam for a given $\delta(\kappa)$ can be written as

$$\mathcal{E}_x(z,\omega) = \mathcal{E}_x(0,\omega)\{1 + \delta(\kappa)\cos(\kappa x)\, e^{\gamma(\kappa)z}\}\, e^{\{ik_{vac}n_{2||,E}|\mathcal{E}(\omega)|^2 z\}}, \qquad (14.14)$$

where a real $\gamma(\kappa)$ indicates that the noise grows exponentially with distance in the small signal gain approximation and, since the noise is amplified, the plane-wave solution is unstable. The trial solution 14.14 is introduced into the nonlinear wave equation that includes diffraction, i.e., into

$$-\frac{i}{2k}\frac{\partial^2 E(z,\omega)}{\partial x^2} + \frac{\partial E(z,\omega)}{\partial z} = ik_{vac}n_{2||,E}(-\omega;\omega)|E(z,\omega)|^2 E(z,\omega) \qquad (14.15)$$

and solved for the exponential growth coefficient $\gamma(\kappa)$. Performing the derivatives in Eq. 14.15 and assuming small signal growth so that $|E_x(z,\omega)|^2 \approx |E_x(0,\omega)|^2$ gives

$$\delta(\kappa)\cos(\kappa x)\, e^{\gamma z}\mathcal{E}_x(0,\omega)\left\{\gamma(\kappa) + ik_{vac}n_{2||,E}(-\omega;\omega)|\mathcal{E}_x(0,\omega)|^2 + i\frac{\kappa^2}{2k}\right\} \qquad (14.16)$$

$$\cong ik_{vac}n_{2||,E}(-\omega;\omega)|\mathcal{E}_x(0,\omega)|^2\mathcal{E}_x(0,\omega)[2\delta(\kappa) + \delta^*(\kappa)]\cos(\kappa x)\, e^{\gamma(\kappa)z}.$$

Including the terms on the right-hand side of Eq. 14.15 and simplifying them gives

$$\delta(\kappa)\left\{\gamma(\kappa) + i\frac{\kappa^2}{2k}\right\} = ik_{vac}n_{2||,E}(-\omega;\omega)|\mathcal{E}_x(0,\omega)|^2\left[\delta(\kappa) + \delta^*(\kappa)\right]. \qquad (14.17)$$

Substituting $\delta = \delta_\Re + i\delta_\Im$ into Eq. 14.17 gives two equations, one for real and one for imaginary quantities, which should be solved separately:

$$\gamma(\kappa)\delta_\Re(\kappa) - \delta_\Im(\kappa)\frac{\kappa^2}{2k} = 0,$$

$$\delta_\Im(\kappa)\gamma(\kappa) + \delta_\Re(\kappa)\frac{\kappa^2}{2k} = 2\delta_\Re(\kappa)k_{vac}n_{2||,E}(-\omega;\omega)|\mathcal{E}_x(0,\omega)|^2. \qquad (14.18)$$

Eliminating $\delta_{\Re}(\kappa)$ and $\delta_{\Im}(\kappa)$ gives

$$\gamma^2(\kappa) = \frac{\kappa^2}{2k}\left\{2k_{\text{vac}}n_{2\|,\text{E}}(-\omega;\omega)|\mathcal{E}_x(0,\omega)|^2 - \frac{\kappa^2}{2k}\right\}. \qquad (14.19)$$

A solution for $\Re\{\gamma(\kappa)\}$ exists only for a self-focusing nonlinearity and only as long as

$$|\mathcal{E}_x(0,\omega)|^2 > \frac{\kappa^2}{4kk_{\text{vac}}n_{2\|,\text{E}}(-\omega;\omega)} \quad \rightarrow \quad I_x(0,\omega) > \frac{\kappa^2 nc\varepsilon_0}{8kk_{\text{vac}}n_{2\|,\text{E}}(-\omega;\omega)}, \qquad (14.20)$$

which means that the noise with spatial κ is amplified once the intensity crosses a threshold value different for every value of κ. Note that the larger the periodicity $\Lambda = 2\pi/\kappa$, the lower the threshold for instability. The maximum gain for a given intensity occurs at $d\gamma/d\kappa = 0$ and the corresponding maximum gain coefficient and the value of κ at which maximum gain occurs are

$$\kappa_{\max} = \sqrt{2k_{\text{vac}}kn_{2\|,\text{E}}(-\omega;\omega)}|\mathcal{E}_x(0,\omega)|, \quad \gamma_{\max} = k_{\text{vac}}n_{2\|,\text{E}}(-\omega;\omega)|\mathcal{E}_x(0,\omega)|^2, \qquad (14.21)$$

respectively. The variation in the gain coefficient with the period is shown in Fig. 14.7. Since the growth is exponential, the periodicity with the largest gain will emerge as

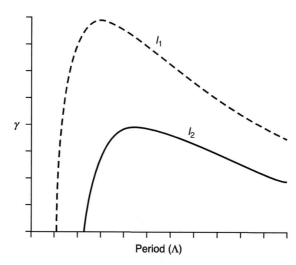

FIGURE 14.7 The gain coefficient versus the period $\Lambda = 2\pi/\kappa$ of the instability for two intensities $I_1 > I_2$. Reproduced with permission from the Optical Society of America (6).

FIGURE 14.8 The evolution of a noisy plane wave with propagation distance above the threshold for modulation instability. (a) $n_{2||,E}(-\omega; \omega) > 0$. (b) $n_{2||,E}(-\omega; \omega) < 0$.

dominant, superimposed on an intensity background as indicated in Fig. 14.8. This growth from noise is called "modulation instability" (MI). Note that below the intensity threshold for a given κ there is no amplification of the noise.

The analysis has been given for a plane wave. However, MI also occurs for wide beams as long as their half-width is much greater than the MI period associated with their peak intensity. An experimental example is shown in Fig. 14.9 for slab AlGaAs waveguides that are effectively a 1D system since diffraction can occur only in the plane of the waveguide (6). For photon energies below one half its bandgap energy, this material exhibits ultrafast Kerr nonlinearities, as discussed in Chapter 11. For intensities above the threshold intensity where the small signal gain approximation breaks down, a beam breaks up into discrete "filaments" in which the energy is progressively more concentrated in the filaments on top of a broad background, as shown in Fig. 14.9b. The spatial profiles shown in Fig. 14.9b are no longer in the small

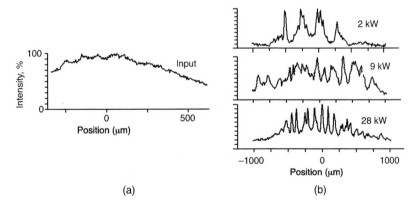

(a)   (b)

FIGURE 14.9 Example of the evolution with the input power of modulation instability in a one-dimensional Kerr medium with a finite width beam. (a) Input beam profile with noise. (b) Output beams at increasing power levels, all above the intensity threshold for modulation instability and no longer in the small signal limit. Reproduced with permission from the Optical Society of America (6).

gain limit in which most of the energy would be in the background instead of the peaks. In this case, saturation of the gain occurs and higher spatial harmonics of κ_{max} appear.

A weak but noisy beam also becomes unstable in the presence of a strong beam of the orthogonal polarization. As discussed in Chapters 9 and 10, the strong (x-polarized) beam leads to a nonlinear phase shift in the weak y-polarized beam so that the solution for plane-wave fields is given by

$$\mathcal{E}_y(z, \omega) = \mathcal{E}_y(0, \omega)\, e^{ik_{vac}n_{2\perp,E}(-\omega;\omega)|\mathcal{E}_x(\omega)|^2 z}. \tag{14.22}$$

An analysis similar to the one above leads to

$$\gamma^2(\kappa) = \frac{\kappa^2}{2k}\left\{ 2k_{vac}n_{2\perp,E}(-\omega;\omega)|\mathcal{E}_x(0,\omega)|^2 - \frac{\kappa^2}{2k} \right\}, \tag{14.23}$$

in which the periodicity $\Lambda = 2\pi/\kappa$ and the gain $\gamma(\kappa)$ are experienced by the weak beam.

Although the analysis above has been for Kerr nonlinearities, MI is a universal phenomenon in nonlinear optics and occurs for all the self-focusing nonlinearities discussed in Chapters 11 and 12. For nonlinearities in which the index change can diffuse some distance in space, the instability period must be greater than the diffusion length. The cascading nonlinearity whose origin lies in second-order nonlinearities also exhibits instabilities. Figure 14.10 shows the numerical simulation of the breakup during second harmonic generation of a noisy fundamental beam using beam propagation techniques based on the coupled wave equations from Chapter 3, with diffraction included. A similar breakup of the second harmonic also occurs. In fact, these effects have been verified experimentally (7).

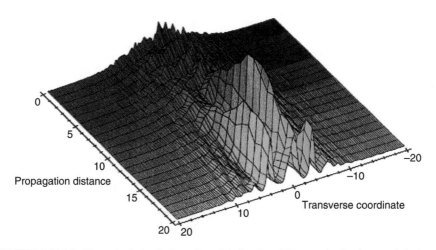

FIGURE 14.10 Numerical simulation of modulation instability on the fundamental during second harmonic generation.

14.3.2 Instabilities in Time

The temporal case is completely analogous to the spatial case. The dispersion in time replaces diffraction in the nonlinear wave equation. However, first a discussion of GVD that broadens temporal pulses is necessary.

14.3.2.1 GVD and Pulse Broadening.
The starting point is the usual wave equation

$$\frac{\partial^2 \vec{E}(\vec{r}, t)}{\partial z^2} - \frac{n^2}{c^2} \frac{\partial^2 \vec{E}(\vec{r}, t)}{\partial t^2} = 0. \tag{14.24}$$

The pulse envelope is described by $\vec{E}(\vec{r}, t)$. Fourier transforming Eq. 14.24 via

$$E(z, \omega) = \frac{1}{2\pi} \int_{-\infty}^{\infty} \mathcal{E}(z, t)\, e^{i(k(\omega_a)z - \omega_a t)}\, e^{i\omega t}\, dt + \text{c.c.}$$

$$\rightarrow \frac{\partial^2 \mathcal{E}(z, \omega - \omega_a)}{\partial z^2} e^{ik(\omega_a)z} + k^2(\omega) \mathcal{E}(z, \omega - \omega_a)\, e^{ik(\omega_a)z} = 0, \tag{14.25}$$

in which $\mathcal{E}(z, \omega - \omega_a)$ describes the frequency spectrum of the pulse around the central frequency ω_a. The slowly varying envelope approximation with

$$\frac{k^2(\omega) - k^2(\omega_a)}{2k(\omega_a)} = \frac{[k(\omega) - k(\omega_a)][k(\omega) + k(\omega_a)]}{2k(\omega_a)} \cong k(\omega) - k(\omega_a)$$

gives, for pulses many cycles long,

$$-\frac{\partial \mathcal{E}(z, \omega - \omega_a)}{\partial z} + i[k(\omega) - k(\omega_a)]\mathcal{E}(z, \omega - \omega_a) = 0. \tag{14.26}$$

Expanding $k(\omega) - k(\omega_a)$, again for pulses many cycles long, gives

$$k(\omega) - k(\omega_a) = \frac{dk(\omega)}{d\omega}\Big|_{\omega_a}(\omega - \omega_a) + \frac{1}{2}\frac{d^2 k(\omega)}{d\omega^2}\Big|_{\omega_a}(\omega - \omega_a)^2 + \cdots, \tag{14.27}$$

where

$$k_1 = \frac{dk(\omega)}{d\omega}\Big|_{\omega_a} = \frac{1}{c}\left(\omega\frac{dn}{d\omega} + n\right) = \frac{n_{\text{group}}}{c} = \frac{1}{v_{\text{group}}}$$

$$k_2 = \frac{d^2 k(\omega)}{d\omega^2}\Big|_{\omega_a} = \frac{d}{d\omega}\left(\frac{1}{v_{\text{group}}}\right) = -\frac{1}{v_{\text{group}}^2}\frac{dv_{\text{group}}}{d\omega}. \tag{14.28}$$

Here k_2 is the GVD with frequency. It quantifies the spread in frequencies around a central frequency associated with a temporal pulse of finite width in time. Each frequency component travels with a different group velocity and the beam spreads in time. The most famous example is that of fused silica, and the dispersion with wavelength of the GVD is shown in Fig. 14.11. Note that at a wavelength of 1.27 μm in fused silica, a pulse will not spread unless it is a very short pulse whose frequency spectrum in the tail extends significantly beyond 1.27 μm.

Note that the fused silica GVD can be either positive or negative. Glass is one of the very few materials in which this sign change occurs in a convenient spectral range for communications applications. In most materials, GVD is positive to beyond wavelengths of 2 μm. Substituting Eq. 14.27 into Eq. 14.26 gives

$$\frac{d\mathcal{E}(z, \omega-\omega_a)}{dz} - ik_1\mathcal{E}(z, \omega-\omega_a)(\omega-\omega_a) - \frac{1}{2}ik_2\mathcal{E}(z, \omega-\omega_a)(\omega-\omega_a)^2 = 0. \quad (14.29)$$

Now applying the inverse Fourier transform to obtain the equivalent equation in the space–time domain gives

$$\frac{\partial\mathcal{E}(z, t)}{\partial z} + k_1\frac{\partial\mathcal{E}(z, t)}{\partial t} + \frac{1}{2}ik_2\frac{\partial^2\mathcal{E}(z, t)}{\partial t^2} = 0, \quad (14.30)$$

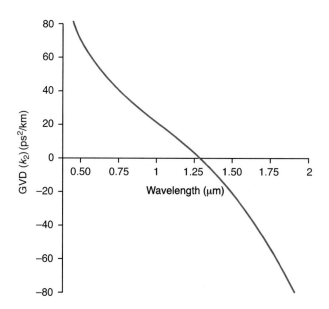

FIGURE 14.11 The dispersion in group velocity dispersion (GVD) with wavelength in fused silica.

and finally eliminating the k_1 term by transforming to the reference frame traveling with the pulse gives

$$\frac{\partial \mathcal{E}(z,T)}{\partial z} + \frac{1}{2} i k_2 \frac{\partial^2 \mathcal{E}(z,T)}{\partial T^2} = 0. \tag{14.31}$$

Note that the pulse spreads in time for both signs of k_2. This can be shown relatively simply (and analytically) for a Gaussian pulse in time. This is achieved by Fourier transforming into the frequency domain, solving for the z-dependence, and transforming back into the time domain via

$$\mathcal{E}(z,T) = \int_{-\infty}^{\infty} \mathcal{E}(0,\omega)\, e^{ik(\omega_a)z + i(k_2/2)(\omega - \omega_a)^2 z - i(\omega - \omega_a)T}\, d(\omega - \omega_a). \tag{14.32}$$

The parameter $\mathcal{E}(0,\omega)$ depends on the detailed pulse shape and for a Gaussian pulse this gives

$$\mathcal{E}(z,T) = \frac{\mathcal{E}(0,T)}{\sqrt{1 + z^2/L_{\text{Dis}}^2}} \exp\left[-\frac{T^2}{2T_1^2} + i \frac{k_2}{|k_2|} \frac{T^2}{T_0^2(1 + z^2/L_{\text{Dis}}^2)} \frac{z}{L_{\text{Dis}}} + i \tan^{-1}\left(\frac{z}{L_{\text{Dis}}}\right) \right]. \tag{14.33}$$

Defining

$$T_1^2 = T_0^2\left[1 + \frac{k_2^2}{T_0^4} z^2\right] = T_0^2\left[1 + \frac{z^2}{L_{\text{Dis}}^2}\right], \qquad L_{\text{Dis}} = \frac{T_0^2}{|k_2|} \tag{14.34}$$

gives

$$\delta\omega = -\frac{\partial\phi(T)}{\partial T} = 2\frac{k_2}{|k_2|} \frac{z/L_{\text{Dis}}}{1 + z^2/L_{\text{Dis}}^2} \frac{T}{T_0^2}, \tag{14.35}$$

where L_{Dis} is called the characteristic dispersion length for pulse broadening. Note that although the broadening is independent of the sign of k_2, the additional phase change does depend on the sign. An important point is that GVD introduces a linear chirp across the pulse. For normal dispersion ($k_2 > 0$, red faster than blue), $\delta\omega$ is negative at the leading edge ($T < 0$) and positive at the trailing edge ($T > 0$). Conversely, for anomalous dispersion ($k_2 < 0$, blue faster than red), $\delta\omega$ is positive at the leading edge ($T < 0$) and negative at the trailing edge ($T > 0$). This is shown in Fig. 14.12.

14.3.2.2 Pulse Instabilities in Time.
Following the procedure for the spatial case, the nonlinear polarization is added to Eq. 14.31; i.e.,

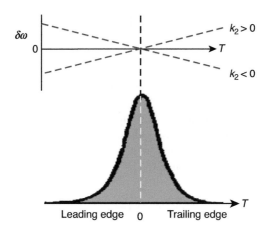

FIGURE 14.12   The frequency chirp along the pulse for different values of group velocity dispersion.

$$\frac{\partial \mathcal{E}(z,T)}{\partial z} + \frac{1}{2}ik_2\frac{\partial^2 \mathcal{E}(z,T)}{\partial T^2} = ik_{\text{vac}}(\omega)n_{2\|,\text{E}}(-\omega;\,\omega)|\mathcal{E}_x(z,T)|^2\mathcal{E}_x(z,T), \quad (14.36)$$

which in the absence of dispersion ($k_2 = 0$) and loss has the solution

$$\mathcal{E}_x(z,T) = \mathcal{E}_x(0,T)\,e^{ik_{\text{vac}}n_{2\|,\text{E}}(-\omega;\omega)|\mathcal{E}_x(0,0)|^2 z}. \quad (14.37)$$

Assuming a pulse envelope with noise that can be described by its Fourier components and the possibility of gain, we obtain

$$\mathcal{E}_x(z,T) = \mathcal{E}_x(0,T)\{1 + \delta(\kappa)\cos(\kappa t)\,e^{\gamma(\kappa)t}\}\,e^{\{ik_{\text{vac}}n_{2\|,\text{E}}|\mathcal{E}(\omega)|^2 z\}}, \quad (14.38)$$

and following the analysis used in the spatial case,

$$\gamma^2(\kappa) = -\frac{\kappa^2 k_2}{2}\left\{2k_{\text{vac}}n_{2\|,\text{E}}(-\omega;\,\omega)|\mathcal{E}_x(0,\omega)|^2 + \frac{\kappa^2 k_2}{2}\right\}. \quad (14.39)$$

In the region of anomalous dispersion ($k_2 < 0$), instabilities occur for glass and grow exponentially in the small gain regime. For fused silica, this is the region with $\lambda_{\text{vac}} > 1.27\,\mu\text{m}$. For shorter wavelengths there is no noise amplification. Note that in a fused silica step-index glass fiber, there is a geometric contribution to the GVD, so that zero GVD is at $1.31\,\mu\text{m}$. Some tuning of the GVD is possible by tailoring the fiber geometry.

Essentially the same comments about response time of the nonlinearity versus pulse width made at the end of Section 14.2 are also valid here.

14.4 SOLITONS (NONLINEAR MODES)

In the preceding section it was shown that the solutions to the nonlinear wave equation for *plane waves* are unstable due to noise. Robust solutions do exist to the nonlinear wave equations that correspond to beams that do not spread or collapse in space or time, or both under appropriate conditions. These are called bright solitons, and they have some very special properties. Spatial solitons exist in all media that exhibit self-focusing. There are also spatial solitons in self-defocusing media, called dark or gray solitons. Mathematically, they consist of a notch in a plane wave. In practice, the wider the extent of the field in which the notch exists, the more stable is the dark soliton. In the linear optics case, solutions to Maxwell's equations are eigenmodes; i.e., they satisfy orthogonality conditions. This is not the case for solitons, although by satisfying the nonlinear wave equation they are nevertheless modes. The only nonlinearity for which realistic analytical solutions exist is the Kerr case. (Other cases need to be analyzed numerically.) The simplest (and only stable) Kerr case is for a single dimension in which light can spread in space or time, i.e., spatial solitons and temporal solitons, respectively. Furthermore, they have very special properties.

14.4.1 Spatial Solitons

Consider the slab (1D) waveguide shown in Fig. 14.13. Light is guided by a thin film of higher refractive index than the surrounding media. The fields decay evanescently into the surrounding media and hence diffraction can occur only in the plane of the film, i.e., one dimension.

The field polarized in the plane of the film (simplest case) is written as

$$\vec{E}(\vec{r}, t) = \frac{1}{2}\hat{e}_y h(x)\mathcal{E}(y, z)\, e^{i(\beta(\omega)z - \omega t)} + \text{c.c.}, \qquad (14.40)$$

where $\beta(\omega)$ is the guided-wave wave vector and $h(x)$ is the field distribution normal to the film, which is oscillatory across the film and decays exponentially into the bounding media (8). The nonlinear wave equation for $\mathcal{E}(y, z)$ in one dimension is

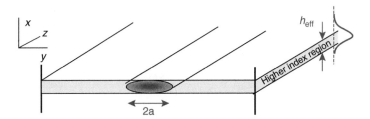

FIGURE 14.13 A slab waveguide with its transverse field distribution along the x-axis with an effective width h_{eff}. Also shown is soliton propagation for which the beam does not spread in the y-dimension.

$$\frac{d^2\mathcal{E}(y,z)}{dy^2} + 2i\beta(\omega)\frac{d\mathcal{E}(y,z)}{dz} = -2k_{\text{vac}}^2(\omega)\overline{n(x)n_{2\|,\text{E}}(-\omega;\,\omega,x)}|\mathcal{E}(y,z)|^2\mathcal{E}(y,z),$$

$$(14.41)$$

where $\beta(\omega) = n_{\text{eff}}(\omega)k_{\text{vac}}(\omega)$ is the guided-wave wave vector, $n_{\text{eff}}(\omega) = \overline{n(x)}$ is the effective index obtained by averaging the refractive index over the intensity distribution along the x-axis, and $\overline{n(x)n_{2\|,\text{E}}(-\omega;\,\omega,x)} = n_{\text{eff}}(\omega)n'_{2\|,\text{E}}(-\omega;\,\omega)$ is the weighted average of the linear index multiplied by the nonlinear index coefficient over the square of the x-dependent intensity distribution. The linear refractive index $n(x)$ and the nonlinear refractive index coefficient $n_{2\|,\text{E}}(-\omega;\,\omega,x)$ are typically different in the three different media constituting a planar waveguide. Writing the field as

$$\mathcal{E}(y,z) = A\rho(y)\,e^{i\Delta\phi^{\text{NL}}(z)} \qquad (14.42)$$

and substituting it into Eq. 14.41 gives

$$\frac{d^2\rho(y)}{dy^2} - 2\beta(\omega)\rho(y)\frac{d\Delta\phi^{\text{NL}}(z)}{dz} = -2k_{\text{vac}}^2(\omega)n_{\text{eff}}(\omega)n'_{2\|,\text{E}}(-\omega;\,\omega)A^2\rho^3(y).$$

$$(14.43)$$

Instead of solving this equation, which is tedious, it is shown that $\rho(y) = \text{sech}(y/a)$ is a valid solution. Substituting this gives

$$\frac{1}{a^2} - \frac{2}{a^2\cosh^2(y/a)} - 2k_{\text{vac}}(\omega)n_{\text{eff}}(\omega)\frac{d\Delta\phi^{\text{NL}}(z)}{dz} = -\frac{2k_{\text{vac}}^2n_{\text{eff}}(\omega)n'_{2\|,\text{E}}(-\omega;\,\omega)}{\cosh^2(y/a)}A^2.$$

$$(14.44)$$

Separating the terms depending on y from those that are independent of y, we obtain

$$\frac{d\Delta\phi^{\text{NL}}(z)}{dz} = \frac{1}{2a^2\beta(\omega)} \quad \rightarrow \quad \Delta\phi^{\text{NL}}(z) = \frac{1}{2a^2\beta(\omega)}z,$$

$$A = \sqrt{\frac{1}{a^2k_{\text{vac}}^2(\omega)n_{\text{eff}}(\omega)n'_{2\|,\text{E}}(-\omega;\,\omega)}}.$$

$$(14.45)$$

Now converting to intensity,

$$I_{\text{sol}} = \frac{c\varepsilon_0 n_{\text{eff}}(\omega)A^2}{2\cosh^2(y/a)} \quad \rightarrow \quad P_{\text{sol}} = \frac{h_{\text{eff}}c\varepsilon_0}{2a^2k_{\text{vac}}^2(\omega)n'_{2\|,\text{E}}(-\omega;\,\omega)}\int_{-\infty}^{\infty}\cosh^{-2}(y/a)\,dy$$

$$= \frac{h_{\text{eff}}c\varepsilon_0}{ak_{\text{vac}}^2(\omega)n'_{2\|,\text{E}}(-\omega;\,\omega)}.$$

$$(14.46)$$

There exists a simple criterion for the existence of spatial solitons of arbitrary order in terms of the diffraction length $L_{\text{Dif}} = z_0 = kw_0^2 \cong k_{\text{vac}}n_{\text{eff}}(\omega)a^2$ and the nonlinear length L_{NL}; i.e., $L_{\text{Dif}} = N^2 L_{\text{NL}}$, with $N = 1$ (9). Noting that $2n_{2\parallel,\text{E}}'(-\omega;\omega) = c\varepsilon_0 n_{\text{eff}}(\omega)n_{2\parallel}'(-\omega;\omega)$ and the effective waveguide cross-sectional area is $A_{\text{eff}} \cong 2ah_{\text{eff}}$, the nonlinear length is defined as

$$L_{\text{NL}} \cong \frac{A_{\text{eff}}}{k_{\text{vac}}n_{2\parallel}'(-\omega;\omega)P_{\text{sol}}}. \qquad (14.47)$$

There exist higher order spatial (and temporal) solitons with the order given by $N = 2, 3, \ldots$, which have progressively higher power requirements. Their field distributions are more complicated than for the $N = 1$ case. The evolution of the $N = 2, 3,$ and 4 soliton fields with distance, for example, is shown in Fig. 14.14. The field is not a constant with z. However, the field reproduces itself periodically with a period given by $z_{\text{SP}} - \pi L_{\text{Dif}}/2$. Higher order solitons are "breather" solitons because of this variation in the field (and intensity) over one period.

It is not possible to create *Kerr* solitons in two dimensions, i.e., in a bulk material— they are not stable. It was shown in Eq. 14.46 that the 1D soliton power at which diffraction balances self-focusing is proportional to h_{eff}/a. In two dimensions, $h_{\text{eff}} \propto a$ and the soliton power is independent of the width of the beam. Therefore the feedback mechanism that allows solitons to compensate for noise, perturbations, and so on in one dimension, i.e., $dP_{\text{sol}}/da < 0$, is not present in two dimensions since

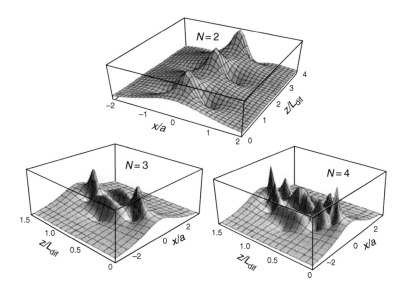

FIGURE 14.14 Evolution of the field profiles of the $N = 2, 3,$ and 4 solitons with propagation distance in diffraction lengths. Courtesy of Prof. Nail Akhmediev, Australia National University.

$dP_{sol}/da = 0$ and hence any noise will destabilize the beam. All other solitons—cascading, photorefractive, liquid crystal, and so on—are stable in two dimensions.

These bright soliton solutions have some unusual properties that are common to all solitons which cannot be understood on the basis of linear optics. The more interesting ones are as follows:

1. The shape and amplitude are either invariant on propagation for $N = 1$ or repeat every soliton period for $N > 1$ because of a balance between self-focusing and diffraction.
2. There is an additional phase shift proportional to z due to an effective nonlinear change in the wave vector.
3. Solitons are robust against fluctuations in width or power; i.e., $aP_{sol} = constant$ for 1D Kerr media. Since $dP_{sol}/da < 0$, $P_{sol} \propto a^{-1}$ and vice versa; i.e., any increase in the soliton power is compensated for by a decrease in the width. For non-Kerr solitons, although $a \times P_{sol}$ may not be a constant, any decrease in P_{sol} is still compensated by an increase in width, and vice versa, with possibly some radiation loss.
4. Solitons have fascinating collision properties. They act like both waves and particles (see Fig. 14.15). Note the interference effects that occur near and on collision—a wave property. Also the number of solitons before collision is equal to the number after collision—a particle property (not always the case for non-Kerr solitons). For the Kerr case only, there are no radiative losses. In phase solitons attract and repel out of phase. At other angles, there is a net energy exchange. For collision in the in-phase case, there is a small lateral shift in

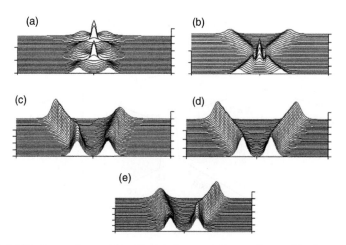

FIGURE 14.15 Interactions and collisions between Kerr solitons: (a) phase difference $\Delta\phi = 0$, parallel soliton incidence; (b) $\Delta\phi = 0$, collision; (c) $\Delta\phi = \pi/2$, parallel input; (d) $\Delta\phi = \pi$, parallel input; (e) $\Delta\phi = 3\pi/2$, parallel input.

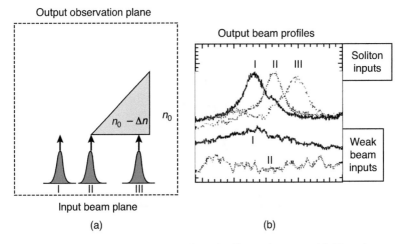

FIGURE 14.16 Experimental demonstration of soliton robustness. (a) Three beams were incident on the prism that changed the guided-wave index by ≈0.03. (b) The input and output beam profiles are shown on the left-hand side. Reproduced with permission from the American Institute of Physics (10).

direction at the output. A similar lateral shift occurs in linear optics when a beam traverses a region of higher refractive index.

5. Kerr solitons are impervious to small perturbations (see Fig. 14.16). In this demonstration of soliton robustness, a "prism" was etched into an AlGaAs slab waveguide, which caused a perturbation for guided waves in the slab geometry shown in Fig. 14.16a (10). Three beams were input at low powers and also as solitons. The beam profiles observed at the output facet are shown in Fig. 14.16b. The weak beams diffract, and in the case of the input aimed at the prism edge, additional edge diffraction occurs. For incident solitons, the soliton in every case survives, even the one incident on the prism edge. The one passing through the prism is deflected like it would be in linear optics by a prism. For non-Kerr solitons, some energy may be lost to radiation but the soliton nature of the output survives. This soliton robustness against perturbations is one of the special properties of solitons.

6. Solitons, except those due to cascading, can form waveguides, which can trap other weaker beams at a different frequency or polarization, or both. Trapping occurs because the soliton creates a refractive index potential well due to $\Delta n_y = n_{2\parallel}(-\omega; \omega)I(y)$ for weak copolarized beams of the same frequency and $\Delta n_x = n_{2\perp}(-\omega; \omega)I(y)$ for orthogonally polarized beams. For beams at a different frequency ω_a, the pertinent nonlinearities are $\Delta n_y = n_{2\parallel}(-\omega_a; \omega)I(y)$ and $\Delta n_x = n_{2\perp}(-\omega_a; \omega)I(y)$, respectively.

Since both the spatial instabilities and solitons have their origin in the same nonlinearity, there clearly must be a connection between them. The fields required

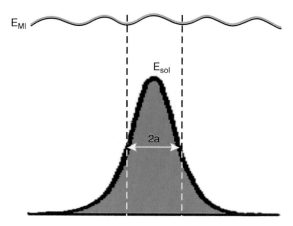

FIGURE 14.17 The fields associated with a soliton of full width 2a and the onset of modulational instability with period $\Lambda = 2a$.

to produce a periodic MI pattern with a period Λ and a soliton of full width $2a \cong \Lambda$ (see Fig. 14.17), given by Eqs 14.20 and 14.45 in a slab waveguide, are given by

$$|E_0|^2_{MI} = \frac{\pi^2}{n'_{2||,E}(-\omega;\omega)n_{eff}(\omega)k^2_{vac}\Lambda^2} \quad \text{and} \quad |E_0|^2_{sol} = \frac{4}{n'_{2||,E}(-\omega;\omega)n_{eff}(\omega)k^2_{vac}(2a)^2},$$
$$(14.48)$$

respectively. As the continuous-wave input intensity is increased, stable solitons appear first and then the onset of modulational instability occurs for wide beams at a higher intensity than for soliton formation.

14.4.2 Temporal Solitons (Temporal Analog of Spatial Solitons)

Temporal solitons were the first optical solitons to be investigated intensively, driven by the possibility of using them for long distance optical communications (9). Digital information in the form of soliton pulses that do not disperse in time was a very attractive concept, but the technical difficulties encountered were too severe for practical applications. Nevertheless, nonlinear optics in fibers does allow the possibility of large nonlinear phase shifts based on the simple Kerr effect, as discussed previously, and many beautiful nonlinear optics experiments were performed, some of which will be discussed in Chapter 17.

   The analysis of this problem and hence many of the unique features of temporal solitons are completely analogous to the spatial case, which has already been discussed in some detail. Self-focusing in the time domain cancels GVD. In the absence of spatial diffraction, typically the case in a fiber, the equation is a form of the nonlinear wave equation, i.e., Eq. 14.41 driven by a nonlinear polarization. Since the index difference between the core and the cladding in fibers is small, the analysis is simplified. In analogy to the spatial case, the solution is of the form $\vec{\mathcal{E}}(z, T) = \hat{e}A \operatorname{sech}(T/T_0) e^{i\Delta\phi^{\mathrm{NL}}(z)}$, with

$$\Delta\phi^{\mathrm{NL}}(z) = \frac{1}{2}\frac{|k_2|}{T_0^2}z, \qquad A = \sqrt{\frac{|k_2|}{T_0^2 k_{\mathrm{vac}} n_{2\|,\mathrm{E}}(-\omega;\omega)}}. \tag{14.49}$$

These temporal pulses exist only in the anomalous dispersion regime and propagate without change in the shape or amplitude with a slow accumulation of an additional nonlinear phase shift. Some typical numbers for silica fibers at $\lambda = 1.55\,\mu\mathrm{m}$ are $k_2 = -20\,\mathrm{ps}^2/\mathrm{km}$, $P_{\mathrm{sol}} = 5\,\mathrm{W}$ for $T_0 = 1\,\mathrm{ps}$, and $P_{\mathrm{sol}} = 50\,\mathrm{mW}$ for $T_0 = 10\,\mathrm{ps}$. The original proposal for communications was to use 10-ps pulses.

The stability, collision, and other properties discussed for temporal solitons are identical to those for spatial solitons. However, there are a few other characteristics that are best illustrated in the temporal domain. One example is the reshaping of the pulse profile from some arbitrary input shape into the classical soliton shape. This is illustrated in Fig. 14.18, which shows the evolution into a soliton of an input Gaussian of approximately the right peak power and temporal width. This reshaping can be adiabatic if the input conditions are very close to the soliton. However, for large differences, energy can be lost via radiation.

An alternate way to discuss the physics of the $N = 1$ soliton is in terms of the nonlinear frequency chirp. For these solitons, not only is the pulse shape constant but there is also no spreading of the frequency spectrum. This requires that the nonlinear frequency chirp be balanced by the chirp due to GVD. These two chirps are reproduced in Fig. 14.19 for self-phase modulation and for GVD. Although the signs of the chirp are opposite due to the two effects, the chirps themselves have different shapes. This pulse reshaping in the previous example and the cancellation of dissimilar chirps are both examples of the robustness of solitons. This robustness is a consequence of the strong interaction between light and matter needed for soliton formation.

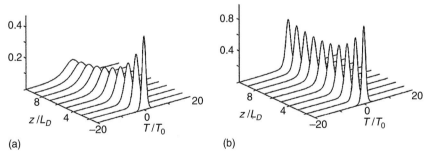

FIGURE 14.18 For Gaussian beam inputs, the reshaping of the temporal pulse profile into that of the $N = 1$ soliton on propagation is shown for (a) an input power below the soliton power ($N = 0.75$) for the input Gaussian width and (b) an input Gaussian at the soliton power ($N = 1$) and width. Courtesy of Dr Alessandro Sandrino, University of Central Florida.

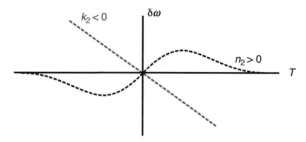

FIGURE 14.19 The frequency chirps due to group velocity dispersion with $k_2 < 0$ and due to self-phase modulation.

14.4.3 Dark Solitons

Temporal dark solitons also exist for $k_2 > 0$ and $n_{2\parallel}(-\omega; \omega) > 0$. They were initially studied in fibers (11). The starting nonlinear wave equations are of course the same, but the solutions for $k_2 > 0$ are different; namely, $\vec{\mathcal{E}}(z, T) = \hat{e} A \tanh(T/T_0) \, e^{i\Delta\phi^{NL}(z)}$ for "dark" solitons with

$$A = \sqrt{\frac{|k_2|}{T_0^2 k_{vac} n_{2\parallel,E}(-\omega; \omega)}}, \qquad \Delta\phi^{NL}(z) = \frac{k_2}{T_0^2} z. \qquad (14.50)$$

Note that the dark soliton is a "notch" in an infinite broad plane-wave background and hence requires, in principle, infinite energy (see Fig. 14.20a). There is a π phase difference between the background fields on opposite sides of the notch. If this phase difference is less than π, the notch does not approach zero (gray soliton) and the notch has a "transverse velocity"; i.e., it travels at an angle (slides sideways along the time axis) in Fig. 14.20b.

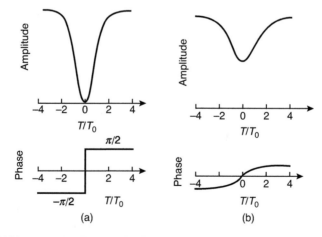

FIGURE 14.20 (a) Dark soliton amplitude and phase distribution. (b) Gray soliton amplitude and phase distribution.

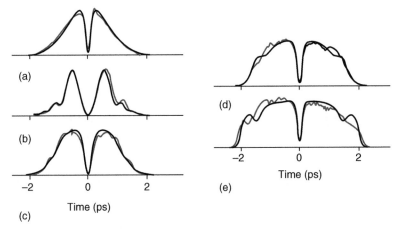

FIGURE 14.21 (a) The theoretical (solid line) and experimental (data points) results for the input intensity profile with a field that changes sign from positive to negative at the minimum dip in the center. Note the very limited extent of the background in which the notch sits. The evolution with an increasing input power of dark solitons as measured at the output end of a fiber. (b)–(e) The output for input powers of 1.5 W (negligible nonlinear effects), 52.5 W, 150 W, and 300 W peak power, respectively. Reproduced with permission from American Physical Society (11).

It is clearly not possible to satisfy the theoretical conditions of an infinitely broad background for the notch to sit in. However, since solitons are robust, solitonic effects should be visible for finite width beams, e.g., the one shown in Fig. 14.21a. The evolution of this input as observed at the output end of the fiber was measured for different input powers, also shown in Fig. 14.21b–e (11). Clearly the notch diffracts initially and then narrows near the soliton power back to the original notch. However, the background diffracts as expected. Additional dips appear in the background. They also evolve with increasing background width. The good agreement between the experiment and the numerical simulation indicates the generation of (imperfect) dark solitons.

Dark spatial solitons cannot exist for self-focusing nonlinearities since diffraction has only one possible sign, namely, negative. In the paraxial approximation $1 \gg \theta$, diffraction is defined as

$$D = \frac{d^2 k_z}{dk_x^2} = \frac{d\sqrt{k^2 - k_x^2}}{dk_x^2} < 0. \qquad (14.51)$$

For dark temporal solitons, the product $n_{2\parallel}(-\omega; \omega)k_2 > 0$ and so dark spatial solitons can also be studied in self-defocusing media as long as $n_{2\parallel}(-\omega; \omega)D > 0$, which can clearly be satisfied, and such solitons have been observed. Dark solitons are stable only for one dimension in Kerr media, but they are stable in two dimensions as well as one dimension for other (than Kerr) defocusing nonlinearities where they appear as nonlinear vortices in a constant field background.

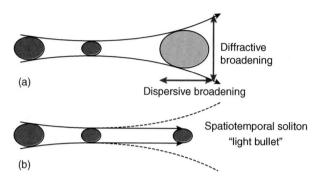

FIGURE 14.22   (a) Normal diffraction and dispersion of a weak intensity, spatiotemporal pulse. (b) Spatiotemporal soliton that neither diffracts in space nor disperses in time. Reproduced with permission from the Optical Society of America (12).

14.4.4   Solitons in Space and Time (Optical Bullets)

The condition for solitons in both space and time, shown in Fig. 14.22, can be written as

$$L > L_{NL} \cong L_{Dif} \cong L_{Dis}. \qquad (14.52)$$

Typically, samples for spatial solitons have lengths of $1-5$ cm. Since glass has a zero dispersion point at $1.27\,\mu m$, glass slab waveguides of the type shown in Fig. 14.13 should be good candidates for space–time solitons. However, at the intensities required for $L_{Dis} \approx$ cm, other nonlinear effects such as multiphoton absorption and ionization occur and for other materials L_{Dis} is much too small due to large GVD.

Quasi-1D spatiotemporal solitons have been demonstrated using the cascading nonlinearity (12). The input beam was elliptical in shape with a large aspect ratio, 4 mm long, 20 μm wide. The beam does not diffract in the long dimension ($L_{Dif} > 1$ m) but does in the transverse dimension ($L_{Dif} \approx 0.7$ mm). The GVD was changed in the transverse dimension by tilting the pulse wave front in that dimension by using novel grating techniques, allowing Eq. 14.53 to be satisfied in the transverse dimension (13). The propagation results for that dimension are shown in Fig. 14.23a. Similarly, the temporal pulse width does not disperse with distance (see Fig. 14.24b). Clearly at the calculated soliton power the beam is a quasi-1D optical bullet. However, as the power input is increased, MI sets in producing multiple filaments along the long axis of the ellipse as predicted in Fig. 14.17 and reproduced in Fig. 14.23c.

14.5   OPTICAL BISTABILITY

This phenomenon was one of the first nonlinear optics effects to be considered for digital, all-optical information processing (14). For an input into an appropriate "cavity," e.g., a Fabry–Perot interferometer and under certain conditions of initial

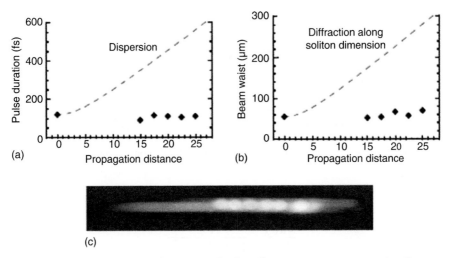

(a)

(b)

(c)

FIGURE 14.23 (a) Pulse duration measured at the soliton power versus propagation distance. (b) Beam width measured along the transverse dimension at the soliton power. (c) The onset of instabilities in the long dimension at powers larger than the soliton power. Reproduced with permission from the Optical Society of America (12).

detuning of the cavity from one of its transmission resonances, the output has two possible states of cavity transmission for a range of input powers. The actual output state obtained depends on whether the input was decreasing or increasing in power approaching these states, i.e., the previous history of the input beam. This behavior can occur either due to nonlinear absorption or nonlinear refraction. The first case is discussed here and the second left as a problem at the end of the chapter.

The simplest "cavity" for understanding this effect is a unidirectional cavity containing a nonlinear medium, as shown in Fig. 14.24. Consider the continuous-wave case, i.e., steady state. Light input through the first beam splitter circulates around the cavity and can exit through either the first or the second beam splitter. It is assumed that when the circulated beam E_I recombines with the incident beam E_{in} via beam splitter 1, $E_R(t) = \sqrt{R}E_I(t) + \sqrt{T}E_{in}(t)$ is reproduced, where R and T are the intensity reflection and transmission coefficients, respectively, identical for both beam splitters 1 and 2. Light passing through the nonlinear medium acquires a nonlinear phase shift $\Delta\phi^{NL} = k_{vac}(\omega)n_{2\|}(-\omega; \omega)LI_R$ per pass, with $TI_R = I_{out}$. ϕ_C is the linear, i.e., low power (linear), round-trip phase accumulation that occurs in a round-trip time t_{RT} and so the total round-trip phase shift is $\phi = \phi_C + \Delta\phi^{NL}$.

After one round trip in the cavity at time t, just before beam splitter 1,

$$\mathcal{E}_I(t) = \sqrt{R}\mathcal{E}_R(t-t_{RT})\, e^{i(\phi_C + \Delta\phi^{NL})}. \tag{14.53}$$

Just after the beam splitter,

$$\mathcal{E}_R(t) = \sqrt{T}\mathcal{E}_{in}(t) + R\mathcal{E}_R(t-t_{RT})\, e^{i(\phi_C + \Delta\phi^{NL})}. \tag{14.54}$$

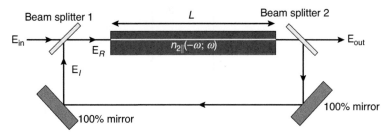

FIGURE 14.24   The unidirectional cavity used for demonstrating bistability.

Since in the steady state $\mathcal{E}_R(t) = \mathcal{E}_R(t - t_{RT})$,

$$\frac{\mathcal{E}_R(t)}{\mathcal{E}_{in}(t)} = \frac{\sqrt{T}}{1 - R\, e^{i(\phi_C + \phi^{NL})}} \tag{14.55}$$

$$\rightarrow \frac{I_{out}}{I_{in}} = \frac{T}{1 + R^2 - 2R\cos\phi}. \tag{14.56}$$

Although this equation cannot be solved analytically, it can be solved numerically, with the results shown in Fig. 14.25 for two different negative detunings from the resonance; i.e., $\phi_C < 0$ (15). The position of the detuning on the transmission curve is shown in Fig. 14.25a. Figures 14.25b–14.25d show the input–output characteristics for the three cases: one with zero detuning from the resonance and two with increasing negative detuning. The plotted curves followed by the output transmission and whether they can be accessed by an increasing or decreasing input intensity are shown by arrows. For a large enough negative detuning (Fig. 14.25d), a hysteresis loop is formed and there are two bistable states for a single input intensity. One can be reached by increasing the intensity (low transmission) and the second by decreasing the intensity (high transmission) from above the hysteresis loop.

This behavior can be explained as follows. Starting on resonance, increasing the input intensity results in a nonlinear phase shift that tunes the total phase off the resonance peak and so the transmission decreases and the output intensity becomes sublinear in the input intensity (Fig. 12.25b). For initially negative detuning, the transmission increases with input intensity as the system moves toward the resonance due to the nonlinear phase shift; i.e., the output increases faster than linear in the input (Fig. 12.25c). In this region, $dI_{out}/dI_{in} > 0$. When resonance is reached, further increases in the input intensity move the system off resonance and the response is similar to Fig. 12.25b. In Fig. 12.25d, a "runaway" effect takes place in which an increase in the input intensity moves the system toward the resonance and reaches a point where the increase in the cavity field and hence the nonlinear phase shift is large enough to continuously move the system to resonance without further increase in the input intensity. This corresponds to a region where $dI_{out}/dI_{in} < 0$ (dashed black line) and stability analysis shows that such regions are unstable (15). Only regions with $dI_{out}/dI_{in} > 0$ are stable. The system "jumps" to the high transmission state.

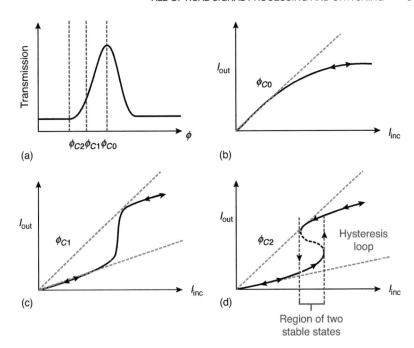

FIGURE 14.25 (a) A low power resonance of the cavity and the detunings from that cavity resonance before the intensity is increased. (b) The input–output for the intensity for zero initial detuning from the cavity resonance peak. (c) The input–output for the intensity for an intermediate input intensity for the initial detuning from the resonance peak ϕ_{C1}. (d) The input–output for a high input intensity for the initial detuning from the resonance peak ϕ_{C2}. The dark dashed line indicates the unstable part of the solution. The green lines indicate the maximum (upper line) and minimum (lower line) transmission possible.

A subsequent increase in the input intensity produces behavior again similar to Fig. 12.25b. When the input intensity decreases from above the hysteresis loop, the cavity field starts in the high transmission state. As the input intensity decreases, the cavity field decreases but remains in the high transmission state until $dI_{out}/dI_{in} < 0$ when the system becomes unstable and can no longer stay on resonance at which point it "jumps" down to the low transmission state. Hence a hysteresis loop is formed due to the feedback provided by the cavity. Which state the system is in depends on the prior history of the input light.

These high and low transmission states have been considered as digital one or zero states, and so in principle digital logic can be performed. Note that a minimum input holding power is needed to maintain each state.

14.6 ALL-OPTICAL SIGNAL PROCESSING AND SWITCHING

Because of the importance of controlled routing of optical signals in communications, a number of schemes have been developed for using light to control light, which in principle can be achieved in shorter times than by electronics, which is limited by electron transit times across some junction. Optically this can, in principle, be implemented by using the intensity-dependent refractive index in control-signal beam geometries. Waveguides, either in fiber or in channel integrated optics form, have the optical cross sections of a few wavelengths that are needed to achieve and maintain the high intensities needed over useful distances. In this section, the classic example of a nonlinear directional (NLDC) coupler in an integrated, channel, optics geometry will be discussed (16).

14.6.1 Linear Coupler

To understand how a nonlinear directional coupler functions, it is necessary to understand the linear coupler first. It consists of two parallel identical channel waveguides (as shown in Fig. 14.26a) in which the optical fields in one waveguide overlap the second waveguide (as shown in Fig. 14.26b). When one waveguide is excited with light, light transfers with propagation distance to the second waveguide (as shown in Fig. 14.26a). The distance required for complete transfer is called the coupling length L_C.

A slab waveguide geometry shown in Fig. 14.27c is used to illustrate the calculation of the coupling length. In the absence of waveguide 2, the film of waveguide 1 is completely surrounded by the cladding of the refractive index n_c. The introduction of the second waveguide introduces an extra region of index n_f so that the field of waveguide 1 introduces a linear polarization

$$P^L(x) = \varepsilon_0(n_f^2 - n_c^2)E_1(x)a_1(z) \tag{14.57}$$

in the n_f region of waveguide 2 where it excites the second waveguide's mode. Inserting this polarization into the slowly varying envelope approximation gives

$$2i\beta\frac{\partial E_2(x)a_2(z)}{\partial z} = -\omega^2\mu_0 P^L(x) = -\omega^2\mu_0\varepsilon_0(n_f^2 - n_c^2)E_1(x)a_1(z). \tag{14.58}$$

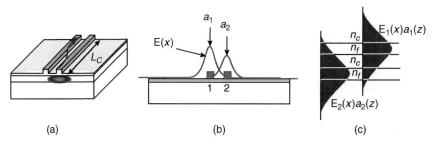

(a)　　　　　　　　　(b)　　　　　　　　　(c)

FIGURE 14.26 (a) A directional coupler of length L_C. (b) Overlap of the fields leading to coupling between the waveguides. (c) Simplified one-dimensional geometry for of the coupling analysis.

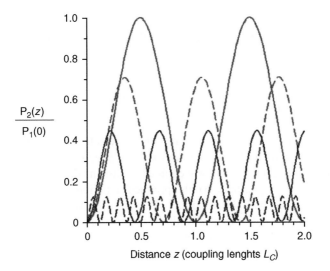

FIGURE 14.27 The fraction of power transferred from the first to the second channel versus propagation distance for different values of the normalized wave-vector mismatch R. Solid red: R = 0. Dashed red: R = 1. Solid blue: R = 2. Dashed blue: R = 8.

Multiplying both sides of the equation by $E_2^*(x)$ and defining the coupling constant κ as

$$\kappa = \frac{\omega(n_f^2 - n_c^2)}{2cn_{\text{eff}}} \frac{\int_1 E_2^*(x)E_1(x)\, dx}{\int_2 |E_2(x)|^2\, dx} \tag{14.59}$$

gives the coupled mode equation

$$\frac{da_2(z)}{dz} = i\kappa a_1(z). \tag{14.60a}$$

Of course waveguide 2 also produces a polarization in Waveguide 1 so that

$$\frac{da_1(z)}{dz} = i\kappa a_2(z). \tag{14.60b}$$

When waveguide 1 is excited, the solution to these equations is trivially found to be

$$a_1(z) = a_1(0)\cos(z\pi/2L_C), \qquad a_2(z) = a_1(0)\sin(z\pi/2L_C), \qquad L_C = \pi/2\kappa. \tag{14.61}$$

For channel waveguides with fields that are functions of x and y, the definition for the coupling constant is more complex. With $E(x, y)$ now normalized so that $|a(z)|^2$ is the power in a waveguide,

$$\kappa = \frac{\omega \varepsilon_0}{4} \int \int_1 [n_1^2(x, y) - n_2^2(x, y)] E_2^*(x, y) E_1(x, y) \, dx \, dy. \tag{14.62}$$

Before adding the nonlinearity to Eqs 14.61, which will mismatch the two waveguides, it will be useful to first consider the effects of linear mismatch due to nonidentical waveguides. The waveguides are assumed to have a wave-vector mismatch $\Delta \beta = \beta_2 - \beta_1$, which would produce a periodic exchange of energy between the channels with the period $L_b = 2\pi/|\beta_2 - \beta_1|$. If the mismatch is small, the modal fields are still essentially the same so that the coupling constant κ does not change and Eqs 14.60 now become

$$\frac{da_1(z)}{dz} = i\kappa a_2(z) \, e^{i(\beta_2 - \beta_1)z}, \qquad \frac{da_2(z)}{dz} = i\kappa a_1(z) \, e^{-i(\beta_2 - \beta_1)z}. \tag{14.63}$$

The solutions are shown graphically in Fig. 14.27. Clearly mismatching the waveguides reduces the coupling significantly. Here $R = 2L_c/L_b$.

14.6.2 Nonlinear Directional Coupler

This device has been used in a number of different configurations for switching and routing (16,17). The simplest excitation case is a single input channel and identical waveguides made from the same nonlinear material. Taking Eq. 14.41 but without diffraction since light is confined in waveguides and adding the coupling from Eqs 14.60 gives

$$i\frac{da_1(z)}{dz} = -\kappa a_2(z) - \gamma |a_1(z)|^2 a_1(z), \qquad i\frac{da_2(z)}{dz} = -\kappa a_1(z) - \gamma |a_2(z)|^2 a_2(z), \tag{14.64}$$

with the normalized γ given by

$$\gamma = \frac{k_{vac} \varepsilon_0}{4\mu_0} \int_{-\infty}^{\infty} \int_{-\infty}^{\infty} n^2(x, y) n_{2||}(-\omega; \omega) |E(x, y)|^4 \, dx \, dy. \tag{14.65}$$

Equations 14.64 have analytical solutions in terms of Jacobi elliptic functions (17). There exists a critical power

$$P_c = \frac{4\kappa}{\gamma} \tag{14.66a}$$

that separates the behavior of the solutions. The solutions for the initial excitation of Channel 1 are given by

$$P_1(z) = \frac{1}{2}P_1(0)\{1 + cn(2\kappa z|m), \qquad P_2(z) = P_1(0) - P_1(z), \qquad m = \frac{P_1(0)}{P_c}.$$

(14.66b)

The solutions are shown in Fig. 14.28. Exactly at P_c in the asymptotic limit of a very long coupler, $L \gg L_C$, the power is split 50–50 between the waveguides. This is an unstable point, and the slightest deviation upward in power leads to oscillations that are double the period of oscillations just below P_c. Far above the critical power, there is essentially no power transfer to Waveguide 2.

Some aspects of the nonlinear behavior can be understood in terms of an intensity-dependent propagation wave vector in the channels, i.e., $\beta_1(P_1(z)) = \beta_1 + n_2(-\omega; \omega)P_1(z)/A_{\text{eff}}$ and $\beta_2(P_2(z)) - \beta_2 + n_2(-\omega; \omega)P_2(z)/A_{\text{eff}}$, where A_{eff} is the effective channel area. This means that optical power detunes the wave vector of the two channels and partially inhibits the power transfer. This results in an increase in the effective coupling length that approaches infinity at the critical power.

The behavior of these devices is also nonreciprocal for initially detuned devices; i.e., $\beta_1 \neq \beta_2$ (18). If $\beta_2 > \beta_1$ and with power incident in Channel 1, the power-dependent increase in $\beta_1(P_1(z))$ initially increases the power transfer to Channel 2, whereas if $\beta_1 > \beta_2$, $\beta_1(P_1(z))$ inhibits the transfer of power to Channel 2 (18).

A potential application of this device to demultiplexing or routing is shown in Fig. 14.29. The strong control pulse in the lower channel detunes the coupler during its passage through the device, as indicated in Fig. 14.29 (19). No signal can cross out of

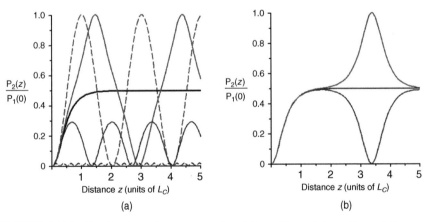

FIGURE 14.28 The fraction of power transferred from Channel 1 to Channel 2 versus propagation distance for different values of $P_1(0)/P_c$. (a) Red curves: dashed, $P_1(0)/P_c = 0.4$; solid, $P_1(0)/P_c = 0.95$. Black curve: $P_1(0)/P_c = 1.0$. Blue curve: dashed, $P_1(0)/P_c = 2$; solid, $P_1(0)/P_c = 1.05$. (b) Red curves: $P_1(0)/P_c = 0.9999$. Black curve: $P_1(0)/P_c = 1.0$. Blue curve: $P_1(0)/P_c = 1.0001$.

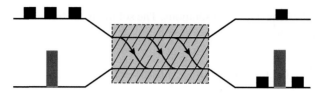

FIGURE 14.29 Nonlinear directional coupler used for routing a single (dark) weak pulse from a pulse train by coincidence in time with a strong control pulse in the lower channel.

the signal channel if it is coincident with the control pulse. The control pulse can be orthogonally polarized, or even be at a different frequency.

PROBLEMS

1. A nonlinear first mode is a solution to a nonlinear wave equation for a plane wave, i. e., no diffraction. Show that for a self-focusing medium, the nonlinear wave equation is the nonlinear Schrödinger equation for the field and can be written as

$$\frac{1}{2}\frac{\partial^2 u(\omega)}{\partial \tau^2} + i\frac{\partial u(\omega)}{\partial \xi} + |u|^2 u = 0, \quad \text{with}$$

$$u = \sqrt{\frac{3}{8}\frac{\tilde{\chi}^{(3)}_{xxxx}(-\omega; \omega, -\omega, \omega)}{n^2}}E(\omega), \quad \tau = kx, \ \xi = kz.$$

2. Often saturation in a nonlinearly induced index change is modeled by an index change of the form $\Delta n = n_{2\parallel}(-\omega; \omega)I - n_3(-\omega; \omega, -\omega, \omega)I^2$. Assuming that a nonlinear phase shift of $\pi/2$ is required for a soliton, show that this form of index change leads to stable spatial solitons with two transverse dimensions up to a threshold power beyond which the solitons are unstable. Evaluate the threshold intensity and interpret it in terms of the index change induced by the nonlinearities n_2 and n_3.

3. For temporal solitons in fibers, one can define the dispersion length L_D as $L_D = T_0^2/|k_2|$, the nonlinear length L_{NL} as $L_{NL} = cA_{eff}/n_2\omega P$, and the fundamental soliton as $L_D/L_{NL} = N^2$ (where $N = 1$ gives the fundamental soliton).
   (a) Show that the nonlinear wave equation can be written as

$$i\frac{\partial u}{\partial \xi} = sign(k_2)\frac{1}{2}\frac{\partial^2}{\partial \tau^2}u - N^2|u|^2 u,$$

with $u = \mathcal{E}/\sqrt{P}$, $\xi = z/L_D$, and $\tau = T/T_0$.

(b) The solution for the $N=2$ soliton is given by

$$u(\xi,\tau) = \frac{4\left[\cosh(3\tau)+3\,e^{4i\xi}\cosh(\tau)\right]e^{i\xi/2)}}{\left[\cosh(4\tau)+4\cosh(2\tau)+3\cos(4\xi)\right]}.$$

This field is a nightmare to verify for the $N=2$ soliton unless one has a complex equation simplifier. Instead, take the required derivatives of $u(\xi,\tau)$, evaluate both sides of the equation, and show that the nonlinear wave equation is satisfied at $\xi=0$ and $\tau=0$.

(c) Plot the intensity $|u(\xi,\tau)|^2$ as a function of ξ and show that the intensity is periodic in ξ with a period of $\xi_0=\pi/2$. Such a soliton is known as a "breather." Thus for a soliton the shape does not have to be invariant on propagation. A periodic reconstruction of the intensity also qualifies as a soliton.

(d) In principle, the $N=2$ soliton can be used for pulse compression. How would you do this?

4. In this chapter a dispersive bistability was considered in which the cavity contained materials whose refractive index changed with intensity. Bistability can also occur due to an intensity-dependent absorption. Consider the cavity in the figure below, in which there is saturation of the absorption as discussed for a two-level system in Chapter 11; i.e., $\alpha=\alpha_0/(1+I/I_{sat})$. Derive the relationship between the input and the output intensity, plot it, and find the range of values of $C_0=R\alpha_0 L/(1-R)$ for which bistable behavior occurs.

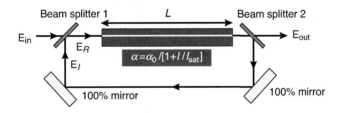

5. The periodicity induced in the frequency spectrum due to self-phase modulation in the time domain has an equivalent effect in the spatial domain for the component of the propagation wave vector orthogonal to the beam propagation axis. Show that a cylindrically symmetric Gaussian beam of finite width traveling in a medium capable of $\Delta n(I)L$ of multiples of π leads to concentric rings. What happens to the ring pattern as the input intensity is increased? [Hint: This phenomenon has been observed in liquid crystals.]

6. Show that the solution to Eqs 14.64 gives the critical power in Eqs 14.60.

REFERENCES

1. J. H. Marburger, "Self-focusing: theory," Prog. Quantum Electron., **4**, 35–110 (1975).

2. G. Fibich and A. L. Gaeta, "Critical power for self-focusing in bulk media and in hollow waveguides," Opt. Lett., **25**, 335–337 (2000).

3. B. Crosignani, A. Yariv, and S. Mookherjea, "Nonparaxial spatial solitons and propagation-invariant pattern solutions in optical Kerr media," Opt. Lett., **29**, 1254–1256 (2004).

4. A. Couairon and A. Mysyrowicz, "Femtosecond filamentation in transparent media," Phys. Rep., **441**, 47–189 (2007).

5. R. H. Stolen and C. Lin, "Self-phase-modulation in silica optical fibers," Phys. Rev. A, **17**, 1448–1453 (1978).

6. R. Malendevich, L. Jankovic, G. Stegeman, and J. S. Aitchison, "Spatial modulation instability in a Kerr slab waveguide," Opt. Lett., **26**, 1879–1881 (2001).

7. R. A. Fuerst, D.-M. Baboiu, B. Lawrence, W. E. Torruellas, G. I. Stegeman, and S. Trillo, "Spatial modulational instability and multisoliton-like generation in a quadratically nonlinear optical medium," Phys. Rev. Lett., **78**, 2760–2763 (1997).

8. G. I. Stegeman and C. T. Seaton, "Nonlinear integrated optics," Appl. Phys. Rev., **58**, R57–R78 (1985).

9. G. Agrawal, Nonlinear Fiber Optics, 3rd Edition (Academic Press, San Diego, 2001).

10. J. U. Kang, G. I. Stegeman, G. Hamilton, and J. S. Aitchison, "Robustness of spatial solitons in AlGaAs waveguides," Appl. Phys. Lett., **70**, 1363–1365 (1997).

11. A. M. Weiner, J. P. Heritage, R. J. Hawkins, R. N. Thurston, E. M. Kirschner, D. E. Leaird, and W. J. Tomlinson, "Experimental observation of the fundamental dark soliton in optical fibers," Phys. Rev. Lett., **61**, 2445–2448 (1988).

12. X. Liu, L. Qian, and F. W. Wise, "Generation of optical spatiotemporal solitons," Phys. Rev. Lett., **82**, 4631 (1999); F. Wise and P. Di Trapani, "The hunt for light bullets—spatiotemporal solitons," Opt. Photonics News, **13**, 28–32 (2002).

13. P. Di Trapani, D. Caironi, G. Valiulis, A. Dubietis, R. Danielius, and A. Piskarskas, "Observation of temporal solitons in second-harmonic generation with tilted pulses," Phys. Rev. Lett., **81**, 570–573 (1998).

14. H. M. Gibbs, G. Khitrova, and N. Peyghambarian, eds, Nonlinear Photonics (Springer-Verlag, Berlin, 1990).

15. A. C. Walker, B. S. Wherrett, and S. D. Smith, "First implementation of optical digital computing circuits using nonlinear devices," in Nonlinear Photonics, edited by H. M. Gibbs, G. Khoitrova, and N. Peyghambarian (Springer-Verlag, Berlin, 1990), Chap. 4.

16. G. I. Stegeman, A. Villeneuve, J. S. Aitchison, and C. N. Ironside, "Nonlinear integrated optics and all-optical switching in semiconductors," in Fabrication Properties and Applications of Low-Dimensional Semiconductors, NATO ASI Series, edited by M. Balkanski and I. Yanchev (Kluwer Academic Publishers, Dordrecht, 1995), pp. 415–449.

17. S. Jensen, "The nonlinear coherent coupler," IEEE J. Quantum Electron., **QE-18** 1580–1583 (1982).

18. S. Trillo and S. Wabnitz, "Nonlinear reciprocity in a coherent mismatched directional coupler," Appl. Phys. Lett., **49**, 752–754 (1986).

19. A. Villeneuve, P. Mamyshev, J. U. Kang, G. I. Stegeman, J. S. Aitchison, and C. N. Ironside, "Efficient time-domain demultiplexing with separate signal and control wavelengths in an AlGaAs nonlinear directional coupler," IEEE J. Quantum Electron., **31**, 2165–2172 (1995).

SUGGESTED FURTHER READING

G. Agrawal, Nonlinear Fiber Optics, 3rd Edition (Academic Press, San Diego, 2001).

R. W. Boyd, S. G. Lukishova, and Y. R. Shen, eds, Self-Focusing, Past and Present: Fundamentals and Prospects (Springer-Verlag, Berlin, 2008).

A. Couairon and A. Mysyrowicz, "Femtosecond filamentation in transparent media," Phys. Rep., **441**, 47–189 (2007).

H. M. Gibbs, G. Khoitrova, and N. Peyghambarian, eds, Nonlinear Photonics (Springer-Verlag, Berlin, 1990).

I. C. Khoo, "Nonlinear optics of liquid crystalline materials," Phys. Rep., **471**, 221–267 (2009).

Y. Kivshar and G. Agrawal, Optical Solitons: From Fibres to Photonic Crystals (Academic Press, San Diego, 2003).

J. H. Marburger, "Self-focusing: theory," Prog. Quantum Electron., **4**, 35–110 (1975).

A. W. Snyder and J. D. Love, Optical Waveguide Theory (Chapman & Hall, London, 1983).

G. I. Stegeman, A. Villeneuve, J. S. Aitchison, and C. N. Ironside, "Nonlinear integrated optics and all-optical switching in semiconductors," in Fabrication, Properties and Applications of Low-Dimensional Semiconductors, NATO ASI Series, edited by M. Balkanski and I. Yanchev (Kluwer Academic Publishers, Dordrecht, 1995), pp. 415–449.

S. Trillo and W. Torruellas, eds, Spatial Soliton (Springer-Verlag, Berlin, 2001).

CHAPTER 15

Multiwave Mixing

The last chapter dealt with various phenomena that occur due to nonlinear refraction (NLR). This chapter discusses a selection of additional phenomena that are based on the mixing of two or more waves to produce a new wave with a different frequency, direction, or polarization. This includes interactions with nonoptical normal modes in matter, such as molecular vibrations that, under appropriate conditions, can be excited optically via nonlinear optics.

In the terminology of nonlinear optics, "degenerate" means that all the beams are at the same frequency ω and "nondegenerate" identifies interactions between waves of different frequencies. The interacting beams can have different polarizations and can travel in different directions. Since the interactions usually occur between coherent waves, the key issue for producing an efficient interaction is wave-vector matching. Because there is dispersion in the refractive index with frequency, collinear wave-vector-matched nondegenerate interactions are not trivial to achieve, especially in bulk media. Of course, the waves can be noncollinear to achieve wave-vector conservation, in which case beam overlap problems play a detrimental role for the efficiency of the interaction. Alternatively, collinear wave-vector matching can be achieved if there is a contribution to the total wave vector from a normal mode of matter, e.g., sound wave, or a vibrational mode such as that occurs in stimulated scattering discussed in Chapter 16.

One of the virtues of waveguides in fiber or integrated optics formats is that they can be engineered to have more than one eigenmode and hence multiple wave vectors available at the same frequency. This enables nondegenerate interactions to occur at more discrete wavelengths than in the bulk case. Furthermore, at the very high intensities achievable in fibers, NLR via $n_{2\|}(-\omega_b; \omega_a)$ and $n_{2\perp}(-\omega_b; \omega_a)$ can tune the refractive index at different frequencies, allowing wave-vector matching that normally could not be wave-vector matched. The important variables are the intensity of the strong pump beams and dispersion in the nonlinearity with frequency. In fact, the combination of many nonlinear effects can lead to coherent continua over octaves in frequency that occurs in fibers and will be discussed in Chapter 17.

15.1 DEGENERATE FOUR-WAVE MIXING

This interaction with all waves of the same frequency is useful primarily for two reasons: One, in the appropriate spectral regions it can be used to measure $n_{2\parallel}(-\omega; \omega)$; and two, the output of the interaction is the phase conjugate of the signal input.

15.1.1 Beam Geometry and Nonlinear Polarization

The most common interaction involves three input beams: two collinear, but traveling in opposite directions, which act as pump beams p1 and p2; and one signal beam labeled s, which can travel in any direction relative to the pumps. Note that optimizing the overlap of beams to increase the conversion efficiencies requires small relative angles between the signal and the pump beam propagation directions. When the two pump beams are properly aligned in opposite directions, a conjugate beam, labeled c, travels backward along the signal beam direction (see Fig. 15.1). Wave-vector conservation means that $\vec{k}_{p1} + \vec{k}_{p2} + \vec{k}_s + \vec{k}_c = 0$. It is important to make the pump beam alignment as precise as possible. This means that $\vec{k}_s = -\vec{k}_c$ and $\vec{k}_{p2} = -\vec{k}_{p1}$. In isotropic materials or high symmetry crystals, $|\vec{k}_{p1}| = |\vec{k}_{p2}| = |\vec{k}_s| = |\vec{k}_c|$ since all the beams are at the same frequency.

For the most general case, the Fourier transform of the nonlinear polarization discussed in Chapter 8 is given by

$$P_i^{(3)}(\omega_4) = \varepsilon_0 \int \int \int \hat{\chi}_{ijkl}^{(3)}(-\omega_4; \omega_3, \omega_2, \omega_1) E_j(\omega_1) E_k(\omega_2) E_l(\omega_3) \delta(\omega - \omega_1 - \omega_2 - \omega_3)$$
$$d\omega_1 \, d\omega_2 \, d\omega_3,$$

$$(15.1)$$

in which each of the fields $E(\omega_m)$ designated are the total fields, i.e., each of the fields consists of the four fields shown in Fig. 15.1:

$$\vec{E}(\omega) = \frac{1}{2} \left\{ \begin{array}{l} \vec{E}^{(p1)}(\omega)\delta(\omega - \omega_{p1}) + \vec{E}^{*(p1)}(\omega)\delta(\omega + \omega_{p1}) \\ +\vec{E}^{(p2)}(\omega)\delta(\omega - \omega_{p2}) + \vec{E}^{*(p2)}(\omega)\delta(\omega + \omega_{p2}) \\ +\vec{E}^{(s)}(\omega)\delta(\omega - \omega_s) + \vec{E}^{*(s)}(\omega)\delta(\omega + \omega_s) \\ +\vec{E}^{(c)}(\omega)\delta(\omega - \omega_c) + \vec{E}^{*(c)}(\omega)\delta(\omega + \omega_c) \end{array} \right\}. \quad (15.2)$$

The frequencies will ultimately be all put equal, but at this stage, subscripts 1, 2, and so on, allow the pertinent beams to be identified. All the fields correspond to different

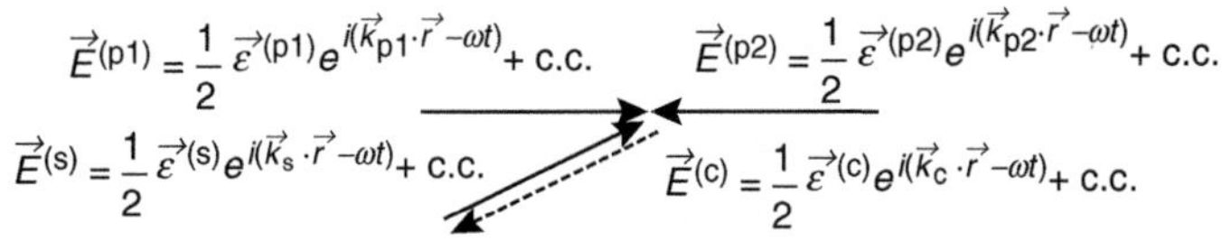

$$\vec{E}^{(p1)} = \frac{1}{2}\vec{\varepsilon}^{(p1)}e^{i(\vec{k}_{p1}\cdot\vec{r} - \omega t)} + \text{c.c.} \qquad \vec{E}^{(p2)} = \frac{1}{2}\vec{\varepsilon}^{(p2)}e^{i(\vec{k}_{p2}\cdot\vec{r} - \omega t)} + \text{c.c.}$$

$$\vec{E}^{(s)} = \frac{1}{2}\vec{\varepsilon}^{(s)}e^{i(\vec{k}_s\cdot\vec{r} - \omega t)} + \text{c.c.} \qquad \vec{E}^{(c)} = \frac{1}{2}\vec{\varepsilon}^{(c)}e^{i(\vec{k}_c\cdot\vec{r} - \omega t)} + \text{c.c.}$$

FIGURE 15.1 Classical four-wave mixing geometry and definitions of the fields.

eigenmodes, and so $\hat{\chi}_{ijkl}^{(3)}(-\omega; \omega, -\omega, \omega)$ contains six contributions in the general case. For simplicity, it will be assumed that the beams are copolarized and that $|E^{(p1)}|^2 \approx |E^{(p2)}|^2 \gg |E^{(s)}|^2$ and $|E^{(c)}|^2$. Since it is the signal and the conjugate beam that are the primary outputs of interest, the nonlinear polarization should consist only of beam products containing both pump beams and the signal or the conjugate beam; therefore,

$$
\begin{aligned}
P_x^{(3)}(\omega_s) = \frac{1}{4}\varepsilon_0 \Big\{ &\hat{\chi}_{xxxx}^{(3)}(-\omega; \omega_{p1}, -\omega_c, \omega_{p2}) + \hat{\chi}_{xxxx}^{(3)}(-\omega; \omega_{p2}, -\omega_c, \omega_{p1}) \\
+ &\hat{\chi}_{xxxx}^{(3)}(-\omega; \omega_{p1}, \omega_{p2}, -\omega_c) + \hat{\chi}_{xxxx}^{(3)}(-\omega; \omega_{p2}, \omega_{p1}, -\omega_c) \\
+ &\hat{\chi}_{xxxx}^{(3)}(-\omega; -\omega_c, \omega_{p1}, \omega_{p2}) + \hat{\chi}_{xxxx}^{(3)}(-\omega; -\omega_c, \omega_{p2}, \omega_{p1}) \Big\} \\
& \mathcal{E}_x^{(p1)} \mathcal{E}_x^{(p2)} \mathcal{E}_x^{*(c)} e^{i\vec{k}_s \cdot \vec{r}}
\end{aligned}
\tag{15.3a}
$$

for the signal and

$$
\begin{aligned}
P_x^{(3)}(\omega_c) = \frac{1}{4}\varepsilon_0 \Big\{ &\hat{\chi}_{xxxx}^{(3)}(-\omega; \omega_{p1}, -\omega_s, \omega_{p2}) + \hat{\chi}_{xxxx}^{(3)}(-\omega; \omega_{p2}, -\omega_s, \omega_{p1}) \\
+ &\hat{\chi}_{xxxx}^{(3)}(-\omega; \omega_{p1}, \omega_{p2}, -\omega_s) + \hat{\chi}_{xxxx}^{(3)}(-\omega; \omega_{p2}, \omega_{p1}, -\omega_s) \\
+ &\hat{\chi}_{xxxx}^{(3)}(-\omega; -\omega_s, \omega_{p1}, \omega_{p2}) + \hat{\chi}_{xxxx}^{(3)}(-\omega; -\omega_s, \omega_{p2}, \omega_{p1}) \Big\} \\
& \mathcal{E}_x^{(p1)} \mathcal{E}_x^{(p2)} \mathcal{E}_x^{*(s)} e^{i\vec{k}_c \cdot \vec{r}}
\end{aligned}
\tag{15.3b}
$$

for the conjugate beam.

Note that various nonlinear susceptibilities can have different combinations of resonances. Assume again for simplicity that the frequency is sufficiently far from any material resonances so that all the $\tilde{\chi}_{xxxx}^{(3)}(-\omega)$ are equal, i.e., near the nonresonant (Kleinman) limit. There are, however, other factors to be considered when there is spatial diffusion of a nonlinearly induced index change, e.g., in semiconductors. This is best understood in terms of a "grating" model in the Kleinman limit.

15.1.2 Grating Model

In the grating model, the products of two fields create an index grating in space and the third input wave is deflected by the grating to form beams s or c. Initially the only two field products of interest in forming the grating are (p1 or p2) × (p2 or p1, or c or s). The possible field products are

$$
\begin{aligned}
\frac{1}{4}\Big[&\{\delta(\omega_1 - \omega)\delta(\omega_2 - \omega)\}\{2\mathcal{E}^{(p1)}\mathcal{E}^{(p2)} + 2\mathcal{E}^{(p1)}\mathcal{E}^{(s)}e^{i(\vec{k}_p + \vec{k}_s)\cdot\vec{r}} \\
&+ 2\mathcal{E}^{(p1)}\mathcal{E}^{(c)}e^{i(\vec{k}_p - \vec{k}_s)\cdot\vec{r}} + 2\mathcal{E}^{(p2)}\mathcal{E}^{(s)i(-\vec{k}_p + \vec{k}_s)\cdot\vec{r}} + 2\mathcal{E}^{(p2)}\mathcal{E}^{(c)i(-\vec{k}_p - \vec{k}_s)\cdot\vec{r}}\} \\
&+ \{\delta(\omega_1 - \omega)\delta(\omega_2 + \omega)\}\{2\mathcal{E}^{(p1)}\mathcal{E}^{*(p2)}e^{2i\vec{k}_p} + 2\mathcal{E}^{(p1)}\mathcal{E}^{*(s)}e^{i(\vec{k}_p - \vec{k}_s)\cdot\vec{r}} \\
&+ 2\mathcal{E}^{(p1)}\mathcal{E}^{*(c)}e^{i(\vec{k}_p + \vec{k}_s)\cdot\vec{r}} + 2\mathcal{E}^{(p2)}\mathcal{E}^{*(s)-i(\vec{k}_p + \vec{k}_s)\cdot\vec{r}} + 2\mathcal{E}^{(p2)}\mathcal{E}^{*(c)i(-\vec{k}_p + \vec{k}_s)\cdot\vec{r}}\} + \text{c.c.}\Big]
\end{aligned}
\tag{15.4}
$$

There are *two* kinds of time dependence present here corresponding to the first two inputs (ω_1, ω_2) in $\tilde{\chi}^{(3)}_{xxxx}(-\omega; \omega_1, \omega_2, \omega_3)$:

1. $\delta(\omega_1 - \omega)\delta(\omega_2 - \omega) + \text{c.c.}$ oscillates at 2ω.
2. $\delta(\omega_1 - \omega)\delta(\omega_2 + \omega) + \text{c.c.}$ is zero frequency (DC) in time.

The first requires a purely electronic nonlinearity (ultrafast); otherwise the contribution of that nonlinearity is zero. There are no time restrictions on the second, the DC grating, and so essentially all the nonlinear processes discussed in Chapters 11 and 12 can contribute. The products of the fields in Eq. 15.3a produce "gratings" in the effective linear susceptibility of the form $\tilde{\chi}^{(1)}_{xx,\text{effective}} = \tilde{\chi}^{(3)}_{xxxx}\mathcal{E}(\omega_1)\mathcal{E}(\omega_2)$, which can be interpreted as a modulation of the refractive index, i.e., $\Delta n \propto \tilde{\chi}^{(1)}_{xx,\text{effective}}$.

In the grating model, the products of two fields—one pump beam and one signal or conjugate beam—create a grating in space and the second pump beam is deflected by the grating to form the fourth beam c or s, respectively. The gratings that do conserve wave vector and the responsible field combinations are illustrated in Fig. 15.2. For example, in Fig. 15.2a, the field combinations $E^{(p1)}E^{(s)}$ and $E^{*(p2)}E^{*(c)}$ lead to the grating wave vector $\vec{k}_p + \vec{k}_s$. This grating then "reflects" (Fig. 15.2b) the "other" pump beam, i.e., $E^{(p2)}$ for $E^{(p1)}E^{(s)}$ or $E^{*(p1)}$ for $E^{*(p2)}E^{*(c)}$ into $E^{*(c)}$ or $E^{(s)}$, respectively. Note that the grating in Fig. 15.2a oscillates at 2ω whereas the grating in Fig. 15.2c is DC in time. For a Kerr medium, both gratings will lead to reflected beams.

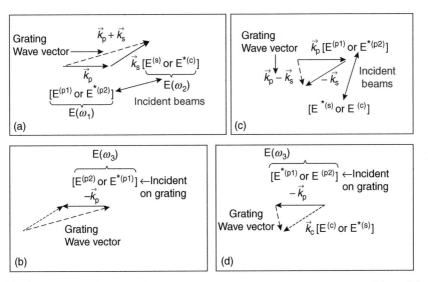

FIGURE 15.2 (a) The small period (large wave vector) grating created by mixing of the specified beams. (b) These gratings then reflect the incident third beam to produce either a signal or a conjugate beam. (c) Same as (a) but for a large period (small wave-vector) grating. (d) Same as (b) but for a large period grating.

There are additional gratings in Eq. 15.4 that are not shown in Fig. 15.2, since they do not participate in degenerate four-wave mixing (D4WM). For example, the grating $\mathcal{E}^{(\text{p}1)}\mathcal{E}^{*(\text{p}2)}\,e^{2i\vec{k}_\text{p}\cdot\vec{r}}$ cannot be wave-vector matched, nor can $\mathcal{E}^{(\text{p}1)}\mathcal{E}^{(\text{p}2)}$.

There are other nonlinearities besides electronic nonlinearities that give rise to D4WM, and it is the details of their physics that determines whether they contribute to both gratings. For example, carrier generation and recombination would not produce a grating in the visible or near-infrared region that needs to respond at 2ω. Furthermore, excited carriers "diffuse" over a distance L_d (Fig. 11.1), which results in the optically induced grating "washing away" if $L_d > \Lambda_g = 2\pi/|\vec{k}_\text{p} - \vec{k}_\text{s}|$. In this case, the grating would not (or only partially) contribute to D4WM.

Assuming that the material response is fast enough to respond to all the wave-vector-matched gratings, i.e., the Kerr case, the pertinent nonlinear polarization for generating the signal and the conjugate beam,

$$P_x^{(3)}(\omega) = \frac{6}{4}\varepsilon_0\tilde{\chi}_{xxxx}^{(3)}(-\omega;\,\omega,\,-\omega,\,\omega)\mathcal{E}_x^{(\text{p}1)}\mathcal{E}_x^{(\text{p}2)}\left\{\mathcal{E}_x^{*(\text{s})}\,e^{-i\vec{k}_\text{s}\cdot\vec{r}} + \mathcal{E}_x^{*(\text{c})}\,e^{+i\vec{k}_\text{s}\cdot\vec{r}}\right\}.$$

$$(15.5)$$

The term containing $\mathcal{E}_x^{*(\text{s})}$ generates the conjugate beam and the one with $\mathcal{E}_x^{*(\text{c})}$ generates the signal. Substituting $\tilde{\chi}_{xxxx}^{(3)}(-\omega;\,\omega,\,-\omega,\,\omega)$ in terms of $n_{2\|}(-\omega;\,\omega)$, the nonlinearity defined for a *single* input beam,

$$P_x^{(3)}(\omega) = 2n^2 c\varepsilon_0^2 n_{2\|}(-\omega;\,\omega)\mathcal{E}_x^{(\text{p}1)}\mathcal{E}_x^{(\text{p}2)}\left\{\mathcal{E}_x^{*(\text{s})}\,e^{-i\vec{k}_\text{s}\cdot\vec{r}} + \mathcal{E}_x^{*(\text{c})}\,e^{+i\vec{k}_\text{s}\cdot\vec{r}}\right\}. \quad (15.6)$$

15.1.3   D4WM Field Solutions

Using the slowly varying envelope approximation (SVEA) in the undepleted pump beam approximation, we obtain

$$-\frac{d\mathcal{E}^{(\text{c})}(z',\omega)}{dz'} = i\omega n\varepsilon_0 n_{2\|}(-\omega;\,\omega)\mathcal{E}^{(\text{p}1)}(\omega)\mathcal{E}^{(\text{p}2)}(\omega)\mathcal{E}^{*(\text{s})}(z',\omega), \qquad (15.7\text{a})$$

$$\frac{d\mathcal{E}^{(\text{s})}(z',\omega)}{dz'} = i\omega n\varepsilon_0 n_{2\|}(-\omega;\,\omega)\mathcal{E}^{(\text{p}1)}(\omega)\mathcal{E}^{(\text{p}2)}(\omega)\mathcal{E}^{*(\text{c})}(z',\omega), \qquad (15.7\text{b})$$

where the z'-axis is the direction along which the signal and conjugate beams travel. Note the minus sign in Eq. 15.7a, which indicates that the conjugate beam grows in the backward direction, i.e., along $-z'$-axis. The z'-axis is normally rotated from the z-axis by a small angle θ to facilitate the separation of the signal and the conjugate beam from the pump beams and to optimize the beam interaction volume. Note that the processes are assumed to be phase matched, i.e., perfect contradirectional alignment of the pump beams. Defining

$$\ell_{\text{4WM}}^{-1} = \omega n\varepsilon_0 n_{2\|}(-\omega;\,\omega)\mathcal{E}^{(\text{p}1)}(\omega)\mathcal{E}^{(\text{p}2)}(\omega), \qquad (15.8)$$

which is a constant for no depletion of the pump beams, Eqs 15.7 simplifies to

$$\frac{d\mathcal{E}^{(c)}(z')}{dz'} = -i\frac{1}{\ell_{4WM}}\mathcal{E}^{*(s)}(z'), \tag{15.9a}$$

$$\frac{d\mathcal{E}^{(s)}(z')}{dz'} = i\frac{1}{\ell_{4WM}}\mathcal{E}^{*(c)}(z'). \tag{15.9b}$$

Taking d/dz' of Eq. 15.9a and substituting from Eq. 15.9b gives

$$\frac{d^2\mathcal{E}^{(c)}(z',\omega)}{dz'^2} = -\frac{1}{\ell_{4WM}^2}\mathcal{E}^{(c)}(z',\omega), \tag{15.10}$$

which has the solutions of the form

$$\mathcal{E}^{(c)}(z') = A^{(c)}\sin\left(\frac{z'}{\ell_{4WM}}\right) + B^{(c)}\cos\left(\frac{z'}{\ell_{4WM}}\right),$$
$$\mathcal{E}^{(s)}(z') = A^{(s)}\sin\left(\frac{z'}{\ell_{4WM}}\right) + B^{(s)}\cos\left(\frac{z'}{\ell_{4WM}}\right). \tag{15.11}$$

Applying the boundary conditions $\mathcal{E}^{(c)}(L',\omega) = 0$ and $\mathcal{E}^{(s)}(0,\omega) \neq 0$, where $L \cong L'$ is the sample length for small angles θ, we obtain

$$\mathcal{E}^{(s)}(z',\omega) = \mathcal{E}^{(s)}(0,\omega)\left[\tan\left(\frac{L'}{\ell_{4WM}}\right)\sin\left(\frac{z'}{\ell_{4WM}}\right) + \cos\left(\frac{z'}{\ell_{4WM}}\right)\right],$$
$$\mathcal{E}^{(c)}(z',\omega) = i\mathcal{E}^{*(s)}(0,\omega)\left[\tan\left(\frac{L'}{\ell_{4WM}}\right)\cos\left(\frac{z'}{\ell_{4WM}}\right) - \sin\left(\frac{z'}{\ell_{4WM}}\right)\right]. \tag{15.12}$$

The output of interest is usually the conjugate beam. The common designation of this process as "phase conjugation" comes from the output of the conjugate beam, which is clearly the conjugate of the input signal; i.e.,

$$\mathcal{E}^{(c)}(0,\omega) = i\mathcal{E}^{*(s)}(0,\omega)\left[\tan\left(\frac{L'}{\ell_{4WM}}\right)\right]. \tag{15.13}$$

It is common to define "reflectivity" as

$$\frac{|\mathcal{E}^{(c)}(0,\omega)|^2}{|\mathcal{E}^{(s)}(0,\omega)|^2} = \tan^2\left(\frac{L'}{\ell_{4WM}}\right), \tag{15.14}$$

which implies that the reflectivity can be greater than unity. This is not a contradiction, since the photons needed come from the pump beams. The "transmission" of the signal beam is given by

$$\frac{|\mathcal{E}^{(s)}(L',\omega)|^2}{|\mathcal{E}^{(s)}(0,\omega)|^2} = \left| \tan\left(\frac{L'}{\ell_{4WM}}\right) \sin\left(\frac{L'}{\ell_{4WM}}\right) + \cos\left(\frac{L'}{\ell_{4WM}}\right) \right|^2 = \sec^2\left(\frac{L'}{\ell_{4WM}}\right).$$

(15.15)

Note that as $L'/\ell_{4WM} \to \pi/2$, $R \to \infty$ and $T \to \infty$, which is of course impossible and is a consequence of the assumption of no pump beam depletion. Note that the signal beam also experiences gain.

15.1.4 Manley–Rowe Relations

These relations describe the power flow in the interaction that must be conserved. Multiplying Eqs 15.7a and 15.7b by $\mathcal{E}^{*(c)}(z',\omega)$ and $\mathcal{E}^{*(s)}(z',\omega)$, respectively, and converting them to intensity equations,

$$-\mathcal{E}^{*(c)}(z,\omega)\frac{d\mathcal{E}^{(c)}(z,\omega)}{dz} = i\omega n \varepsilon_0 n_2 \mathcal{E}^{(p1)}(\omega)\mathcal{E}^{(p2)}(\omega)\mathcal{E}^{*(s)}(z,\omega)\mathcal{E}^{*(c)}(z,\omega),$$

$$\mathcal{E}^{*(s)}(z,\omega)\frac{d\mathcal{E}^{(s)}(z,\omega)}{dz} = i\omega n \varepsilon_0 n_2 \mathcal{E}^{(p1)}(\omega)\mathcal{E}^{(p2)}(\omega)\mathcal{E}^{*(c)}(z,\omega)\mathcal{E}^{*(s)}(z,\omega)$$

(15.16)

$$\to \quad -\frac{dI^{(c)}(z)}{dz} = \frac{dI^{(s)}(z)}{dz}. \tag{15.17}$$

Since the two waves travel in opposite directions, they both grow with their respective propagation distances. This energy must come from the pump beams. Assuming that pump depletion does occur and following the procedure that led to Eq. 15.3a, it can be used to find the nonlinear polarization driving the pump beams. The grating formed by, e.g., $E^{(p1)}E^{(s)}$ is then used to reflect $E^{(c)}$ into $E^{(p2)}$ and so on. Inserting the nonlinear polarization for the pump beams into the SVEA for small angles θ gives

$$-\frac{d\mathcal{E}^{*(p1)}(z,\omega)}{dz} = i\omega n \varepsilon_0 n_2 \mathcal{E}^{*(s)}(z,\omega)\mathcal{E}^{*(c)}(z,\omega)\mathcal{E}^{(p2)}(z,\omega),$$

$$\frac{d\mathcal{E}^{*(p2)}(z,\omega)}{dz} = i\omega n \varepsilon_0 n_2 \mathcal{E}^{*(s)}(z,\omega)\mathcal{E}^{*(c)}(z,\omega)\mathcal{E}^{(p1)}(z,\omega).$$

(15.18)

These equations are also converted into intensity relations by multiplying the first of Eqs 15.18 by $\mathcal{E}^{(p1)}(z,\omega)$ and the second by $\mathcal{E}^{(p2)}(z,\omega)$, giving

$$-\mathcal{E}^{(p1)}(z,\omega)\frac{d\mathcal{E}^{*(p1)}(z,\omega)}{dz} = i\omega n \varepsilon_0 n_2 \mathcal{E}^{*(s)}(z,\omega)\mathcal{E}^{*(c)}(z,\omega)\mathcal{E}^{(p1)}(z,\omega)\mathcal{E}^{(p2)}(z,\omega),$$

$$\mathcal{E}^{(p2)}(z,\omega)\frac{d\mathcal{E}^{*(p2)}(z,\omega)}{dz} = i\omega n \varepsilon_0 n_2 \mathcal{E}^{*(s)}(z,\omega)\mathcal{E}^{*(c)}(z,\omega)\mathcal{E}^{(p1)}(z,\omega)\mathcal{E}^{(p2)}(z,\omega).$$

(15.19)

Noting that the left-hand side of Eqs 15.16 and 15.19 are identical, we obtain

$$\frac{dI^{(\text{p2})}(z)}{dz} = -\frac{dI^{(\text{c})}(z)}{dz}, \qquad -\frac{dI^{(\text{p1})}(z)}{dz} = \frac{dI^{(\text{s})}(z)}{dz}. \tag{15.20}$$

As the forward traveling pump beam depletes, the forward traveling signal beam grows, and as the backward traveling pump beam depletes, the backward traveling conjugate beam grows. Energy is conserved, and both signal and conjugate beams experience gain.

15.1.5 Wave-Vector Mismatch

The case discussed here is for a small misalignment of the two pump beams (Fig. 15.3) that results in the same small beam misalignment of the conjugate and signal beams.
This introduces a wave-vector mismatch $\Delta k = 2k[1 - \cos(\theta/2)] \cong k\theta^2/2$ along the z-axis into the equations that become (in the undepleted pump beam approximation)

$$\frac{d\mathcal{E}^{(\text{c})}(z)}{dz} = -i\frac{1}{\ell_{4\text{WM}}}\mathcal{E}^{*(\text{s})}(z)\,e^{i\Delta kz}, \qquad \frac{d\mathcal{E}^{(\text{s})}(z)}{dz} = i\frac{1}{\ell_{4\text{WM}}}\mathcal{E}^{*(\text{c})}(z)\,e^{i\Delta kz}. \tag{15.21}$$

Defining $\mathcal{E}^{(i)}(z) = B^{(i)}(z)\exp[i\Delta kz/2]$ and substituting it into Eqs 15.21 results in

$$\frac{d^2 B^{(i)}(z)}{dz^2} = -\left(\frac{\Delta k^2}{4} + \frac{1}{\ell_{4\text{WM}}^2}\right) B^{(i)}(z) = -\mu^2 B^{(i)}(z), \qquad \mu^2 = \frac{\Delta k^2}{4} + \frac{1}{\ell_{4\text{WM}}^2}, \tag{15.22}$$

and the solutions are of the form

$$B^{(\text{s})}(z) = C^{(\text{s})}\sin(\mu z) + D^{(\text{s})}\cos(\mu z), \qquad B^{(\text{c})}(z) = C^{(\text{c})}\sin(\mu z) + D^{(\text{c})}\cos(\mu z). \tag{15.23}$$

The details of the calculation of the reflectivity are left as a problem at the end of the chapter. Applying the usual boundary conditions $\mathcal{E}^{(\text{c})}(L',\omega) = 0$ and $\mathcal{E}^{(\text{s})}(0,\omega) \neq 0$ gives

$$R = \frac{\sin^2(\mu L)}{\cos^2(\mu L) + [\Delta k\ell_{4\text{WM}}/2]^2} \quad \rightarrow \quad R_{\text{max}}\left(\mu L = \frac{\pi}{2}\right) = \left(\frac{2}{\Delta k\ell_{4\text{WM}}}\right)^2 \tag{15.24}$$

FIGURE 15.3 Pump beams misaligned by a small angle θ.

Clearly the misalignment leads to a reduction in the conjugate signal that can be even larger if beam overlap also becomes a problem.

15.1.6 Linear Absorption

Absorption can be included approximately if there is no significant pump depletion due to conversion into the conjugate and signal beams. Clearly the absorption affects all four waves.

$$\frac{d\mathcal{E}^{(p1)}(z)}{dz} = -\frac{\alpha}{2}\mathcal{E}^{(p1)}(z) \quad \rightarrow \quad \mathcal{E}^{(p1)}(z) = \mathcal{E}^{(p1)}(0)\,e^{-(\alpha/2)z},$$

$$\frac{d\mathcal{E}^{(p2)}(z)}{dz} = \frac{\alpha}{2}\mathcal{E}^{(p2)}(z) \quad \rightarrow \quad \mathcal{E}^{(p2)}(z) = \mathcal{E}^{(p2)}(L)\,e^{(\alpha/2)(z-L)}. \tag{15.25}$$

Including these expressions for the pump loss gives for the signal and conjugate beams

$$\frac{d\mathcal{E}^{(s)}(z)}{dz} = \frac{i}{\ell_{4WM}}\mathcal{E}^{*(c)}(z) - \frac{\alpha}{2}\mathcal{E}^{(s)}(z),$$

$$\frac{d\mathcal{E}^{(c)}(z)}{dz} = -\frac{i}{\ell_{4WM}}\mathcal{E}^{*(s)}(z) + \frac{\alpha}{2}\mathcal{E}^{(c)}(z). \tag{15.26}$$

Redefining $\frac{1}{\ell_{4WM}} = \omega n\varepsilon_0 n_2 \mathcal{E}^{(p1)}(0)\mathcal{E}^{(p2)}(L)\,e^{-\alpha L/2}$ and $\mu^2 = \frac{1}{\ell_{4WM}^2} - \left[\frac{\alpha}{2}\right]^2$ gives

$$R = \frac{\sin^2(\mu L)}{\ell_{4WM}^2 [\mu\cos(\mu L) + \{\alpha/2\}\sin(\mu L)]^2}, \tag{15.27}$$

which is valid as long as $\frac{1}{\ell_{4WM}^2} - \left[\frac{\alpha}{2}\right]^2 \gg 0$, which again reduces the conjugate signal, as expected.

15.1.7 Material Characterization via D4WM

This application was introduced briefly in Chapter 13, but the detailed equations were not discussed. It was shown in Eq. 15.14 that the reflectivity $\propto \tan^2(L/\ell_{4WM})$. For small reflectivities, $\tan(L/\ell_{4WM}) \cong L/\ell_{4WM}$ and so

$$R = [\omega n\varepsilon_0 n_{2\parallel}(-\omega;\omega)L]^2 |E^{(p1)}(0)|^2 |E^{(p2)}(L)|^2 = 4[k_{vac}Ln_{2\parallel}(-\omega;\omega)]^2 I^{(p1)}(0)I^{(p2)}(L). \tag{15.28}$$

Hence the nonlinearity $n_{2\parallel}(-\omega;\omega)$ can be evaluated from the conjugate output. Note, however, that it is assumed explicitly that two-photon absorption can be neglected, i.e., far from any two-photon resonances. The effect of two-photon absorption associated with a complex $\chi^{(3)}$ will be discussed in Section 15.1.8.

Experiments are usually done with pulsed beams, and Eq. 15.28 is for the continuous-wave case with plane waves. The simplest case is for lossless, Gaussian input beams collinear in space.

$$I^{(p1)}(0, r_\perp, t) = I^{(p1)}(0, 0, 0) \exp\left[-\frac{r_\perp^2}{w_{p1}^2} - \frac{t^2}{\tau_{p1}^2} \right],$$

$$I^{(p2)}(L, r_\perp, t) = I^{(p2)}(L, 0, 0) \exp\left[-\frac{r_\perp^2}{w_{p2}^2} - \frac{t^2}{\tau_{p2}^2} \right], \qquad (15.29)$$

$$I^{(s)}(0, r_\perp, t) = I^{(s)}(0, 0, 0) \exp\left[-\frac{r_\perp^2}{w_s^2} - \frac{t^2}{\tau_s^2} \right].$$

Assuming no beam spreading due to diffraction, for the spatial part of the problem,

$$I^{(c)}(0, 0, t) = 4[k_{vac} L n_{2\parallel}(-\omega; \omega)]^2 I^{(p1)}(0, 0, t) I^{(p2)}(L, 0, t) I^{(s)}(0, 0, t)$$

$$\times \exp\left[-r_\perp^2 \left\{ \frac{1}{w_{p1}^2} + \frac{1}{w_{p2}^2} + \frac{1}{w_s^2} \right\} \right]$$

$$\to I^{(c)}(0, r_\perp, t) = I^{(c)}(0, 0, t) \exp\left[-\frac{r_\perp^2}{w_c^2} \right], \qquad \frac{1}{w_c^2} = \frac{1}{w_{p1}^2} + \frac{1}{w_{p2}^2} + \frac{1}{w_s^2},$$

$$(15.30)$$

which indicates that the conjugate pulse is narrower in space than are the input beams (see Fig. 15.4)—another example of the beam narrowing implicit with wave mixing in nonlinear optics.

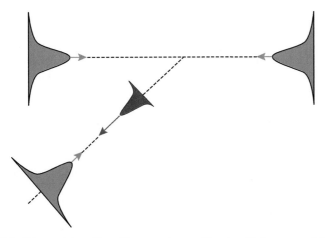

FIGURE 15.4 The spatial profiles of the pump beams. The signal is in green and the conjugate output beam in red.

In the usual case, all three input beams have the same beam width w, which gives for the conjugate power

$$P^{(c)}(0,t) = 2\pi \int_0^\infty I^{(c)}(0,r_\perp,t)r_\perp \, dr_\perp = \pi\frac{w^2}{3}I^{(c)}(0,0,t)$$

$$\to \quad P^{(c)}(0,t) = \frac{4}{3}\left(\frac{k_{vac}Ln_{2\parallel}(-\omega;\omega)}{\pi w^2}\right)^2 P^{(p1)}(0,t)P^{(p2)}(L,t)P^{(s)}(0,t). \tag{15.31}$$

For temporal pulses, the conjugate signal is also narrowed in time to a pulse width τ_c, similar to the spatial case in Fig. 15.4:

$$P^{(c)}(0,t) = \frac{4}{3}\left(\frac{k_{vac}Ln_{2\parallel}(-\omega;\omega)}{\pi w^2}\right)^2 P^{(p1)}(0,0)P^{(p2)}(L,0)P^{(s)}(0,0)\, e^{-t^2/\tau_c^2},$$

$$\frac{1}{\tau_c^2} = \frac{1}{\tau_{p1}^2} + \frac{1}{\tau_{p2}^2} + \frac{1}{\tau_s^2} \xrightarrow{\tau_{p1}^2=\tau_{p2}^2=\tau_s^2=\tau} \frac{3}{\tau^2}. \tag{15.32}$$

The analysis is considerably simplified if the pulse width $\tau \gg nL/c$ (i.e., sample transit time) and all the input beams have the same τ. Since it is the pulse energy that is most frequently measured, integrating over time gives

$$\Delta E_c = \sqrt{\frac{\pi}{2}}\tau_c P_c(0) \quad \to \quad \Delta E_c = \frac{8\tau_c}{3\sqrt{3}\pi^3 w^4\tau^2}[k_{vac}Ln_{2\parallel}(-\omega;\omega)]^2 \Delta E_{p1}\Delta E_{p2}\Delta E_s. \tag{15.33}$$

15.1.8 Complex $\chi^{(3)}$

In the previous analysis it was assumed that only index modulation occurs via $n_{2\parallel}(-\omega;\omega)$. But in general, $\hat{\chi}^{(3)} = \Re\text{eal}\{\hat{\chi}^{(3)}\} + i\,\Im\text{mag}\{\hat{\chi}^{(3)}\}$ and the absorption change associated with the imaginary component can also contribute to the degenerate four-wave mixing conjugate and signal waves. The general definition of ℓ_{4WM} for real susceptibilities was

$$\frac{1}{\ell_{4WM}} = \frac{3k_{vac}}{4n}\hat{\chi}_{xxxx}^{(3)}(-\omega;\omega,-\omega,\omega)\mathcal{E}^{(p1)}(\omega)\mathcal{E}^{(p2)}(\omega), \tag{15.34}$$

and, for complex $\hat{\chi}_{xxxx}^{(3)}(-\omega;\omega,-\omega,\omega)$, ℓ_{4WM} is also a complex quantity given by

$$\frac{1}{\ell_{4WM}} = \frac{3k_{vac}}{4n}\left\{\Re\text{eal}\left[\hat{\chi}_{xxxx}^{(3)}(-\omega;\omega,-\omega,\omega) + i\,\Im\text{mag}\left[\hat{\chi}_{xxxx}^{(3)}(-\omega;\omega,-\omega,\omega)\right]\right]\right\}$$
$$\mathcal{E}^{(p1)}(\omega)\mathcal{E}^{(p2)}(\omega). \tag{15.35}$$

In this case the reflectivity R is given by

$$R = \tan^2\left(\frac{L}{|\ell_{4\text{WM}}|}\right), \qquad \frac{1}{|\ell_{4\text{WM}}|} = \frac{3k_{\text{vac}}}{4n}\left(\frac{[\Re\text{eal}\{\hat{\chi}^{(3)}_{xxxx}(-\omega;\omega,-\omega,\omega)\}]^2}{+[\Im\text{mag}\{\hat{\chi}^{(3)}_{xxxx}(-\omega;\omega,-\omega,\omega)\}]^2}\right)^{1/2}$$

$$\mathcal{E}^{(\text{p1})}(\omega)\mathcal{E}^{(\text{p2})}(\omega). \tag{15.36}$$

Therefore the absorption gratings formed by the imaginary part of the third-order susceptibility also contribute to the four-wave mixing signal. The absorption gratings limit the value of D4WM for measuring $n_{2\|}(-\omega;\omega)$ to regions of weak two-photon absorption. This is the main reason why this process is no longer used to obtain reliable data near resonance. The reflectivity can now be expressed as

$$R = 4(k_{\text{vac}}L)^2\left\{n_{2\|}^2(-\omega;\omega) + \frac{\alpha_{2\|}^2(-\omega;\omega)}{4k_{\text{vac}}^2}\right\}I^{(\text{p1})}(0)I^{(\text{p2})}(L),$$

$$R = 4(k_{\text{vac}}L)^2\left\{\Delta n_{\text{p1}}^{\text{NL}}\Delta n_{\text{p2}}^{\text{NL}} + \frac{\Delta\alpha_{\text{p1}}^{\text{NL}}\Delta\alpha_{\text{p2}}^{\text{NL}}}{4k_{\text{vac}}^2}\right\}, \tag{15.37}$$

where $\Delta n_{\text{p1}}^{\text{N}}$ and $\Delta n_{\text{p2}}^{\text{N}}$ are the index changes and $\Delta\alpha_{\text{p1}}^{\text{NL}}$ and $\Delta\alpha_{\text{p2}}^{\text{NL}}$ are the absorption changes due to the two pump beams separately, not just for the Kerr effect but also for any third-order nonlinearity, subject of course to the time and space response considerations of the different gratings induced. Furthermore, any process that modulates the refractive index or the absorption will produce D4WM, subject to spatial diffusion of gratings.

15.1.9 D4WM Including NLR

At the intensity levels required for efficient four-wave mixing, NLR can also contribute to and complicate the phase-matching process. All four waves can contribute to cross-NLR, and this problem is complicated further if pump depletion occurs. The discussion here is limited to the "no pump depletion" case and the conjugate and signal waves too weak to contribute NLR.

Assuming no pump depletion, the phase-mismatched case, and that the interaction takes place near the nonresonant regime, the equations describing the pump beams, including self- and cross-NLR, are

$$\frac{d\mathcal{E}^{(\text{p1})}}{dz} = ik_{\text{vac}}n_{2\|}(-\omega;\omega)(I_1 + 2I_2)\mathcal{E}^{(\text{p1})}(0)$$

$$\Rightarrow \quad \mathcal{E}^{(\text{p1})}(z) = \mathcal{E}^{(\text{p1})}(0)\,e^{ik_{\text{vac}}n_{2\|}(-\omega;\omega)(I_1+2I_2)z},$$

$$-\frac{d\mathcal{E}^{(\text{p2})}(z)}{dz} = ik_{\text{vac}}n_{2\|}(-\omega;\omega)(2I_1 + I_2)\mathcal{E}^{(\text{p2})}(z) \tag{15.38}$$

$$\Rightarrow \quad \mathcal{E}^{(\text{p2})}(z) = \mathcal{E}^{(\text{p2})}(L)\,e^{-ik_{\text{vac}}n_{2\|}(-\omega;\omega)(2I_1+I_2)(z-L)}.$$

Note that for beam p2 the nonlinear phase due to cross-NLR accumulates, starting at $z = L$. Therefore the product of the pump fields is

$$\mathcal{E}^{(p1)}(z)\mathcal{E}^{(p2)}(z) = \mathcal{E}^{(p1)}(0)\mathcal{E}^{(p2)}(L)\, e^{ik_{vac}n_{2||}(-\omega;\,\omega)(I_2-I_1)z+i\Psi}, \tag{15.39}$$

and $\psi = ik_{vac}n_{2||}(-\omega;\,\omega)(2I_1+I_2)L$. The cross-NLR due to the pump beams must now be included in Eq. 15.7a for both the signal and the conjugate beam. Defining the parameters

$$\delta = 2k_{vac}n_{2||}(-\omega;\,\omega)(I_1+I_2), \qquad \Gamma = \omega n\varepsilon_0 n_{2||}(-\omega;\,\omega)\mathcal{E}^{(p1)}(\omega)\mathcal{E}^{(p2)}(\omega)\, e^{i\psi},$$
$$\kappa = \Delta k + k_{vac}n_{2||}(-\omega;\,\omega)(I_2-I_1) \tag{15.40}$$

gives for Eqs 15.7a

$$\frac{d\mathcal{E}^{(s)}(z)}{dz} = i\delta\mathcal{E}^{(s)}(z) + i\Gamma\mathcal{E}^{(c)*}(z)\, e^{i\kappa z},$$
$$-\frac{d\mathcal{E}^{(c)}(z)}{dz} = i\delta\mathcal{E}^{(c)}(z) + i\Gamma\mathcal{E}^{(s)*}(z)^{i\kappa z}. \tag{15.41}$$

These equations can now be simplified with the following substitution:

$$\mathcal{E}^{(s)}(z) = B_s(z)\, e^{i\delta z}, \qquad \mathcal{E}^{(c)}(z) = B_c(z)\, e^{-i\delta z}, \tag{15.42}$$

and so Eqs 15.41 become

$$\frac{dB_c(z)}{dz} = -i\Gamma\, e^{i\kappa z}B_s^*(z),$$
$$\frac{dB_s(z)}{dz} = i\Gamma\, e^{i\kappa z}B_c^*(z). \tag{15.43}$$

These equations have simple solutions of the form

$$B_s(z) = e^{i\kappa z/2}[C_s\sin(\mu z) + C_c\cos(\mu z)],$$
$$B_c(z) = e^{i\kappa z/2}[D_s\sin(\mu z) + D_c\cos(\mu z)], \qquad \mu^2 = \frac{\kappa^2}{4} + |\Gamma|^2. \tag{15.44}$$

With the boundary conditions $\mathcal{E}^{(c)}(L) = 0$ and $\mathcal{E}^{(s)}(0) = C_c$, R is

$$R = \frac{|\mathcal{E}^{(c)}(0)|^2}{|\mathcal{E}^{(s)}(0)|^2} = \frac{\sin^2(\mu L)}{\cos^2(\mu L) + (\kappa^2/4|\Gamma|^2)}. \tag{15.45}$$

In the limit $\Delta k = 0$, the NLR effects detune the four-wave mixing process and reduce the intensity of the conjugate beam. However, if $\Delta k = -k_{vac}n_{2||}(-\omega; \omega)(I_2 - I_1)$ so that $\kappa = 0$ and the pump beams are not of equal intensity, it is possible to use the NLR at a specific intensity difference to cancel out the wave-vector mismatch.

15.2 DEGENERATE THREE-WAVE MIXING

The designation three-wave mixing is somewhat misleading since the number of photons involved in energy and wave-vector conservation is actually 4. This is a geometry used for thin samples with two pump beams incident symmetrically displaced by small angles θ from the normal to the surface of the sample, as shown in Fig. 15.5.

For simplicity it is assumed that the two pump beams are copolarized along the x-axis. The two pump beam wave vectors are $\vec{k}_{p1} = (k_x, k_y, k_z) = k(0, -\sin\theta, \cos\theta)$ and $\vec{k}_{p2} = k(0, \sin\theta, \cos\theta)$. The nonlinear mixing terms of interest are $[E^{(p1)}]^2 E^{*(p2)}$ and $[E^{(p2)}]^2 E^{*(p1)}$, and so there are two input eigenmodes and assuming a Kerr nonresonant response the nonlinear polarization is

$$P^{(3)}(z) = \frac{3}{4}\varepsilon_0\tilde{\chi}^{(3)}_{xxxx}(-\omega; \omega, -\omega, \omega)$$
$$\{[\mathcal{E}^{(p1)}]^2\mathcal{E}^{*(p2)}\, e^{i(2\vec{k}_{p1}-\vec{k}_{p2})\cdot\vec{r}} + [\mathcal{E}^{(p2)}]^2\mathcal{E}^{*(p1)}\, e^{i(2\vec{k}_{p2}-\vec{k}_{p1})\cdot\vec{r}}\} \tag{15.46}$$

Assuming for simplicity no pump depletion and approximately the nonresonant limit and substituting for $\tilde{\chi}^{(3)}_{xxxx}(-\omega; \omega, -\omega, \omega)$ in terms of $n_{2||}(-\omega; \omega)$ in regions where two-photon absorption is negligible, the nonlinear polarizations for the signal beams are

$$P^{(s1)}(z) = [n\varepsilon_0]^2 cn_{2||}(-\omega; \omega)[\mathcal{E}^{(p2)}]^2\mathcal{E}^{*(p1)}\, e^{i(2\vec{k}_{p2}-\vec{k}_{p1})\cdot\vec{r}},$$
$$P^{(s2)}(z) = [n\varepsilon_0]^2 cn_{2||}(-\omega; \omega)[\mathcal{E}^{(p1)}]^2\mathcal{E}^{*(p2)}\, e^{i(2\vec{k}_{p1}-\vec{k}_{p2})\cdot\vec{r}}. \tag{15.47}$$

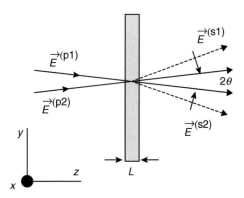

FIGURE 15.5 The beam–sample geometry for three-wave mixing.

Inserting these into the SVEA gives

$$\frac{d\mathcal{E}^{(s1)}(z)}{dz} = i\kappa_{s1}\, e^{i(2\vec{k}_{p2}-\vec{k}_{p1}-\vec{k}_{s1})\cdot\vec{r}}, \quad \kappa_{s1} = \frac{\omega n \varepsilon_0 n_{2\|}(-\omega;\,\omega)}{2}[\mathcal{E}^{(p2)}]^2 \mathcal{E}^{*(p1)},$$

$$\frac{d\mathcal{E}^{(s2)}(z)}{dz} = i\kappa_{s2}\, e^{i(2\vec{k}_{p1}-\vec{k}_{p2}-\vec{k}_{s2})\cdot\vec{r}}, \quad \kappa_{s2} = \frac{\omega n \varepsilon_0 n_{2\|}(-\omega;\,\omega)}{2}[\mathcal{E}^{(p1)}]^2 \mathcal{E}^{*(p2)}.$$

$$(15.48)$$

Clearly these two interactions cannot be exactly wave-vector matched. Defining

$$2\vec{k}_{p2} - \vec{k}_{p1} = k(0,\ 3\sin\theta,\ \cos\theta), \qquad 2\vec{k}_{p1} - \vec{k}_{p2} = k(0,\ -3\sin\theta,\ \cos\theta)$$

$$(15.49)$$

and assuming that the beams are much wider in the x–y plane than a wavelength, wave vector is conserved in the x–y plane. For the signal field $E^{(s1)}$, which must be a solution to the wave equation,

$$k_z^2 = k^2 - k_y^2 = k^2 - 9k^2\sin^2\theta = k^2[1 - 9\sin^2\theta]$$

$$\rightarrow \quad k_z = k\left(1 - \frac{9}{2}\sin^2\theta\right) \cong k\left(1 - \frac{9}{2}\theta^2\right). \qquad (15.50)$$

Similarly for $E^{(s2)}$, $k_z \cong k(1 - \frac{9}{2}\theta^2)$. Therefore the wave-vector mismatch along the z-axis is given by

$$\Delta k_z = 2k_{z,p2} - k_{z,p1} - k_{z,s1} = 2k_{z,p1} - k_{z,p2} - k_{z,s2} = k\left(\cos\theta - 1 + \frac{9}{2}\theta^2\right) = 4k\theta^2,$$

$$(15.51)$$

and the solution to Eqs 15.48 are

$$\mathcal{E}^{(s1)}(L) = i\kappa_{s1}L\, e^{i\Delta k_z L/2} \operatorname{sinc}\left(\frac{\Delta k_z L}{2}\right), \qquad \mathcal{E}^{(s2)}(L) = i\kappa_{s2}L\, e^{i\Delta k_z L/2} \operatorname{sinc}\left(\frac{\Delta k_z L}{2}\right)$$

$$(15.52)$$

for the field amplitudes and

$$I^{(s1)}(L) = \left[k_{vac}Ln_{2\|}(-\omega;\,\omega)\operatorname{sinc}\left(\frac{\Delta k_z L}{2}\right)I^{(p2)}(0)\right]^2 I^{(p1)}(0),$$

$$I^{(s2)}(L) = \left[k_{vac}Ln_{2\|}(-\omega;\,\omega)\operatorname{sinc}\left(\frac{\Delta k_z L}{2}\right)I^{(p1)}(0)\right]^2 I^{(p2)}(0).$$

$$(15.53)$$

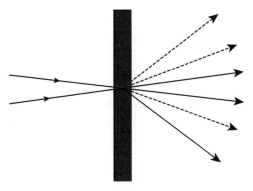

FIGURE 15.6   Multiple output beams in degenerate three-wave mixing.

for the intensities. For small angles θ, the sinc^2 term becomes approximately unity and

$$
\begin{aligned}
I^{(\mathrm{s}1)}(L) &= [k_{\mathrm{vac}}Ln_{2\|}(-\omega;\,\omega)I^{(\mathrm{p}2)}(0)]^2 I^{(\mathrm{p}1)}(0), \\
I^{(\mathrm{s}2)}(L) &= [k_{\mathrm{vac}}Ln_{2\|}(-\omega;\,\omega)I^{(\mathrm{p}1)}(0)]^2 I^{(\mathrm{p}2)}(0).
\end{aligned}
\tag{15.54}
$$

If there is also a significant two-photon absorption, similar to the D4WM case, then

$$
\frac{I^{(\mathrm{s}1)}(L)}{I^{(\mathrm{p}2)}(0)} = \frac{I^{(\mathrm{s}2)}(L)}{I^{(\mathrm{p}1)}(0)} = [k_{\mathrm{vac}}L]^2 \left\{ n_{2\|}^2(-\omega;\,\omega) + \frac{\alpha_{2\|}^2(-\omega;\,\omega)}{4k_{\mathrm{vac}}^2} \right\} I^{(\mathrm{p}1)}(0) I^{(\mathrm{p}2)}(0)
$$

$$
= [k_{\mathrm{vac}}L]^2 \left\{ \Delta n_{\mathrm{p}1}^{\mathrm{NL}} \Delta n_{\mathrm{p}2}^{\mathrm{NL}} + \frac{\Delta \alpha_{\mathrm{p}1}^{\mathrm{NL}} \Delta \alpha_{\mathrm{p}2}^{\mathrm{NL}}}{4k_{\mathrm{vac}}^2} \right\}.
\tag{15.55}
$$

The signal beams can then mix with the pump beams to produce new signal beams at even larger angles from the surface normal etc. (as shown in Fig. 15.6), with successive beams being much weaker. Occasionally in the literature these extra beams have erroneously been interpreted as due to higher order nonlinearities.

15.3   NONDEGENERATE WAVE MIXING

This section discusses the wave mixing interactions in which some of the input and output frequencies are not the same. This invariably makes wave-vector matching a much more difficult process.

15.3.1 All-Optical Nondegenerate Wave Mixing

One of the most attractive features of the degenerate four-wave mixing process just discussed above is that it is automatically wave-vector matched if the oppositely propagating pump beam geometry is correct. Because the frequencies are equal, the tolerances on a pump beam misalignment by $\Delta\theta$ gives a wave-vector mismatch proportional to $\Delta\theta^2/2$, as discussed previously. In the most general nondegenerate case with frequency inputs ω_1, ω_2, ω_3, and ω_4 in which ω_1 and ω_2 are the pump beams, for wave-vector and frequency conservation,

$$\vec{k}(\omega_1) + \vec{k}(\omega_2) + \vec{k}(\omega_3) + \vec{k}(\omega_4) = \frac{1}{c}\left\{ \omega_1 n(\omega_1) \frac{\vec{k}(\omega_1)}{|k(\omega_1)|} + \omega_2 n(\omega_2) \frac{\vec{k}(\omega_2)}{|\vec{k}(\omega_2)|} \right.$$

$$\left. +\omega_3 n(\omega_3) \frac{\vec{k}(\omega_3)}{|k(\omega_3)|} + \omega_4 n(\omega_4) \frac{\vec{k}(\omega_4)}{|\vec{k}(\omega_4)|} \right\} = 0 \quad (15.56a)$$

and

$$\omega_1 + \omega_2 = \omega_3 + \omega_4, \quad (15.56b)$$

respectively, must be satisfied. Satisfying both conditions is, in principle, possible in a bulk sample for arbitrary frequencies, but for efficient interactions the allowed wave-vector tolerances on the angles for wave-vector matching will be very small unless all the frequencies are almost equal.

Most of the observations of nondegenerate mixing with all four optical beams have been achieved with a special condition on the frequencies and wave vectors, namely, only one input frequency so that all the beams travel in the same direction and

$$2\omega_1 = \omega_3 + \omega_4, \quad (15.57)$$

$$2\omega_1 n(\omega_1) = \omega_3 n(\omega_3) + \omega_4 n(\omega_4). \quad (15.58)$$

If there were no dispersion in the refractive index with frequency, satisfying Eq. 15.58 would trivially be the same as satisfying Eq. 15.57, but of course that is not the case.

Now a specific case is assumed that will show how NLR can be used to wave-vector match a nondegenerate wave mixing process. All the plane-wave optical fields are assumed to propagate along the z-axis (see Fig. 15.7) and are written as

$$\vec{E}^{(1)} = \frac{1}{2}\vec{\mathcal{E}}(\omega_1)\,e^{i(k_1 z - \omega_1 t)} + \text{c.c.}, \qquad \vec{E}^{(3)} = \frac{1}{2}\vec{\mathcal{E}}(\omega_3)\,e^{i(k_3 z - \omega_3 t)} + \text{c.c.},$$

$$\vec{E}^{(4)} = \frac{1}{2}\vec{\mathcal{E}}(\omega_3)\,e^{i(k_4 z - \omega_4 t)} + \text{c.c.} \quad (15.59)$$

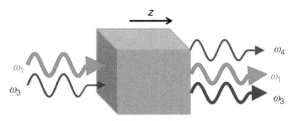

FIGURE 15.7   The geometry for the nondegenerate three-wave mixing interaction.

Assuming copolarized beams, the Kerr effect near the nonresonant regime, and cross-NLR only due to the pump beams, the signal and idler ($\equiv$ conjugate) beam nonlinear polarizations are, respectively, given by

$$\mathcal{P}^{(\mathrm{NL})}(\omega_3) = 2\frac{6}{4n_1c}\tilde{\chi}^{(3)}_{xxxx}(-\omega_3;\omega_1,-\omega_1,\omega_3)I_1\mathcal{E}(\omega_3)$$

$$+\frac{3}{4}\varepsilon_0\tilde{\chi}^{(3)}_{xxxx}(-\omega_3;\omega_1,\omega_1,-\omega_4)[\mathcal{E}(\omega_1)]^2\mathcal{E}^*(\omega_4)\,e^{i[\Delta k+2n_{2\|}(-\omega_1;\omega_1)k_{\mathrm{vac}}(\omega_1)I_1]z}\Big\},$$

$$(15.60)$$

$$\mathcal{P}^{(\mathrm{NL})}(\omega_4) = 2\frac{6}{4n_1c}\tilde{\chi}^{(3)}_{xxxx}(-\omega_4;\omega_1,-\omega_1,\omega_4)I_1\mathcal{E}(\omega_4)$$

$$+\frac{3}{4}\varepsilon_0\tilde{\chi}^{(3)}_{xxxx}(-\omega_4;\omega_1,\omega_1,-\omega_3)[\mathcal{E}(\omega_1)]^2\mathcal{E}^*(\omega_3)\,e^{i[\Delta k+2n_{2\|}(-\omega_1;\omega_1)k_{\mathrm{vac}}(\omega_1)I_1]z}\Big\},$$

$$(15.61)$$

where $\Delta k = 2k(\omega_1) - k(\omega_3) - k(\omega_4) = 2k_{\mathrm{vac}}(\omega_1)n_1 - k_{\mathrm{vac}}(\omega_3)n_3 - k_{\mathrm{vac}}(\omega_4)n_4$, setting the input pump beam phase to zero so that $[\mathcal{E}(\omega_1)]^2 = |\mathcal{E}(\omega_1)|^2$, and I_1 is the intensity of the input pump mode. Defining the parameters along the lines of Eq. 15.40, namely,

$$\delta_{3,4} = 2k_{\mathrm{vac}}(\omega_{3,4})\frac{n_1}{n_{3,4}}n_{2\|}(-\omega_1;\omega_1)I_1,\quad \kappa = \Delta k + 2k_{\mathrm{vac}}(\omega_1)n_1n_{2\|}(-\omega_1;\omega_1)I_1,$$

$$\Gamma_{3,4} = k_{\mathrm{vac}}(\omega_{3,4})\frac{n_1}{n_{3,4}}n_{2\|}(-\omega_1;\omega_1)I_1\,e^{2i\psi(\omega_1)}, \tag{15.62}$$

and inserting the nonlinear polarizations into the SVEA, we obtain

$$\frac{d\mathcal{E}(\omega_3,z)}{dz} = i\delta_3\mathcal{E}(\omega_3,z) + i\Gamma_3\mathcal{E}^*(\omega_4,z)e^{i\kappa z},$$

$$\frac{d\mathcal{E}(\omega_4,z)}{dz} = i\delta_4\mathcal{E}(\omega_4,z) + i\Gamma_4\mathcal{E}^*(\omega_3,z)e^{i\kappa z}. \tag{15.63}$$

Equations (15.63) are quite similar (except for a minus sign) to Eqs 15.41. Again using a similar substitution for the fields, namely,

$$\mathcal{E}(\omega_3, z) = B_3(z)e^{i\delta_3 z}, \quad \mathcal{E}(\omega_4, z) = B_4(z)e^{i\delta_4 z}, \tag{15.64}$$

gives

$$\left\{\frac{dB_3}{dz} + i\delta B_3\right\}e^{i\delta_3 z} = i\delta_3 B_3 e^{i\delta_3 z} + i\Gamma_3 B_4^* e^{-i\delta_4 z + i\kappa z} \quad \rightarrow \quad \frac{dB_3}{dz} = i\Gamma_3 B_4^* e^{i[\kappa - (\delta_3 + \delta_4)]z},$$

$$\tag{15.65a}$$

$$\left\{\frac{dB_4}{dz} + i\delta_4 B_4\right\}e^{i\delta_4 z} = i\delta_4 B_4 e^{i\delta_4 z} + i\Gamma_4 B_3^* e^{i\kappa z - i\delta_3 z} \quad \rightarrow \quad \frac{dB_4}{dz} = i\Gamma_4 B_3^* e^{i[\kappa - (\delta_4 + \delta_3)]z}.$$

$$\tag{15.65b}$$

Taking the d/dz derivative of Eq. 15.65a and substituting Eq. 15.65b for dB_4^*/dz and Eq. 15.65a for B_4^* gives

$$\frac{d^2 B_3}{dz^2} = i\Gamma_3 \left\{\frac{dB_4^*}{dz}e^{i[\kappa - (\delta_3 + \delta_4)]z} + i[\kappa - (\delta_3 + \delta_4)]B_4^* e^{i[\kappa - (\delta_3 + \delta_4)]z}\right\}$$

$$\tag{15.66a}$$

$$\rightarrow \quad \frac{d^2 B_3}{dz^2} = \Gamma_3 \Gamma_4 B_3 + i[\kappa - (\delta_3 + \delta_4)]\frac{dB_3}{dz}.$$

Similarly for B_4,

$$\frac{d^2 B_4}{dz^2} = \Gamma_3 \Gamma_4 B_3 + i[\kappa - (\delta_3 + \delta_4)]\frac{dB_3}{dz}. \tag{15.66b}$$

The roots of these differential equations are

$$\mu_\pm = \frac{i(\kappa + \delta_3 - \delta_4)}{2} \pm \frac{1}{2}\sqrt{-[\kappa - (\delta_3 + \delta_4)]^2 + 4\Gamma_3\Gamma_4} = \frac{i(\kappa + \delta_3 - \delta_4)}{2} \pm g. \tag{15.67}$$

The signal and idler grow exponentially for

$$4\Gamma_3\Gamma_4 > [\kappa - (\delta_3 + \delta_4)]^2 \tag{15.68}$$

and for $[\kappa - (\delta_3 + \delta_4)]^2 > 4\Gamma_3\Gamma_4$, where g is imaginary and the solutions are oscillatory. The solutions to Eq. 15.66a for B_3 and B_4 including the boundary conditions $\mathcal{E}(\omega_4, 0) = 0, \mathcal{E}(\omega_3, 0) \neq 0$ for the condition defined by Eq. 15.68 are

$$\mathcal{E}(\omega_3, z) = \mathcal{E}(\omega_3, 0)\left[\cosh(zg) - i\frac{\kappa - (\delta_3 + \delta_4)}{2g}\sinh(gz)\right]e^{-i\{[\kappa - (\delta_3 + \delta_4)]/2\}z},$$

$$\tag{15.69}$$

$$\mathcal{E}(\omega_4, z) = i\frac{\Gamma_4}{g}\mathcal{E}^*(\omega_3, 0)\sinh(gz)e^{-i\{[\kappa - (\delta_3 + \delta_4)]/2\}z}.$$

The calculation of the actual gain coefficient is complicated and tedious and does not lead to simple analytical results. Assuming that $\Delta\omega = \omega_4 - \omega_1 = \omega_1 - \omega_3 \ll \omega_1$, a number of simplifications can be made that give insights into the conditions for gain. Expanding Δk around $k(\omega_1)$ using Eq. 13.27 gives

$$2k(\omega_1) - k(\omega_3) - k(\omega_4) \cong -k_2(\omega_1)\Delta\omega^2 \tag{15.70}$$

and so the sign of Δk is negative in the normal dispersion region and positive in the anomalous dispersion region. The condition for gain can be written as

$$\left[2\frac{n_1}{\sqrt{n_3 n_4}}\sqrt{k_{\text{vac}}(\omega_3)k_{\text{vac}}(\omega_4)}n_{2\parallel}(-\omega_1;\omega_1)I_1\right]^2$$
$$-\left\{\Delta k - n_{2\parallel}(-\omega_1;\omega_1)I_1\left[2k_{\text{vac}}(\omega_1)-\frac{n_1}{n_3}k_{\text{vac}}(\omega_3)-\frac{n_1}{n_4}k_{\text{vac}}(\omega_3)\right]\right\}^2 > 0. \tag{15.71}$$

Expanding further the right-hand side gives

$$2k_{\text{vac}}(\omega_1)-\frac{n_1}{n_3}k_{\text{vac}}(\omega_3)-\frac{n_1}{n_4}k_{\text{vac}}(\omega_4)$$
$$\cong k_{\text{vac}}(\omega_1)\left[2-n_1\frac{n_3+n_4}{n_3 n_4}\right]+\frac{1}{2}k_1 n_1\frac{n_4-n_3}{n_3 n_4}\Delta\omega-\frac{1}{6}k_2 n_1\frac{n_3+n_4}{n_3 n_4}\Delta\omega^2, \tag{15.72}$$

and the left-hand side

$$\sqrt{k_{\text{vac}}(\omega_3)k_{\text{vac}}(\omega_4)}\cong\sqrt{\left[k_{\text{vac}}(\omega_1)-\frac{\Delta\omega}{c}\right]\left[k_{\text{vac}}(\omega_1)+\frac{\Delta\omega}{c}\right]}=\sqrt{k_{\text{vac}}^2(\omega_1)-\frac{\Delta\omega^2}{c^2}}\cong k_{\text{vac}}(\omega_1) \tag{15.73}$$

and $n_1/\sqrt{n_3 n_4} \cong 1$, which gives for Eq. 15.71

$$[2k_{\text{vac}}(\omega_1)n_{2\parallel}(-\omega_1;\omega_1)I_1]^2-\left\{-\frac{1}{2}k_2\Delta\omega^2-n_{2\parallel}(-\omega_1;\omega_1)I_1\right.$$
$$\left.\left[k_{\text{vac}}(\omega_1)\left\{2-\left(\frac{n_1}{n_3}+\frac{n_1}{n_4}\right)\right\}+\left(\frac{n_1}{n_3}-\frac{n_1}{n_4}\right)k_1\Delta\omega-\frac{1}{2}k_2\left(\frac{n_1}{n_3}+\frac{n_1}{n_4}\right)\Delta\omega^2\right]\right\}^2>0.$$
$$\tag{15.74}$$

Noting that

$$2k_{vac}(\omega_1) \gg \left[k_{vac}(\omega_1) \left\{ 2 - \left(\frac{n_1}{n_3} + \frac{n_1}{n_4} \right) \right\} + \left(\frac{n_1}{n_3} - \frac{n_1}{n_4} \right) k_1 \Delta\omega - \frac{1}{2}k_2 \left(\frac{n_1}{n_3} + \frac{n_1}{n_4} \right) \Delta\omega^2 \right] \right\},$$

Eq. 15.74 can be written as

$$\left[2k_{vac}(\omega_1)n_{2\parallel}(-\omega_1;\omega_1)I_1 - \frac{1}{2}k_2\Delta\omega^2 \right] \left[2k_{vac}(\omega_1)n_{2\parallel}(-\omega_1;\omega_1)I_1 + \frac{1}{2}k_2\Delta\omega^2 \right] > 0.$$

$$(15.75)$$

Gain occurs for both signs of the GVD and the nonlinearity, provided that the intensity exceeds the threshold value

$$2k_{vac}(\omega_1)|n_{2\parallel}(-\omega_1;\omega_1)|I_1 = |\Delta k| = \frac{1}{2}|k_2|\Delta\omega^2. \qquad (15.76)$$

Since in the nonresonant limit $2n_{2\parallel}(-\omega_1;\omega_1) = n_{2\parallel}(-\omega_3;\omega_1) = n_{2\parallel}(-\omega_4;\omega_1)$, the interpretation of Eq. 15.76 is obvious; i.e., the cross-phase NLR due to the pump beam must exceed the index detuning from the resonance for gain to occur.

15.3.1.1 Nondegenerate Two-Photon Vibrational Resonance.

The wave interaction case considered here involves not only optical fields but also a normal mode of the medium, namely, the vibration of a molecule. How such a vibration changes the polarizability of a molecule was discussed in Section 12.3. The wave mixing of interest here is a two-photon absorption process (sum frequency resonance) within the vibrational manifold of the ground state of a CO_2 molecule (same as in Section 12.3) and requires incident beams in the infrared. As will be clear, this is a multiwave mixing process.

The incident optical field is written as

$$\vec{E}_{inc}(\vec{r}, t) = \frac{1}{2}\vec{\mathcal{E}}(\omega_a) e^{i(\vec{k}_a \cdot \vec{r} - \omega_a t)} + \text{c.c.} \qquad (15.77)$$

Following the procedure outlined in Section 12.3, the oxygen displacements are given by

$$\ddot{q} + 2\bar{\tau}_v^{-1}\dot{q} + \bar{\Omega}_v^2 \bar{q} = \frac{1}{2\bar{m}_{eff}} \frac{\partial \bar{\alpha}}{\partial \bar{q}} \bigg|_{\bar{q}=0} [f^{(1)}(\omega_a)]^2 \vec{E} \cdot \vec{E}$$

$$\rightarrow \quad \bar{Q}(\vec{K}, \Omega) e^{i(\vec{K} \cdot \vec{r} - \Omega t)} + \text{c.c} = \frac{1}{\bar{m}_{eff}} \frac{\partial \bar{\alpha}}{\partial \bar{q}} \bigg|_{\bar{q}=0} [f^{(1)}(\omega_a)]^2 \frac{\vec{E} \cdot \vec{E}}{D(2\omega_a)},$$

$$D(2\omega_a) = \bar{\Omega}_v^2 - 4\omega_a^2 - i4\omega_a\bar{\tau}_v^{-1}. \qquad (15.78)$$

The field product $\vec{E} \cdot \vec{E}$ produces a response at zero and at $2\omega_a$ frequencies, and the harmonic term responsible for two-photon absorption is

$$\vec{E}(\vec{r}, t) \cdot \vec{E}(\vec{r}, t) = \frac{1}{4} \mathcal{E}^2(\omega_a) \, e^{2i(\vec{k}_a \cdot \vec{r} - \omega_a t)} + \text{c.c.}, \qquad (15.79)$$

which gives for the optically driven vibration

$$\bar{q}(\vec{r}, t) = \frac{1}{2} \left[\frac{\mathcal{E}^2(\omega_a)}{4\bar{m}_{\text{eff}} D(2\omega_a)} [f^{(1)}(\omega_a)]^2 \frac{\partial \bar{\alpha}}{\partial \bar{q}} \Big|_{\bar{q}=0} e^{2i(\vec{k}_a \cdot \vec{r} - \omega_a t)} + \text{c.c.} \right]. \qquad (15.80)$$

Note that although the vibrations are local to single molecule, there is a spatial periodicity (π/k_a) imposed on the vibrational displacements of individual molecules by the mixing of the light beam with itself. The nonlinear macroscopic polarization is

$$\vec{P}^{\text{NL}}(\vec{r}, t) = N\vec{q}(\vec{r}, t) \cdot \frac{\partial \overset{\leftrightarrow}{\bar{\alpha}}}{\partial \bar{q}} \Big|_{\bar{q}=0} \cdot \vec{E}(\vec{r}, t) f^{(1)}(\omega_a)$$

$$= N\bar{q}(\vec{r}, t) \cdot \frac{\partial \overset{\leftrightarrow}{\bar{\alpha}}}{\partial \bar{q}} \Big|_{\bar{q}=0} \cdot \frac{1}{2} \{ \mathcal{E}(\omega_a) \, e^{i(\vec{k}_a \cdot \vec{r} - \omega_a t)} + \vec{\mathcal{E}}^*(\omega_a) \, e^{-i(\vec{k}_a \cdot \vec{r} - \omega_a t)} \} f^{(1)}(\omega_a)$$

$$\propto \bar{Q}(\omega_a) \vec{\mathcal{E}}^*(\omega_a) \, e^{-i(\vec{k}_a \cdot \vec{r} - \omega_a t)} + \text{c.c.} \qquad (15.81)$$

Finally this gives

$$\mathcal{P}^{\text{NL}}(\omega_a) = \frac{N}{8\bar{m}_{\text{eff}}} \left[\frac{\partial \bar{\alpha}}{\partial \bar{q}} \Big|_{\bar{q}=0} \right]^2 f^{(3)} \frac{|\mathcal{E}(\omega_a)|^2 \mathcal{E}(\omega_a)}{\bar{\Omega}_v^2 - 4\omega_a^2 - 4i\omega_a \bar{\tau}_v^{-1}}, \quad f^{(3)} = [f^{(1)}(\omega_a)]^4. \quad (15.82)$$

Inserting this polarization into the SVEA yields

$$\frac{d\mathcal{E}(\omega_a)}{dz} = i \frac{\omega_a N}{16 n_a \bar{m}_{\text{eff}} \varepsilon_0 c} \left[\frac{\partial \bar{\alpha}}{\partial \bar{q}} \Big|_{\bar{q}=0} \right]^2 f^{(3)} \frac{|\mathcal{E}(\omega_a)|^2 \mathcal{E}(\omega_a)}{D(2\omega_a)}. \qquad (15.83)$$

Near resonance,

$$\frac{1}{D(2\omega_a)} = \frac{\bar{\Omega}_v^2 - 4\omega_a^2 + i4\omega_a \bar{\tau}_v^{-1}}{[\bar{\Omega}_v^2 - 4\omega_a^2]^2 + (4\omega_a \bar{\tau}_v^{-1})^2} \underset{\text{near resonance}}{\simeq} \frac{\bar{\Omega}_v - 2\omega_a + i\bar{\tau}_v^{-1}}{(\bar{\Omega}_v - 2\omega_a)^2 + \bar{\tau}_v^{-2}} \frac{1}{4\omega_a} \qquad (15.84)$$

$$\Rightarrow \frac{dI(\omega_a)}{dz} = -\alpha_{2||}(-\omega_a; \omega_a)I^2(\omega_a),$$

$$\alpha_{2||}(-\omega; \omega) = \frac{N\bar{\tau}_v^{-1}}{16 n_a^2 \bar{m}_{\text{eff}} \varepsilon_0^2 c^2} \left[\frac{\partial\bar{\alpha}}{\partial\bar{q}}\bigg|_{\bar{q}=0}\right]^2 f^{(3)} \frac{1}{(\bar{\Omega}_v - 2\omega_a)^2 + \bar{\tau}_v^{-2}} \qquad (15.85a)$$

and

$$\frac{d\mathcal{E}(\omega_a)}{dz} = i k_{\text{vac}} n_{2||}(-\omega_a; \omega_a)I(\omega_a)\mathcal{E}(\omega_a),$$

$$\rightarrow n_{2||}(-\omega_a; \omega_a) = \frac{N}{32 n_a^2 \varepsilon_0^2 \bar{m}_{\text{eff}} c\omega_a} \left[\frac{\partial\bar{\alpha}}{\partial\bar{q}}\bigg|_{\bar{q}=0}\right]^2 f^{(3)} \frac{\bar{\Omega}_v - 2\omega_a}{(\bar{\Omega}_v - 2\omega_a)^2 + \bar{\tau}_v^{-2}}. \qquad (15.85b)$$

It is important to point out that for this two-photon resonance, there is also dispersion in the nonlinear index coefficient, as discussed in Chapter 10.

15.3.1.2 Nonlinear Raman Spectroscopy.

The processes described next involve the difference frequency created by the mixing of two input beams being approximately equal to the vibration frequency of a molecule. The resulting resonance at the difference frequency can be used as a tool to identify vibrational modes since it leads to an enhancement of the appropriate nonlinear susceptibility. Because the process is nonlinear, it can attain monolayer sensitivity with intense inputs (1). The input fields are written as

$$\vec{E}_{\text{inc}} = \frac{1}{2}\left(\vec{\mathcal{E}}(\omega_a)\, e^{i(\vec{k}_a \cdot \vec{r} - \omega_a t)} + \text{c.c.} + \vec{\mathcal{E}}(\omega_b)\, e^{i(\vec{k}_b \cdot \vec{r} - \omega_b t)} + \text{c.c.}\right). \qquad (15.86)$$

15.3.1.2.1 Raman-Induced Kerr Effect Spectroscopy.

This process leads to a Raman-induced birefringence in an isotropic medium with enhancement occurring when the difference frequency matches a vibrational mode of the medium. The pump field $\vec{E}(\omega_a)$ polarized along the x-axis makes the medium birefringent for the field $\vec{E}(\omega_b)$ polarized along the y-axis with $\omega_a > \omega_b$. Note that the process uses an off-diagonal element of the hyperpolarizability tensor. The molecular and macroscopic nonlinear polarizability is given by

$$\bar{p}_i = \left[\bar{\alpha}_{ij}^L + \bar{q}_n \frac{\partial\bar{\alpha}_{ijn}}{\partial\bar{q}_n}\bigg|_{\bar{q}_n=0}\right] E_{\text{loc},j}(\omega_b)$$

$$\Rightarrow P_i^{\text{NL}} = N \frac{\partial\bar{\alpha}_{ijn}}{\partial q_n}\bigg|_{\bar{q}_n=0} \bar{q}_n f^{(1)}(\omega_{\text{NL}}) f^{(1)}(\omega_b) E_j(\omega_b). \qquad (15.87)$$

The interaction potential

$$\bar{V}_{\text{int}} \propto -\int \bar{p}_i \cdot dE_{i,\text{loc}}(\omega_a) = -\bar{q}_n \frac{\partial \bar{\alpha}_{ijn}}{\partial \bar{q}_n}\Big|_{\bar{q}_n=0} f^{(1)}(\omega_a) f^{(1)}(\omega_b) E_i(\omega_a) E_j(\omega_b)$$

$$(15.88)$$

gives the nonlinear force on the vibrating masses

$$\bar{F}_n = -\frac{\partial \bar{V}_{\text{int}}}{\partial \bar{q}_n} = N \frac{\partial \bar{\alpha}_{ijn}}{\partial \bar{q}_n}\Big|_{\bar{q}_n=0} f^{(1)}(\omega_b) f^{(1)}(\omega_a) E_i(\omega_a) E_j(\omega_b), \qquad (15.89)$$

which drives the vibrational modes via

$$\ddot{\bar{q}}_n + 2\bar{\tau}_v^{-1}\dot{\bar{q}}_n + \bar{\Omega}_v^2 \bar{q}_n = \frac{1}{\bar{m}_{\text{eff}}} \frac{\partial \bar{\alpha}_{ijn}}{\partial \bar{q}_n}\Big|_{\bar{q}_n=0} f^{(1)}(\omega_b) f^{(1)}(\omega_a) E_i(\omega_a) E_j(\omega_b). \qquad (15.90)$$

Solving for the resonantly enhanced amplitude of the vibration gives

$$\bar{q}_n(\vec{r},t) = \frac{1}{4\bar{m}_{\text{eff}}} \frac{\partial \bar{\alpha}_{xyn}}{\partial \bar{q}_n}\Big|_{\bar{q}_n=0} f^{(1)}(\omega_a) f^{(1)}(\omega_b) \times \frac{\mathcal{E}_x^*(\omega_a)\mathcal{E}_y(\omega_b)}{D^*(\omega_a - \omega_b)} e^{i[(\vec{k}_b - \vec{k}_a)\cdot\vec{r} - (\omega_b - \omega_a)t]} + \text{c.c.},$$

$$(15.91)$$

which is substituted into Eq. 15.89, giving

$$P_y^{\text{NL}}(\vec{r},t) = \frac{1}{4} N \frac{\partial \bar{\alpha}_{yxn}}{\partial \bar{q}_n}\Big|_{\bar{q}_n=0} f^{(1)}(\omega_{\text{NL}} = \omega_b) f^{(1)}(\omega_a) \left[\bar{Q}_n^*(\omega_a - \omega_b) e^{-i[(\vec{k}_a - \vec{k}_b)\cdot\vec{r} - (\omega_a - \omega_b)t]} + \text{c.c.} \right]$$

$$\times [\mathcal{E}_x(\omega_a) e^{i[\vec{k}_a \cdot \vec{r} - \omega_a t]} + \text{c.c.}]. \qquad (15.92)$$

Writing

$$P_y^{\text{NL}}(\vec{r},t) = \frac{N}{8\bar{m}_{\text{eff}}} \frac{\partial \bar{\alpha}_{yxn}}{\partial \bar{q}_n}\Big|_{\bar{q}_n=0} \times \frac{\partial \bar{\alpha}_{xyn}^*}{\partial \bar{q}_n}\Big|_{\bar{q}_n=0} f^{(3)} \frac{|\mathcal{E}_x(\omega_a)|^2 \mathcal{E}_y(\omega_b)}{D^*(\omega_a - \omega_b)} e^{i[\vec{k}_b \cdot \vec{r} - \omega_b t]} + \text{c.c}$$

$$(15.93)$$

with $f^{(3)} = [f^{(1)}(\omega_a)]^2 [f^{(1)}(\omega_b)]^2$ and inserting this polarization into the SVEA gives

$$\frac{d\mathcal{E}_y(\omega_b)}{dz} = i \frac{N\omega_b}{8n_a n_b \bar{m}_{\text{eff}} \varepsilon_0^2 c^2} \left| \frac{\partial \bar{\alpha}_{yxn}}{\partial \bar{q}_n}\Big|_{\bar{q}_n=0} \right|^2 f^{(3)} \frac{I(\omega_a)\mathcal{E}_y(\omega_b)}{D^*(\omega_a - \omega_b)}. \qquad (15.94)$$

There are two contributions implicit in Eq. 15.94:

1. An index change produced at frequency ω_b by the beam of frequency ω_a.
2. Nonlinear gain or loss induced in beam b by beam a. For difference frequencies $(\omega_a - \omega_b) \approx \bar{\Omega}_v$,

$$
\frac{1}{D^*(\omega_a - \omega_b)} = \frac{\bar{\Omega}_v^2 - (\omega_a - \omega_b)^2 - 2i(\omega_a - \omega_b)\bar{\tau}_v^{-1}}{[\bar{\Omega}_v^2 - (\omega_a - \omega_b)^2]^2 + (2[\omega_a - \omega_b]\bar{\tau}_v^{-1})^2}
$$
$$
\cong \frac{\bar{\Omega}_v - [\omega_a - \omega_b] - i\bar{\tau}_v^{-1}}{(\bar{\Omega}_v - [\omega_a - \omega_b])^2 + \bar{\tau}_v^{-2}} \frac{1}{2\bar{\Omega}_v}. \tag{15.95}
$$

After substituting Eq. 15.95 into Eq. 15.94, the resulting imaginary part leads to

$$
\frac{d\mathcal{E}_y(\omega_b)}{dz} = i\frac{\omega_b}{c} n_{2\perp}(-\omega_b; \omega_a) I_x(\omega_a)\mathcal{E}_y(\omega_b)
$$
$$
\rightarrow \quad n_{2\perp}(-\omega_b; \omega_a) = \frac{Nf^{(3)}}{8n_b n_a \bar{m}_{\text{eff}}\varepsilon_0^2 c\bar{\Omega}_v} \left|\frac{\partial\bar{\alpha}_{yxn}}{\partial\bar{q}_n}\right|_{\bar{q}_n=0}^2 \frac{\bar{\Omega}_v - [\omega_a - \omega_b]}{(\bar{\Omega}_v - [\omega_a - \omega_b])^2 + \bar{\tau}_v^{-2}}. \tag{15.96}
$$

The real component results in nonlinear gain at ω_b and nonlinear loss at ω_a; i.e., a photon from beam a breaks up into a b photon and an optical phonon.

$$
\frac{d\mathcal{E}_y(\omega_b)}{dz} = \frac{N\omega_b\bar{\tau}_v^{-1}f^{(3)}}{8n_b n_a \bar{m}_{\text{eff}}\varepsilon_0^2 c^2\bar{\Omega}_v} \left|\frac{\partial\bar{\alpha}_{yxn}}{\partial\bar{q}_n}\right|_{\bar{q}_n=0}^2 \frac{1}{(\bar{\Omega}_v - [\omega_a - \omega_b])^2 + \bar{\tau}_v^{-2}} I_x(\omega_a)\mathcal{E}_y(\omega_b). \tag{15.97}
$$

The field gain coefficient is given by

$$
\gamma_\perp(-\omega_b; \omega_a) = \frac{N\omega_b\bar{\Gamma}_v f^{(3)}}{8n_b n_a \bar{m}\varepsilon_0^2 c^2\bar{\Omega}_v} \left|\frac{\partial\bar{\alpha}_{yxn}}{\partial\bar{q}_n}\right|_{\bar{q}_n=0}^2 \frac{1}{(\bar{\Omega}_v - [\omega_a - \omega_b])^2 + (\bar{\Gamma}_v/2)^2}. \tag{15.98}
$$

Hence by modulating the intensity of beam a on and off, the intensity of beam b is also modulated (see Fig. 15.8):

$$
\frac{I(\omega_b, L)}{I(\omega_b, 0)} = e_\perp^{(-\omega_b; \omega_a)I(\omega_a)L} \cong 1 + \gamma_\perp(-\omega_b; \omega_a)I(\omega_a)L
$$
$$
\rightarrow \quad \frac{I(\omega_b, L) - I(\omega_b, 0)}{I(\omega_b, 0)} = \gamma_\perp(-\omega_b; \omega_a)I(\omega_a)L. \tag{15.99}
$$

Varying $\omega_a - \omega_b$ through Ω_v gives a resonance in the transmission of the ω_b beam.

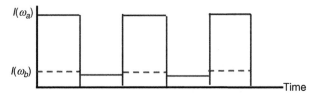

FIGURE 15.8 The modulation of the beam at frequency ω_b achieved by turning the beam a on and off.

15.3.1.2.2 Coherent Anti-Stokes Raman Scattering.

In this process there are two incident copolarized optical pump beams of frequency ω_a and ω_b ($\omega_b > \omega_a$) and a third beam of frequency $2\omega_b - \omega_a$ is generated. The physics of this process is shown in Fig. 15.9. Varying $\omega_b - \omega_a$ through the vibrational frequency Ω_v produces a sufficiently strong signal that monolayer thick films produce measurable signals (1).

The analysis is the same as in the two previous examples. The pertinent nonlinear polarization is $P^{NL}(\vec{r}, 2\omega_b - \omega_a) \propto \bar{Q}(\vec{r}, \omega_b - \omega_a)E(\vec{r}, \omega_b) \propto e^{-i(2\omega_b - \omega_a)t}$. The excited vibration for this case is given by

$$\bar{q}_n(\vec{r}, t) \propto \frac{1}{4\bar{m}_{\text{eff}}} \frac{\partial \bar{\alpha}_{iin}}{\partial \bar{q}_n}\Big|_{\bar{q}_n=0} \frac{E_i(\vec{r}, \omega_b)E_i^*(\vec{r}, \omega_a)}{D(\omega_b - \omega_a)} f^{(1)}(\omega_a)f^{(1)}(\omega_b) \, e^{-i(\omega_b - \omega_a)t} + \text{c.c.}$$

$$(15.100)$$

The nonlinear polarization is

$$\mathcal{P}_i^{NL}(2\omega_b - \omega_a) = \frac{N}{4\bar{m}_{\text{eff}}} \left|\frac{\partial \bar{\alpha}_{iin}}{\partial \bar{q}_n}\Big|_{\bar{q}_n=0}\right|^2 f^{(3)} \frac{\mathcal{E}_i^2(\omega_b)\mathcal{E}_i^*(\omega_a)}{D(\omega_b - \omega_a)}, \qquad (15.101)$$

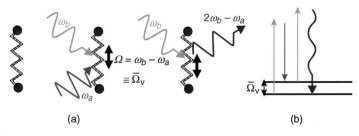

(a)   (b)

FIGURE 15.9 (a) Schematic of the coherent anti-Stokes Raman scattering process. Beams with frequencies ω_a and ω_b mix in a molecule with the difference frequency $\omega_b - \omega_a$ and excite a vibrational normal mode at this difference frequency. The ω_b beam scatters from (is modulated by) this vibration, and a new beam at the frequency $2\omega_b - \omega_a$ is generated and the vibration energy is reduced by $\hbar(\omega_b - \omega_a)$. (b) The energy level diagram showing the excitation of the molecule to its first vibrational level and then the scattering of the second beam to a higher energy photon.

which, when inserted into the SVEA formalism, gives

$$\frac{d\mathcal{E}_i(2\omega_b - \omega_a)}{dz} = i\frac{N\omega_c}{8n(\omega_c)c\varepsilon_0\bar{m}_{\mathrm{eff}}}\left|\frac{\partial\bar{\alpha}_{iin}}{\partial\bar{q}_n}\right|_{\bar{q}_n=0}^2 f^{(3)}\frac{\mathcal{E}_i^2(\omega_b)\mathcal{E}_i^*(\omega_a)}{D(\omega_b - \omega_a)}e^{i\Delta\vec{k}\cdot\vec{r}},$$

with $f^{(3)} = f^{(1)}(\omega_a)[f^{(1)}(\omega_b)]^2 f^{(1)}(2\omega_b - \omega_a).$ $\qquad$ (15.102)

The field at $\omega_c = 2\omega_a - \omega_b$ is written as $\frac{1}{2}\mathcal{E}_i(\omega_c)e^{i[k(\omega_c)z - \omega_c t]} + \mathrm{c.c.}$, and the output signal in the negligible pump beam depletion regime is

$$\mathcal{E}_i(L,\omega_c) = i\frac{N\omega_c L}{8n_c c\varepsilon_0\bar{m}_{\mathrm{eff}}}\left|\frac{\partial\bar{\alpha}_{iin}}{\partial\bar{q}_n}\right|_{\bar{q}_n=0}^2 f^{(3)}\frac{\mathcal{E}_i^2(\omega_b)\mathcal{E}_i^*(\omega_a)}{D(\omega_b - \omega_a)}e^{i\Delta k(L/2)}\mathrm{sinc}\left(\Delta k\frac{L}{2}\right)$$

$$I(L,\omega_c) = \frac{N^2\omega_c^2 L^2}{64n_c n_b^2 n_a c^4 \varepsilon_0^4 \bar{m}_{\mathrm{eff}}^2 \bar{\Omega}_v^2}\left|\frac{\partial\bar{\alpha}_{iin}}{\partial\bar{q}_n}\right|_{\bar{q}_n=0}^4 [f^{(3)}]^2\frac{\mathrm{sinc}^2(\Delta k\frac{L}{2})I^2(\omega_b)I(\omega_a)}{[\bar{\Omega}_v - (\omega_b - \omega_a)]^2 + \bar{\tau}_v^{-2}}.$$

$$(15.103)$$

As indicated by the last two equations, this is a coherent process that needs to be phase matched for optimum efficiency. Because of the normal dispersion in the refractive index in the visible region (see Fig. 15.10a), codirectional beams do not result in phase matching. However, phase matching is possible with non-codirectional beams (see Fig. 15.10b).

There is also a "background" nonlinear contribution due to the usual electronic nonlinearity, i.e., $\hat{\chi}_{iiii}^{(3)}(-[2\omega_b - \omega_a]; \omega_b, -\omega_a, \omega_b)$. The typical signal observed for *comparable* contributions of the resonant and background components is shown in Fig. 15.10c.

Finally, note there is also an equivalent process, called coherent Stokes Raman scattering, by which the output signal occurs at a lower frequency ($\omega_c = 2\omega_a - \omega_b$) than does either input signal. The wave-vector matching conditions are different from those for coherent anti-Stokes Raman scattering and so only one process can occur at a time.

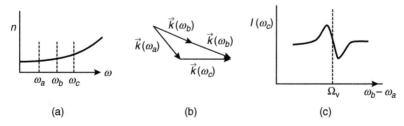

(a) $\qquad$ (b) $\qquad$ (c)

FIGURE 15.10 (a) Normal dispersion in the refractive index. Because of the curvature, codirectional wave-vector matching is not possible. (b) Noncollinear wave-vector matching for coherent anti-Stokes Raman scattering (CARS). (c) CARS signal when the background electronic nonlinearity is comparable to the vibrational resonance nonlinearity.

PROBLEMS

1. Consider a degenerate four-wave mixing experiment with the signal beam incident at an angle θ to the direction of one of the pump beams. The signal beam is orthogonally polarized to the two copolarized pump beams. Assuming an isotropic medium, calculate the reflectivity, i.e., the ratio of the conjugate beam intensity to the incident signal beam intensity, for the case that the pump beams are polarized

   (a) in the plane of wave vectors and

   (b) perpendicular to the plane of wave vectors.

   For what crystal classes can three polarizations be the same, and the fourth orthogonal to them?

2. Assume that the pump beams in D4WM are slightly misaligned to give along the z-axis a wave-vector mismatch $\Delta k = -(k_{p1} + k_{p2} + k_s + k_c)$.

   (a) Show that the evolution of the signal and conjugated beams is given by

   $$\frac{d\mathcal{E}^{(c)}(z)}{dz} = -i\frac{1}{\ell_{4WM}}\mathcal{E}^{*(s)}(z)\,e^{i\Delta kz}, \qquad \frac{d\mathcal{E}^{(s)}(z)}{dz} = i\frac{1}{\ell_{4WM}}\mathcal{E}^{*(c)}(z)\,e^{i\Delta kz}.$$

   (b) Show that the (intensity) degenerate four-wave mixing reflection coefficient is given by

   $$R = \frac{\sin^2(\mu L)}{\cos^2(\mu L) + [\Delta k\ell_{4WM}/2]^2}.$$

3. Show in D4WM that if linear absorption α (intensity) exists at the degenerate four-wave mixing frequency, the degenerate four-wave mixing reflectivity is given by

   $$R = \frac{\sin^2(\mu L)}{\ell_{4WM}^2[\mu \cos(\mu L) + \{\alpha/2\}\sin(\mu L)]^2},$$

   $$\frac{1}{\ell_{4WM}} = \omega n\varepsilon_0 n_2 \mathcal{E}^{(p1)}(0)\mathcal{E}^{(p2)}(L)\,e^{-\alpha L/2}, \qquad \mu^2 = \frac{1}{\ell_{4WM}^2} - \left[\frac{\alpha}{2}\right]^2.$$

4. Consider three-wave mixing in a medium of width L, nonlinearity $n_{2\parallel}(-\omega; \omega)$, and an intensity absorption coefficient α at the frequency of the two input beams that are incident at angles $\pm\theta$ relative to the normal to the surface. Derive an expression for the intensity of either one of the new beams at the exit facet.

5. Identify and use the gratings created by the mixing of two fields in a three-wave mixing geometry. Consider degenerate (same frequency for input and output beams) three-wave mixing in the small conversion limit (*no depletion of input fields*) via the nonresonant third-order susceptibility $\tilde{\chi}^{(3)}_{xxxx}(-\omega; \omega, -\omega, \omega)$ for

the geometry shown in the figure below (angles θ are small). The beams are all z-polarized, and the medium is isotropic with $|\vec{E}_1| = |\vec{E}_2|$ at the input.

(a) Identify all the nonlinear gratings formed by the product of two input fields (including both the products of the field with itself and the products with the other input field), separating them into the ones that respond at 2ω and those that respond at zero frequency.

(b) From (a), identify and discard the field combinations that lead to SPM or XPM since they do not lead to energy exchange.

(c) Identify the products of three fields that do not lead to new fields being generated at the input frequency ω and give reasons why they do not give rise to new fields that can approximately satisfy wave-vector conservation.

(d) Write down the coupled wave equations for both first-order (at $\pm 3\theta$) scattered beams. Are the two input beams always equally depleted on propagation? Why/why not?

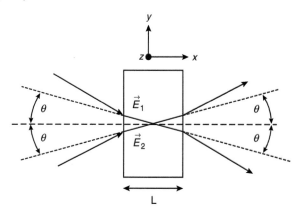

6. Consider the *electrostrictive* effect in an isotropic medium.

(a) Find $P(3\omega)$ for THG (Third Harmonic Generation).

(b) Find $P(\omega)$ for degenerate four-wave mixing when all three input beams are copolarized.

7. Consider the Raman-induced Kerr effect spectroscopy process with a right circularly polarized pump beam at frequency ω_a and a left circularly polarized weak beam at frequency ω_b propagating along the z-axis. Find the nonlinear gain (or loss) in the left circularly polarized beam intensity due to the presence of the pump beam. The only nonzero terms in the hyperpolarizability tensor are $\partial \bar{\alpha}_{xyn}/\partial \bar{q}_n = \partial \bar{\alpha}_{yxn}/\partial \bar{q}_n$.

8. Consider the Raman-induced Kerr effect spectroscopy process with a right circularly polarized pump beam at frequency ω_a and a left circularly polarized weak beam at frequency ω_b propagating along the z-axis. Find the nonlinear gain (or loss) in the left circularly polarized beam intensity due to the presence of the

pump beam. The only nonzero terms in the hyperpolarizability tensor are $\partial \bar{\alpha}_{xxn}/\partial \bar{q}_n = \partial \bar{\alpha}_{yyn}/\partial \bar{q}_n$.

9. Consider the Raman-induced Kerr effect spectroscopy process with a left circularly polarized pump beam at frequency ω_a and a linearly polarized weak beam at frequency ω_b propagating along the z-axis. Find the nonlinear gain (or loss) in the linearly polarized beams due to the presence of the pump beam. The only nonzero terms are

   (a) $\partial \bar{\alpha}_{xyn}/\partial \bar{q}_n = \partial \bar{\alpha}_{yxn}/\partial \bar{q}_n$ and

   (b) $\partial \bar{\alpha}_{xxn}/\partial \bar{q}_n = \partial \bar{\alpha}_{yyn}/\partial \bar{q}_n$.

10. Find the contribution to $n_{2\parallel}(-\omega; \omega)$ and $\alpha_{2\parallel}(-\omega; \omega)$ due to the symmetric "breathing" mode in methane.

11. In coherent anti-Stokes Raman scattering, a nonlinear polarization at frequency $2\omega_b - \omega_a$ also occurs due to electronic processes and due to the "nonresonant" tails of other vibrational contributions. Assuming that this extra contribution is nonresonant, i.e., independent of frequency, and can be written as $3\tilde{\chi}_{iiii}^{(3)}(2\omega_b - \omega_a; \omega_b, \omega_b, -\omega_a)$, derive the line shape for $I(2\omega_b - \omega_a)$ in the frequency range $10\,2\bar{\tau}_v^{-1} + \omega_b - \omega_a > \omega_b - \omega_a > \omega_b - \omega_a - 10\,2\bar{\tau}_v^{-1}$ for the ratio of the maximum vibrational contribution to the background amplitude contribution of 0.33, 1.0, and 3.0.

12. Derive the formula for a plane-polarized output coherent anti-Stokes Raman scattering beam at frequency ω_c with a left circularly polarized beam at frequency ω_a and a linearly polarized beam at frequency ω_b (same polarization as the output beam) propagating along the z-axis for the two cases.

    (a) $\partial \bar{\alpha}_{xyn}/\partial \bar{q}_n = \partial \bar{\alpha}_{yxn}/\partial \bar{q}_n$.

    (b) $\partial \bar{\alpha}_{xxn}/\partial \bar{q}_n = \partial \bar{\alpha}_{yyn}/\partial \bar{q}_n$.

REFERENCE

1. W. M. Hetherington III, Z. Z. Ho, E. W. Koenig, G. I. Stegeman, and R. M. Fortenberry, "CARS spectroscopy of the surface phonons on ZnO," Chem. Phys. Lett., **128**, 150–152 (1986).

SUGGESTED FURTHER READING

R. W. Boyd, Nonlinear Optics, 3rd Edition (Academic Press, Burlington, MA, 2008).

M. D. Levenson and S. S. Kano, Introduction to Nonlinear Laser Spectroscopy (Academic Press, San Diego, 1988).

S. H. Lin, A. A. Villaeys, and Yuichi Fujimura, Advances in Multiphoton Processses and Spectroscopy (World Scientific Press, Singapore, 1998).

Y. R. Shen, Principles of Nonlinear Optics (John Wiley & Sons, New York, 1984).

CHAPTER 16

Stimulated Scattering

Stimulated scattering is a fascinating process that requires a strong coupling between light and matter. The properties of matter that interact with light are vibrational and rotational modes, concentrations of different species, spin, sound waves, and in general any property that can undergo fluctuations in its population and couples to light. Typically the output light is shifted down in frequency from a pump beam frequency, and in general the interaction leads to a growth equation for the shifted light intensity that is proportional to the intensity itself. This in turn leads to an exponential growth of the signal before saturation sets in due to pump beam depletion. Furthermore, the matter modes experience gain. For example, the amplitude of the vibration also grows with distance. These stimulated processes have a threshold because gain must exceed loss.

The two most common processes are stimulated Raman scattering (SRS) and stimulated Brillouin scattering (SBS), which involve vibrational normal modes of molecules (optical phonons) and traveling sound waves (acoustic phonons), respectively. Although both are stimulated processes, the output properties (direction, magnitude of the frequency shift, and so on) are very different in the two cases. For SRS and associated processes, such as anti-Stokes beam generation, a number of beams can be generated with frequencies shifted up or down from the pump frequency by multiples of the vibrational mode frequency. Furthermore, the gain spectrum leads to useful processes such as amplification and lasing. For SBS, because sound waves are propagating modes and there is a specific relation between their wave vector and frequency, wave-vector conservation leads to a single output beam traveling in the direction opposite to that of the input beam.

The sequence of events and energy flow in all stimulated scattering processes, made specific first for SRS, is shown in Fig. 16.1. The SRS process is initiated by noise, thermally induced fluctuations in the optical fields, and Raman-active vibrational modes. An incident pump field (ω_P) interacts with vibrational fluctuations, in the process losing a photon that is downshifted in frequency to produce a Stokes wave (ω_S, Stokes photon) by the vibrational frequency (Ω) and also emitting an optical phonon (quantum of vibrational energy $\hbar\omega_v$) (see Fig. 16.2). These in turn also

Nonlinear Optics: Phenomena, Materials, and Devices, George I. Stegeman and Robert A. Stegeman.
© 2012 John Wiley & Sons, Inc. Published 2012 by John Wiley & Sons, Inc.

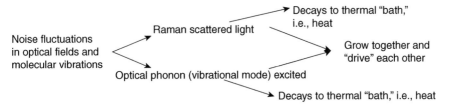

FIGURE 16.1 The sequence of events that leads to growth of the Stokes photon and the optical phonon populations.

stimulate further breakup of pump photons in the classical exponential population dynamics process, in which the more you have, the more you get. Therefore the pump decays with propagation distance, and both the phonon population and the Stokes wave grow together. If the generation rate of Stokes light exceeds the loss rate, stimulated emission occurs and the Stokes beam grows exponentially. All the interacting species also decay due to their usual decay mechanisms, typically leading to an increase in the temperature of the medium. (This loss must be taken into account explicitly when formulating the Manley–Rowe relations.) Any Raman-active mode will modulate the polarizability and therefore participate in this process.

Nonlinear optics in fibers has been progressively more important for many applications, especially frequency shifting and amplification, and a number of these will also be discussed in this chapter.

16.1 STIMULATED RAMAN SCATTERING

Just as discussed in Chapter 15, it is the difference frequency between the optical fields that excites coherently the phonon modes. Starting from "noise" requires a quantum mechanical treatment, the details of which are beyond the scope of this textbook. Here only the classical case is considered, i.e., both the pump and a small amount of the Stokes beams already exist.

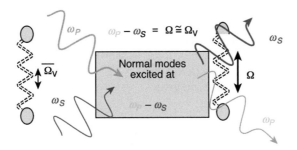

FIGURE 16.2 Schematic of the stimulated Raman scattering process.

The first step is to calculate the response in a molecule's frame of reference. The total optical field present consists of both the pump and the Stokes copolarized fields; i.e.,

$$\vec{E}_T(\vec{r}, t) = \frac{1}{2}\hat{e}_x\left\{\mathcal{E}_P\, e^{i(\vec{k}_P\cdot\vec{r}-\omega_P t)} + \text{c.c.} + \mathcal{E}_S\, e^{i(\vec{k}_S\cdot\vec{r}-\omega_S t)} + \text{c.c.}\right\}. \tag{16.1}$$

The Raman hyperpolarizability couples the light waves to the vibrations as discussed in Chapter 15; i.e.,

$$\bar{\alpha}_{ij} = \left[\bar{\alpha}_{ij}^L + \bar{q}_n\frac{\partial\bar{\alpha}_{ijn}}{\partial\bar{q}_n}\bigg|_{\bar{q}_n=0}\right], \tag{16.2}$$

which leads to the nonlinear polarization induced in a molecule of the form

$$\vec{p}^{\,\text{NL}}(\vec{r}, t, \omega_{\text{NL}}) = \vec{q}\cdot\frac{\partial\vec{\bar{\alpha}}}{\partial\vec{q}}\bigg|_{\bar{q}=0} : \vec{E}(\vec{r}, t, \omega_P)f^{(1)}(\omega_{\text{NL}})f^{(1)}(\omega_P), \tag{16.3}$$

where $f^{(1)}(\omega_{\text{NL}})$ and $f^{(1)}(\omega_P)$ are the local field corrections at the nonlinear frequency-shifted beam (the Stokes beam in the present case) and the pump beam, respectively. Simplifying the analysis to the most common case of the coupling between the light waves copolarized along the direction of the vibrations, the Raman nonlinear polarization induced is

$$\frac{\partial\bar{\alpha}_{ijk}}{\partial\bar{q}_k} = \frac{\partial\bar{\alpha}_{xxx}}{\partial\bar{q}_x} = \frac{\partial\bar{\alpha}}{\partial\bar{q}}$$

$$\rightarrow\quad \bar{p}^{\text{NL}}(\omega_S) = \bar{q}\frac{\partial\bar{\alpha}}{\partial\bar{q}}\bigg|_{\bar{q}=0} f^{(1)}(\omega_S)f^{(1)}(\omega_P)E(\omega_P), \tag{16.4}$$

$$\bar{p}^{\text{NL}}(\omega_P) = \bar{q}\frac{\partial\bar{\alpha}}{\partial\bar{q}}\bigg|_{\bar{q}=0} f^{(1)}(\omega_P)f^{(1)}(\omega_S)E(\omega_S).$$

To find the all-optical force driving the vibrations, the interaction potential between the induced dipoles and the incident fields is calculated and the gradient of the potential is taken to find the nonlinear force, as shown previously in Chapter 12. The vibrations are written as $\bar{q} = \frac{1}{2}\bar{Q}(\vec{K}, \Omega)\, e^{i(\vec{K}\cdot\vec{r}-\Omega t)} + \text{c.c.}$, and solving the driven simple harmonic oscillator problem gives

$$\bar{Q}(\vec{K}, \Omega)\, e^{i(\vec{K}\cdot\vec{r}-\Omega t)} + \text{c.c.} = \frac{2}{m_{\text{eff}}}\frac{\partial\bar{\alpha}}{\partial\bar{q}}\bigg|_{\bar{q}=0} f^{(2)}\frac{\vec{E}(\omega_S)\cdot\vec{E}(\omega_P)}{D(\Omega)}, \tag{16.5}$$

$$f^{(2)} = f^{(1)}(\omega_P)f^{(1)}(\omega_S), \quad D(\Omega) = D(\omega_P - \omega_S) = \Omega^2 - (\omega_P - \omega_S)^2 - 2i\bar{\tau}_v^{-1}(\omega_P - \omega_S),$$

with $\bar{\tau}_v$ being the optical phonon decay time at $\Omega \cong \Omega_v$. Therefore

$$\bar{q} \propto \frac{1}{4\bar{m}_{eff}} \frac{\partial \tilde{\alpha}}{\partial \bar{q}}\bigg|_{\bar{q}=0} f^{(1)}(\omega_P)f^{(1)}(\omega_S) \left\{ \frac{E_P E_S^*}{D(\omega_P - \omega_S)} e^{-i(\omega_P - \omega_S)t} + \frac{E_P^* E_S}{D^*(\omega_P - \omega_S)} e^{-i(\omega_S - \omega_P)t} \right\},$$

(16.6)

which gives for the "Maxwell" nonlinear polarizations (including the "extra" local field correction discussed in Chapter 8)

$$\boldsymbol{P}_S^{NL}(z) e^{i(k_S z - \omega_S t)} = \frac{N}{4\bar{m}_{eff} D^*(\omega_P - \omega_S)} \left[\frac{\partial \tilde{\alpha}}{\partial \bar{q}}\bigg|_{\bar{q}=0} \right]^2$$

$$\times \left[f^{(1)}(\omega_P)f^{(1)}(\omega_S) \right]^2 |\boldsymbol{\mathcal{E}}_P(z)|^2 \boldsymbol{\mathcal{E}}_S(z) e^{i(k_S z - \omega_S t)},$$

(16.7)

$$\boldsymbol{P}_P^{NL}(z) e^{i(k_P z - \omega_P t)} = \frac{N}{4\bar{m}_{eff} D(\omega_P - \omega_S)} \left[\frac{\partial \tilde{\alpha}}{\partial \bar{q}}\bigg|_{\bar{q}=0} \right]^2$$

$$\times \left[f^{(1)}(\omega_P)f^{(1)}(\omega_S) \right]^2 |\boldsymbol{\mathcal{E}}_S(z)|^2 \boldsymbol{\mathcal{E}}_P(z) e^{i(k_P z - \omega_P t)}.$$

Note that both polarizations $\boldsymbol{P}_S^{NL}(z)$ and $\boldsymbol{P}_P^{NL}(z)$ have exactly the correct wave vector for phase matching to the Stokes and pump fields, respectively. Also, for simplicity in the analysis, assume that the laser and Stokes beams are collinear. However, note that SRS can occur in all directions since $\boldsymbol{P}_S^{NL}(z)$ is independent of $\vec{k}_P$.

The calculation now follows the usual procedure of substituting into the slowly varying envelope approximation (SVEA). This gives

$$k^{NL} = k_S \quad \rightarrow \quad \frac{d\boldsymbol{\mathcal{E}}_S(z)}{dz} = i \frac{N\omega_S}{8\bar{m}_{eff} n_S \varepsilon_0 c D^*(\omega_P - \omega_S)} \left[\frac{\partial \tilde{\alpha}}{\partial \bar{q}}\bigg|_{\bar{q}=0} \right]^2 f^{(3)} |\boldsymbol{\mathcal{E}}_P(z)|^2 \boldsymbol{\mathcal{E}}_S(z),$$

(16.8a)

$$k^{NL} = k_P \quad \rightarrow \quad \frac{d\boldsymbol{\mathcal{E}}_P(z)}{dz} = i \frac{N\omega_P}{8\bar{m}_{eff} n_P \varepsilon_0 c D(\omega_P - \omega_S)} \left[\frac{\partial \tilde{\alpha}}{\partial \bar{q}}\bigg|_{\bar{q}=0} \right]^2 f^{(3)} |\boldsymbol{\mathcal{E}}_S(z)|^2 \boldsymbol{\mathcal{E}}_P(z),$$

(16.8b)

with $f^{(3)} = [f^{(1)}(\omega_P)f^{(1)}(\omega_S)]^2$. Multiplying Eq. 16.8a by $\boldsymbol{\mathcal{E}}_S^*(z)$ to obtain an equation for the intensity gives

$$\frac{dI_S(z)}{dz} = \frac{N\omega_S}{4\bar{m}_{eff} n_S n_P c^2 \varepsilon_0^2} \left[\frac{\partial \tilde{\alpha}}{\partial \bar{q}}\bigg|_{\bar{q}=0} \right]^2 f^{(3)} I_P(z) I_S(z) \left\{ \frac{i}{D^*(\omega_P - \omega_S)} + c.c. \right\}. \quad (16.9)$$

Since

$$\frac{i}{D^*(\omega_P - \omega_S)} + c.c. = \frac{4\bar{\tau}_v^{-1}(\omega_P - \omega_S)}{\left(\bar{\Omega}_v^2 - [\omega_P - \omega_S]^2\right)^2 + 4\bar{\tau}_v^{-2}[\omega_P - \omega_S]^2}, \quad (16.10)$$

we obtain

$$\frac{dI_S(z)}{dz} = g_R I_S(z) I_P(z) - \alpha_S I_S(z),$$

$$g_R = \frac{N\omega_S}{\bar{m}_{\text{eff}} n_S n_P c^2 \varepsilon_0^2} \left[\frac{\partial \bar{\alpha}}{\partial \bar{q}} \Big|_{\bar{q}=0} \right]^2 f^{(3)} \frac{(\omega_P - \omega_S) \bar{\tau}_v^{-1}}{\left(\bar{\Omega}_v^2 - [\omega_P - \omega_S]^2 \right)^2 + 4\bar{\tau}_v^{-2} [\omega_P - \omega_S]^2}.$$

$$(16.11)$$

Note that the linear damping of the Stokes wave has been added phenomenologically via α_S. Here g_R is defined as the Raman gain coefficient with a maximum at $\Omega_v = \omega_P - \omega_S$ of the magnitude

$$g_{R,\text{max}} = \frac{N\omega_S}{4\bar{m}_{\text{eff}} n_S n_P c^2 \varepsilon_0^2 \bar{\Omega}_v \bar{\tau}_v^{-1}} \left[\frac{\partial \bar{\alpha}}{\partial \bar{q}} \Big|_{\bar{q}=0} \right]^2 f^{(3)}. \qquad (16.12)$$

For the pump beam, Eq. 16.8b leads to

$$\frac{dI_P(z)}{dz} = \frac{N\omega_P}{4\bar{m}_{\text{eff}} n_P n_S c^2 \varepsilon_0^2} \left[\frac{\partial \bar{\alpha}}{\partial \bar{q}} \Big|_{\bar{q}=0} \right]^2 f^{(3)} I_S(z) I_P(z) \left\{ \frac{i}{D(\omega_P - \omega_S)} + \text{c.c.} \right\}, \quad (16.13)$$

which gives

$$\frac{dI_P(z)}{dz} = -g_R \frac{\omega_P}{\omega_S} I_S(z) I_P(z) - \alpha_P I_P(z). \qquad (16.14)$$

A number of important properties of this process are now clear. It is automatically wave-vector matched. Also, Eqs 16.13 and 16.14 immediately lead to

$$-\frac{d}{dz} \left[\frac{I_P(z)}{\hbar \omega_P} \right] = \frac{d}{dz} \left[\frac{I_S(z)}{\hbar \omega_S} \right] \quad \rightarrow \quad -\frac{dN_P}{dz} = \frac{dN_S}{dz} \qquad (16.15)$$

for the photon flux. Therefore for every Stokes photon generated, a pump photon is annihilated. Furthermore, assuming no pump depletion (which is of course valid only in the early stages of SRS), Eq. 16.11 gives

$$I_S(z) = I_S(0) \, e^{[g_R I_P(0) - \alpha_S] z}. \qquad (16.16)$$

This equation shows that above threshold, i.e., $g I_P(0) > \alpha_S$, an *exponential* growth of the Stokes wave takes place. The variation of the Raman gain coefficient with $\omega_P - \omega_S$ is shown in Fig. 16.3. Note that the Raman gain process occurs over the full frequency range $\omega_P > \omega_S$ and that there is loss for $\omega > \omega_P$. Since the gain coefficient peaks at $\bar{\Omega}_v$ for the usual case that $\bar{\Omega}_v \bar{\tau}_v \gg 1$, the frequency $\omega_S = \omega_P - \bar{\Omega}_v$ grows the

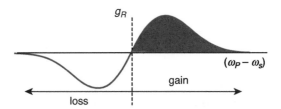

FIGURE 16.3   The variation of the Raman gain coefficient with frequency $\omega_P - \omega_S$.

fastest, and in this scenario, the stimulated Stokes signal narrows with propagation distance to less than the pump beam linewidth and the phase of the SRS signal is independent of the laser phase. However, if the temporal coherence of the input laser is too short, the $\delta\omega_P$ may be larger than τ_v^{-1} and it is necessary to average over $\delta\omega_P$ to get net gain. Finally, note that SRS can occur in all directions since the gain of Stokes wave does not depend on $\vec{k}_P$. However, only in the forward (or backward) direction of the Stokes wave is the beam overlap optimized and therefore the growth is exponential over long distances. Even though SRS also occurs in the backward direction, the threshold (discussed in Section 16.1.1.2) turns out to be higher for this case, and in these "winner-take-all" scenarios, the forward SRS dominates.

There is a simple reason why there is no gain for an anti-Stokes beam in this region of early growth of the Stokes beam. Anti-Stokes beam generation would require the annihilation of an optical phonon and the only phonon population available initially is due to thermal fluctuations. (Subsequently it will be shown here later that an anti-Stokes wave is generated via coupling to the pump beam and the phonons created via the Stokes process, but this turns out to *not* be a stimulated process and to require phase matching.)

16.1.1   Raman Amplification

The most obvious application of SRS is to the all-optical amplification of optical signals, as illustrated in Fig. 16.4a.

16.1.1.1   *Optimum Conversion Efficiency.*   It was shown previously that when $I_S(z)$ grows by one photon, $I_P(z)$ decreases by one photon and $\hbar(\omega_P - \omega_S)$ of energy is lost to the vibrational mode, and eventually ends up as heat. For optimum conversion, $I_P(L) = 0$ and $I_S(0) \cong 0$ and so

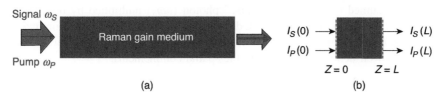

FIGURE 16.4   (a) The amplification of a signal by the stimulated Raman scattering process. (b) Boundary conditions for optimum amplification.

$$\frac{1}{\omega_P}[I_P(L) - I_P(0)] = -\frac{1}{\omega_S}[I_S(L) - I_S(0)] \quad \rightarrow \quad \frac{I_P(0)}{\omega_P} = \frac{I_S(L)}{\omega_S}. \quad (16.17)$$

16.1.1.2 *Raman Amplification Properties: Attenuation, Saturation, Pump Depletion, Threshold.* With no pump depletion (small signal gain) to the Stokes beam, the coupled intensity equations can be solved analytically. Integrating

$$\frac{dI_P}{dz} = -\alpha_P I_P \rightarrow I_P(z) = I_P(0)\, e^{-\alpha_P z}$$

$$\rightarrow \quad \frac{dI_S(z)}{dz} = g_R I_S(z) I_P(0)\, e^{-\alpha_P z} - \alpha_S I_S(z). \quad (16.18)$$

This gives

$$\int_0^L \frac{dI_S(z)}{I_S(z)} = \int_0^L [g_R I_P(0)\, e^{-\alpha_P z} - \alpha_S]\, dz$$

$$\Rightarrow \quad I_S(z) = I_S(0)\, e^{g_R I_P(0) L_{\text{eff}} - \alpha_S L}, \quad \text{with } L_{\text{eff}} = \frac{1 - \exp(-\alpha_P L)}{\alpha_P}. \quad (16.19)$$

The unsaturated (no pump depletion) amplifier gain is defined as $G_A = \exp[g_R I_P(0) L_{\text{eff}}]$. It is normally a useful approximation that $\alpha_P = \alpha_S = \alpha$, and in this case, Eqs 16.14 and 16.16 give

$$G_S = \frac{(1 + r_0)\, e^{-\alpha L}}{r_0 + G_A^{-(1+r_0)}}, \quad \text{with } r_0 = \frac{\omega_P}{\omega_S} \frac{I_S(0)}{I_P(0)}, \quad (16.20)$$

where G_S is the saturated gain shown in Fig. 16.5 as a function of the parameter $G_A r_0$, which contains the ratio of the input intensities of the pump and Stokes beams (1). Note that the higher the input power, the faster the saturation occurs, as expected.

To calculate approximately the threshold intensity, the Stokes beam intensity $I_S^{\text{eff}}(0)$ is assumed to be a single "noise" photon ($\hbar\omega_S$) multiplied by the Stokes frequency bandwidth of the unsaturated gain profile, assumed to have a Lorentzian line shape (2). Mathematically, for the most important case of a single-mode fiber (which has well-defined modes due to the spatial confinement),

$$I_S^{\text{eff}}(0) = A_{\text{eff}} \hbar\omega_S \sqrt{\frac{2\pi}{I_P(0) L_{\text{eff}}} \left| \frac{\partial^2 g_R}{\partial\omega^2} \right|_{\omega=\omega_S}^{-1/2}}, \quad (16.21)$$

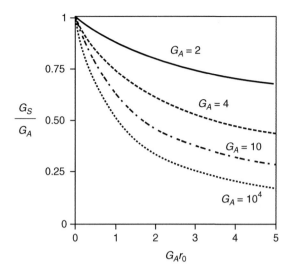

FIGURE 16.5   The variation in the normalized saturated gain versus a parameter that contains the ratio of the intensities of the input signal to the pump beam.

where A_{eff} is the effective core area of the fiber. The stimulated Raman "threshold" pump intensity $I_P^{\text{th}}(0)$ is defined as *the input pump intensity for which the output pump intensity equals the Stokes output intensity*; i.e.,

$$I_S(L) = I_S^{\text{eff}}(0) \exp[g_R I_P(0) L_{\text{eff}}] = I_P(L) = I_P^{\text{th}}(0) e^{-\alpha_P L}$$
$$\Rightarrow \quad g_R I_P^{\text{th}}(0) L_{\text{eff}} \approx 16. \tag{16.22}$$

The parameter $\sqrt{\partial^2 g_R / \partial \omega^2}|_{\omega = \omega_s}$ is shown approximately in Fig. 16.6 for silica glass. For the backward propagating Stokes beam, the definition for threshold is $I_P^{\text{th}}(0) e^{-\alpha_P L} = I_S(0)$, and so

$$I_S^{\text{eff}}(L) \exp[g_R I_P(0) L_{\text{eff}}] = I_P(L)$$
$$\Rightarrow \quad g_R I_P^{\text{th}}(0) L_{\text{eff}} \approx 20. \tag{16.23}$$

The threshold for the backward propagating Stokes beam is higher than that for the forward propagating Stokes beam. Therefore, the forward propagating Stokes beam starts first and typically grows so fast that it depletes the pump, and so the backward propagating Stokes beam never really grows significantly.

16.1.1.3 *Raman Amplification—Pulses.* Consider the real case in which the Stokes and pump beams propagate with different group velocities $v_g(\omega_S)$ and $v_g(\omega_P)$ so that the pulses separate in space. Hence the interaction efficiency is greatly reduced when the walk-off time is greater than or equal to pump pulse width Δt. This results in a Stokes signal spread in both time and space, as shown in Fig. 16.7.

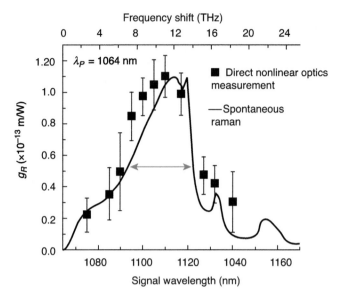

FIGURE 16.6 The measured Raman gain spectrum (data points) and the Raman gain deduced from the spontaneous Raman scattering spectrum (solid line). The bandwidth parameter $\sqrt{\partial^2 g_R / \partial \omega^2 |_{\omega = \omega_s}}$ is shown by the arrow. Reproduced with permission from the Optical Society of America (3).

The situation is even worse for SRS in the backward direction. Because of small overlap, amplification is very small.

16.1.1.4 *Stimulated Raman Gain in Glasses.* The derivation of g_R was for the simplest case, i.e., a single isolated vibrational mode that decays exponentially

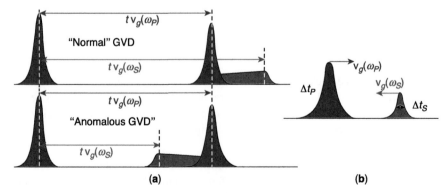

FIGURE 16.7 The effect of group velocity dispersion (GVD) on stimulated Raman scattering in the two regions of GVD for (a) copropagating and (b) counterpropagating pump and Stokes beams (same in both dispersion regions).

with time once excited. This does not give enough bandwidth for the gain to be useful. In practice, inhomogeneously broadened glass media give much bigger bandwidths, the most common being *silica* glass for which the Raman gain spectrum is shown in Fig. 16.6. Glasses are amorphous media in which the response of various molecular vibrations is broadened due to a number of mechanisms and gives rise to a continuous Raman gain spectrum. Clearly the frequency dependence of $g_R(\omega_P - \omega_S)$ is non-analytical and quite complicated.

The value of g_R in fused silica is not large. However, the figure of merit for Raman gain is g_R/α_1, i.e., the product of the magnitude of the gain and the distance over which it can be maintained in the presence of loss. The extremely low loss in silica fibers ($\alpha_1 < 0.2$ dB/km) is the key factor that makes silica glass attractive for gain applications. Note that since $g_R^{max} \propto \omega_S$, $g_R^{max}(\text{visible}) > g_R^{max}(1.55 \, \mu m)$, which means that it is smaller at communications wavelengths than in the visible region, where many spontaneous Raman scattering experiments have been performed to identify promising glasses. For example, at $\lambda_P = 1.55 \, \mu m$, in silica-based fibers with $A_{eff} = 50 \, \mu m^2$ and $\alpha = 0.2$ dB/km, the required threshold power for SRS is $P_P^{th} = 600$ mW. Nevertheless, with the development of high power diode pumps, such pump powers are now available and Raman gain in glasses provides an alternative to erbium-doped glass amplifiers for optical communications in the wavelength range of 1.3–1.6 μm. As a result, there has been recent interest in new glasses with larger bandwidths and higher gains (4). Figure 16.8 shows typical Raman gain coefficients and current losses in different glass families. As was initially the case for silica fibers, focused efforts including more refined glass purification and manufacture techniques will be needed to reduce losses to more acceptable levels in glasses identified for further technological applications. Figure 16.9 shows two examples of the glasses with larger gain coefficients and broader gain spectra relative

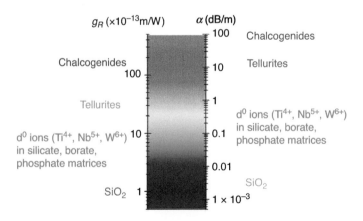

FIGURE 16.8 Raman gain and current status (as of 2011) of losses for different glass families. It is expected that these loss figures will be reduced substantially as these glasses become processed into fibers and their optimum processing conditions established. Courtesy of Dr Clara Rivero, Lockheed Martin.

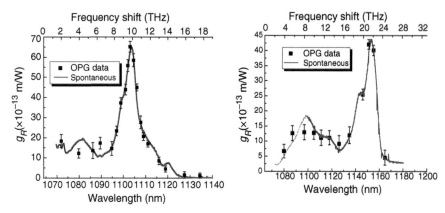

FIGURE 16.9 Raman gain spectra of two glasses pumped at $\lambda_P = 1.064\,\mu$m. The percentages of the constituents are (a) 18Ge–5Ga–7Sb–70S and (b) 59.5TeO$_2$–25.5TlO$_{0.5}$–15PbO. Note the increased bandwidth and increased gain coefficient relative to fused silica. The single data points were obtained with a direct measurement of Raman gain, whereas the continuous red lines are obtained from spontaneous Raman scattering. Reproduced with permission from the Optical Society of America (5,6).

to fused silica (5,6). Most specifically, chalcogenide glasses with a typical Raman gain spectrum shown in Fig. 16.9a may provide better alternatives in the communications regime for limited bandwidth applications (6).

The Raman gain spectrum of these glasses, including silica, clearly contains a number of peaks due to individual vibrational modes ($\bar{\Omega}_\beta$) whose spectra are broadened by the random orientation of individual molecules, intermolecular interactions, and disorder in the medium. These effects lead to a distribution of vibrational frequencies $F(\bar{\Omega}_\beta - \bar{\Omega}_{\beta 0})$ normalized as follows: $\int_{-\infty}^{\infty} F(\bar{\Omega}_\beta - \bar{\Omega}_{\beta 0})\, d\bar{\Omega}_\beta = 1$ (discussed in Appendix 12.1). For notational convenience, $\bar{\Omega}_v$ and $\bar{\tau}_v$ for the βth mode are written as $\bar{\Omega}_\beta$ and $\bar{\tau}_\beta$. The Raman gain for a single vibrational mode of an isolated molecule was given by Eq. 16.11. Averaging over all molecular orientations and including all the Raman-active modes $\vec{q}_\beta$ gives

$$
g_R = \sum_\beta \frac{N\omega_S \pi}{\bar{m}_{\beta 0,\text{eff}}\, n_S n_P c^2 \varepsilon_0^2} \left\langle \left[\frac{\partial \bar{\alpha}^\beta}{\partial \bar{q}_\beta}\Big|_{\bar{q}_\beta = 0} \right]^2 \right\rangle_{\text{angles}} f^{(3)}
$$

$$
\times \int_0^\infty \frac{F(\bar{\Omega}_\beta - \bar{\Omega}_{\beta 0})(\omega_P - \omega_S)(1/\bar{\tau}_\beta \pi)}{(\bar{\Omega}_\beta^2 - [\omega_P - \omega_S]^2)^2 + \bar{\tau}_\beta^{-2}[\omega_P - \omega_S]^2}\, d\bar{\Omega}_\beta. \tag{16.24}
$$

Near resonances,

$$
\frac{(\omega_P - \omega_S)\left(\bar{\tau}_\beta^{-1}/\pi\right)}{\left(\bar{\Omega}_\beta^2 - [\omega_P - \omega_S]^2\right)^2 + 4\bar{\tau}_\beta^{-2}[\omega_P - \omega_S]^2} \approx \frac{\left(\bar{\tau}_\beta^{-1}/\pi\right)}{\left(\bar{\Omega}_\beta - [\omega_P - \omega_S]\right)^2 + \bar{\tau}_\beta^{-2}} \times \frac{1}{4[\omega_P - \omega_S]}, \tag{16.25}
$$

and noting that typically the spectral width of the distribution function $F(\bar{\Omega}_\beta - \bar{\Omega}_{\beta 0})$ is much broader than the Lorentzian line shape characterized by $\bar{\tau}_\beta$, we obtain

$$\frac{\bar{\tau}_\beta^{-1}/\pi}{(\bar{\Omega}_\beta - [\omega_P - \omega_S])^2 + \bar{\tau}_\beta^{-2}} \rightarrow \delta(\bar{\Omega}_\beta - [\omega_P - \omega_S]). \qquad (16.26)$$

Inserting Eqs 16.25 and 16.26 into Eq. 16.24 and integrating it over $d\bar{\Omega}_\beta$ gives

$$g_R(\omega_S) \cong \sum_\beta \frac{N\omega_P\pi}{4\bar{m}_{\beta 0,\text{eff}}\bar{\Omega}_{\beta 0}n^2c^2\varepsilon_0^2}\left\langle\left[\left.\frac{\partial\bar{\alpha}^\beta}{\partial\bar{q}^\beta}\right|_{\bar{q}^\beta=0}\right]^2\right\rangle f^{(3)}F(\omega_P - \omega_S - \bar{\Omega}_{\beta 0}). \quad (16.27)$$

Substituting Eq. A.12.1.29 from Appendix 12.1 on Raman scattering into Eq. 16.27 gives the Raman gain spectrum in terms of the spontaneous Raman scattering spectrum as

$$g_R(\omega_S) \cong \frac{2c^2\pi(4\pi)^2}{n^2k_BT\omega_P^3}\sum_\beta \bar{\Omega}_{\beta 0}\frac{P_{x,\text{dm}}^{RS,\beta}(\omega_S)}{\Delta\Omega I_{P,x}}. \qquad (16.28)$$

It is important to note that the Raman scattering experiments need to be completely calibrated to obtain numerical estimates for the gain and not just its spectrum; i.e., the absolute Raman cross sections need to be measured from the experimental geometry, pump laser power, and the measured (calibrated) scattered intensity (4).

16.1.2   Raman Laser

Because the SRS process arises from stimulated emission of the Stokes beam, placing it in a cavity leads to a Raman laser, as illustrated in Fig. 16.10. The threshold condition is $R\,e^{[g_{R,\text{max}}I_P - \alpha_S]L} > 1$, with $g_{R,\text{max}}$ given by Eq. 16.12. For $L \approx 10\,\text{m}$, $P_P^{\text{th}} = 1\,\text{mW}$ for lasing in silica fibers, which explains why to date Raman lasers have been restricted primarily to fibers (7).

There are a number of applications that require lasers in the mid-infrared (mid-IR) region of the frequency spectrum, and fiber lasers are emerging as the preferred

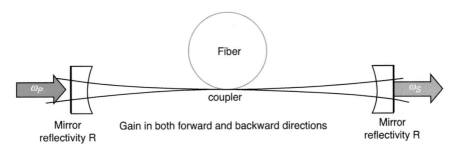

FIGURE 16.10   Geometry of a Raman laser.

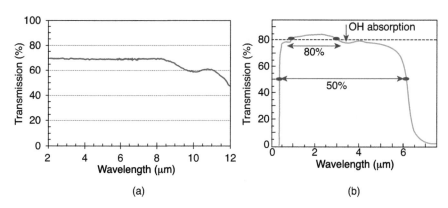

FIGURE 16.11 Mid-infrared band edge due to multiphonon absorption for (a) As_2S_3 chalcogenide glass (courtesy of Dr Brandon Shaw at Naval Research Laboratory) and (b) $80TeO_2$–$10Na_2O$–$10ZnO$ tellurite glass. Reproduced with permission from the Optical Society of America (9).

solution. For example, vibrational absorption signatures for important gases, such as CO_2, occur in this region of the spectrum (as shown previously in Fig. 6.20) for the atmospheric transmittance in the near- and mid-IR windows (8). Conveniently, the chalcogenide and tellurite glasses possess large transparency windows that extend further into the mid-IR region than do fused silica, which makes them attractive candidates as core technologies for mid-IR lasers. An example of the multiphonon "vibrational" edge in the mid-IR region for arsenic trisulfide (As_2S_3) and a tellurite glass containing $80TeO_2$–$10Na_2O$–$10ZnO$ is illustrated in Fig. 16.11 (9).

Multiwavelength mid-IR sources with power at discrete wavelengths up to 5 μm are also needed, with appropriate pump lasers restricted to commonly available 2-μm sources, such as thulium-doped fiber lasers or indium phosphide (InP) semiconductor lasers. Although mid-IR semiconductor lasers based on gallium arsenide quantum cascade technology and InP are available for laser emission in the 2–10-μm range, fiber lasers using nonlinear optics offer an alternative solution for two reasons: First, their self-guiding properties are more resistant to temperature and vibration extremes for operation in harsh environments that have large temperature and vibration extremes. For stable operation, monolithic fiber lasers using internal gratings offer alternatives to solid-state lasers and nonlinear crystals that use free space optics (10). Second, the long interaction lengths and small effective areas offered in fibers allow the modest nonlinear coefficients of glasses to be efficiently used.

For example, one fiber solution would be to use Raman shifting in nonlinear fibers to progressively shift the energy from 2 μm out to the desired mid-IR wavelengths (11). A suitable nonlinear fiber could be fabricated from a chalcogenide glass or a tellurite glass, both of which exhibit good transmission into the mid-IR region yet have significant differences in the peak intensity and spectral bandwidth of their Raman gain curves. The Raman gain curves for two candidates that are being actively pursued are 18Ge–5Ga–7Sb–70S chalcogenide glass and $59.5TeO_2$–$25.5TlO_{0.5}$–15PbO tellurite glass, and their Raman gain spectra are shown in Fig. 16.9. The main

FIGURE 16.12   Example of a Raman laser fiber for mid-infrared wavelength generation. HR, high reflectance; OC, output coupler.

structures of the Raman gain spectra are located near a frequency shift $\Delta v \sim 10\,\text{THz}$ for the chalcogenide glass and $\Delta v \sim 22\,\text{THz}$ for the tellurite glass. One of the many trade-offs is the peak Raman gain coefficient and its spectral shift. To shift energy from 2 to $5\,\mu\text{m}$, a large frequency shift of $\Delta v \sim 90\,\text{THz}$ is desired. The tellurite composition can come close to a frequency shift of $\Delta v \sim 90\,\text{THz}$ via a cascade of four Raman shifts at the peak Raman gain coefficient of $\Delta v \sim 22\,\text{THz}$, whereas the chalcogenide composition would need nine successive Raman shifts of the peak Raman gain coefficient of $\Delta v \sim 10\,\text{THz}$. Therefore, the tellurite glass would be the more attractive option in this case.

It is instructive to consider a mid-IR laser design that could be implemented as shown in Fig. 16.12. A 2-μm pump source, in the form of a thulium-doped fiber laser or a fiber-coupled InP semiconductor laser, can be spliced onto an appropriate length and size of tellurite fiber. At the input to the Raman fiber laser (RFL) cavity, a series of fiber Bragg gratings (FBGs) can be either spliced into the laser between the pump source and the Raman-active fiber or inscribed directly into the Raman-active tellurite fiber by a femtosecond laser (10). The output of the RFL cavity also contains a series of FBGs, of which the spectra and reflectivity coefficients are to be defined. In general, the FBGs at the input end of the RFL cavity should reflect as much energy back as possible into the cavity such that $R(v_{p+j\Delta v}) \approx 100\%$, where subscripts p and j denote the pump and number of the Raman shifts, respectively. At the output end of the RFL cavity, the reflectivities of FBGs are defined such that $0 \leq R(v_{p+j\Delta v}) \leq 100\%$. For optimum conversion efficiency into the longer wavelengths of interest, it would be optimum to define the center wavelengths of FBGs at the peak of the Raman gain spectrum of $\Delta v \sim 22\,\text{THz}$ to generate as much power at the longest wavelengths as possible. If specific wavelengths of emission that do not coincide near successive Raman shifts of $\Delta v \sim 22\,\text{THz}$ are desired, then the respective FBG spectrum can be defined for those wavelengths, although the overall conversion efficiency of the laser will decrease due to the lower Raman gain coefficients available off the peak of the Raman gain spectrum.

16.1.3   Coherent Anti-Stokes Beam Generation

Figure 16.13 shows an easily observed output from silica fibers as obtained with pulsed lasers. There are clearly many output frequencies, including frequencies upshifted from the pump frequency. Multiple Stokes beams occur when an initial Stokes beam becomes the pump for a second Stokes beam and so on. Anti-Stokes lines are also visible in Fig. 16.13b. The origin of these anti-Stokes lines is discussed in this section.

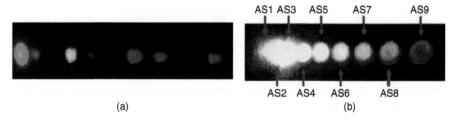

FIGURE 16.13 (a) Stokes signals obtained with a doubled YAG laser. (b) Anti-Stokes signals obtained with mid-infrared-sensitive film from a fiber with excitation at 1.064 μm.

To this point the focus has been on terms such as $|E_P|^2 E_S$ and $|E_S|^2 E_P$, which corresponded to the Stokes signal at $\omega_S = \omega_P - \bar{\Omega}_v$. But clearly many anti-Stokes lines are also observed experimentally, as indicated in Fig. 16.13b. This requires tracking the population of phonons since a phonon must be destroyed to upshift the pump frequency to the anti-Stokes frequency. An anti-Stokes field of the form

$$\frac{1}{2}\mathcal{E}_A\, e^{i(\vec{k}_A\cdot\vec{r}-\omega_A t)} + \text{c.c.}, \quad \text{with } \omega_A = \omega_P + \bar{\Omega}_v, \tag{16.29}$$

is added to the previous analysis. Once a phonon is created in the Stokes process, it can mix with a pump (laser) photon via the anti-Stokes process, as illustrated in Fig. 16.14.

The solution for $\bar{q}$ was already given in Eq. 16.6 and inserting it into

$$\vec{P}^{\text{NL}}(\vec{r}, t) = N\bar{q}\, \frac{\partial\bar{\alpha}}{\partial\bar{q}}\bigg|_{\bar{q}=0}\, \vec{E}_{\text{total}} \tag{16.30}$$

now gives

$$\begin{aligned}
\vec{P}^{\text{NL}}(\vec{r}, t) = \frac{N}{4}\bigg\{ &\frac{\partial\bar{\alpha}}{\partial\bar{q}}\bigg|_{\bar{q}=0}\, \bar{Q}(\vec{k}_P - \vec{k}_S, \omega_P - \omega_S)\, e^{i([\vec{k}_P-\vec{k}_S]\cdot\vec{r}-[\omega_P-\omega_S]t)} \\
&+ \frac{\partial\bar{\alpha}}{\partial\bar{q}}\bigg|_{\bar{q}=0}\, \bar{Q}^*(\vec{k}_P - \vec{k}_S, \omega_P - \omega_S)\, e^{-i([\vec{k}_P-\vec{k}_S]\cdot\vec{r}-[\omega_P-\omega_S]t)}\bigg\} \\
\times\big\{ &\mathcal{E}_P\, e^{i(\vec{k}_P\cdot\vec{r}-\omega_P t)} + \mathcal{E}_P^*\, e^{-i(\vec{k}_P\cdot\vec{r}-\omega_P t)} \\
&+ \mathcal{E}_S\, e^{i(\vec{k}_S\cdot\vec{r}-\omega_S t)} + \mathcal{E}_S^*\, e^{-i(\vec{k}_S\cdot\vec{r}-\omega_S t)} + \mathcal{E}_A\, e^{i(\vec{k}_A\cdot\vec{r}-\omega_A t)} + \mathcal{E}_A^*\, e^{-i(\vec{k}_A\cdot\vec{r}-\omega_A t)}\big\}.
\end{aligned} \tag{16.31}$$

FIGURE 16.14 The two-step process that leads to an anti-Stokes beam.

The product $\mathcal{E}_P\, e^{i(\vec{k}_P\cdot\vec{r}-\omega_P t)}\bar{Q}^*(\vec{k}_P-\vec{k}_S,\omega_P-\omega_S)\,e^{-i([\vec{k}_P-\vec{k}_S]\cdot\vec{r}-[\omega_P-\omega_S]t)}$ generates the polarization

$$\mathcal{P}_S^{\mathrm{NL}}=\frac{1}{2}N\frac{\partial\bar{\alpha}}{\partial\bar{q}}\bigg|_{\bar{q}=0}f^{(1)}(\omega_P)f^{(1)}(\omega_S)\bar{Q}^*(\vec{K},\Omega)\mathcal{E}_P. \tag{16.32}$$

The products $\mathcal{E}_S\, e^{i(\vec{k}_S\cdot\vec{r}-\omega_S t)}\bar{Q}(\vec{k}_P-\vec{k}_S,\omega_P-\omega_S)\,e^{i([\vec{k}_P-\vec{k}_S]\cdot\vec{r}-[\omega_P-\omega_S]t)}$ and $\mathcal{E}_A\, e^{i(\vec{k}_A\cdot\vec{r}-\omega_A t)}\bar{Q}^*(\vec{k}_P-\vec{k}_S,\omega_P-\omega_S)\,e^{-i([\vec{k}_P-\vec{k}_S]\cdot\vec{r}-[\omega_P-\omega_S]t)}$ generate the polarization

$$\mathcal{P}_P^{\mathrm{NL}}=\frac{1}{2}N\frac{\partial\bar{\alpha}}{\partial\bar{q}}\bigg|_{\bar{q}=0}f^{(1)}(\omega_P)\Big[f^{(1)}(\omega_S)\bar{Q}(\vec{K},\Omega)\mathcal{E}_S+f^{(1)}(\omega_A)\bar{Q}^*(\vec{K},\Omega)\mathcal{E}_A\,e^{-i[\vec{k}_P-\vec{k}_S-\vec{k}_A]\cdot\vec{r}}\Big].$$

$$\tag{16.33}$$

The product $\mathcal{E}_P\, e^{i(\vec{k}_P\cdot\vec{r}-\omega_P t)}\bar{Q}(\vec{k}_P-\vec{k}_S,\omega_P-\omega_S)\,e^{i([\vec{k}_P-\vec{k}_S]\cdot\vec{r}-[\omega_P-\omega_S]t)}$ generates the polarization

$$\mathcal{P}_A^{\mathrm{NL}}=\frac{1}{2}N\frac{\partial\bar{\alpha}}{\partial\bar{q}}\bigg|_{\bar{q}=0}f^{(1)}(\omega_P)f^{(1)}(\omega_A)\bar{Q}(\vec{K},\Omega)\mathcal{E}_P\,e^{i[2\vec{k}_P-\vec{k}_S]\cdot\vec{r}}. \tag{16.34}$$

Note that $\Delta\vec{k}=2\vec{k}_P-\vec{k}_S-\vec{k}_A$ for the nonlinear anti-Stokes polarization and the dispersion in refractive index means that the waves are not collinear for efficient generation on phase match, similar to the coherent anti-Stokes Raman scattering case discussed previously in Chapter 15. Thus the anti-Stokes process is a coherent process and requires phase matching (not automatic) and hence will not experience exponential gain. Since the anti-Stokes process involves the annihilation of optical phonons that are generated by the Stokes process, the required large population of optical phonons requires the generation of a strong Stokes signal.

Applying the SVEA to the nonlinear polarizations in Eqs 16.32–16.34 gives

$$\frac{d\mathcal{E}_S(z)}{dz}=i\frac{N\omega_S}{4n_S\varepsilon_0 c}\frac{\partial\bar{\alpha}}{\partial\bar{q}}\bigg|_{\bar{q}=0}f^{(1)}(\omega_P)f^{(1)}(\omega_S)\bar{Q}^*(z,\vec{K},\Omega)\mathcal{E}_P(z),$$

$$\frac{d\mathcal{E}_P(z)}{dz}=i\frac{N\omega_P}{4n_P\varepsilon_0 c}\frac{\partial\bar{\alpha}}{\partial\bar{q}}\bigg|_{\bar{q}=0}f^{(1)}(\omega_P)\Big[f^{(1)}(\omega_S)\bar{Q}(z,\vec{K},\Omega)\mathcal{E}_S(z)$$

$$+f^{(1)}(\omega_A)\bar{Q}^*(z,\vec{K},\Omega)\mathcal{E}_A(z)\,e^{-i\Delta\vec{k}\cdot\vec{r}}\Big], \tag{16.35}$$

$$\frac{d\mathcal{E}_A(z)}{dz}=i\frac{N\omega_A}{4n_A\varepsilon_0 c}\frac{\partial\bar{\alpha}}{\partial\bar{q}}\bigg|_{\bar{q}=0}f^{(1)}(\omega_P)f^{(1)}(\omega_A)\bar{Q}(z,\vec{K},\Omega)\mathcal{E}_P(z)\,e^{i\Delta\vec{k}\cdot\vec{r}}.$$

Converting Eqs 16.35 into intensity equations by multiplying them by $\mathcal{E}_S^*,\mathcal{E}_P^*$, and $\mathcal{E}_A^*$, respectively, gives

$$\frac{1}{\omega_S}\frac{dI_S(z)}{dz}+\frac{1}{\omega_A}\frac{dI_A(z)}{dz}=-\frac{1}{\omega_P}\frac{dI_P(z)}{dz}, \tag{16.36}$$

which indicates that for every Stokes photon created one pump photon is destroyed *and* for every anti-Stokes photon created another pump photon is destroyed. Also, for every Stokes photon created an optical phonon is also created, and for every anti-Stokes photon created an optical phonon is destroyed.

What is missing in the conservation of energy is the flow of mechanical energy $E_{mech}(t)$ into the optical phonon modes via the nonlinear mixing interaction, and its subsequent decay into heat. The time average (continuous bar above the quantities) of the total mechanical energy per unit volume associated with a molecular vibration is

$$\overline{E_{mech}} = \overline{\text{Kinetic Energy} + \text{Potential Energy}} = \frac{1}{2}N\bar{m}_{eff}\overline{[\dot{\bar{q}}^2 + \bar{\Omega}_v^2\bar{q}^2]} \simeq \frac{1}{2}N\bar{m}_{eff}\bar{\Omega}_v^2|\bar{Q}|^2.$$

$$(16.37)$$

For the SVEA in the time domain, $\bar{Q} \to \bar{Q}(t)$ and $E_{mech} \to E_{mech}(t)$ and using Eq. 15.90 to substitute for $\ddot{\bar{q}} + \bar{\Omega}_v^2\bar{q}$, we obtain

$$\overline{\frac{dE_{mech}}{dt}} = N\bar{m}_{eff}\overline{\dot{\bar{q}}[\ddot{\bar{q}} + \bar{\Omega}_v^2\bar{q}]} = N\bar{m}_{eff}\overline{\dot{\bar{q}}\left[-2\bar{\tau}_v^{-1}\dot{\bar{q}} + \frac{1}{\bar{m}_{eff}}F_{\bar{q}}(z,t)\right]}.$$

$$(16.38)$$

Here $2\bar{m}_{eff}N\overline{\dot{\bar{q}}^2}\bar{\tau}_v^{-1} = N\bar{m}_{eff}\bar{\tau}_v^{-1}\bar{\Omega}_v^2\bar{Q}(t)\bar{Q}^*(t) = 2\bar{\tau}_v^{-1}\overline{E_{mech}(t)}$ is the dissipative term by which the vibration decays, losing its energy to the thermal bath. $N\overline{\dot{\bar{q}}F_{\bar{q}}(z,t)}$ is the all-optical driving term due to the mixing of optical fields. This yields the energy flow balance for the vibrations

$$\overline{\frac{dE_{mech}(t)}{dt}} + 2\bar{\tau}_v^{-1}\overline{E_{mech}(t)} = N\overline{\dot{\bar{q}}F_{\bar{q}}(z,t)}.$$

$$(16.39)$$

To obtain a *nonzero time average* for the driving term (right-hand side of Eq. 16.39),

$$F_{\bar{q}} = \frac{1}{2}F_{\bar{q}}e^{-i\Omega t} + \text{c.c.} \quad \to \quad \overline{\dot{\bar{q}}F_{\bar{q}}} = -\frac{1}{4}i\Omega\bar{Q}F_{\bar{q}} - \frac{1}{4}i\Omega Q F_{\bar{q}}^* + \text{c.c.}$$

$$(16.40)$$

Therefore

$$\overline{\frac{dE_{mech}}{dt}} + 2\bar{\tau}_v^{-1}\overline{E_{mech}} = i\frac{N\Omega}{8}\frac{\partial\bar{\alpha}}{\partial\bar{q}}\bigg|_{\bar{q}=0}$$

$$\times \left[\bar{Q}^*f^{(1)}(\omega_S)f^{(1)}(\omega_P)\mathcal{E}_S^*\mathcal{E}_P - \bar{Q}f^{(1)}(\omega_A)f^{(1)}(\omega_P)\mathcal{E}_P\mathcal{E}_A^*e^{i\Delta kz}\right] + \text{c.c.}$$

$$(16.41)$$

Substituting for $\bar{Q}^*$ and $\bar{Q}$ into Eq. 16.41 finally gives

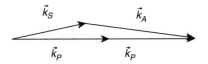

FIGURE 16.15 The wave-vector matching condition for the anti-Stokes beam generation.

$$\frac{1}{\hbar\Omega}\left\{\frac{\overline{dE_{mech}}}{dt}+2\bar{\tau}_v^{-1}\overline{E_{mech}}\right\}=\frac{1}{\hbar\omega_S}\frac{dI_S}{dz}-\frac{1}{\hbar\omega_A}\frac{dI_A}{dz}, \qquad (16.42)$$

which means that an optical phonon is created when a Stokes photon is annihilated and a phonon is eliminated when an anti-Stokes photon is created. Furthermore, the phonon flux loses energy due to decay to the thermal bath. Since Stokes photons are generated in all directions, the generation of anti-Stokes photons "eat out" a cone in the Stokes generation because wave vector must be conserved in the anti-Stokes generation (see Fig. 16.15).

Multiple anti-Stokes waves can be generated in a way similar to the multiple Stokes case, with the difference that the anti-Stokes waves need to be wave-vector matched. The sequence is shown in Fig. 16.16. In the first step a phonon and a Stokes photon are emitted. In the second, the phonon interacts with the pump beam to create anti-Stokes photons ($\omega_A = \omega_P + \Omega_v$). Finally, another phonon interacts with the first anti-Stokes wave to generate photons in the direction of the second anti-Stokes wave ($\omega_{2A} = \omega_A + \Omega_v$) and so on.

16.2 STIMULATED BRILLOUIN SCATTERING

The second commonly encountered stimulated scattering phenomenon, especially in fibers, is SBS. The material normal modes involved are acoustic phonons with displacements along the light propagation axis. In contrast to optical phonons that are stationary, acoustic waves *travel* at the velocity of sound. The geometry of the interacting waves in fibers is shown in Fig. 16.17. It will be shown that for the Stokes case for which exponential growth occurs, acoustic waves traveling in the direction backward to that of the pump supply the $\vec{k}$ vector to conserve wave vector in the light–sound interaction. It will also turn out that an anti-Stokes wave will be self-damping.

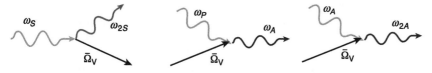

FIGURE 16.16 The three-step process necessary for the generation of second anti-Stokes photons.

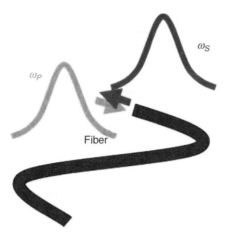

FIGURE 16.17   The geometry of the stimulated Brillouin scattering in a fiber.

The interacting optical waves are potentially

$$\vec{E}(\vec{r}, t) = \frac{1}{2}\hat{e}_x \left[\mathcal{E}_P \, e^{i(k_P z - \omega_P t)} + \mathcal{E}_S \, e^{i(-k_S z - \omega_S t)} + \mathcal{E}_A \, e^{i(-k_A z + -_A t)} + \text{c.c.} \right], \quad (16.43)$$

and the sound waves must be collinear with the pump and Stokes waves for efficient SBS,

$$q = \frac{1}{2} \left[Q_+ \, e^{i(Kz - \Omega_S t)} + Q_+^* \, e^{i(-Kz + \Omega_S t)} + Q_- \, e^{i(-Kz - \Omega_S t)} + Q_-^* \, e^{i(Kz + \Omega_S t)} \right], \quad (16.44)$$

where subscripts "+" and "−" refer to sound waves traveling along the $+z$-axis and $-z$-axis, respectively.

There are good reasons why it was assumed in Eq. 16.44 that only backward traveling Stokes and anti-Stokes waves can occur. Since the speed of sound is orders of magnitude ($\approx 10^5$) less than that of light, $\omega/k = c \gg \Omega_s/K = v_s$. Furthermore, since $\omega_S = \omega_P \pm \Omega_s$ and wave vector must be conserved, the maximum K possible is $\approx 2k_P$ and therefore ω_S equals ω_P and k_S equals k_P to within the ratio v_s/c. Note that if the Stokes or anti-Stokes waves propagate in the same direction as does the pump, then wave-vector conservation requires that $K = 0$ and hence $\Omega_s = 0$, i.e., no interaction. Just as in SRS, it is the Stokes beam that can grow exponentially and so $\omega_S = \omega_P - \Omega_S$ since the acoustic phonons needed for the interaction are generated in this process and can stimulate further conversion to the Stokes beam.

There are only two possible field products for generating the Stokes wave, namely, $\mathcal{E}_S \propto \mathcal{E}_P Q_+^*$ and $\mathcal{E}_S \propto \mathcal{E}_P Q_-^*$ and their complex conjugates. For the first product, $-k_S = k_p - K$, and for the second, $-k_S = k_p + K$. Clearly the second product cannot be wave-vector matched. Hence the pump waves interact with phonons traveling in the same direction. For the anti-Stokes case, $\omega_A = \omega_P + \Omega_S$, the possibilities are $\mathcal{E}_A \propto \mathcal{E}_P Q_+$ and $\mathcal{E}_A \propto \mathcal{E}_P Q_-$, with wave vectors $-k_A = k_p + K$ and $-k_A = k_p - K$, respectively. Only the second case can be wave-vector matched. Note that the Stokes and anti-Stokes waves are generated via interactions with *different* phonon populations: forward traveling for Stokes waves and backward propagating for anti-Stokes waves. Since the anti-Stokes process annihilates phonons, the only population of these phonons available is due to thermal fluctuations and hence this process is quickly damped.

Proceeding with the Stokes case, the nonlinear polarization due to the acousto-optic interaction is written for the gas and condensed matter cases, respectively, as

$$\text{Gas/liquid}: \quad P_i^{\text{NL}}(\vec{r}, t) = -\varepsilon_0(n^2 - 1)[\vec{\nabla} \cdot \vec{q}]E_i(\vec{r}, t),$$
$$\text{Solid}: \quad P_i^{\text{NL}}(\vec{r}, t) = -\varepsilon_0 n_i^2 n_j^2 p_{iijj}^{\text{AO}}[\vec{\nabla} \cdot \vec{q}]_{jj} E_i(\vec{r}, t). \tag{16.45}$$

The coefficient p_{iijj}^{AO} is the acousto-optics coefficient, also known as the Pockels coefficient.

16.2.1 Equation of Motion for Sound Waves in a Gas or Liquid and SVEA

The equation of motion for driven sound waves is very similar to that of light; namely,

$$\rho\left(\ddot{\vec{q}} + 2\tau_s^{-1}\dot{\vec{q}}\right) - \rho v_s^2 \frac{\partial^2 \vec{q}}{\partial z^2} = \vec{F}_q, \tag{16.46}$$

where ρ is the mass density, $\vec{q}$ is the displacement of the medium from its equilibrium, τ_s is the decay time for the sound wave, and $\vec{F}_q$ is the usual driving force that comes from the mixing of optical fields. The acousto-optic interaction for light collinear with the sound waves can occur only for longitudinal (compressional) sound waves, as shown in Fig. 16.18. This modulation of the density generates a modulation of the refractive index, which in turn affects light.

The calculation now proceeds in the way adopted previously for SRS. The acousto-optic interaction potential is given by

$$V_{\text{int}} = -\int \vec{P}^{\text{NL}} \cdot d\vec{E} = \frac{1}{2}\varepsilon_0(n^2 - 1)\left[\vec{\nabla} \cdot \vec{q}\right]\vec{E}_{\text{total}} \cdot \vec{E}_{\text{total}}. \tag{16.47}$$

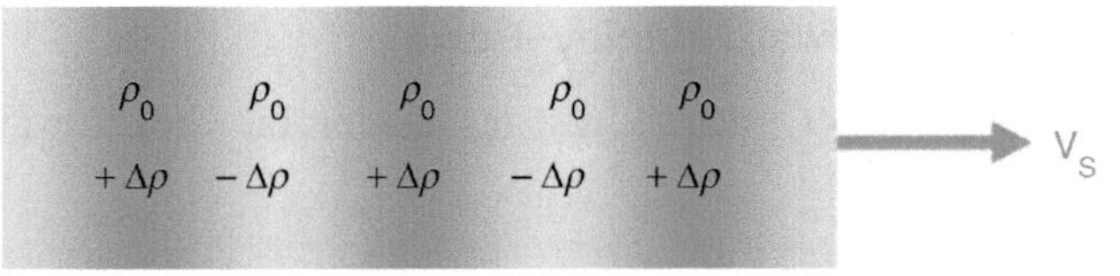

FIGURE 16.18 Longitudinal phonon (acoustic wave) modulating the density and hence the refractive index of a medium.

It is useful to replace the term $[\vec{\nabla} \cdot \vec{q}]\vec{E}_{total} \cdot \vec{E}_{total}$ as follows:

$$\vec{E}_{total} \cdot \vec{E}_{total}\frac{\partial q_z}{\partial z} = \frac{\partial(q_z\vec{E}_{total} \cdot \vec{E}_{total})}{\partial z} - q_z\frac{\partial(\vec{E}_{total} \cdot \vec{E}_{total})}{\partial z}, \qquad (16.48a)$$

and for the Stokes case

$$q_z\vec{E}_{total} \cdot \vec{E}_{total} \propto \mathcal{E}_S^*\mathcal{E}_P Q_+^* \exp\{-i[k_P + k_S - K]z\} = \mathcal{E}_S^*\mathcal{E}_P Q_+^* \;\to\; \frac{d\{q_z\vec{E}_{total} \cdot \vec{E}_{total}\}}{dz} = 0,$$

$$\Rightarrow \vec{E}_{total} \cdot \vec{E}_{total}\frac{\partial q_z}{\partial z} = -q_z\frac{\partial(\vec{E}_{total} \cdot \vec{E}_{total})}{\partial z} = -i\frac{1}{2}q_z(k_P + k_S)\mathcal{E}_S^*\mathcal{E}_P e^{i(k_P + k_S)z} + \text{c.c.}$$

$$(16.48b)$$

Therefore V_{int} can now be written as

$$V_{int} = \frac{1}{4}i\varepsilon_0(n^2 - 1)q_z(k_P + k_S)\mathcal{E}_S^*\mathcal{E}_P e^{i(k_P + k_S)z - i[\omega_P - \omega_S]t} + \text{c.c.} \qquad (16.49)$$

Finding the force $\vec{F}_q$ from $\partial V_{int}/\partial q_z$ and substituting it into Eq. 16.46 gives

$$\rho\left(\ddot{q}_z + 2\tau_s^{-1}\dot{q}_z\right) - \rho v_s^2\frac{\partial^2 q_z}{\partial z^2} = \frac{1}{4}i\varepsilon_0(n^2 - 1)(k_P + k_S)\mathcal{E}_S^*\mathcal{E}_P e^{i(k_P + k_S)z - i[\omega_P - \omega_S]t} + \text{c.c.} \qquad (16.50)$$

The damping of acoustic phonons at the frequencies typical of SBS (tens of gigahertz) is large with decay lengths less than $100\,\mu m$. This limits (saturates) the growth of the phonons and in the SBS case the phonons are damped as fast as they are created after $\approx 100\,\mu m$. Thus solving in this *steady-state* case for the phonons, we obtain

$$Q_+^* = i\frac{\varepsilon_0(n^2 - 1)}{2\rho}\frac{1}{(\omega_P - \omega_S)^2 - \Omega_S^2 - 2i(\omega_P - \omega_S)\tau_s^{-1}}\left[K\mathcal{E}_P^*(z)\mathcal{E}_S(z)\right], \qquad (16.51)$$

which implies that $\mathcal{E}_P^*(z)\mathcal{E}_S(z) = \text{constant}$ when the acoustic phonon flux saturates.

16.2.2 Power Flow (Toward Manley–Rowe)

In Eq. 16.48b, it is the Q_+^* that is responsible for the Stokes wave, and so $\vec{\nabla} \cdot \vec{q} \to -iKQ_+^*$ and

$$\mathcal{P}_S^{NL} = i\frac{1}{2}\varepsilon_0(n_S^2 - 1)(KQ_+^*)\mathcal{E}_P. \qquad (16.52)$$

Noting that $Q_+E_S \propto e^{-i(\omega_P - \omega_S)t - i\omega_S t} \propto e^{-i\omega_P t}$, we obtain

$$\mathcal{P}_P^{\text{NL}} = -i\frac{1}{2}\varepsilon_0(n^2 - 1)(KQ_+)\mathcal{E}_S. \tag{16.53}$$

Applying the SVEA gives

$$-\frac{d\mathcal{E}_S(z)}{dz} = \frac{\omega_S(n^2 - 1)KQ_+^*}{4n_S c}\mathcal{E}_P(z), \qquad \frac{d\mathcal{E}_P(z)}{dz} = -\frac{\omega_P(n^2 - 1)KQ_+}{4n_P c}[\mathcal{E}_S(z)]. \tag{16.54}$$

Converting to intensities by multiplying by the appropriate complex conjugate fields gives

$$-\frac{dI_S(z)}{dz} = \frac{\omega_S(n^2 - 1)K\varepsilon_0}{8}\{Q_+^*\mathcal{E}_P(z)\mathcal{E}_S^*(z) + \text{c.c.}\}, \tag{16.55a}$$

$$\frac{dI_P(z)}{dz} = -\frac{\omega_P(n^2 - 1)K\varepsilon_0}{8}\{Q_+\mathcal{E}_P^*(z)\mathcal{E}_S(z) + \text{c.c.}\} \tag{16.55b}$$

$$\rightarrow \qquad -\frac{1}{\omega_P}\frac{dI_P(z)}{dz} = -\frac{1}{\omega_S}\frac{dI_S(z)}{dz}. \tag{16.55c}$$

Therefore the pump beam that travels along the $+z$-axis depletes and the Stokes beam that travels along the $-z$-axis grows as it travels; i.e., when the pump beam loses a photon, the Stokes beam gains a photon.

To get power flux balance, the equivalent of the SVEA for acoustic waves and power flow are needed. Applying the SVEA assumptions to Eq. 16.46 gives

$$\frac{dQ_+(z)}{dz} + \gamma_S Q_+(z) + \text{c.c.} = i\frac{1}{2\rho v_S^2 K}F_q(z, t)$$

$$\rightarrow \qquad \frac{dQ_+(z)}{dz} = -\left\{\frac{1}{4\rho v_S^2}\varepsilon_0(n^2 - 1)\mathcal{E}_S^*\mathcal{E}_P + \gamma_S Q_+(z)\right\}, \tag{16.56}$$

with $\gamma_s = \tau_s^{-1}v_s^{-1}$. For acoustic power flow, the time average potential energy = time average kinetic energy = $\rho\dot{q}_z^2/2$ and the total energy per unit volume is

$$(U_+) = \overline{\rho\dot{q}_z^2} = \frac{1}{4}\rho[-\Omega_S^2 Q_+^2 \, e^{2i(Kz - \Omega t)} + \text{c.c.} + 2\Omega_S^2 Q_+ Q_+^*] = \frac{1}{2}\rho\Omega_S^2 Q_+ Q_+^*. \tag{16.57}$$

The total acoustic intensity is given by the product of the velocity of sound and (U_+); i.e.,

$$\overline{I_+(\Omega)} = \frac{1}{2}\rho v_s \Omega_s^2 |Q_+|^2 \quad \rightarrow \quad \frac{dI_+(\Omega_s,z)}{dz} = \frac{1}{2}\rho v_s \Omega_s^2 \left(Q_+ \frac{dQ_+^*}{dz} + Q_+^* \frac{dQ_+}{dz}\right).$$

$$(16.58)$$

Substituting Eq. 16.56 gives

$$\frac{dI_+(\Omega_s,z)}{dz} + 2\frac{\tau_s^{-1}}{v_s} I_+(\Omega_s,z) = -\frac{\Omega_s \varepsilon_0 (n^2 - 1)K}{8} \{Q_+^* \mathcal{E}_P(z)\mathcal{E}_s^*(z) + \text{c.c.}\}. \quad (16.59)$$

Note that the decay coefficient for the acoustic phonon *energy* is $\alpha_s = 2\tau_s^{-1}v_s^{-1}$.

The acoustic (Eq. 16.59) and optical power (Eq. 16.55) flows are now combined to give the complete Manley–Rowe relations. Therefore

$$\frac{1}{\Omega_s}\left[\frac{dI_+(\Omega_s,z)}{dz} + 2\frac{\tau_s^{-1}}{v_s} I_+(\Omega_s,z)\right] = -\frac{1}{\omega_P}\frac{dI_P(z)}{dz} = -\frac{1}{\omega_S}\frac{dI_S(z)}{dz}. \quad (16.60)$$

The phonon beam grows in the forward direction by picking up energy from the pump beam. The energy associated with $\hbar[\omega_P - \omega_S] = \hbar\Omega_s$, i.e., the sound waves, also decays into heat. The Stokes wave grows in the backward direction because it also picks up energy from the pump.

16.2.3 Exponential Growth

To this point, exponential growth has not been demonstrated. When the growth of the acoustic phonons is balanced by their attenuation, steady state is reached for Q_+^*. In that case the solution given by Eq. 16.51 can be substituted into Eq. 16.55a to give

$$\frac{dI_S(z)}{dz} = -\frac{\omega_S(n^2 - 1)^2\Omega_s}{4\rho v_s^2 \tau_s^{-1}c^2 n^2}\frac{\tau_s^{-2}}{(\omega_P - \omega_S - \Omega_S)^2 + \tau_s^{-2}}I_S(z)I_P(z) = -g_B I_S(z)I_P(z)$$

$$\Rightarrow \quad g_B = \frac{\omega_S(n^2 - 1)^2\Omega_s}{4\rho v_s^2 \tau_s^{-1}c^2 n^2}\frac{\tau_s^{-2}}{(\omega_P - \omega_S - \Omega_S)^2 + \tau_s^{-2}}$$

$$\Rightarrow \quad g_B^{\max} = \frac{\omega_S(n^2 - 1)^2\Omega_s}{4\rho v_s^2 \tau_s^{-1}c^2 n^2}.$$

$$(16.61)$$

Also for the pump beam,

$$\frac{dI_P(z)}{dz} = -g_B\frac{\omega_P}{\omega_S}I_S(z)I_P(z) = -g_B I_S(z)I_P(z). \quad (16.62)$$

The loss can be added phenomenologically to Eqs 16.61 and 16.62; therefore,

$$\frac{dI_S(z)}{dz} = -g_B I_S(z) I_P(z) + \alpha_S I_S(z), \quad \frac{dI_P(z)}{dz} = -g_B \frac{\omega_P}{\omega_S} I_S(z) I_P(z) - \alpha_P I_P(z).$$

$$(16.63)$$

Clearly Eq. 16.61 gives an exponential growth of the Stokes beam along the $-z$-axis; i.e., it occurs only once the acoustic phonon population saturates. At this point from Eq. 16.56, we obtain

$$\frac{dQ_+(z)}{dz} = 0 \quad \rightarrow \quad Q_+(z) = -\frac{\varepsilon_0 (n^2 - 1)}{2\gamma_s \rho v_s^2} (\mathcal{E}_P \mathcal{E}_S^*). \qquad (16.64)$$

Since $\omega_P = \omega_S$ (to parts per 10^5), the pump beam decays with the same exponent as the Stokes beam grows and hence $\mathcal{E}_P \mathcal{E}_S^* = \text{constant}$, as discussed previously.

The evolution of the pump beam in the forward direction and the Stokes beam in the backward direction is shown in Fig. 16.19a.

The amplifier Brillouin gain given by $[I_S(0) - I_S(L)]/I_S(L)$ for the fixed pump input is shown in Fig. 16.19b as a function of the input pump to Stokes output conversion efficiency. Note that the deviation from the curve occurs only after 90% conversion efficiency, after which it falls to zero.

If instead of a gas, the sample is an isotropic solid, e.g., a fiber, Eq. 16.45 has the form

$$\vec{P}^{\text{NL}}(\vec{r}, t) = -\varepsilon_0 n^4 p_{12} [\vec{\nabla} \cdot \vec{q}(\vec{r}, t)] \vec{E}(\vec{r}, t). \qquad (16.65)$$

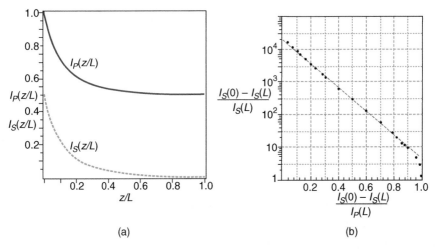

(a)   (b)

FIGURE 16.19 (a) The evolution of the pump beam in the forward direction and the Stokes beam in the backward direction for the value of $I_S(L)/I_P(0) = 0.169$, which produces 50% conversion for $g_B L I_P(0) = 10$. (b) Brillouin gain as a function of conversion efficiency from the pump to the Stokes beams for the gain parameter $g_B L I_P(0) = 10$.

In this case

$$g_B = \frac{\omega_S n^6 p_{12}^2 \Omega_s}{4\rho v_s^2 \tau_s^{-1} c^2} \frac{\tau_s^{-2}}{(\omega_P - \omega_S - \Omega_S)^2 + \tau_s^{-2}}, \qquad g_B^{max} = \frac{\omega_S n^6 p_{12}^2 \Omega_s}{4\rho v_s^2 \tau_s^{-1} c^2}. \qquad (16.66)$$

The acousto-optic constant for materials typically falls in the range $1 \geq p_{12} \geq 0.1$.

16.2.4 SBS Threshold

The analysis up to this point assumes no pump depletion. Threshold and saturation effects are similar to those discussed previously for Raman gain. Since $\omega_S, \omega_P \gg \Omega_s$, $\alpha_S \cong \alpha_P = \alpha$ is an excellent approximation. For no depletion of pump *except* for absorption,

$$\frac{dI_S(z)}{dz} = g_B I_S(z) I_P(z) - \alpha I_S(z), \qquad \frac{dI_P(z)}{dz} = -\alpha I_P(z), \qquad L_{eff} = \frac{1 - \exp(-\alpha L)}{\alpha}$$

$$\Rightarrow \quad I_P(z) = I_P(0)\, e^{-\alpha L_{eff}}, \qquad I_S(0) = I_S(L)\, e^{g_B I_P(0) L_{eff} - \alpha L}, \qquad (16.67)$$

where $I_S(0)$ is the Stokes output intensity and $I_S(L)$ is the Stokes input intensity.

The Brillouin threshold pump intensity is defined as $I_P^{th}(0)$ for which $I_P(0)\, e^{-\alpha L} = I_S(0)$ (2). Assuming unsaturated gain with a Lorentzian line shape for g_B (Eq. 16.61), we obtain

$$g_B I_P^{th}(0) L_{eff} \approx 21. \qquad (16.68)$$

SBS has been seen in fibers at milliwatt power levels for continuous-wave single-frequency inputs. It is the dominant nonlinear effect for continuous-wave beams. For example, for fused silica, $\lambda_P = 1.55\,\mu m$, $n = 1.45$, $v_s = 6\,km/s$, $\Omega_S/2\pi = 11$ GHz, $1/\pi\tau_s \approx 17\,MHz$, which gives $g_B \approx 5 \times 10^{-11}$ m/W. This value is ≈ 500 times larger than the g_R value. However, $2/\tau_s$ is quite small and requires stable single-frequency input to use the larger gain. For non-single-frequency pumps with $\Delta\omega > 1/\pi\tau_s$ (tens to hundreds of megahertz), the effective Brillouin gain is reduced by $\approx 1/\pi\tau_s \Delta\omega_p$.

16.2.5 Pulsed Pump and Stokes Beams

As illustrated in Fig. 16.20, the Stokes and pump beams travel in opposite directions and the overlap with a growing Stokes beam is very small. Hence the Stokes amplification is very small. The shorter the pump pulse, the lesser the Stokes beam is generated; thus, this is a very inefficient process. As a result, SRS dominates for pulses when pulse width is $\ll Ln/c$.

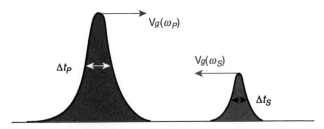

FIGURE 16.20   The geometry for oppositely propagating pulsed pump and Stokes beams.

PROBLEMS

1. In a liquid made up of anisotropic molecules, the rotation of a molecule is strongly hindered by viscosity. Normally the molecules have random orientations. Any induced orientation of the molecules decays back to a random distribution via a time $\bar{\tau}_D$ that is the Debye relaxation time. Local reorientation of the molecules can be induced by surfaces, by linear optics if the molecules have a permanent dipole moment, and by nonlinear optics via the modulation of the molecular polarizability at the difference frequency between two optical fields. If these optical fields are strong enough, this process can become stimulated, e.g., like stimulated Raman scattering. Consider cigar-shaped molecules (e.g., CS_2) with polarizability $\bar{\alpha}_{\parallel}$ along the long cigar axis and $\bar{\alpha}_{\perp}$ in the plane orthogonal to the long axis. The detailed description of this process requires using the "Debye rotational diffusion equation for the orientational distribution function," which is the probability of finding molecular axis ($\bar{\alpha}_{\parallel}$) in a cone at an angle θ_0 to the x-axis (field direction). Consider a simplified version of this equation, in which the evolution of an angle change $\Delta\theta$ due to the mixing of the optical fields for molecules initially oriented at an angle θ_0 relative to the direction of the optical fields is given by

$$\Delta\dot{\theta} + \frac{\Delta\theta}{\bar{\tau}_D} = -\frac{\partial V_{\text{int}}}{\partial \theta}.$$

(Note that this derivative is the torque on the molecule.)

(a) In analogy to stimulated Raman scattering, we assume that there are two copolarized fields present: a pump field at frequency ω_a and a stimulated field at frequency ω_b. Show that

$$\Delta\theta(\omega_a - \omega_b) = -\frac{1}{4}\left[\bar{\alpha}_{\parallel} - \bar{\alpha}_{\perp}\right]\sin(2\theta_0)\frac{\bar{\tau}_D E(\omega_a)E^*(\omega_b)}{1 - i(\omega_a - \omega_b)\bar{\tau}_D}\left(\frac{\varepsilon_r(\omega_a) + 2}{3}\right)\left(\frac{\varepsilon_r(\omega_b) + 2}{3}\right).$$

(b) Show that in the limit of small $\Delta\theta$ the induced *nonlinear* polarization in a molecule along the incident field polarization is given by

$$p_x^{NL}(\omega_b) = \frac{1}{4}[\bar{\alpha}_{||} - \bar{\alpha}_\perp]^2 \sin^2(2\theta_0) f^{(3)} \frac{\bar{\tau}_D |E(\omega_a)|^2 E(\omega_b)}{1 + i(\omega_a - \omega_b)\bar{\tau}_D},$$

$$\text{with } f^{(3)} = \left(\frac{\varepsilon_r(\omega_a) + 2}{3}\right)^2 \left(\frac{\varepsilon_r(\omega_b) + 2}{3}\right)^2.$$

(c) The macroscopic nonlinear polarization $p_x^{NL}(\omega_b)$ is given by $N\langle p_x^{NL}(\omega_b)\rangle$, where $\langle\rangle$ denotes an average over all orientations θ_0, weighted by the appropriate probability. Show that the slowly varying envelope approximation gives

$$\frac{dE(\omega_b)}{dz} = i\frac{\omega_b N\bar{\tau}_D}{15n_b\varepsilon_0 c}[\bar{\alpha}_{||} - \bar{\alpha}_\perp]^2 \frac{1 - i(\omega_a - \omega_b)\bar{\tau}_D}{1 + (\omega_a - \omega_b)^2\bar{\tau}_D^2} f^{(3)} |E(\omega_a)|^2 E(\omega_b).$$

(d) Find the stimulated Rayleigh wing gain spectrum and show that it peaks at $\omega_a - \omega_b \cong \bar{\tau}_D^{-1}$.

2. Derive the effective frequency spectrum of the Raman Stokes signal when the laser input intensity has a Gaussian frequency response shape and the input Stokes field has single frequency for the two cases

   (a) $\omega_{P0} - \omega_{S0} \gg \bar{\tau}_v^{-1} \gg \Delta\omega_P$ and

   (b) $\omega_{P0} - \omega_{S0} \gg \Delta\omega_P \gg \bar{\tau}_v^{-1}$.

   By taking the ratio of Raman gain coefficients found in (a) and (b), show that a pump beam whose bandwidth is much greater than the optical phonon lifetime greatly reduces the Raman gain.

3. Consider stimulated Raman scattering in the approximation that $\alpha_P = \alpha_S$. Show that the nonlinear coupled wave equations

   $$\frac{dI_S(z)}{dz} = g_R I_S(z)I_P(z) - \alpha I_S(z), \qquad \frac{dI_P(z)}{dz} = -g_R\frac{\omega_P}{\omega_S}I_S(z)I_P(z) - \alpha I_P(z)$$

   can be solved to give the saturated gain as

   $$\Rightarrow \quad G_S = \frac{(1 + r_0)e^{-\alpha L}}{r_0 + G_A^{-(1+r_0)}}, \quad \text{with } r_0 = \frac{\omega_P}{\omega_S}\frac{I_S(0)}{I_P(0)} \text{ and } G_A = e^{g_R I_P(0)L_{eff}}.$$

4. Consider stimulated Brillouin scattering in the approximation that $\alpha_P = \alpha_S = 0$. Show that the nonlinear coupled wave equations

   $$\frac{dI_S(z)}{dz} = g_B I_S(z)I_P(z), \qquad \frac{dI_P(z)}{dz} = -g_B I_S(z)I_P(z)$$

can be solved to give the saturated gain as

$$G_S = \frac{I_S(0)}{I_S(L)} = \frac{e^{[(1-b_0)g_B I_P(0)L]} - b_0}{1 - b_0},$$

$$\text{with } b_0 = \frac{I_S(0)}{I_P(0)} \text{ and } G_A = g_B I_P(0)L (\text{unsaturated gain}).$$

5. Consider a system made of two species A and B that are perfectly miscible in one another. The relative dielectric constant is given by $\varepsilon_r = C_A \varepsilon_A + C_B \varepsilon_B$. This system, like any other, is subject to statistical fluctuations that can produce local fluctuations in the concentration of species A and B. A fluctuation ΔC in species A results in a fluctuation in the dielectric constant given by $\Delta \varepsilon_r \approx \Delta C [\varepsilon_A - \varepsilon_B]$. The concentration fluctuation decays exponentially in space and time according to the diffusion equation

$$\left(\frac{\partial}{\partial t} - D \nabla^2 \right) \Delta C = 0.$$

The "nonlinear force" driving the fluctuations can be written as

$$\left(\frac{\partial}{\partial t} - D \nabla^2 \right) \Delta C = -\frac{D(\partial \varepsilon / \partial C) \nabla^2 (\vec{E}_1 \cdot \vec{E}_2^*)}{2 \rho_0 (\partial \mu / \partial C)},$$

where μ is the chemical potential. Show that stimulating scattering can occur from these concentration fluctuations and derive the Stokes intensity.

6. Derive the effective frequency spectrum of the Brillouin Stokes signal when the laser input intensity has a Gaussian frequency response shape and the input Stokes field has single frequency for the two cases
   (a) $\omega_{P0} - \omega_{S0} \gg \tau_s^{-1} \gg \Delta \omega_P$ and
   (b) $\omega_{P0} - \omega_{S0} \gg \Delta \omega_P \gg \tau_s^{-1}$.
   By taking the ratio of Brillouin gain coefficients found in (a) and (b), show that a pump beam whose bandwidth is much greater than the acoustic phonon lifetime greatly reduces the Brillouin gain.

7. Show that the gain coefficient for the backward propagating Raman Stokes beam is the same as for the forward propagating Stokes beam.

REFERENCES

1. G. Agrawal, Nonlinear Fiber Optics, 3rd Edition (Academic Press, San Diego, 2001).
2. R. G. Smith, "Optical power handling capacity of low loss optical fibers as determined by stimulated Raman and Brillouin scattering," Appl. Opt., **11**, 2489–2494, (1972).

3. R. Stegeman, C. Rivero, G. Stegeman, K. Richardson, P. Delfyett, Jr., L. Jankovic, and H. Kim, "Raman gain measurements in bulk glass samples," J. Opt. Soc. Am. B, **22**, 1861–1867 (2005).

4. C. A. Rivero, High Gain/Broadband Oxide Glasses for Next Generation Raman Amplifiers, PhD Thesis (University of Central Florida, 2005).

5. R. Stegeman, G. Stegeman, P. Delfyett, L. Petit, N. Carlie, K. Richardson, and M. Couzi, "Raman gain measurements and photo-induced transmission effects of germanium and arsenic-based chalcogenide glasses," Opt. Express, **14**, 11702–11708 (2006).

6. R. Stegeman, C. Rivero, K. Richardson, G. Stegeman, P. Delfyett, Y. Guo, A. Pope, A. Schulte, T. Cardinal, P. Thomas, and J.-C. Champarnaud-Mesjard, "Raman gain measurements of thallium–tellurium oxide glasses," Opt. Express, **13**, 1144–1149 (2005).

7. R. H. Stolen, C. Lin, and R. K. Jain, "A time-dispersion-tuned fiber Raman oscillator," Appl. Phys. Lett., **30**, 340–342 (1977).

8. http://en.wikipedia.org/wiki/Infrared_window

9. A. Lin, A. Zhang, E. Bushong, and J. Toulouse, "Solid-core tellurite glass fiber for infrared and nonlinear applications," Opt. Express, **17**, 16716–16721 (2009).

10. S. J. Mihailov, Ch. W. Smelser, P. Lu, R. B. Walker, D. Grobnic, H. Ding, G. Henderson, and J. Unruh, "Fiber Bragg gratings made with a phase mask and 800-nm femtosecond radiation," Opt. Lett., **28**, 995–997 (2003).

11. M. Rini, I. Cristiani, V. Degiorgio, A. S. Kurkov, and V. M. Paramonov, "Experimental and numerical optimization of a fiber Raman laser," Opt. Commun., **203**, 139–144 (2002).

SUGGESTED FURTHER READING

G. Agrawal, Nonlinear Fiber Optics, 3rd Edition (Academic Press, San Diego, 2001).

R. W. Boyd, Nonlinear Optics, 3rd Edition (Academic Press, Burlington MA, 2008).

M. J. Damzen, V. Vlad, A. Mocofanescu, and V. Babin, Stimulated Brillouin Scattering: Fundamentals and Applications, Institute of Physics: Series in Optics and Optoelectronics (Taylor & Francis, Boca Raton Florida, 2003).

S. H. Lin, A. A. Villaeys, and Y. Fujimura, Advances in Multiphoton Processses and Spectroscopy (World Scientific Press, Singapore, 1998).

Y. R. Shen, Principles of Nonlinear Optics (John Wiley & Sons, New York, 1984).

CHAPTER 17

Ultrafast and Ultrahigh Intensity Processes

With the exception of Chapters 5, 13 and 14, the nonlinear optical processes analyzed in the preceding chapters were either for continuous waves or for $>$1-ps-long pulses. However in Chapter 13, specifically Fig. 13.1, it was shown that there is a time (pulse width) dependence to the nonlinear phenomena such as the vibrational (Raman) contribution below 1 ps. This was not taken into account when taking the time derivative of the nonlinear polarization $\left(\partial^2 \vec{P}^{(NL)}(\vec{r}, t)/\partial t^2\right)$ in the right-hand side of Eq. 2.21 and clearly must be included when dealing with the Raman process. Furthermore, the spectral breadth of ultrafast processes, especially near the zero group velocity dispersion (GVD) point, requires the next term (third derivative) in the expansion of the wave vector around a central frequency. One of the goals of this chapter is to include these terms and show how they lead to new phenomena, especially at ultrahigh intensities. In this chapter the nonlinear wave equation will be expanded to include higher order dispersion and noninstantaneous response of the nonlinearity.

Although the nonlinearities of fused silica glass are quite small, optical fibers do have a number of advantages for nonlinear optics. The two-dimensional confinement to areas of the order of the squared wavelength allows high intensities and the low losses in the mid-infrared region allow nonlinear effects to accumulate over long distances. In addition, the GVD is small and changes sign in this region of the spectrum. With the invention of photonic crystal fibers, the GVD can be controlled, which allows phenomena such as continuum generation to occur over very wide spectral ranges. Finally, if a parametric interaction that requires wave-vector matching takes place in a multimode fiber waveguide, there exist multiple guided modes with which wave-vector matching can be achieved. A three-wave-mixing process will be used to illustrate this flexibility. However there are complexities in the analysis of fiber phenomena.

In this chapter, there will also be a brief introduction to photonic crystal fibers and an example of their role in nonlinear optics. Three examples of unusual phenomena in fibers will be discussed: self-steepening of pulses, soliton self-frequency shift, and continuum generation. The last topic to be discussed—high harmonic generation—is fascinating, because it cannot be understood in terms of the usual expansion of a

Nonlinear Optics: Phenomena, Materials, and Devices, George I. Stegeman and Robert A. Stegeman.
© 2012 John Wiley & Sons, Inc. Published 2012 by John Wiley & Sons, Inc.

nonlinear polarization in terms of the products of the fields. It is the first example of the new field of extreme nonlinear optics.

17.1 EXTENDED NONLINEAR WAVE EQUATION

The starting point for extending the nonlinear wave equation is the usual wave equation driven by a nonlinear polarization field

$$\nabla^2 E(\vec{r}, t) - \frac{n^2}{c^2} \frac{\partial^2}{\partial t^2} E(\vec{r}, t) = \mu_0 \frac{\partial^2 P^{(\text{NL})}(\vec{r}, t)}{\partial t^2}, \tag{17.1}$$

with fields given by

$$E(\vec{r}, t) = \frac{1}{2} \mathcal{E}(\vec{r}, t) \, e^{i(k_0 z - \omega_0 t)} + \text{c.c.}, \qquad P(\vec{r}, t) = \frac{1}{2} \mathcal{P}(\vec{r}, t) \, e^{i(k_0 z - \omega_0 t)} + \text{c.c.} \tag{17.2}$$

Following the procedures mentioned in Section 14.3.2.1, the Fourier transform of Eq. 17.1 is given by

$$\nabla_\perp^2 \mathcal{E}(\vec{r}, \omega) + \frac{\partial^2 \mathcal{E}(\vec{r}, \omega)}{\partial z^2} + 2ik_0 \frac{\partial \mathcal{E}(\vec{r}, \omega)}{\partial z} + (k^2 - k_0^2) \mathcal{E}(\vec{r}, \omega)$$
$$= -\omega_0^2 \mu_0 \left[1 + i \frac{\omega - \omega_0}{\omega_0} \frac{\partial}{\partial \omega} \right]^2 \mathcal{P}^{(\text{NL})}(\vec{r}, \omega). \tag{17.3}$$

From Eqs 14.27 and 14.28, the $k^2 - k_0^2$ term can be expanded as

$$k^2 - k_0^2 = G(\omega - \omega_0) \cong 2k_0 k_1 (\omega - \omega_0) + k_0 k_2 (\omega - \omega_0)^2 + k_1^2 (\omega - \omega_0)^2$$
$$+ \frac{1}{3} k_0 k_3 (\omega - \omega_0)^3 + \cdots, \tag{17.4}$$

which includes k_3, the dispersion in the GVD for dealing with interactions at and near the GVD $= 0$ point in fibers. Keeping only the leading terms on the left-hand side, we obtain

$$\nabla_\perp^2 \mathcal{E}(\vec{r}, \omega) + \frac{\partial^2 \mathcal{E}(\vec{r}, \omega)}{\partial z^2} + 2ik_0 \frac{\partial \mathcal{E}(\vec{r}, \omega)}{\partial z} + G(\omega - \omega_0) \mathcal{E}(\vec{r}, \omega). \tag{17.5}$$

Returning to the time domain, Eq. 17.3 becomes

$$\nabla_\perp^2 \mathcal{E}(\vec{r}, t) + \frac{\partial^2 \mathcal{E}(\vec{r}, t)}{\partial z^2} + k_0 \left\{ 2i \frac{\partial}{\partial z} + 2ik_1 \frac{\partial}{\partial t} - k_2 \frac{\partial^2}{\partial t^2} - \frac{k_1^2}{k_0} \frac{\partial^2}{\partial t^2} - \frac{1}{3} ik_3 \frac{\partial^3}{\partial t^3} \right\} \mathcal{E}(\vec{r}, t)$$

$$= -\omega_0^2 \mu_0 \left[1 + i \frac{1}{\omega_0} \frac{\partial}{\partial t} \right]^2 \mathcal{P}^{(\text{NL})}(\vec{r}, t). \tag{17.6}$$

The discussion is now limited to nonlinear refraction effects for a single input beam sufficiently far from the material resonances that the imaginary part of the nonlinearity associated with loss can be neglected. Dealing next with the nonlinear polarization term that contains the information about the specific form of the nonlinearity that will be assumed to include both the Kerr and the vibrational Raman contribution, we obtain for sub-picosecond pulses

$$
\mathcal{P}^{(NL)}(\vec{r}, t) = \frac{3}{4}\varepsilon_0 \int_{-\infty}^{t}\int_{-\infty}^{t}\int_{-\infty}^{t} \Re\text{eal}\left\{\hat{\chi}^{(3)}_{\text{Kerr}}(-t; t - t_1, t - t_2, t - t_3)\right.
$$

$$
\left. + \hat{\chi}^{(3)}_{\text{RAM}}(-t; t - t_1, t - t_2, t - t_3)\right\} \times \mathcal{E}(\vec{r}, t_1)\mathcal{E}^*(\vec{r}, t_2)\mathcal{E}(\vec{r}, t_3)\, dt_1\, dt_2\, dt_3.
$$

$$(17.7)$$

It has been shown that the vibrational contribution $\hat{\chi}^{(3)}_{\text{RAM}}(-t; t - t_1, t - t_2, t - t_3)$ can be written as

$$
\hat{\chi}^{(3)}_{\text{RAM}}(-t; t - t_1, t - t_2, t - t_3) = R(t - t_1)\delta(t - t_2)\delta(t - t_3), \qquad (17.8)
$$

where $R(t - t_1)$ is a normalized Raman response function with $\int_{-\infty}^{\infty} R(t - t_1)\, dt_1 = 1$ (1). For the instantaneous Kerr nonlinearity $R(t - t_1) = \delta(t - t_1)$ and for the vibrational Raman contribution, $R(t - t_1)$ depends on the molecule. Thus after conversion to the frequency domain (see Section 14.3.2.1 for details), allowing for the nonlinear susceptibility to vary over the bandwidth of the pulse by expanding it about the carrier frequency as

$$
\tilde{\chi}^{(3)}(-\omega; \omega, -\omega, \omega) = \tilde{\chi}^{(3)}(-\omega_0; \omega_0, -\omega_0, \omega_0) + (\omega - \omega_0)\left[\frac{d\tilde{\chi}^{(3)}(-\omega; \omega, -\omega, \omega)}{d\omega}\right]\Bigg|_{\omega_0},
$$

$$(17.9)$$

and a simplification of the notation $\tilde{\chi}^{(3)}(-\omega_0) \equiv \tilde{\chi}^{(3)}(-\omega_0; \omega_0, -\omega_0, \omega_0)$, the right-hand side of Eq. 17.6 becomes

$$
-\frac{3}{4}\omega_0^2\varepsilon_0\mu_0\left\{\tilde{\chi}^{(3)}_{\text{Kerr}}(-\omega_0) \times \left[1 + \frac{\omega - \omega_0}{\tilde{\chi}^{(3)}_{\text{Kerr}}(-\omega_0)}\left\{\frac{\partial}{\partial\omega}\tilde{\chi}^{(3)}_{\text{Kerr}}(-\omega)\right\}\Bigg|_{\omega_0}\right]\right.
$$

$$
\times\left[1 + 2\frac{(\omega - \omega_0)}{\omega_0}\right]\mathcal{E}(\vec{r}, \omega)|\mathcal{E}(\vec{r}, \omega)|^2\bigg]
$$

$$
+ \hat{\chi}^{(3)}_{\text{RAM}}(-\omega) \times \left[1 + \frac{\omega - \omega_0}{\hat{\chi}^{(3)}_{\text{RAM}}(-\omega)}\left\{\frac{\partial}{\partial\omega}\hat{\chi}^{(3)}_{\text{RAM}}(-\omega)\right\}\Bigg|_{\omega_0}\right]\left[1 + 2\frac{(\omega - \omega_0)}{\omega_0}\right]
$$

$$
\times \int\int_{-\infty}^{\infty} R(\omega - \omega_1)\mathcal{E}(\vec{r}, \omega_1)\mathcal{E}(\vec{r}, \omega_2)\mathcal{E}^*(\vec{r}, \omega_1 + \omega_2 - \omega)\, d\omega_1\, d\omega_2\bigg\}. \quad (17.10)
$$

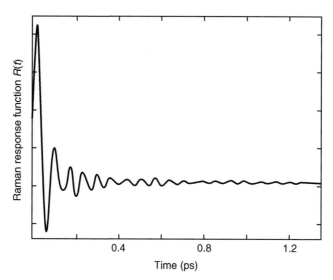

FIGURE 17.1 The temporal response function $R(t)$ for silica glass. Reproduced with permission from the Optical Society of America (1).

$R(t)$ can be evaluated by Fourier transforming the Raman gain or the spontaneous Raman scattering spectrum of the material. An example of $R(t)$ obtained in this way for silica glass is shown in Fig. 17.1. The principal contribution to the dispersion of $\hat{\chi}^{(3)}(-\omega; \omega, -\omega, \omega)$ occurs over the first few hundred femtoseconds in silica glass.

Converting Eqs 17.10 back to the time domain and including Eq. 17.6, we obtain

$$
\nabla_\perp^2 \mathcal{E}(\vec{r},t) + \frac{\partial^2 \mathcal{E}(\vec{r},t)}{\partial z^2} + 2ik_0 \frac{\partial \mathcal{E}(\vec{r},t)}{\partial z} + k_0 \left\{ 2ik_1 \frac{\partial}{\partial t} - \left(k_2 + \frac{k_1^2}{k_0} \right) \frac{\partial^2}{\partial t^2} - \frac{1}{3} ik_3 \frac{\partial^3}{\partial t^3} \right\} \mathcal{E}(\vec{r},t)
$$

$$
\cong -\frac{3}{4} \omega_0^2 \varepsilon_0 \mu_0 \left\{ \tilde{\chi}_{\text{Kerr}}^{(3)}(-\omega_0) \left[1 + i \frac{1}{\tilde{\chi}_{\text{Kerr}}^{(3)}(-\omega)} \left. \frac{\partial \tilde{\chi}_{\text{Kerr}}^{(3)}(-\omega)}{\partial \omega} \right|_{\omega_0} \frac{\partial}{\partial t} + 2i \frac{1}{\omega_0} \frac{\partial}{\partial t} \right] \right.
$$

$$
\mathcal{E}(\vec{r},t) |\mathcal{E}(\vec{r},t)|^2 + \hat{\chi}_{\text{RAM}}^{(3)}(-\omega_0) \left[1 + i \frac{1}{\hat{\chi}_{\text{RAM}}^{(3)}(-\omega)} \left. \frac{\partial \hat{\chi}_{\text{RAM}}^{(3)}(-\omega)}{\partial \omega} \right|_{\omega_0} \frac{\partial}{\partial t} + 2i \frac{1}{\omega_0} \frac{\partial}{\partial t} \right]
$$

$$
\left. \mathcal{E}(\vec{r},t) \int_{-\infty}^{\infty} R(t'-t) |\mathcal{E}(\vec{r},t'-t)|^2 \, dt' \right\}, \tag{17.11}
$$

which contains all the relevant effects that are valid down to tens of femtoseconds. The first term on the right-hand side of Eq. 17.11 describes the Kerr response, and the second term gives the vibrational contribution. Assuming that the pulse envelope still contains many cycles so that the pulse evolves slowly over a wavelength, a Taylor expansion for $|\mathcal{E}(\vec{r}, t'-t)|^2$ can be used, which gives

$$|\mathcal{E}(\vec{r}, t'-t)|^2 \approx |\mathcal{E}(\vec{r}, t)|^2 \delta(t'-t) - t' \frac{\partial |\mathcal{E}(\vec{r}, t)|^2}{\partial t}\bigg|_t \qquad (17.12)$$

$$\rightarrow \int_{-\infty}^{\infty} R(t'-t)|E(\vec{r}, t'-t)|^2 \, dt' = |E(\vec{r}, t)|^2 \left[1 - \int_{-\infty}^{\infty} t' R(t') \, dt' \right] = [1 - T_R] |E(\vec{r}, t)|^2. \qquad (17.13)$$

Hence the Raman part of the response is characterized by its first moment in time. The slowly evolving field approximation is invoked; i.e., $\partial^2 \mathcal{E}(\vec{r}, t)/\partial z^2$ is neglected. The frame of reference from the laboratory frame of reference to one traveling with the pulse is made with the transformations $z = z'$ and $T = t - v_g z = t - k_1 z$, with

$$\frac{\partial}{\partial z} = \frac{\partial}{\partial z'} - k_1 \frac{\partial}{\partial T}, \quad \frac{\partial}{\partial t} = \frac{\partial}{\partial T} \qquad (17.14)$$

to give

$$\nabla_\perp^2 \mathcal{E}(\vec{r}, T) + 2ik_0 \frac{\partial \mathcal{E}(\vec{r}, T)}{\partial z} - k_0 \left\{ \left(k_2 + \frac{k_1^2}{k_0} \right) \frac{\partial^2}{\partial T^2} + \frac{1}{3} ik_3 \frac{\partial^3}{\partial T^3} \right\} \mathcal{E}(\vec{r}, T)$$

$$\cong -\frac{3}{4} \omega_0^2 \varepsilon_0 \mu_0 \left\{ \tilde{\chi}_{\text{Kerr}}^{(3)}(-\omega_0) \times \left[1 + i \frac{1}{\tilde{\chi}_{\text{Kerr}}^{(3)}(-\omega)} \frac{\partial \tilde{\chi}_{\text{Kerr}}^{(3)}(-\omega)}{\partial \omega} \bigg|_{\omega_0} + 2i \frac{1}{\omega_0} \frac{\partial}{\partial T} \right] \mathcal{E}(\vec{r}, T)|\mathcal{E}(\vec{r}, T)|^2 \right.$$

$$\left. + \hat{\chi}_{\text{RAM}}^{(3)}(-\omega_0) \times \left[1 + i \frac{1}{\hat{\chi}_{\text{RAM}}^{(3)}(-\omega)} \frac{\partial \hat{\chi}_{\text{RAM}}^{(3)}(-\omega)}{\partial \omega} \bigg|_{\omega_0} - 2i \frac{T_R}{\omega_0} \frac{\partial}{\partial t T} \right] \mathcal{E}(\vec{r}, T)|\mathcal{E}(\vec{r}, T)|^2 \right\}. \qquad (17.15)$$

It is useful to write this equation in terms of $n_{2\|}(-\omega_0; \omega_0)$ and $n_{2\|, \text{vib}}(-\omega_0; \omega_0)$ as either one of

$$\nabla_\perp^2 \mathcal{E}(\vec{r}, T) + 2ik_0 \frac{\partial \mathcal{E}(\vec{r}, T)}{\partial z} - k_0 \left\{ \left(k_2 + \frac{k_1^2}{k_0} \right) \frac{\partial^2}{\partial T^2} + \frac{1}{3} ik_3 \frac{\partial^3}{\partial T^3} \right\} \mathcal{E}(\vec{r}, T) \cong -n_0^2 k_{0, \text{vac}}^2 c \varepsilon_0$$

$$\times \left\{ n_{2\|}(-\omega_0; \omega_0) \left[1 + i \frac{1}{n_{2\|}(-\omega_0; \omega_0) \partial \omega} n_{2\|}(-\omega; \omega) \bigg|_{\omega_0} + 2i \frac{1}{\omega_0} \frac{\partial}{\partial T} \right] \mathcal{E}(\vec{r}, T)|\mathcal{E}(\vec{r}, T)|^2 \right.$$

$$+n_{2\|,\text{vib}}(-\omega_0;\omega_0)\left[1+i\frac{1}{n_{2\|,\text{vib}}(-\omega_0;\omega_0)\partial\omega}n_{2\|,\text{vib}}(-\omega;\omega)\Big|_{\omega_0}-2i\frac{T_R}{\omega_0}\frac{\partial}{\partial T}\right]$$

$$\mathcal{E}(\vec{r},T)|\mathcal{E}(\vec{r},T)|^2\Bigg\}$$

(17.16a)

or

$$\nabla_\perp^2\mathcal{E}(\vec{r},T)+2ik_0\frac{\partial\mathcal{E}(\vec{r},T)}{\partial z}-k_0\left\{\left(k_2+\frac{k_1^2}{k_0}\right)\frac{\partial^2}{\partial T^2}+\frac{1}{3}ik_3\frac{\partial^3}{\partial T^3}\right\}\mathcal{E}(\vec{r},\ T)\cong-2n_0k_{0,\text{vac}}^2$$

$$\times\left\{n_{2\|}(-\omega_0;\omega_0)\left\{1+i\frac{1}{n_{2\|}(-\omega_0;\omega_0)}\frac{\partial n_{2\|}(-\omega;\omega)}{\partial\omega}\Bigg|_{\omega_0}+2i\frac{1}{\omega_0}\frac{\partial}{\partial T}\right\}\mathcal{E}(\vec{r},T)I(T)\right.$$

$$+n_{2\|,\text{vib}}(-\omega_0;\omega_0)\left[1+i\frac{1}{n_{2\|,\text{vib}}(-\omega_0;\omega_0)}\frac{\partial n_{2\|,\text{vib}}(-\omega;\omega)}{\partial\omega}\Bigg|_{\omega_0}-2i\frac{T_R}{\omega_0}\frac{\partial}{\partial T}\right]\mathcal{E}(\vec{r},T)I(T)\Bigg\}.$$

(17.16b)

The physical significance of most of the new terms in Eqs 17.16a is relatively straightforward. Under normal circumstances, the k_2 term dominates the GVD. However, near the GVD $=0$ point, k_3 becomes the dominant term, describing the effects of dispersion on the temporal pulse envelope. The relative importance of the term k_1^2/k_0 will depend on the material for very short pulses. On the right-hand side, the frequency dispersion in the third-order susceptibility will be important in the off-resonance regime. The $\left[\frac{\partial}{\partial T}\right]\mathcal{E}(\vec{r},T)|\mathcal{E}(\vec{r},T)|^2$ term essentially introduces an intensity dependence to the photon flux; i.e., the group velocity and the last (Raman) term introduces the effect of coupling to the vibration modes.

17.2 FORMALISM FOR ULTRAFAST FIBER NONLINEAR OPTICS

This section establishes the equations needed to analyze nonlinear phenomena in fibers. There is essentially no new physics other than the two-dimensional confinement of the beam so that spatial diffraction does not occur.

Here a brief summary of the relevant properties of fiber propagation is given. In the most general case, there exist multiple guided modes with different wave vectors $\beta^{(s,m)}(\omega)=n_{\text{eff}}^{(s,m)}k_{\text{vac}}(\omega)$ at a given frequency, where $n_{\text{eff}}(\omega)$ is the effective guided wave refractive index of the fiber modes obtained by solving the wave equation in all the guiding media and applying the usual boundary conditions at the interfaces between the media. The parameters s and m take on integer values and characterize the radial and angular dependence of the fields, respectively. In step index fibers (see Fig. 17.2), $n_{\text{eff}}(\omega)$ lies between $n_{\text{core}}(\omega)$ and n_{clad}, the core and cladding refractive indices, respectively. Although the field distributions can be quite complicated, only

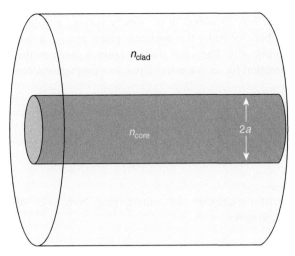

FIGURE 17.2 Step index fiber of core radius a, with core index n_{core} and cladding index n_{clad}.

those polarized along the x- or y-axis are considered in the present discussion. A specific guided field is written as

$$\vec{E}^{(1)}(\vec{r}, t) = \frac{1}{2} a_1^{(s_1, m_1)}(z, T) C^{(s_1, m_1)} \vec{\mathcal{E}}^{(s_1, m_1)}(x, y) \, e^{i(\beta_1^{(s_1, m_1)} z - \omega_1 T)} + \text{c.c.} \qquad (17.17)$$

The fields $C^{(s_1, m_1)} \vec{\mathcal{E}}^{(s_1, m_1)}(x, y)$ are normalized so that $|a_1^{(s_1, m_1)}(z, T)|^2$ is the power $P(z, T)$ carried in the mode and $\vec{\mathcal{E}}^{(s_1, m_1)}(x, y)$ is the normalized field distribution with $\vec{\mathcal{E}}^{(s_1, m_1)}(0, 0) = 1$. The normalization constant $C^{(s_1, m_1)}$ is given by

$$\frac{1}{2} c \varepsilon_0 \left| C^{(s_1, m_1)} \right|^2 \int_{-\infty}^{\infty} \int_{-\infty}^{\infty} n(x, y) |\vec{\mathcal{E}}^{*(s_1, m_1)}(\vec{r}) \cdot \vec{\mathcal{E}}^{(s_1, m_1)}(\vec{r})|^2 \, dx \, dy = 1$$

$$\rightarrow \quad \left| C^{(s_1, m_1)} \right|^2 = \frac{2}{c \varepsilon_0} \left(\int_{-\infty}^{\infty} \int_{-\infty}^{\infty} n(x, y) \left| \vec{\mathcal{E}}^{*(s_1, m_1)}(\vec{r}) \cdot \vec{\mathcal{E}}^{(s_1, m_1)}(\vec{r}) \right|^2 dx \, dy \right)^{-1}.$$

$$(17.18)$$

The number of modes, $M(\omega)$, in a step index fiber given by

$$M(\omega) = \Im\text{nteger} \left\{ a \frac{4}{\pi} \frac{\omega}{c} \sqrt{n_{\text{core}}^2 - n_{\text{clad}}^2} \right\} \qquad (17.19)$$

grows with core fiber radius a and frequency ω. The terminology $\Im\text{nteger}\{ \cdot \}$ means truncation to the next lower integer since the number of modes must have integer values.

As an example of how problems in the slowly varying envelope approximation are dealt with in fibers, consider the nonlinear phase change imparted to a guided wave field at high intensity. The usual starting point is given by the slowly varying envelope approximation for an intensity-dependent propagation constant as

$$2i\beta_1^{(s_1,m_1)} C^{(s_1,m_1)} \mathcal{E}^{(s_1,m_1)}(x,y) \frac{\partial a_1^{(s_1,m_1)}(z,T)}{\partial z} = -c\varepsilon_0 n^2(x,y)k_{vac}^2 n_{2||}(-\omega_0; \omega_0 : x, y)$$

$$\times C^{(s_1,m_1)} \left|C^{(s_1,m_1)}\right|^2 \mathcal{E}^{(s_1,m_1)}(x,y) \left|\mathcal{E}^{(s_1,m_1)}(x,y)\right|^2 \left|a_1^{(s_1,m_1)}(z,T)\right|^2 a_1^{(s_1,m_1)}(z,T).$$

$$(17.20)$$

Following the usual prescription, i.e., multiplying both sides of Eq. 17.20 by $C^{*(s_1,m_1)} \mathcal{E}^{*(s_1,m_1)}(x,y)$ gives

$$\left|C^{(s_1,m_1)}\right|^2 \left|\mathcal{E}^{(s_1,m_1)}(x,y)\right|^2 \frac{\partial a_1^{(s_1,m_1)}(z,T)}{\partial z} = i\frac{c\varepsilon_0 k_{vac}^2}{2\beta_1^{(s_1,m_1)}} \left|C^{(s_1,m_1)}\right|^4$$

$$\times n^2(x,y)n_{2||}(-\omega_0; \omega_0 : x, y)\left|\mathcal{E}^{(s_1,m_1)}(x,y)\right|^4 \left|a_1^{(s_1,m_1)}(z,T)\right|^2 a_1^{(s_1,m_1)}(z,T).$$

$$(17.21)$$

Simplifying to the weakly guiding case in which the optical properties of the core and cladding are almost identical, i.e., $n_{core} - n_{clad} \ll n_{core}$, $n_{2||}(-\omega_0; \omega_0 : x, y) \cong n_{2||}(-\omega_0; \omega_0)$, $\beta_1^{(s_1,m_1)} \cong nk_{vac}$, integrating both sides over the fiber's cross-sectional area, and substituting for $|C^{(s_1,m_1)}|^2$ from Eq. 17.18 gives

$$\frac{\partial a_1^{(s_1,m_1)}(z,T)}{\partial z} = ik_{vac}n_{2||}(-\omega_0; \omega_0) \frac{\int_{-\infty}^{\infty}\int_{-\infty}^{\infty} \left|\mathcal{E}^{(s_1,m_1)}(x,y)\right|^4 dx\, dy}{[\int_{-\infty}^{\infty}\int_{-\infty}^{\infty} \left|\mathcal{E}^{(s_1,m_1)}(x,y)\right|^2 dx\, dy]^2}$$

$$\times |a_1^{(s_1,m_1)}(z,T)|^2 a_1^{(s_1,m_1)}(z,T).$$

$$(17.22)$$

The ratio of integrals has dimension of 1/area, leading to the definition

$$\frac{\int_{-\infty}^{\infty}\int_{-\infty}^{\infty} \left|\mathcal{E}^{(s_1,m_1)}(x,y)\right|^4 dx\, dy}{[\int_{-\infty}^{\infty}\int_{-\infty}^{\infty} \left|\mathcal{E}^{(s_1,m_1)}(x,y)\right|^2 dx\, dy]^2} = \frac{1}{A_{eff}}$$

$$(17.23)$$

for the effective interaction area and so finally

$$\frac{\partial a_1^{(s_1,m_1)}(z,T)}{\partial z} = ik_{vac}n_{2||}(-\omega_0; \omega_0)a_1^{(s_1,m_1)}(z,T)\frac{|a_1^{(s_1,m_1)}(z,T)|^2}{A_{eff}}.$$

$$(17.24)$$

For a fiber, there is no spatial diffraction due to the two-dimensional confinement and it is straightforward to show that Eq. 17.16a can be rewritten as

$$2ik_0 \frac{\partial a^{(s_1,m_1)}(z,T)}{\partial z} - k_0 \left\{ \left(k_2 + \frac{k_1^2}{k_0} \right) \frac{\partial^2}{\partial T^2} + \frac{1}{3} ik_3 \frac{\partial^3}{\partial T^3} \right\} a^{(s_1,m_1)}(z,T) \cong -2n_0 k_{0,\text{vac}}^2$$

$$\times \left\{ n_{2\|}(-\omega_0;\omega_0) \left[1 + i \frac{1}{n_{2\|}(-\omega_0;\omega_0)} \left. \frac{\partial n_{2\|}(-\omega;\omega)}{\partial\omega} \right|_{\omega_0} + 2i\frac{1}{\omega_0}\frac{\partial}{\partial T} \right] a^{(s_1,m_1)}(z,T) \right.$$

$$\times \frac{P^{(s_1,m_1)}(z,T)}{A_{\text{eff}}} + n_{2\|,\text{vib}}(-\omega_0;\omega_0) \left[1 + i\frac{1}{n_{2\|,\text{vib}}(-\omega_0;\omega_0)} \left. \frac{\partial n_{2\|,\text{vib}}(-\omega;\omega)}{\partial\omega} \right|_{\omega_0} \right.$$

$$\left. \left. -2i\frac{T_R}{\omega_0}\frac{\partial}{\partial T} \right] a^{(s_1,m_1)}(z,T) \frac{P^{(s_1,m_1)}(z,T)}{A_{\text{eff}}} \right\}. \tag{17.25}$$

The discovery of photonic crystal fibers has had a huge impact on nonlinear optics in fibers. The cross section of an early photonic crystal fiber is shown in Fig. 17.3 (2). A naive explanation of the function of the holes is that they reduce the refractive index in the vicinity of the core region and act as a cladding. The size, shape, and distribution pattern of the holes all play key roles in the optical properties of the photonic crystal fiber, including the GVD. The effective nonlinearity $n_{2\|}(-\omega;\omega)$ depends on the "filling fraction" of the glass. The key property that is affected by the photonic crystal

FIGURE 17.3 Example of the cross section of a typical photonic crystal fiber. Reproduced with permission from the Optical Society of America (2).

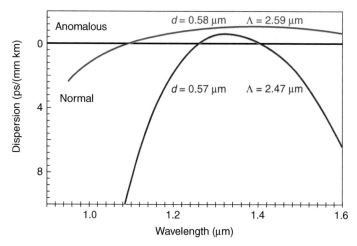

FIGURE 17.4 The dispersion $(D = -2\pi ck_2/\lambda^2)$ with wavelength of two similarly engineered photonic crystal fibers with 11 periods. Red curve: pitch $\Lambda = 2.59\,\mu\text{m}$; hole diameter $d = 0.58\,\mu\text{m}$. Blue curve: $\Lambda = 2.47\,\mu\text{m}$; $d = 0.57\,\mu\text{m}$. Reproduced with permission from the Optical Society of America (3).

structure is the GVD, which can be shifted in wavelength, e.g., into the visible region, and whose dispersion can be controlled. An example of design and fabrication to produce a flat GVD response and the tolerances required is shown in Fig. 17.4 (3).

17.3 EXAMPLES OF NONLINEAR OPTICS IN FIBERS

Many interesting nonlinear phenomena have been observed in fibers at high intensities and long propagation distances. A comprehensive discussion can be found in the book *Nonlinear Fiber Optics* by Agrawal (4). A few examples are presented here.

17.3.1 Nondegenerate Four-Wave Mixing

The analysis of the case with one input at frequency (ω_1) generating two side bands at frequencies $\omega_{3,4} = \omega_1 \pm \Delta\omega$ previously analyzed in bulk media will be reconsidered in a fiber geometry. The wave-vector-matching condition becomes

$$2\omega_1 n_{\text{eff}}^{(s_1,m_1)}(\omega_1) - \omega_3 n_{\text{eff}}^{(s_3,m_3)}(\omega_3) + \omega_4 n_{\text{eff}}^{(s_4,m_4)}(\omega_4) = 0. \qquad (17.26)$$

The multiplicity of guided waves in a multimode fiber enhances the probability of obtaining wave-vector matching, and this phenomenon has indeed been observed for different mode combinations. The modal fields are written as

$$\vec{E}^{(1)}(\vec{r}, T) = \frac{1}{2} C^{(m_1,s_1)} \vec{\mathcal{E}}^{(s_1,m_1)}(x,y) a_1^{(m_1,s_1)}(z, T) \, e^{i(\beta^{(s_1,m_1)}(\omega_1)z - \omega_1 t)} + \text{c.c.},$$

$$\vec{E}^{(3)}(\vec{r}, T) = \frac{1}{2} C^{(m_3,s_3)} \vec{\mathcal{E}}^{(s_3,m_3)}(x,y) a_3^{(m_3,s_3)}(z, T) \, e^{i(\beta^{(s_3,m_3)}(\omega_3)z - \omega_3 t)} + \text{c.c.}, \quad (17.27)$$

$$\vec{E}^{(4)}(\vec{r}, T) = \frac{1}{2} C^{(m_4,s_4)} \vec{\mathcal{E}}^{(s_4,m_4)}(x,y) a_4^{(m_4,s_4)}(z, T) \, e^{i(\beta^{(s_4,m_4)}(\omega_4)z - \omega_4 t)} + \text{c.c.}$$

The drawback in using multimode fibers, however, is that the efficiency is proportional to the spatial integral over the overlap product of the fields and the third-order susceptibility, e.g.,

$$\int\int \tilde{\chi}_{iijj}^{(3)}(-[2\omega_1 - \omega_{3,4}]; \omega_1, \omega_1, \omega_{3,4}) \left[\mathcal{E}_i^{*(s_1,m_1)}(x,y)\right]^2 \mathcal{E}_j^{*(s_3,m_3)} \mathcal{E}_j^{*(s_4,m_4)}(x,y) \, dx \, dy,$$

$$(17.28)$$

where $i = j$ for copolarized waves and $i \neq j$ for signal and idler orthogonally polarized to the pump. Since the fields in the core are given by the product of Bessel functions of order $s - 1$, s, or $s + 1$ and either $\sin(m\phi)$ or $\cos(m\phi)$, only fortuitous or carefully engineered circumstances lead to large values of the overlap integral and hence efficient processes.

Assuming copolarized beams, the Kerr effect near the nonresonant regime, weak guiding and no pump beam attenuation, the nonlinear refraction due to the pump beam product based on Eq. 17.24 produces the nonlinear phase shown below:

$$a_1^{(s_1,m_1)}(z, T) = a_1^{(s_1,m_1)}(0, T) \, e^{ik_{\text{vac}}(\omega_1)\overline{n_{2\|}(-\omega_1; \omega_1)}\frac{[a_1^{(s_1,m_1)}(0, T)]^2}{A_{\text{eff}}}z}. \quad (17.29)$$

The coupled wave equations are similar to those of the bulk case; i.e.,

$$\frac{da_3^{(s_3,m_3)}(z, T)}{dz} = i\delta_3 a_3^{(s_3,m_3)}(z, T) + i\Gamma_3 a_4^{*(s_4,m_4)}(z, T) \, e^{i\kappa z},$$

$$(17.30)$$

$$\frac{da_4^{(s_4,m_4)}(z, T)}{dz} = i\delta_4 a_4^{(s_4,m_4)}(z, T) + i\Gamma_4 a_3^{*(s_3,m_4)}(z, T) \, e^{i\kappa z},$$

with the following definitions in which the cross-nonlinear refraction term $n_{2\|}(-\omega_{3,4}; \omega_1) \cong 2n_{2\|}(-\omega_1; \omega_1)$, the nonlinear coefficient for wave mixing between the signal and the idler term $\tilde{\chi}^{(3)}(-\omega_{3,4}; \omega_1, -\omega_{4,3}, \omega_1)$ is approximated by $\tilde{\chi}^{(3)}(-\omega_1; \omega_1, -\omega_1, \omega_1)$ (i.e., the frequencies are far off resonance), and the product $[a_1^{(s_1,m_1)}(0, T)]^2 = |a_1^{(s_1,m_1)}(0, T)|^2 \, e^{2i\psi}$ (where ψ is the phase of the input beam, which is set to zero). Furthermore,

$$\kappa = \Delta\beta + 2k_{\text{vac}}(\omega_1)\overline{n_{2\|}(-\omega_1; \omega_1)} \frac{\left|a_1^{(s_1,m_1)}(z, T)\right|^2}{A_{\text{eff}}^{(1,1,1)}}, \quad (17.31a)$$

$$\Delta\beta = 2\beta^{(s_1,m_1)}(\omega_1) - \beta^{(s_3,m_3)}(\omega_3) - \beta^{(s_4,m_4)}(\omega_4),$$

$$\delta_3 = 2k_{vac}(\omega_3)\frac{n_1}{n_3}\overline{n_{2\|}}(-\omega_1;\,\omega_1)\frac{\left|a_1^{(s_1,m_1)}(0,T)\right|^2}{A_{eff}^{(3,3,1,1)}},$$

$$\delta_4 = 2k_{vac}(\omega_4)\frac{n_1}{n_4}\overline{n_{2\|}}(-\omega_1;\,\omega_1)\frac{\left|a_1^{(s_1,m_1)}(0,T)\right|^2}{A_{eff}^{(4,4,1,1)}},$$

$$\frac{1}{A_{eff}^{(3,3,1,1)}} = \frac{\int_{-\infty}^{\infty}\int_{-\infty}^{\infty}[\mathcal{E}^{(s_1,m_1)}(x,y)]^2[\mathcal{E}^{*(s_3,m_3)}(x,y)]^2\,dx\,dy}{[\int_{-\infty}^{\infty}\int_{-\infty}^{\infty}|\mathcal{E}^{(s_3,m_3)}(x,y)|^2\,dx\,dy]^2},$$

$$\frac{1}{A_{eff}^{(4,4,1,1)}} = \frac{\int_{-\infty}^{\infty}\int_{-\infty}^{\infty}[\mathcal{E}^{(s_1,m_1)}(x,y)]^2[\mathcal{E}^{*(s_4,m_4)}(x,y)]^2\,dx\,dy}{[\int_{-\infty}^{\infty}\int_{-\infty}^{\infty}|\mathcal{E}^{(s_4,m_4)}(x,y)|^2\,dx\,dy]^2},$$

(17.31b)

$$\Gamma_3 = k_{vac}(\omega_3)\frac{n_1}{n_3}\overline{n_{2\|}}(-\omega_1;\,\omega_1)\frac{\left|a_1^{(s_1,m_1)}(0,T)\right|^2}{A_{eff}^{(3,4,1,1)}}e^{2i\psi(\omega_1)},$$

$$\Gamma_4 = k_{vac}(\omega_4)\frac{n_1}{n_4}\overline{n_{2\|}}(-\omega_1;\,\omega_1)\frac{\left|a_1^{(s_1,m_1)}(0,T)\right|^2}{A_{eff}^{(4,3,1,1)}}e^{2i\psi(\omega_1)},$$

$$\frac{1}{A_{eff}^{(4,3,1,1)}} = \frac{\int_{-\infty}^{\infty}\int_{-\infty}^{\infty}[\mathcal{E}^{(s_1,m_1)}(x,y)]^2\mathcal{E}^{*(s_3,m_3)}(x,y)\mathcal{E}^{*(s_4,m_4)}(x,y)\,dx\,dy}{[\int_{-\infty}^{\infty}\int_{-\infty}^{\infty}|\mathcal{E}^{(s_4,m_4)}(x,y)|^2\,dx\,dy]^2},$$

$$\frac{1}{A_{eff}^{(3,4,1,1)}} = \frac{\int_{-\infty}^{\infty}\int_{-\infty}^{\infty}[\mathcal{E}^{(s_1,m_1)}(x,y)]^2\mathcal{E}^{*(s_3,m_3)}(x,y)\mathcal{E}^{*(s_4,m_4)}(x,y)\,dx\,dy}{[\int_{-\infty}^{\infty}\int_{-\infty}^{\infty}|\mathcal{E}^{(s_3,m_3)}(x,y)|^2\,dx\,dy]^2}.$$

(17.31c)

Unless the mode numbers s_i, $i = 1-4$, are all equal and the mode numbers m_j, $j = 1-4$, are also all equal, the effective areas $A_{eff} \equiv A_{eff}^{(1,1,1,1)}$, $A_{eff}^{(4,3,1,1)}$, and $A_{eff}^{(3,4,1,1)}$ can be quite different from one another. If the fiber is single mode at all frequencies, then the mode numbers are indeed all equal, and so the areas are roughly equal and can be replaced by just A_{eff} in Eq. 17.23 and

$$\delta_{3,4} = 2k_{vac}(\omega_{3,4})\frac{n_1}{n_{3,4}}\overline{n_{2\|}}(-\omega_1;\,\omega_1)\frac{P_1^{(s_1,m_1)}(z,T)}{A_{eff}} = 2\Gamma_{3,4}.$$ (17.32)

Making the same substitutions as in Section 15.3.1, the solutions are

$$a_3^{(s_3,m_3)}(z,T) = a_3^{(s_3,m_3)}(0,T)\left[\cosh(zg) - i\frac{\kappa - (\delta_3 + \delta_4)}{2g}\sinh(gz)\right]e^{-i\frac{\kappa - (\delta_3 + \delta_4)}{2}z},$$

$$a_4^{(s_4,m_4)}(z,T) = i\frac{\Gamma_4}{g}a_3^{*(s_3,m_3)}(0,T)\sinh(gz)\,e^{-i\frac{\kappa - (\delta_3 + \delta_4)}{2}z},$$

(17.33)

and the gain condition is again

$$4\Gamma_3\Gamma_4 > [\kappa - (\delta_3 + \delta_4)]^2. \tag{17.34}$$

The conclusion is that the *analysis* of guided-wave interactions is considerably more complex than that in bulk media. But the advantages of high intensity over long distances and ease of experimental implementation have resulted in many experiments being performed in fibers (4).

17.3.2 Pulse Self-Steepening

This phenomenon was first considered to explain effects observed in liquids (5). Adapting this theory to high intensity propagation in fibers requires the general equation 17.25, which contains many different phenomena that can occur simultaneously. Stripping away all the terms unrelated to pulse self-steepening, we obtain

$$\frac{\partial a_1^{(s_1,m_1)}(z,T)}{\partial z} \cong i\zeta_1 a_1^{(s_1,m_1)}(z,T)\left|a_1^{(s_1,m_1)}(z,T)\right|^2 - \zeta_2\frac{\partial\left[a_1^{(s_1,m_1)}(z,T)\left|a_1^{(s_1,m_1)}(z,T)\right|^2\right]}{\partial T}, \tag{17.35a}$$

with

$$\zeta_1 = \frac{k_{\mathrm{vac}}(\omega_0)\overline{n_{2\|}(-\omega_0;\,\omega_0)}}{A_{\mathrm{eff}}}, \qquad \zeta_2 = 2\frac{k_{\mathrm{vac}}(\omega_0)\overline{n_{2\|}(-\omega_0;\,\omega_0)}}{\omega_0 A_{\mathrm{eff}}}. \tag{17.35b}$$

Clearly the solution depends on the pulse shape $a^{(s_1,m_1)}(z,T)$ due to the term proportional to ζ_2. Writing $a_1^{(s_1,m_1)} = \sqrt{P_1(z,t)}\,e^{i\varphi_1(z,T)}$, Eq. 17.35a separates into an equation for power and one for the phase:

$$\frac{\partial\varphi(z,T)}{\partial z} = \zeta_1 P_1(z,T) - \zeta_2 P_1(z,T)\frac{\partial\varphi_1(z,T)}{\partial T}, \tag{17.36a}$$

$$\frac{\partial P_1(z,T)}{\partial z} = -3\zeta_2 P_1(z,T)\frac{\partial P_1(z,T)}{\partial T}. \tag{17.36b}$$

Assuming a Gaussian pulse shape $P_1(z,T) \propto P_1(z,0)\exp[-T^2/T_0^2]$ and using the method of characteristics to solve Eq. 17.36b, the solution for fibers is (6)

$$P_1(z,T) \propto e^{-\left\{\frac{T}{T_0} - \zeta_2 P_1(z,T)z\right\}^2}. \tag{17.37}$$

The pulse shape evolution with increasing distance is shown in Fig. 17.5. The trailing edge becomes progressively steeper. This arises due to an intensity-dependent group velocity.

A different effect occurs for solitons due to Eq. 17.35a, where it is manifest by a shift in the soliton peak with propagation distance. The soliton "robustness" leaves the soliton's shape essentially preserved. Physically this effect is due to a shift in the carrier frequency with increasing distance and hence the group velocity.

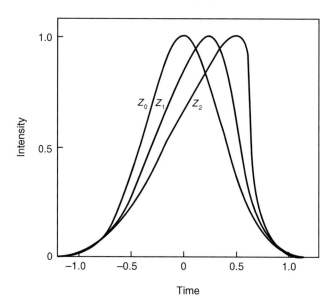

FIGURE 17.5 Self-steepening of the trailing edge of a Gaussian pulse for increasing propagation distance z. Reproduced with permission from the American Physical Society (5).

17.3.3 Soliton Self-Frequency Shift

A self-frequency shift also occurs due to the Raman effect for short pulses in fibers. It was shown in Chapter 16 that the Raman gain always occurs for any intense waveform. The spectral dependence of the Raman gain is reproduced in Fig. 17.6 along with the "pump" (soliton) pulse that produces it. This effect is described by the following terms from Eq. 17.25:

$$\frac{\partial a_1^{(s_1,m_1)}(z,T)}{\partial z} + i\zeta_4 \frac{\partial^2 a_1^{(s_1,m_1)}(z,T)}{\partial T^2} \cong i\zeta_1 a_1^{(s_1,m_1)}(z,T)\left|a_1^{(s_1,m_1)}(z,T)\right|^2$$

$$+\zeta_3 \frac{\partial\left\{a_1^{(s_1,m_1)}(z,T)\left|P_1(z,T)\right|\right\}}{\partial T}, \tag{17.38a}$$

with the substitutions

$$\zeta_1 = \frac{k_{\text{vac}}(\omega_0)\overline{n_{2\|}(-\omega_0;\omega_0)}}{A_{\text{eff}}}, \quad \zeta_3 = -2T_R\frac{\overline{n_{2\|,\text{vib}}(-\omega_0;\omega_0)}}{cA_{\text{eff}}}, \quad \zeta_4 = \frac{1}{2}k_2. \tag{17.38b}$$

To isolate this effect, the term responsible for "self-steepening" was set to zero. The Raman gain amplifies the low frequency tail of an arbitrary pulse, distorting the

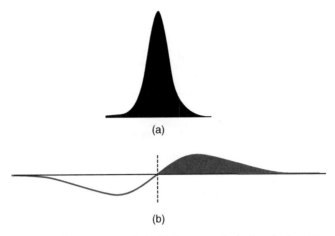

FIGURE 17.6 (a) Soliton (pump pulse). (b) Raman gain (red region) and loss (no color) produced by the soliton due to the stimulated Raman process.

pulse shape and shifting its center of energy toward lower frequencies. On the high frequency side, the pulse is attenuated. This process channels power progressively into lower frequencies, making the frequency spectrum asymmetric.

A fascinating example of this phenomenon occurs when the "pump" beam is a temporal soliton. The physics is essentially the same. Energy is lost to the vibrational modes, and the center of energy of the pulse moves toward lower frequencies. However, because of soliton robustness, the soliton retains its shape and properties as it moves toward lower frequencies. This case can be treated by a perturbation method for solitons which is based on the Euler–Lagrangian equation of classical mechanics, which is beyond the scope of this textbook. This approach has been shown to give the frequency shift for a temporal soliton (see Section 14.4.2 for details about temporal solitons) as (7)

$$\Delta\omega_{\text{sol}}(z) = -\frac{8}{15}\frac{|k_2|T_R}{T_0^4}z. \tag{17.39}$$

The typical evolution shown in Fig. 17.7 is another example of soliton robustness.

17.3.4 Continuum Generation

There are a number of effects that lead to the power of an input beam being spread over a large range of frequencies. These include self- and cross-nonlinear refraction, four-wave mixing of various kinds, the self-frequency shifts just discussed, and higher harmonic generation. Although normally many of these interactions are not phase matched, it was shown in Chapter 16 that nonlinear refraction can be used to achieve phase matching at sufficiently high intensities. Using multimode fibers enhances the number of possible wave-mixing interactions. Furthermore, each new frequency that is created can then act as a pump for the subsequent creation of more new

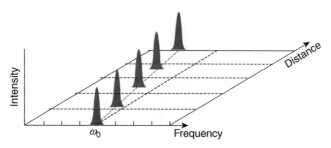

FIGURE 17.7 The soliton self-frequency shift on propagation toward lower frequencies.

frequencies. All these effects are further enhanced in photonic crystal fibers in which the GVD can be engineered to be small over large spectral regions (3). In fibers, both step index and photonic crystal, this has resulted in the generation of broad frequency continua. Although the physics of the contributing phenomena are understood, a simple theory for predicting the general shape of the continuum without extensive numerical computing just does not exist in general.

The best results in terms of bandwidth have been obtained in photonic crystal fibers because their properties can be engineered so precisely. A picture taken of the side scattered light in the visible region for such a photonic crystal fiber excited at 1550 nm is shown in Fig. 17.8 (8). The evolution of the visible spectrum with distance is clear,

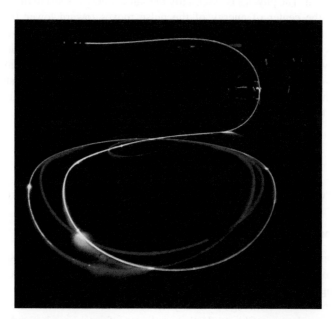

FIGURE 17.8 Continuum of light scattered by imperfections inside a SF6 photonic crystal fiber, a few meters long, which was excited at 1550 nm. Note the progression of color from the deep red (near-infrared) to purple (ultraviolet) at the output. Reproduced with permission from the Optical Society of America (8).

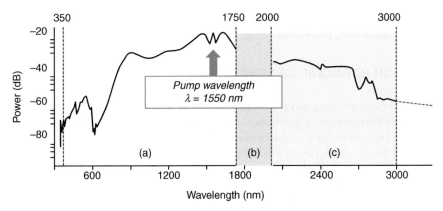

FIGURE 17.9   The continuum generated in a SF6 photonic crystal fiber, 5.7 mm long, excited at 1550 nm. The gap between 1750 and 2000 nm is due to the lack of overlap of the two detectors used. Reproduced with permission from the Optical Society of America (9).

from red to purple. The best results have shown a measured output spectrum that extends from 375 nm to more than 3000 nm (9). Note that the region where no spectrum appears in Fig. 17.9 does not mean that there are no frequencies generated in this range. The gap is actually due to the lack of an appropriate detector for this range when the experiments were done.

Continuum generation has been obtained from ~0.5 µm all the way out to in excess of 4 µm by using a nonsilica fiber with good transmission characteristics out to these important long wavelengths. Using a standard single-mode fiber as seed followed by a ZBLAN fluoride fiber, an average power of 23 mW was demonstrated out to 4.5 µm (see Fig. 17.10) (10). The authors obtained reasonable agreement between experiment

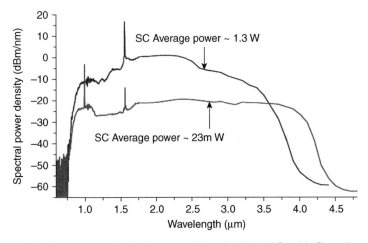

FIGURE 17.10   Continuum of radiation out to 4.5 µm in ZBLAN fluoride fibers. Reproduced with permission from the Optical Society of America (10).

and theoretical calculations, which included dispersion out to the fourth-order, self-nonlinear refraction, pulse steepening, and the Raman terms.

17.4   HIGH HARMONIC GENERATION

One of the most amazing and provocative results in nonlinear optics was reported in the early 1980s in which the generation of a large number of odd order harmonics of an incident CO_2 laser beam was observed at very high intensities from electron plasmas generated from carbon targets (11). Carman et al. reported that instead of continuously decreasing with decreasing wavelength, the harmonic intensity plateaued between the 16th and 42nd harmonics corresponding to 662 and 252 nm. Reproduced in Fig. 17.11 are the results of Carman et al., which show the plateau and the subsequent drop-off to zero after the plateau (11). This result signaled the birth of a new field now called "extreme nonlinear optics." A few years later, Ferray et al. observed a similar plateau in the harmonic generation efficiency with a Nd:YAG laser out to 32 nm in the atomic noble gas Xe (12). This effect typically requires intensities of $10^{13} - 10^{15}$ W/cm^2, which lead to the ionization of these gases, resulting in free electrons and ionized atoms. Subsequent work showed indications of excitation of up to almost 300 harmonics with the wavelengths out to 2.7 nm in the X-ray region (13).

The existence of the plateaus shown in Fig. 17.11 ruled out the possibility that these results could be explained in terms of an expansion of the nonlinear polarization in powers of the field, e.g., Eq. 1.4. The general physics of the generation of the harmonics and their maximum number is now reasonably well understood as a three-step process, but many important details are still emerging over 25 years later (14–17). (In fact, the same scientific community that investigated this phenomenon pioneered the generation of sub-femtosecond pulses, partly in response to this problem (15).) As the intensity of a beam input into a noble gas is increased, an electron can be ionized by a number of different processes, starting with multiphoton absorption, tunneling out of a reduced potential well of the atom, and then finally ejecting directly, as shown in

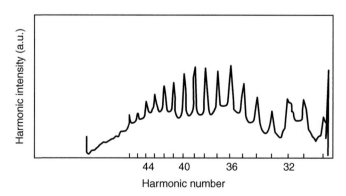

FIGURE 17.11   The results of Carman et al. on the measured harmonic intensity versus harmonic number. Reproduced with permission from American Physical Society (11).

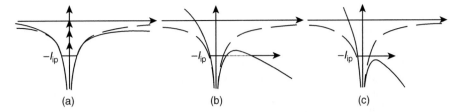

FIGURE 17.12 Ionization of electrons for increasing light intensity. The dashed line is the unperturbed potential well with small fields. The solid line is the potential well perturbed by the strong optical field. I_{ip} is the ionization potential. (a) The mechanism is multiphoton absorption. (b) Tunneling through the potential well side lowered by the strong field. (c) Direct emission due to strong distortion of the barrier. Reproduced with permission from American Physical Society (16).

Fig. 17.12 (16,18). Mechanisms 2 and 3 require optical fields strong enough to tilt the atomic potential well in individual atoms and molecules (see the figure).

The ionization of the electrons is the first step in the three-step process shown in Fig. 17.13, which is the accepted simplified version of the physics involved (14–17). In Step 2, the strong fields drive away the electrons from their parent ions into "orbits" (see Fig. 17.13), which depend on the intensity of the pulse (picoseconds to sub-femtoseconds), its polarization, its phase at the time of ionization, as well as other variables. When the oscillating optical field reverses its sign, the orbit returns the electron to the vicinity of the ion, where it can recombine for appropriate orbits that depend primarily on the phase of the optical field at the time of electron emission (15–17). The largest deceleration of the electrons occurs due to recombination with the ion that occurs twice in every cycle of the optical field. As a result, the electron loses the kinetic energy acquired by the electron in the field and the ionization energy due to

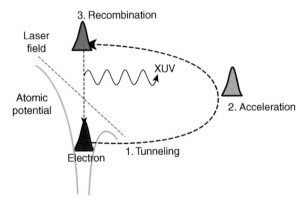

FIGURE 17.13 The three-step process for high harmonic generation (18). In Step 1, an electron tunnels through the atomic potential barrier, which is reduced by the strong field. In Step 2, the electron is accelerated by the strong field, gaining energy and brought back to the vicinity of the parent ion when the field reverses. In Step 3, the electron recombines with the ionized atom releasing its energy, including the ionization potential.

recombination. The radiation contains many sidebands at frequencies $\omega_L + q\omega_L$, with q being an odd integer. The electrons whose orbits do not result in recombination form an electron plasma, which ultimately decays due recombination with ions on time-scales depending on the density of ions, ranging from picoseconds to nanoseconds (19).

The maximum value of q_{max} is determined by the energy lost in the recombination process. Detailed calculations for the recombination process in the presence of the strong laser field have given

$$q_{max}\hbar\omega_L = 3.17\,K + 1.32 I_{ip}, \qquad (17.40)$$

with K being the "jitter" energy of an electron in an oscillating field and I_{ip} being the ionization potential of the atom (14). This agrees remarkably well with experiments. Step 3 requires detailed quantum mechanics to explain the emission process itself since the wave functions of the electrons and the ions are still partially coherent with each other when the electron returns (14). The calculation is beyond the scope of this textbook. Although new details available with femtosecond sources are continuously emerging (at the time of writing of this textbook), the general features of this process have been confirmed.

The "best" fractional transfer of energy from the pump laser to harmonics with wavelengths in the plateau region is about 10^{-7}, and there are continuing efforts to increase this value (20). One approach to maintain the high intensities required is to use a capillary to "guide" the light. Another approach is to wave-vector match the fundamental laser wavelength to a chosen range of harmonic wavelengths. Of course, since the wavelength range of the harmonics is so large, enhancing all the harmonics simultaneously is not possible. Even selectively enhancing a few neighboring harmonics is not trivial since there are so many secondary contributions to the index dispersion over this intensity range such as the density of the atoms, the ions, and the electrons, geometrical effects due to capillary, and the usual nonlinear effects such as self-refraction of the pump beam. Two quasi-phase-matching approaches have resulted in enhancing the harmonic intensities, e.g., by modulating periodically the width of the capillary tube or by optically producing a standing wave pattern in the electron and ion density with two counterpropagating beams in the capillary.

It is quite probable that this field is just the beginning of extreme nonlinear optics.

REFERENCES

1. R. H. Stolen, J. P. Gordon, W. J. Tomlinson, and H. A. Haus, "Raman response function of silica-core fibers," J. Opt. Soc. Am. B, **6**, 1159–1166 (1989).

2. T. A. Birks, J. C. Knight, and P. St. J. Russell, "Endlessly single-mode photonic crystal fiber," Opt. Lett., **22**, 961–963 (1997).

3. W. H. Reeves, J. C. Knight, P. St. J. Russell, and P. J. Roberts, "Demonstration of ultra-flattened dispersion in photonic crystal fibers," Opt. Expr., **10**, 609–613 (2002).

4. G. P. Agrawal, Nonlinear Fiber Optics, 3rd Edition (Academic Press, Burlington, MA, 2008).

5. F. DeMartini, C. H. Townes, T. K. Gustafson, and P. L. Kelley, "Self-steepening of light pulses," Phys. Rev., **164**, 312–323 (1967).

6. N. Tzoar and M. Jain, "Self-phase modulation in long-geometry optical waveguides," Phys. Rev. A, **23**, 1266–1270 (1981).

7. Y. Kodama and A. Hasegawa, "Nonlinear pulse propagation in a monomode dielectric guide," IEEE J. Quantum Electron., **QE-23**, 510–524 (1987).

8. V. V. R. Kumar, G. A., Reeves, J. Knight, P. St. J. Russell, F. Omenetto, and A. Taylor, "Extruded soft glass photonic crystal fiber for ultrabroad supercontinuum generation," Opt. Express, **10**, 1520–1525 (2002).

9. F. G. Omenetto, N. A. Wolchover, M. R. Wehner, M. Ross, A. Efimov, A. J. Taylor, V. V. R. K. Kumar, A. K. George, J. C. Knight, N. Y. Joly, and P. St. J. Russell, "Spectrally smooth supercontinuum from 350 nm to 3 μm in sub-centimeter lengths of soft-glass photonic crystal fibers," Opt. Expr., **14**, 4928–4934, (2006).

10. C. Xia, M. Kumar, M.-Y. Cheng, R. S. Hegde, M. N. Islam, A. Galvanauskas, H. G. Winful, and F. L. Terry, Jr., "Power scalable mid-infrared supercontinuum generation in ZBLAN fluoride fibers with up to 1.3 watts time-averaged power," Opt. Expr., **15**, 865 871 (2007).

11. R. L. Carman, C. K. Rhodes, and R. F. Benjamin, "Observation of harmonics in the visible and ultraviolet created in CO_2-laser-produced plasmas," Phys. Rev. A, **24**, 2649–2663 (1981).

12. M. Ferray, A. L'Huillier, X. F. Li, L. A. Lompré, G. Mainfray, and C. Manus, "Multiple-harmonic conversion of 1064 nm radiation in rare gases," J. Phys. B At. Mol. Opt., **21**, L31–L35 (1988).

13. Z. Chang, A. Rundquist, H. Wang, M. M. Murnane, and H. C. Kapteyn "Generation of coherent soft X rays at 2.7 nm using high harmonics," Phys. Rev. Lett., **79**, 2967–2970 (1997).

14. M. Lewenstein, Ph. Balcou, M. Yu. Ivanov, A. L'Huillier, and P. B. Corkum, "Theory of high-harmonic generation by low frequency laser fields," Phys. Rev. A, **49**, 2117–2132 (1994).

15. F. Krausz and M. Ivanov, "Attosecond physics," Rev. Mod. Phys., **81**, 163–234 (2009).

16. T. Brabec and F. Krausz, "Intense few-cycle laser fields: frontiers of nonlinear optics," Rev. Mod. Phys., **72**, 545–591 (2000).

17. C. Winterfeldt, Chr. Spielmann, and G. Gerber, "Colloquium: Optimal control of high-harmonic generation," Rev. Mod. Phys., **80**, 117–140 (2008).

18. http://en.wikipedia.org/wiki/High_Harmonic_Generation.

19. S. Suntsov, D. Abdollahpour, D. G. Papazoglou, and S. Tzortzakis, "Efficient third-harmonic generation through tailored IR femtosecond laser pulse filamentation in air," Opt. Expr., **17**, 3190–3195 (2009).

20. A. L. Lytle, Phase Matching and Coherence of High-Order Harmonic Generation in Hollow Waveguides, PhD Thesis (University of Colorado, Colorado, 2008). Available online at http://jila01.colorado.edu/pubs/thesis/lytle/.

SUGGESTED FURTHER READING

G. P. Agrawal, Nonlinear Fiber Optics, 3rd Edition (Academic Press, Burlington, MA, 2008).

R. W. Boyd, Nonlinear Optics, 3rd Edition (Academic Press, Burlington, MA, 2008).

F. Krausz and M. Ivanov, "Attosecond physics," Rev. Mod. Phys., **18**, 163 (2009).

F. Poli, A. Cucinotta, and S. Selleri, Photonic Crystal Fibers, Springer Series in Materials Science, Volume **102** (Springer, Heidelberg, 2007).

P. St. J. Russell, "Photonic crystal fibers," J. Lightwave Technol., **24**, 4729–4749 (2006).

A. W. Snyder and J. D. Love, Optical Waveguide Theory (Chapman & Hall, London, 1983).

C. Winterfeldt, C. Spielmann, and G. Gerber, "Colloquium: optimal control of high-harmonic generation," Rev. Mod. Phys., **80**, 117 (2008).

Units, Notation, and Physical Constants

A.1. UNITS OF THIRD-ORDER NONLINEARITY

It is unfortunate that the literature contains values of nonlinearities in two different systems of units. Although the most common system—the SI system—is used in this textbook, it is important to be able to translate measured values of nonlinear susceptibilities between the two systems of units to take advantage of all the published measurements of nonlinearity.

The SI standard of units was adopted in 1948 through a Resolution of the 9th Conférence Générale des Poids et Mesures, the CGPM (known in English as the General Conference on Weights and Measures) (1). Before this year, at least three different sets of units were used in the optics literature, which resulted in some confusion. Unfortunately, researchers in the field of nonlinear optics who started in the 1960s did not take advantage of this standardization and the proliferation of units continued. In the SI system, the standard units of length, mass, and time are meters, kilograms, and seconds—"mks." There is a second system of units called "cgs," which stands for centimeter–gram–second with the centimeter, gram, and second being the standard units of length, mass, and time, respectively. This second system of units was adapted to electromagnetism in the time of Gauss called the "Gaussian" system, which was adapted to nonlinear optics and is used in the nonlinear optics literature even to the present day. Gaussian units are a subsystem of the "cgs" units. The conversion can be adapted from Tables A1.1 and A1.2 (1).

For convenience, limiting the discussion to the nonresonant limit of the susceptibilities, the total polarization, including the nonlinear part, induced in a material is written as

$$P_i(t) = \varepsilon_0 \left[\chi_{ij}^{(1)} E_j(t) + \chi_{ijk}^{(2)} E_j(t) E_k(t) + \chi_{ijk\ell}^{(3)} E_j(t) E_k(t) E_\ell(t) + \cdots \right] \quad (A1.1)$$

and

$$P_i(t) = \left[\chi_{ij}^{(1)} E_j(t) + \chi_{ijk}^{(2)} E_j(t) E_k(t) + \chi_{ijk\ell}^{(3)} E_j(t) E_k(t) E_\ell(t) + \cdots \right] \quad (A1.2)$$

in the mks and Gaussian systems of units, respectively. Substituting the units for ε_0 and the field in Eq. A1.1 shows that $\chi_{ij}^{(1)}$ is dimensionless for both cases. Furthermore,

Nonlinear Optics: Phenomena, Materials, and Devices, George I. Stegeman and Robert A. Stegeman.
© 2012 John Wiley & Sons, Inc. Published 2012 by John Wiley & Sons, Inc.

TABLE A1.1 Unit Conversion between the "mks" and "cgs" Units for Mechanical Parameters

Quantity	Symbol	cgs Unit	cgs Unit Abbreviation	Definition	Equivalent in SI Units
Length, position	L, x	Centimeter	cm	1/100 of m	10^{-2} m
Mass	M	Gram	g	1/1000 of kg	10^{-3} kg
Time	T	Second	s	1 s	1 s
Velocity	V	Centimeter per second	cm/s	cm/s	10^{-2} m/s
Acceleration	A	Gallon	gal	cm/s^2	10^{-2} m/s^2
Force	F	Dyne	dyn	g cm/s^2	10^{-5} N
Energy	E	Erg	erg	g cm^2/s^2	10^{-7} J
Power	P	Erg per second	erg/s	g cm^2/s^3	10^{-7} W
Wave number	K	Kayser	cm^{-1}	cm^{-1}	100 m^{-1}

in the mks system, $\chi_{ijk..}^{(n)}$ has the units of (meter/volt)$^{n-1}$. From the equivalence of Eqs A1.1 and A1.2, the nonlinear polarization per unit volume induced in Gaussian units by a field of 1 V/m is given as

$$P_{\text{Gaussian}}^{(n)} = \chi_{\text{Gaussian}}^{(n)} \left(\frac{1}{2.998 \times 10^4} \right)^n E_{\text{mks}}^n \text{ statvolt/cm}. \tag{A1.3}$$

Defining 1 statV/cm = 1 dyn/esu and 1 dyn = 1 esu^2/cm^2 gives 1 statV/cm = 1 esu/cm^2. And so the units for P_{Gaussian} are esu/cm^2. The units of polarization in the mks system are Coulombs/meter^2. Noting again from Table A1.2 that 1 C = 2.998 × 10^9 esu, the polarization A1.3 can be rewritten in mks units as

$$P_{\text{mks}}^{(n)} = \chi_{\text{Gaussian}}^{(n)} \left(\frac{1}{2.998 \times 10^4} \right)^n E_{\text{mks}}^n \frac{\text{esu}}{\text{cm}^2} \times \frac{1\,\text{C}}{2.998 \times 10^9\,\text{esu}} \times \frac{10^4\,\text{cm}^2}{1\,\text{m}^2}$$

$$= \chi_{\text{Gaussian}}^{(n)} \left(\frac{10^{-5-4n}}{2.998^{n+1}} \right) E_{\text{mks}}^n, \tag{A1.4}$$

TABLE A1.2 Conversion between mks Units and Gaussian Units for Electromagnetic Parameters

Quantity	Symbol	SI Unit	Gaussian Unit
Electric charge	q	1 C	$(10^{-1}c)$ statC (esu)
Electric current	I	1 A	$(10^{-1}c)$ Fr s^{-1}
Electric potential (voltage)	Φ (V)	1 V	$(10^8 c^{-1})$ statV
Electric field	E	1 V/m	$(10^6 c^{-1})$ statV/cm
Magnetic field	B	1 T	(10^4) G
Magnetic field	H	1 A/m	$(4\pi \times 10^{-3})$ Oe

Here "c" refers to the speed of light (2.998 × 10^{10} in cm/s). Furthermore, the units of the induced polarization are Coulombs/meter in SI and are the same as the electric field in the Gaussian system. From the second table, a field of 1 statvolt-cm gives a field of 1/(3 × 10^4) in volts/m and a field of 1 V/m gives in Gaussian units a field of 3 × 10^4 statvolt/cm.

which is now in Coulombs/meter^2. From Eq. A1.1

$$P_{\text{mks}}^{(n)} = \chi_{\text{Gaussian}}^{(n)} \left(\frac{1}{2.998 \times 10^4}\right)^n E_{\text{mks}}^n = \varepsilon_0 \chi_{\text{mks}}^{(n)} E_{\text{mks}}^n, \qquad (A1.5)$$

and substituting $\varepsilon_0 = 4\pi \times 2.998^2 \times 10^9$ gives

$$\chi_{\text{mks}}^{(n)} = \frac{4\pi \times 10^{-4(n-1)}}{(2.998)^{n-1}} \chi_{\text{Gaussian}}^{(n)}. \qquad (A1.6)$$

Next we consider the conversion between the nonlinear refractive index coefficients defined as Eq. 10.29. Following the same procedure as outlined above, we find

$$n_{2\|,E,\text{mks}} = \frac{n_{2\|,E,\text{Gaussian}}}{(2.998 \times 10^4)^2}. \qquad (A1.7)$$

A.2. VALUES OF USEFUL CONSTANTS

$$c = 2.998 \times 10^8 \text{ m/s}$$

$$\mu_0 (\text{permeability of vacuum}) = 4\pi \times 10^{-7} \text{ Wb/(Am)}$$

$$\varepsilon_0 (\text{permittivity of vacuum}) = 8.854 \times 10^{-12} \text{ C}^2/(\text{N} - \text{m})^2$$

$$e \text{ (charge on electron)} = 1.602 \times 10^{-19} \text{ C}$$

$$m_p \text{ (proton mass)} = 1.673 \times 10^{-27} \text{ kg}$$

$$m_e \text{ (electron mass)} = 9.109 \times 10^{-31} \text{ kg}$$

$$a_B \text{ (Bohr radius)} = 5.292 \times 10^{-11} \text{ m}$$

$$k_B \text{ (Boltzmann constant)} = 1.381 \times 10^{-23} \text{ m}^2 \text{ kg}/(\text{s}^2 \text{ K})$$

$$h \text{ (Planck's constant)} = 6.626 \times 10^{-34} \text{ J/s}$$

$$N_A \text{ (Avogadro's number)} = 6.022 \times 10^{23} \text{ particles/mol}$$

$$\alpha \text{ (fine structure constant)} = 137.035 \text{ (dimensionless)}$$

REFERENCE

1. http://en.wikipedia.org/wiki/Centimetre_gram_second_system_of_units.

Nonlinear Optics: Phenomena, Materials, and Devices, George I. Stegeman and Robert A. Stegeman.
© 2012 John Wiley & Sons, Inc. Published 2012 by John Wiley & Sons, Inc.

Printed and bound by CPI Group (UK) Ltd, Croydon, CR0 4YY

16/04/2025

14658355-0001